LEAD-FREE ELECTRONICS

IEEE Press
445 Hoes Lane
Piscataway, NJ 08854

Technical Reviewers
Dave Hillman, Rockwell Collins, Inc.
John W. Stafford, Director, JWS Consulting, P.L.C.

LEAD-FREE ELECTRONICS

iNEMI PROJECTS LEAD TO SUCCESSFUL MANUFACTURING

Edited by

Edwin Bradley
Carol A. Handwerker
Jasbir Bath
Richard D. Parker
Ronald W. Gedney

WILEY-INTERSCIENCE
A JOHN WILEY & SONS, INC., PUBLICATION

For general information on our other products and services or for technical support, please contact our Customer Care Department within the United States at (800) 762-2974, outside the United States at (317) 572-3993 or fax (317) 572-4002.

Wiley also publishes its books in a variety of electronic formats. Some content that appears in print may not be available in electronic formats. For more information about Wiley products, visit our web site at www.wiley.com.

Library of Congress Cataloging-in-Publication Data is available.

ISBN 978-0-471-44887-7

Printed in the United States of America

10 9 8 7 6 5 4 3 2 1

CONTENTS

PREFACE

The International Electronics Manufacturing Initiative (iNEMI) is an industry-led consortium whose mission is to assure leadership of the global electronics manufacturing supply chain. With a membership that includes many large electronics manufacturers, suppliers, associations, government agencies, and universities, iNEMI provides an environment in which partners and competitors alike can collectively anticipate future technology and business needs and effectively develop collaborative courses of action to meet those needs.

Deployment of new technologies requires extensive evaluation and characterization of new materials and processes as well as demonstration of reliability. iNEMI members use existing resources to develop and deploy new manufacturing technologies and efficient business practices necessary to maintain a responsive supply chain infrastructure. Combined, these companies have sufficient critical mass to make an impact, whether it is in influencing development of industry standards or creating consensus requirements to reduce risk for users and suppliers.

The movement to lead (Pb)-free electronic assembly represented one of the largest challenges ever to the electronics industry. For well over 50 years, eutectic lead–tin (Pb–Sn) solder has been studied, categorized, and optimized for electronics manufacturing applications. In a few short years, Pb-free solder assembly would have to be put into wide-scale production, disrupting the status quo. Much of the work to rally the electronics industry to prepare for Pb-free assembly is described in this book.

The first chapter describes the search for a Pb-free replacement solder and the reasoning behind the alloy formulation ultimately recommended. Characterization on the recommended solder was carried out and reported in Chapter 2, while Pb-free solder paste requirements and evaluations are described in Chapter 3. The effect of Pb-free assembly on components is reported in Chapter 4. Chapters 5 and 6 report on iNEMI efforts to characterize the reliability of the new materials, and they present a literature review comparing Sn–Pb and Pb-free solder reliability. Chapter 7 describes the present understanding of a specific reliability risk—tin whiskers—that may arise with the move to Pb-free assembly. Chapters 8 and 9 summarize iNEMI projects on assembly and rework with the chosen Pb-free solder. Chapter 10 discusses the work of iNEMI members on the infrastructure required to implement Pb-free assembly in high-volume manufacturing.

Several hundred researchers from more than 100 companies, universities, and government agencies have contributed to the material in this book. The editors would like to thank them for providing the hardware, facilities, and data generation that has made this book possible. We hope the material herein will help the electronics industry to move forward.

RONALD W. GEDNEY

July 2007

CONTRIBUTORS

R. J. Arvikar, Vectron International-Hudson, Hudson, New Hampshire

Jasbir Bath, Solectron Corporation, Milpetas, California

Elizabeth Benedetto, Hewlett Packard, Houston, Texas

Edwin Bradley, Motorola, Plantation, Florida

Quyen Chu, Jabil Circuit, Saint Petersburg, Florida

Jean-Paul Clech, EPSI Inc., Montclair, New Jersey

Jerry Gleason, Hewlett Packard, Palo Alto, California

Carol A. Handwerker, Purdue University, West Lafayette, Indiana

Ursula Kattner, National Institute of Standards and Technology, Gaithersburg, Maryland

Matt Kelly, IBM, Markham, Ontario, Canada

C. J. Lee, Cisco Systems, San Jose, California

Ning-Cheng Lee, Indium Corporation of America, Clinton, New York

Nick Lycoudes, Freescale, Austin, Texas

Ken Lyjak, IBM, Research Triangle Park, North Carolina

Jack McCullen, Intel, Chandler, Arizona

Jim McElroy, iNEMI, Herndon, Virginia

Kilwon Moon, National Institute of Standards and Technology, Gaithersburg, Maryland

Richard D. Parker, Delphi Electronics & Safety, Kokomo, Indiana

Charlie Reynolds, IBM, Hopewell Junction, New York

Heidi L. Reynolds, Sun Microsystems, Menlo Park, California

Patrick Roubaud, Hewlett Packard, Grenoble, France

Joe Smetana, Alcatel-Lucent, Plano, Texas

Polina Snugovsky, Celestica, Ontario, Canada

John Sohn, Lucent (retired)

Cynthia Williams, iNEMI, Herndon, Virginia

LEAD-FREE ELECTRONICS

INTRODUCTION

JASBIR BATH and CAROL A. HANDWERKER

In 1999, electronics companies in Japan were driving green consumer products for the 2001–2004 market. Primarily this was achieved by eliminating tin–lead (Sn–Pb) solder in the assembly process. At the same time the European Union was proposing legislation to ban Sn–Pb solder in electronic products by 2008 (which was later moved to 2006). Dr. Iwona Turlik, Motorola, a member of the Board of Directors of the National Electronics Manufacturing Initiative (NEMI), now the International Electronics Manufacturing Initiative (iNEMI), foresaw the dramatic impact this initiative would have on electronics manufacturing. She convinced the iNEMI Board that projects had to be undertaken immediately, because it would take several years to adequately address a new solder technology. The iNEMI Lead-Free Assembly Project was established to begin the work to implement Pb-free soldering into electronics manufacturing. As the project developed, several goals emerged: (1) Choose, if possible, a single Pb-free solder that could be recommended as an industry standard; (2) provide a set of manufacturing processes and tools that would enable a participating company to quickly implement lead-free soldering if it decided to do so; and (3) provide sufficient data to demonstrate manufacturability and reliability of the alloy and processes chosen.

It was recognized that one project would not be able to address all the issues that would need to be solved with the introduction of a lead-free soldering process. Follow-on projects were undertaken to fill in the gaps. Results from several of these projects are also provided in this book.

The over-arching goal of the iNEMI program was to provide the necessary processes and tools (to be as compatible as possible with existing assembly infrastructure) so that each member company could decide on implementation for its own needs and on its own schedule.

Lead-Free Electronics. Edited by Bradley, Handwerker, Bath, Parker, and Gedney

LEAD-FREE ASSEMBLY PROJECT

The primary objective of the iNEMI Lead-Free Assembly Project initiated in 1999 and completed in 2002 was to demonstrate the capability to deliver products with Pb-free interconnects in volume, utilizing, as much as possible, processes similar to Sn–Pb solders and existing assembly tools. To realize this objective, the project team set out to:

- Facilitate a common Pb-free solder alloy composition for electronics assembly.

After a thorough literature survey and consultation with six solder suppliers, there were a small number of solders that might be acceptable replacements for the Sn–Pb eutectic. Picking one of these and concentrating efforts on that alloy allowed a more thorough study to be undertaken and began the effort toward a global industry standard.

- Work with component and PCB suppliers to develop specifications necessary to meet the higher temperature reflow conditions required for the new alloy.

Through the IPC, input from a number of printed wiring board companies was collected on which board materials would be acceptable, as would interconnect finishes. The test boards used in this study were a result of the best inputs available at the time.

- Develop criteria that industry could use to evaluate Pb-free processes.
- Monitor environmental legislation to adjust activities if necessary.
- Share information in a timely manner to promote a successful, common path to Pb-free assembly.

The project participants included OEMs/EMS companies such as Agilent Technologies, Alcatel Canada, Celestica, HP, Delphi, IBM, Intel, Eastman Kodak, Lucent, Motorola, Sanmina-SCI, Solectron, and StorageTek (now SUN); solder suppliers such as Alpha Metals, Heraeus, Indium Corp., Kester Solder, and Johnson Manufacturing, component suppliers such as ChipPac, Intel, Motorola, TI, and FCI; and government; other institutions such as the National Institute of Standards and Technology (NIST), SUNY Binghamton, ITRI (US), and IPC and equipment suppliers such as BTU, Universal Instruments Corporation, DEK, Vitronics-Soltec, Orbotech, Sonoscan, and VJ Technologies.

The Pb-Free Assembly Project was broken down into four specific group efforts as shown:

- Alloy
  - Development of alloy material properties and databases for modeling
  - Interface with academia, professional societies, and government agencies

- Components/PCBs
  - Effect of high-temperature reflow
  - Pb-free terminations
- Process Development
  - Generic processes for the reliability test boards
  - Process characterization benchmarking
- Solder Reliability
  - Transparent test procedure
  - Common data to share with the industry

ALLOY GROUP

1. Responsibilities of the Alloy Group

The primary responsibility of the iNEMI Pb-Free Assembly Alloy Group was to provide the Task Force with critical data and analyses needed for making decisions with respect to solder alloys, manufacturing, and assembly reliability. This responsibility required the Alloy Group to provide assessments of candidate solder systems to allow the group to make an informed choice of industry standard lead-free alloys for reflow and wave soldering.

- Generate key data for decision making if not available in the literature.
- Develop recommended practices and experimental procedures to measure the mechanical, thermal, electrical, and wetting properties of lead-free solders.
- Develop public domain solder databases for properties and literature references for lead-free alloys.
- Promote modeling for solder joint reliability through generation of best possible data and modeling methods.

2. Assessment of Candidate Solder Systems

The first task of this group was to recommend a standard lead-free solder alternative. Industry could benefit significantly by focusing on one alloy for replacing the common Sn–Pb solder:

- Electronic Manufacturing Service (EMS) companies, in particular, would not have to have multiple manufacturing lines to handle a variety of solder alloys.
- By concentrating available resources on one solution, data could be gathered more quickly, speeding up introduction of lead-free soldering to manufacturing.
- Component and/or board lead finishes would only have to be compatible with one new alloy.

By cooperatively developing a single alloy solution, it was recognized that it would be possible to implement a replacement sooner, avoid multiple manufacturing processes, and enhance basic understanding of the material while assuring its reliability.

3. Selection Criteria

The Alloy Group, led by Jasbir Bath and Carol Handwerker, began with a literature review, in order to identify and use as much existing data as possible for the alloy choice. A general call went out to the electronic packaging community for data, published and unpublished, on the properties, manufacturability, and reliability of Pb-free solder alloys. Members of the task group also sought opinions from experts in the field, such as NIST in the United States and Soldertec [formerly the International Tin Research Institute (ITRI)] in the United Kingdom, obtained a patent search, and sought the advice of six North American solder manufacturers. These data were gathered and distributed electronically by NIST and were reviewed and discussed in an open forum at the IPC Works Conference in 1999.

Based on the findings of these initial investigations, the industrial members of the alloy selection group defined the following criteria for alloy selection:

1. If possible, stay with ternary alloys (or less). Quaternary alloys can present control difficulties.
2. The new alloy should be near-eutectic (e.g., no large pasty range during cool-down).
3. The new alloy should be as close as possible to eutectic Pb–Sn in melting point and manufacturability (in order to be able to use existing manufacturing tools where possible).
4. The new alloy should be equal to or better than eutectic Pb–Sn in reliability (when used in electronic assemblies).
5. The new alloy should create minimal cost impact over eutectic Pb–Sn (understanding that the solder cost is a very small part of assembly cost).
6. Avoid using a patented alloy if possible, so industry freedom of action is guaranteed.
7. Using the best knowledge available, do not choose an alloy that will have environmental issues in the future.

4. Candidate Alloys

A key report used by the iNEMI selection group in identifying potential alloy replacements came from a three-year study by the National Center for Manufacturing Sciences (NCMS) which evaluated over 79 solder alloys. Based on this study, input from the alloy selection group, and other information including oral and written reports from the EU DTI and IDEALS consortia, a short list of solders was chosen as follows:

1. Sn–58Bi eutectic alloy
2. Sn–Zn–Bi system

3. Sn–Ag–Bi system
4. Sn–Ag–Cu system
5. Sn–3.5Ag eutectic alloy
6. Sn–0.7Cu eutectic alloy

These solders were evaluated by the alloy selection group to determine the relative advantages and disadvantages of each, and details are presented in the Alloy Selection chapter.

In November 1999, the iNEMI Task Force announced its recommendations for lead-free solder. For reflow applications (which represent at least 70% of all board assembly production), iNEMI recommended the use of Sn–3.9Ag–0.6Cu, a predominantly tin-rich alloy with 3.9% silver and 0.6% copper (percentages are by weight). For wave solder production (which requires larger amounts of solder), the group recommended Sn–0.7Cu, a less expensive tin–copper alloy (tin with 0.7% copper), or an alternative standard Sn–3.5Ag (tin with 3.5% silver). The Sn–3.9Ag–0.6Cu ternary alloy could also be used for wave soldering; the other two alloys were recommended because many of the project participants wanted a lower-cost alternative to Sn–3.9Ag–0.6Cu for wave soldering.

Following iNEMI's recommendation of specific lead-free alloys, the Alloy Group, then led by Carol Handwerker, concentrated on developing data needed for modeling alloy thermodynamics, mechanical properties, and solder joint reliability and on assisting the other groups in interpretation of the manufacturing and reliability data being developed.

PROCESS GROUP

The task of the process group was to determine if the chosen Sn–Ag–Cu alloy would be manufacturable on current assembly manufacturing lines. The process development was led by Jasbir Bath at the Solectron facility in Milpitas, CA. To ensure manufacturability, the process was transferred to an assembly facility at Universal Instruments in Binghamton, NY, for the reliability test hardware build.

A tin–lead (Sn–37Pb) no-clean solder paste and a lead-free Sn–3.9Ag–0.6Cu no-clean solder paste were selected for the iNEMI tin–lead and lead-free reliability test board builds based on evaluations on printability, solderability after reflow, and X-ray inspection after reflow. The selected tin–lead and Sn–Ag–Cu pastes were used to successfully assemble components and boards with both tin–lead and lead-free finishes for accelerated thermal cycle (ATC) reliability testing using existing manufacturing tooling. The six types of lead-free and tin–lead components assembled were CSP169, CSP208, PBGA256, CBGA256, TSOP48, and 2512 chip resistor. Differences in the visual appearance between the tin–lead and lead-free solder paste assembled boards were noted. From a reliability perspective, these visual differences were not found to be significant in subsequent ATC testing.

COMPONENT GROUP

The Component group, led by Richard Parker of Delphi, worked on identifying and recommending the best materials for the supplier industry to use, in delivering compatible components and printed wiring boards (PWBs) that met the Pb-free requirements. Recommendations were made for PWBs and terminal finishes for the reliability testing.

Surface Finishes for IC Lead Frames

A number of component lead finishes appeared to be satisfactory [i.e., nickel–palladium–gold (Ni–Pd–Au), Ni–Pd, tin–bismuth (Sn–Bi)], but the predominant solution being offered by industry was the pure tin (Sn) finish. However, the use of high-percentage Sn alloys or pure Sn coatings have renewed concerns regarding Sn whiskers, which is discussed in more detail in Chapter 7 of this book.

Moisture Sensitivity for Plastic Packages

Molded plastic packaged IC components were believed to be the most sensitive components, as a family, to the increased temperature exposure that would result from the industry transition to lead-free solders. The main area of concern was the moisture sensitivity level (MSL) of these packages. A desire to limit peak reflow temperatures (PRTs) led to several studies to help understand the thermal mass effects on real circuit boards. This resulted in new temperature testing parameters being proposed for the IPC/JEDEC MSL standard specification (J-STD-020).

RELIABILITY GROUP

The Reliability Group, led by John Sohn of Lucent, was charged with evaluating Pb-free solder utilizing appropriate Pb-free component and board finishes against a standard Sn–Pb control. The group devised the experimental matrix, lined up individuals and companies to do the actual testing, and analyzed the results. A thorough experimental matrix was devised covering various components, solder–component combinations (including current Pb-containing components assembled using Pb-free paste), and printed wiring board materials and finishes. The reliability tests chosen were:

- Thermal cycling (0°C to 100°C and −40°C to 125°C)
- Three-point bend testing of assembled BGAs
- Electrochemical migration testing of Sn–Ag–Cu soldered no-clean pastes

NIST provided a thorough failure analysis (metallurgical cross-section and analysis) of all test cells to understand the root cause of thermal cycling and three-point bend test failures. In addition, red-dye-penetrant testing was carried out on ball-grid-array (BGA) parts. Several companies provided statistical analysis of the resulting data, which was essential to developing and determining the project conclusions.

The reliability of the Pb-free solder joints was found to be equivalent or superior to the reliability of the Pb-containing joints made using current material sets and assessed by thermal cycling and three-point bend testing. No electrochemical migration issues were identified for the Sn–3.9Ag–0.6Cu no-clean solder paste reflowed alloy.

FOLLOW-ON PROJECTS/WORK

The electronics assembly industry has accumulated some 50 years of experience in processing Sn–Pb solder. A single project could not begin to address all the issues associated with such a huge undertaking as Pb-free assembly. A number of follow-on iNEMI projects were initiated, many of which are still ongoing.

Tin Whiskers

The formation and growth of tin whiskers pose potentially significant long-term reliability issues for electronic components with pure tin or high tin content Pb-free finishes. Reducing the risk of tin-whisker-related failures involves a combination of choosing an effective mitigation strategy and conducting tin whisker acceptance testing and process control of the tin plating process. This chapter deals with the first two aspects of this threefold approach. Recommended mitigation strategies and tin whisker test development are discussed in detail. The importance of ongoing process monitoring is mentioned, although the topic of tin-plating process control is beyond the scope of this work.

Lead-Free Reflow and Rework

This chapter provides an overview and assessment of the printability of lead-free solder pastes together with reflow, rework, and inspection of lead-free solder in surface mount technology compared with tin–lead. The effect of the wetting, temperature profile, and solder joint peak temperature of the solder joints is discussed in relation to reliability, visual appearance, and associated assembly issues. Studies determining the temperatures that will likely occur on board and components during lead-free reflow and rework are reviewed. An assessment is made on the need for equipment changes (if any) for lead-free reflow and rework soldering and any adjustments required in inspection equipment/criteria for X-ray and AOI inspection. Acoustic microscopy of lead-free soldered parts is also reviewed.

Case Study: Pb-Free Assembly, Rework, and Reliability Analysis of IPC Class 2 Assemblies

A team of iNEMI companies collaborated for three years to develop Pb-free assembly and rework processes for double-sided, 14-layer, printed circuit boards (PCB) in two thicknesses (0.093 in. and 0.135 in.) with electrolytic Ni–Au and immersion Ag board surface finishes. This extended the work carried out by the first iNEMI Pb-free development team (1999–2002) to large, thicker boards. All SMT assembly, PTH wave assembly, and component rework processes were carried out on production equipment. Various test vehicles including the reliability test board were used in a multiple-phase development project to develop Pb-free assembly and rework parameters and temperature profiles prior to a 100-board process technology verification build. Following the double-sided SMT and wave assembly build, half of the printed circuits assemblies were passed through a series of representative component rework protocols. Each build group was then subjected to a series of mechanical and thermal reliability stress tests followed by failure analysis. A special reliability test board was designed utilizing a high-temperature laminate designed for Pb-free soldering. Approximately 30% of the assemblies were Sn–Pb control samples. The rework development process used the NEMI Sn–3.9Ag–0.6Cu solder. The rework of large, thick PCBs with Pb-free solder poses a significant challenge to the industry. The lessons learned and recommendations for future work are discussed.

Implementing RoHS and WEEE-Compliant Products

There is much more to the conversion to Pb-free electronics than the resolution of technology gaps. In today's distributed manufacturing environment, cooperation across the value chain is a necessity—from product design through to end-of-life disposition—in order to achieve RoHS compliance. While the focus of this chapter is on the deployment issues associated with a particular set of regulations, the concepts described here would generally apply to any major regulation-driven technology change that is broadly adopted by industry, and thus these observations will remain relevant for future applications.

This work has resulted in a solid first step for the successful introduction of lead-free soldering by the North America electronics industry and has been referred to on numerous occasions globally as a model/benchmark for successful company collaboration and important lead-free development work. Although the basics are complete, the engineering work to improve reliability, cost, and manufacturing yield is ongoing with ever more follow-on projects adding to the platform of knowledge. An engineer's work is never done.

We gratefully acknowledge the efforts of Iwona Turlik and Ron Gedney, the iNEMI program manager for Pb-free projects, in making this project a success: Iwona Turlik for having the vision and drive to initiate the project and Ron Gedney for leading us to completion of the project and this book. We dedicate this book to them.

CHAPTER 1

Alloy Selection

CAROL A. HANDWERKER, URSULA KATTNER, KILWON MOON, JASBIR BATH, EDWIN BRADLEY, and POLINA SNUGOVSKY

1.1. INTRODUCTION

Between 1991 and 2003, national and international research projects in the United States, the European Union, and Japan were formed to examine lead (Pb)-free alternatives to tin–lead eutectic solder and to understand the implications of such a change before it became required by law, by tax, or by market pressure [1–18]. The UK Department of Trade and Industry (DTI) developed a comprehensive report on the major results of these lead-free solder research projects [18]. Parallel to these multi-participant studies were similar investigations by individual companies and research organizations into Pb-free alternatives to Sn–Pb eutectic and near-eutectic solders. All of these studies determined that there was no "drop-in" replacement for Sn–Pb eutectic solder.

In 1999, with the proposed ban on lead in the European Union combined with the substantial Pb-free solder development efforts by Japanese manufacturers, the International Electronics Manufacturing Initiative (iNEMI) formed its Lead-Free Task Force with the goal of helping the North American electronics industry develop the capability to produce lead-free products by 2001. The first task of this group was to recommend a "standardized" lead-free solder alternative [9–11]. In approaching the overall issue of lead-free solders, the iNEMI team members realized that they could make a major contribution to the industry if they could recommend a single solder solution to replace the tin–lead eutectic paste used for high-volume surface-mount component assembly. This is of particular importance to the electronic manufacturing service (EMS) providers, for minimizing their investment in equipment and process optimization required for solders with different assembly behavior, and for components with different moisture sensitivity levels. This became the overriding goal of the project.

Lead-Free Electronics. Edited by Bradley, Handwerker, Bath, Parker, and Gedney

In making an alloy selection, the iNEMI team carried out a thorough literature review and patent review and gathered all available data that 30 member companies, including five solder manufacturers, could bring to the table. The NCMS and IDEALS Project Reports were particularly helpful in narrowing the decision [1–8, 13, 14]. The NCMS work, for example, demonstrated that a solder with a large "pasty" range leads to stresses in through-hole joints during the cool-down phase and, in many cases, to separation of the solder fillet along its interface with the printed wiring board (PWB) copper land (also known as "fillet lifting") or to pad delamination [1, 2]. Solder manufacturers generally recommended selection of an alloy with no more than three elements for ease of solder manufacturing. Analysis of the available data led to the following criteria for selecting a new "standard" solder alloy for board assembly:

1. Melting point should be as close to Sn–Pb eutectic as possible.
2. Alloy must be eutectic or very close to eutectic.
3. There should be no more than three elements (ternary composition).
4. Avoid using existing patents, if possible (for ease of implementation).
5. Potential for reliability should be equal to or better than Sn–Pb eutectic.

Application of these criteria led directly to the iNEMI choice of the Sn–Ag–Cu system, and the specific alloy Sn–3.9Ag–0.6Cu ($\pm 0.2\%$) in the Sn–Ag–Cu (SAC) family of alloys as the most promising solution.

In this chapter, the key results and analyses leading to the choice of SAC alloys by iNEMI are discussed in detail. These include data on phase transformations in solders (including melting behavior, solidification pathways, and interface reactions with substrate and lead materials), on wetting behavior, and on mechanical properties (including thermomechanical fatigue). The materials science issues are illustrated using data from a wide range of sources, including the NCMS Lead-Free Solder Projects (US) [1–4, 13, 14], the IDEALS Lead-Free Solder Project (UK) [3, 6–9], the iNEMI Pb-Free Assembly Project (US) [10–12], various Japanese consortia [15–17], the National Institute of Standards and Technology (NIST) [19–23], and the open literature. Based on the choice of a single SAC alloy, the iNEMI Lead-Free Project could begin to address lead-free assembly, including manufacturing yield, process windows for complex boards, component survivability, and assembly reliability, as described in other chapters in this book.

In the last five years since the iNEMI alloy selection was performed, a worldwide consensus has developed that the general-purpose lead-free alloy should be from the Sn–Ag–Cu family. In Europe, Soldertec, the lead-free solder research arm of Tin Technology, selected the range of compositions Sn–(3.4–4.1)Ag–(0.5–0.9)Cu [9, 18], while the IDEALS consortium recommended Sn–3.8Ag–0.7Cu [5–8]. (Note that all compositions are expressed as Sn–vX–yZ, where the X and Z are alloying elements in Sn, with the composition being v mass fraction $\cdot$ 100 of element X, y mass fraction $\cdot$ 100 of element Z, and remainder being Sn; mass fraction $\cdot$ 100 is also abbreviated as wt%.) While numerous lead-free alloys, including

Sn–Ag–Bi–Cu, Sn–8Zn–3Bi, and Sn–58Bi, were investigated by large Japanese OEMs, the Japanese industry has moved over time toward Sn–Ag–Cu alloys. JEITA (Japan Electronics and Information Technology Industries Association) has recommended the Sn–3.0Ag–0.5Cu alloy, partly due also to concerns over patent issues [15–17]. However, widespread cross-licensing of nearly all the tin–silver–copper family of solder alloys by the solder manufacturers means that alloy selection within the SAC system should be driven primarily by overall performance in product applications and other issues, such as cost, rather than by patent issues. Furthermore, the differences among this range of SAC alloys in terms of manufacturing and reliability are generally believed to be small, based on available melting and reliability data. Additional results and analyses on SAC alloys that have emerged since the iNEMI selection of Sn–3.9Ag–0.6Cu as the standard alloy are also discussed and the differences between SAC alloys are examined.

1.2. LEAD-FREE ALLOYS CONSIDERED BY iNEMI IN 1999 AS REPLACEMENTS FOR TIN–LEAD EUTECTIC SOLDER

Based on input from the alloy selection group, the following short list of Pb-free solders considered as replacements for Sn–Pb eutectic was developed:

1. Sn–58Bi eutectic alloy
2. Sn–Zn–Bi system
3. Sn–Ag–Bi system
4. Sn–Ag–Cu system
5. Sn–3.5Ag eutectic alloy
6. Sn–0.7Cu eutectic alloy

Note that all the Pb-free solders considered were tin-rich solders, with the exception of Sn–58Bi eutectic. These solders were compared by the iNEMI alloy selection group to determine the relative advantages and disadvantages of each. A summary of the group's evaluation is presented below. (For additional discussion of the properties of lead-free alloys, see Refs. 1–9.)

1.2.1. Sn–58Bi Eutectic Alloy

The Sn–58Bi eutectic alloy has a melting temperature of 138°C (eutectic temperature) and has been shown to be resistant to fillet lifting and to outperform eutectic Pb–Sn in the NCMS thermal cycling tests for a range of components [1–4]. Its significantly lower melting temperature than eutectic Sn–Pb will preclude its use in applications where the upper use temperature is close to 138°C. For example, the majority of automotive assemblers are looking toward a higher melting point alloy than eutectic Sn–Pb for under-the-hood applications at 150–175°C. During the transition to lead-free solders, there will be components containing lead from the tin–lead surface finishes for some period of time. The Sn–58Bi eutectic solder will react with the Pb to form some fraction of the Sn–Bi–Pb ternary eutectic phases with a eutectic temperature of 96°C. The possibility of a very large "pasty" range and potentially poor solder joints is considered a manufacturing process issue and potential reliability exposure. A detailed analysis of the melting behavior of Sn–Bi–Pb alloys was performed by NIST as part of this project, as described below [19].

An analysis by NCMS determined that there are also issues of cost and continued availability of Bi and other alloying elements for use in such high concentrations. There are approximately 60 million kilograms of tin–lead solder used in electronics per year. Up to 50 million kilograms are used in wave soldering with up to 10 million kilograms in solder paste applications per year. Considering current production and spare capacity, sufficient bismuth to supply the whole electronics solder market would only support a solder containing up to 6 wt% Bi. When additional sources of Bi are considered, the NCMS Lead-Free Project estimated that the Bi composition of a solder completely replacing eutectic Sn–Pb could be as high as 20 wt% Bi, still lower than Sn–58Bi. The eutectic alloy Sn–58Bi may end up being used for some consumer products with low use temperatures and for temperature-sensitive components and substrates [24]. The consumption and availability issue, and its low-melting eutectic formation with lead (Pb) will limit its widespread adoption, particularly until Pb is eliminated from board and component surface finishes.

1.2.2. Sn–Zn–Bi System

A promising alloy in this system (Sn–8Zn–3Bi) has a melting range of 189–199°C, thus having a slightly higher melting temperature than Sn–37Pb (183°C). [The term "melting range" means that the alloy begins to melt at 189°C (solidus temperature) and finishes melting at 199°C (liquidus temperature). The term "melting range" is synonymous with "pasty range."] This temperature range has an obvious advantage over other high-Sn alloys with liquidus temperatures as high as 227°C. However, zinc-containing alloys oxidize easily, showing severe drossing in wave solder pots, are prone to corrosion and have a paste shelf life that is measured in terms of days or weeks compared to months for eutectic Sn–Pb. The bismuth is added to improve the wettability, reduce the liquidus temperature, and reduce corrosion compared with binary Sn–Zn alloys. The presence of bismuth may also result in the formation of low-melting-point eutectic in contact with Sn–Pb-coated components and boards, affecting the reliability of the assembly as in the case of Sn–58Bi. Due to the manufacturing control difficulties, all six of the solder suppliers consulted recommended strongly against adoption of a zinc alloy, as the standard alloy. Given these drawbacks, the suitability of Sn–Zn–Bi as a general replacement for eutectic Sn–Pb is limited.

1.2.3. Sn–Ag–Bi System

The melting range of this alloy family is 210°C to 217°C with bismuth compositions ranging from 3 to 5 wt% and Ag compositions ranging from 2 to 4 wt% [22, 23].

The alloy Sn–3.4Ag–4.8Bi has been shown to outperform eutectic Pb–Sn in thermal cycling tests for all components examined by NCMS [1–4] and by Sandia National Laboratories, which carried out 0–100°C thermal cycling experiments for up to 10,000 cycles on chip capacitors, SOIC gull-wings, and PLCC-J-lead solder joints [25].

In spite of its excellent performance in SMT applications, there are several issues with this alloy. One issue is again the possibility of the formation of the low-melting-point Sn–Pb–Bi eutectic when combined with Sn–Pb-coated components [19]. With low Bi additions, reliability may not be an issue for consumer products: Panasonic has manufactured a consumer product with this type of alloy paste and Pb-containing component finishes and did not detect the presence of lower-melting eutectic in their testing [26]. Alloys of Sn–Ag–Bi have been found to have a severe problem with fillet lifting in through-hole joints with the tendency toward fillet lifting increasing with Bi concentration to a maximum in the range of 5–10% Bi [1–4]. When these alloys are used with tin–lead-coated components and boards, the tendency toward fillet lifting may be increased. All of the other issues noted above for Bi-containing solders also apply to these alloys.

1.2.4. Sn–Ag–Cu System

Alloys in this family with melting ranges near 217–227°C have the most promise as the main replacement for tin–lead solder. The alloys Sn–3.5Ag, Sn–2.6Ag–0.8Cu–0.5Sb, and other high-Sn alloys containing Ag and Cu with small additions of other elements were shown to perform as well as eutectic Pb–Sn for BQFP, PLCC, and 1206 capacitors in thermal cycling tests by NCMS [1–4].

The Sn–3.8Ag–0.7Cu alloy was recommended by the EU IDEALS consortium as the best lead-free alloy for reflow as a result of reliability testing from −20°C to 125°C for up to 3000 cycles and power cycling from 25°C to 110°C for 5000 cycles [5–8]. In these tests, the reliability of Sn–3.8Ag–0.7Cu was equivalent to or better than eutectic Sn–Pb and Sn–Pb–Ag. The lowest eutectic in the system when lead contamination is present is close to the Sn–Pb eutectic. The 7°C higher temperature compared to Sn–Ag–Bi alloys may be a small price to pay to ensure good reliability of through-hole joints. These alloys have an approximately 4°C lower melting temperature than the Sn–3.5Ag eutectic alloy (221°C) with a potential improvement in solderability and reliability.

At the time of the alloy selection, there were three readily available commercial Sn–Ag–Cu solders with "melting" temperatures near 217°C. These are Sn–3.5Ag–0.7Cu, which is available in Japan, and Sn–3.8Ag–0.7Cu and Sn–4Ag–0.5Cu, which are available in North America and Europe. All these have similar wetting characteristics, mechanical properties, and melting behavior. The NEMI lead-free group decided on the Sn–3.9Ag–0.6Cu as the alloy to recommend to the industry, a composition midway between Sn–3.8Ag–0.7Cu and Sn–4Ag–0.5Cu. The ANSI J-STD-006 specifies that an alloying element less than 5 wt% can vary in composition by ±0.2 wt% so the Sn–3.9Ag–0.6Cu alloy would cover both these compositions and ±0.2 wt% is the usual tolerance that a solder manufacturer gives when manufacturing a particular solder alloy.

NIST [21] used a variety of Sn–Ag–Cu alloy compositions to compare to data from Marquette University [27] and Northwestern University [28] to determine that the ternary eutectic had a melting temperature of 216°C to 217°C with a composition of approximately Sn–3.6Ag–0.9Cu. Alloys with compositions within the range Sn–(3.5–4)Ag–(0.5–1)Cu are close enough to the eutectic to have a liquidus temperature between 217°C and 220°C with similar microstructures and mechanical properties, as described below. The literature indicates that the solderability of Sn–Ag–Cu alloys is adequate. The melting behavior of Sn–Ag–Cu alloys is described in greater detail below.

The patented alloy Sn–2.6Ag–0.8Cu–0.5Sb (CASTIN™) is in the same Sn–Ag–Cu family with similar melting temperature range, solderability, and reliability as the alloys discussed above [1]. Additions of <1% antimony do not degrade solderability and only slightly change the melting point. Antimony is considered to be toxic by some companies, but at this low concentration it is not clear whether it would be a major problem.

iNEMI's patent review found many patents in the Sn–Ag–Cu system (Table 1.1) but with considerable overlap. The alloy Sn–4Ag–0.5Cu was reported in a German thesis and a corresponding paper [29] 50 years ago as the ternary peritectic/eutectic, and some solder companies were producing this alloy without any licensing. In the United States, both Sn–3.8Ag–0.7Cu and Sn–4Ag–0.5Cu formulations are available from the main solder manufacturers. Since the selection of the Sn–3.9Ag–0.6Cu alloy, another alloy Sn–3.0Ag–0.5Cu alloy has been used widely in Japan. It appears to have similar characteristics to the other commercially available Pb–free Sn–Ag–Cu alloys.

1.2.5. Sn–3.5Ag Eutectic Alloy

Sn–3.5Ag has been used in the industry for many years in module assembly. Ford (Visteon Automotive Systems) has reported that they have used Sn–3.5Ag solder successfully in production for wave soldering since 1989 [30, 31]. There are no patent issues regarding its use, and it is already available from most of the solder manufacturers in bar, wire, and paste form. The reliability of the alloy is similar to Sn–37Pb [1–4, 30, 31], and the primary difference between the Sn–3.5Ag and Sn–Ag–Cu alloys is the addition of the copper, which lowers the melting temperature by 4°C [16].

1.2.6. Sn–0.7Cu Eutectic Alloy

The eutectic alloy Sn–0.7Cu with a melting temperature of 227°C was another alloy evaluated for reflow and wave soldering. Its melting temperature, which is 10°C higher than the eutectic temperature of Sn–Ag–Cu, makes it undesirable for reflow applications. In wave soldering applications, the temperatures that the boards and components reach are much lower than in reflow soldering. There is a

TABLE 1.1. Relevant Lead-Free Solder Patents

SnAgCu	Patent No.	Assigned to:	Sn	Ag	Cu	Bi	Sb	Zn	In	Other
USA	4,879,096	Oatey Company	88–99.35	0.05–3	0.5–6	0.1–3	—	—	—	
USA	—	Kester Solder	90–93.5	2.0–5	0.3–2	0.5–7	—	—	—	
Japan	08-132277	Ishikawa Kinzoku	Balance	1.0–3	0.5–2	1.0–10	—	—	—	
Japan	08-206874	Matsushita	Balance	0.1–20	—	0.1–25	—	—	0.1–20	Add Cu 0.1–3 or Zn 0.1–15
USA	4,778,733	Engelhard Corporation	92–99	0.05–3	0.7–6	—	—	—	—	
USA	5,527,628	Iowa State University; Sandia	Balance	3.5–7.7	1.0–4.0	0.0–10.0	—	0.0–1.0	—	
Japan	08-215880	Ishikawa Kinzoku	Balance	0.5–3.5	0.5–2.0	—	—	—	—	
Japan	05-050286	Senju/ Matsushita	Balance	3.0–5	0.5–3		0–5			

problem with respect to using Sn–Ag, Sn–Cu, or Sn–Ag–Cu alloys for wave soldering with Pb-containing surface finishes. The alloys themselves show good resistance to fillet lifting; however, additions of Pb cause an increase in the tendency for fillet lifting, as reported by NCMS and others [1–4]. In 2000, the IDEALS project reported that the reliability of Sn–0.7Cu in early screening trials for plated-through-hole solder joints was poor and Sn–0.7Cu was eliminated as a candidate solder for wave soldering applications [5–8].

The most significant advantage of Sn–0.7Cu over lower melting alloys for wave soldering is the cost of bar solder. Because it does not contain Ag or Bi, Sn–0.7Cu is one of the cheapest lead-free solder alloys available. This may be a key criterion for alloys to fill wave solder pots, which have capacities as large as 730 kg (1600 lb). In contrast to wave soldering, the price of the various alloying elements for paste is less of a consideration compared with solder bar because the metal costs account for much less than 50% of the cost of the paste. So the differences in cost among Sn–Ag–Cu, Sn–Cu, Sn–Ag, and Sn–37Pb pastes due to metal cost will be small.

Based on these analyses by the iNEMI Task Force, the member companies chose Sn–Ag–Cu as the alloy system to replace Sn–Pb eutectic, with Sn–3.9Ag–0.6Cu as the specific alloy chosen as the iNEMI standard alloy, with Sn–0.7Cu as a possible alternative alloy for wave soldering when solder cost dominates alloy choice. The materials science concepts behind these choices are described in detail below.

1.3. FUNDAMENTAL PROPERTIES OF LEAD-FREE SOLDER ALLOYS AFFECTING MANUFACTURING AND RELIABILITY

The essential characteristics of an acceptable Pb-free solder alloy are related to PWB assembly and reliability. The liquidus temperature of the solder must be sufficiently low during reflow or wave soldering to avoid damage to the board and components, but high enough to form a solid joint and operate in normal field conditions. The solder joint must solidify without formation of defects that undermine joint integrity; these defects can become evident during and immediately after solidification, as well as during use. The solder joint must be able to withstand the mechanical stresses imposed by use, including thermomechanical fatigue, thermal shock, vibration, and impact.

As noted by all of the consortia, the choices of easily processed metals with liquidus temperatures close to 183°C and a small pasty range are limited. A simple analysis of the Periodic Table of the Elements and the ASM Binary Alloy Phase Diagrams [32] yields a short list of binary systems that could form the basis for acceptable binary or ternary solder alloys.

Laboratory tests for identifying phase transformations, wetting behavior, and mechanical properties can be quite successful in reducing the number of Pb-free alloys to those most likely to be acceptable as replacements for eutectic Sn–Pb in circuit board assembly. Beyond an initial down-selection process based on "pass–fail"-type criteria, no suite of laboratory experiments has yet been identified which can provide an accurate ranking of possible Pb-free alloys. The challenges in using laboratory test results to identify the "perfect" solder alloy to replace Sn–Pb eutectic are illustrated by examining the formal, quantitative ranking process used by the NCMS Pb-Free Solder Project [1].

The NCMS Lead-Free Solder Project developed three sets of materials property criteria for eliminating alloys from further consideration and for ranking the remaining alloys relative to Sn–Pb eutectic in order to include only the most promising alloys in full manufacturing and reliability trials. This "down-selection" process involved tradeoffs in laboratory-test-based properties selected as surrogates for manufacturing and reliability performance. Pass–fail down-selection criteria listed in Table 1.2 were used to reduce the number of alloys on the initial list of

TABLE 1.2. Pass–Fail Down-Selection Criteria

Solder Property	Definition	Acceptable Levels
Liquidus temperature	Temperature at which solder alloy is completely molten.	$<225°C$
Pasty range	Temperature difference between solidus and liquidus temperatures. Represents the temperature range where the alloy is part solid and part liquid.	$<30°C$
Wettability	A wetting balance test assesses the force resulting when a copper wire is immersed in and wetted by a molten solder bath. A large force indicates good wetting, as does a short time to attain a wetting force of zero and a short time to attain a value of two-thirds of the maximum wetting force.	$F_{max} > 300\ \mu N$ $t_0 < 0.6$ s $t_{2/3} < 1$ s
Area of coverage	Assesses the coverage of the solder on Cu after a typical dip test.	>85% coverage
Drossing	Assesses the amount of oxide formed in air on the surface of molten solder after a fixed time at the soldering temperature.	Qualitative scale
Thermomechanical fatigue (TMF-1)	Cycles-to-failure for a given percent failed of a test population based on a specific solder-joint and board configuration, as compared to eutectic Sn–Pb.	Some percentage, usually >50%
Coefficient of thermal expansion (CTE)	Thermal expansion coefficient of the solder alloy is the fraction change of length per °C temperature change. Value used for comparison was CTE of solder alloy at room temperature.	$<2.9 \times 10^{-5}/°C$
Creep	Stress required at room temperature to cause failure in 10,000 minutes.	>3.4 MPa
Elongation	Total percent elongation of material under uniaxial tension at room temperature.	$\gg 10\%$

candidate alloys. The remaining alloys were grouped by alloy composition, and at most one alloy was selected based on the primary phase field in the binary and ternary phase diagrams. (For example, Sn–3Ag–2Bi was selected as representative of Sn–Ag–Bi alloys with beta-Sn being the first phase to solidify during cooling.) Finally, a decision matrix was used to rank the remaining alloys, based on alloy pasty range, on wetting balance values, and on the results of an accelerated thermal cycling (ATC) test using a printed circuit board test vehicle. A full description of the decision matrix methodology, the test methods, and how the decision matrix was applied in the NCMS Project can be found in the NCMS Pb-Free Project Final Report and CD [1].

The problem with this decision matrix approach for ranking alloys lies in the lack of a simple quantitative measure of the solder joint reliability relative to Sn–Pb eutectic using either laboratory tests of materials properties or a limited set of accelerated thermal cycling (ATC) experiments on PWBs. The manufacturing behavior of Pb-free solder alloys is well described by laboratory measurements of their thermodynamic properties and wetting. In contrast, ATC results depend on component type and thermal cycling conditions. This means that the ranking of Pb-free alloys based on a mechanical property measurement or performance in a single-component ATC test would change if different mechanical property tests or ATC test conditions were used. For Sn–Pb eutectic solders, the relationship between ATC test results under different ATC test conditions and product reliability, expressed as an "acceleration factor," is only qualitative though it is generally imagined that it is based on more quantitive data than it is. The acceleration factors for Pb-free solders are not known and are expected to be a function of alloy composition, component type, and thermal cycling conditions. As discussed below, determining the ATC thermal cycling conditions that accurately predict the thermomechanical fatigue life of Pb-free solders for the full range of currently used components, circuit boards, and product conditions remains to be done.

1.3.1. Phase Transformations in Solder Alloys

In terms of phase transformations, solder alloys undergo numerous changes as they melt, come into contact with other materials, and become solid again. Solder alloys melt and react with the board and lead materials while the solder is in the molten state. Solder solidification depends on the ease of nucleation, precipitation on preexisting phases, metastable phase formation, interdiffusion, coarsening, and reactions with substrates and lead materials in the solid state. During use, the solubilities and the distribution of phases change as a result of thermomechanical fatigue. Phase changes may also include "tin pest," the transformation of beta to alpha tin at low temperatures, leading to a volume expansion of 23% and catastrophic disintegration of solder joints [33]. In terms of analyzing the effect of solder alloy composition on manufacturing, some of these are clearly important and straightforward to analyze, such as melting and solidification behavior. The relationship between these properties and solder joint reliability are dependent on board and component materials, including surface finishes, thermal history in processing, and thermomechanical history in use, and are discussed briefly in the reliability section below.

1.3.1.1. Melting Behavior. Choosing a Pb-free solder as a replacement for Sn–Pb eutectic begins with evaluation of alloy melting behavior. Since the behavior of lead-free solder alloys is judged against Sn–Pb eutectic, it is useful to begin with an examination of the Sn–Pb phase diagram (Figure 1.1) and the melting behavior of Sn–Pb alloys. The Sn–Pb phase diagram is characterized by a liquid phase and two solid phases, each with substantial solid solubility. Furthermore, the system is characterized by a simple eutectic with a significant depression of the liquidus temperature (T_ℓ) by almost 50°C, from pure Sn at 232°C to the binary eutectic (Sn–37Pb) at 183°C. The microstructure on solidification is a mixture of Sn and Pb solid solution phases that constitute the "classic" eutectic microstructure.

The Sn–Bi, Sn–Ag, and Sn–Sb systems are typical of the types of melting behavior for Sn–based Pb-free alloys [18–21]. In the Sn–Bi diagram (Figure 1.2), there is significant solid solubility of Bi in Sn, up to 22% Bi in Sn at the eutectic temperature, 139°C. The liquidus temperature decreases with increasing Bi concentration, from 232°C at pure Sn to 139°C at 58% Bi. The solidus temperature decreases with increasing Bi concentration, from 232°C at pure Sn to 139°C at 22% Bi. In the Sn–Ag diagram (Figure 1.3), there is negligible solid solubility of Ag in Sn. The liquidus temperature decreases from 232°C to 221°C at 3.5% Ag. The Sn–Sb

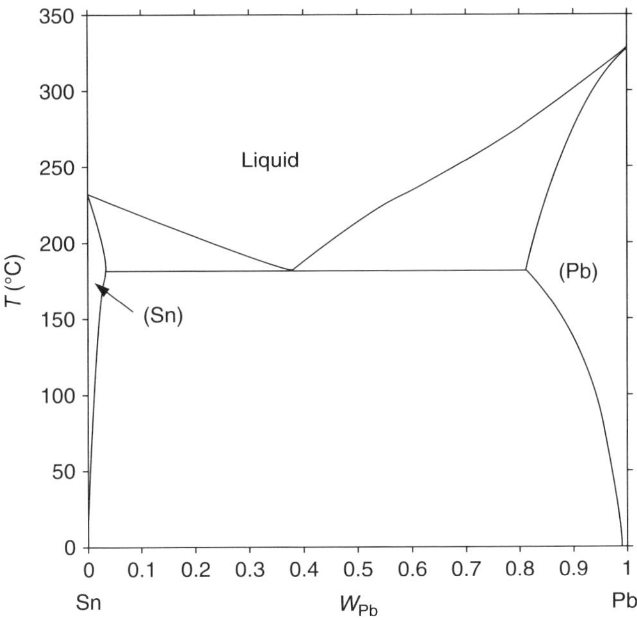

Figure 1.1. Sn–Pb phase diagram.

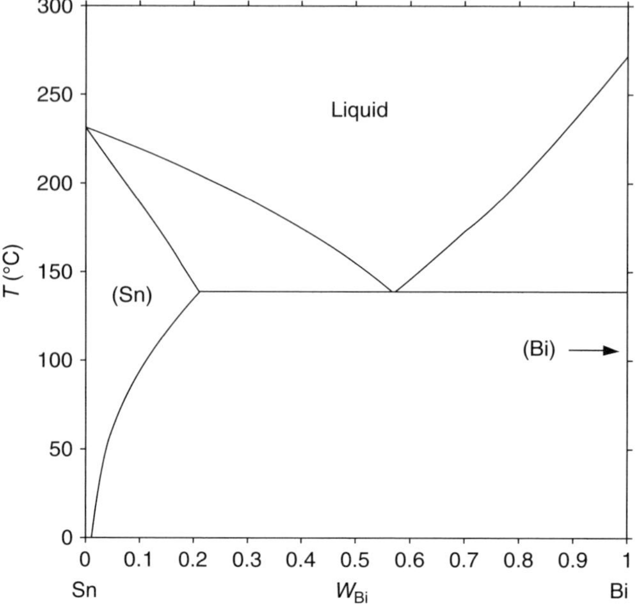

Figure 1.2. Sn–Bi phase diagram.

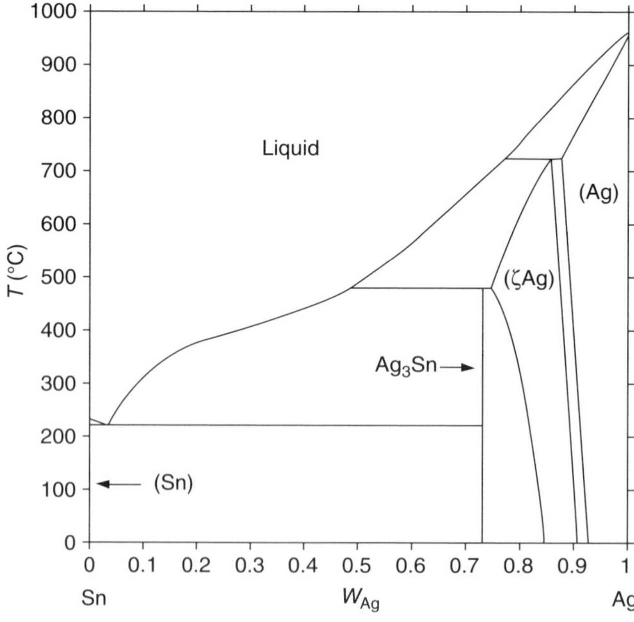

Figure 1.3. Sn–Ag phase diagram.

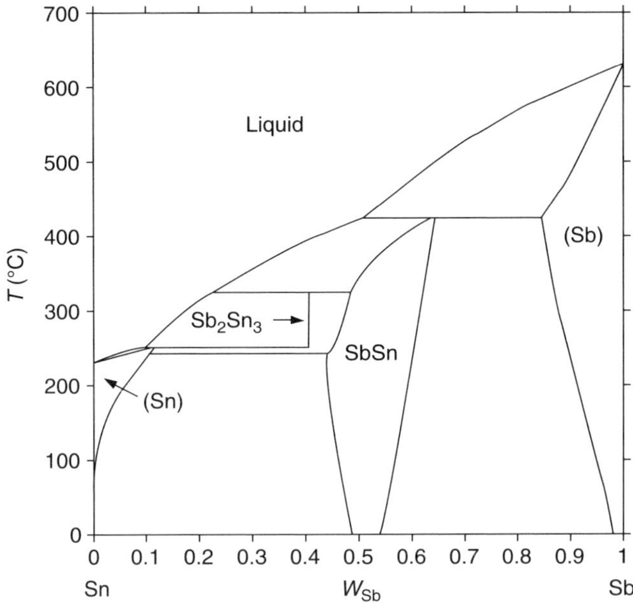

Figure 1.4. Sn–Sb phase diagram.

system (Figure 1.4) contains a peritectic at the Sn-rich side of the phase diagrams, leading to an increase in liquidus temperature with increasing Sb concentration.

There has been widespread desire on the part of the microelectronics industry: (1) to keep the liquidus temperature as close as possible to 183°C, in order to avoid changing manufacturing processes, materials, and infrastructure, (2) to keep the solidus temperature as close as possible to the liquidus temperature, to avoid fillet lifting, and (3) to keep the solidus temperature significantly higher than the solder joint's maximum operating temperature. Eutectics obviously meet the second criterion; however, eutectic Sn-based alloys tend to fall into two temperature regimes with respect to the other two criteria. The high-temperature, Sn-rich eutectics are Sn–0.9Cu (227°C), Sn–3.5 Ag (221°C), Sn–9Zn (T_m = 199°C), and Sn–3.5Ag–0.9Cu (217°C). The low-temperature eutectic solders are Sn–58Bi (139°C), Sn–59Bi–1.2Ag (138°C), and Sn–50.9In (T_m = 120°C). (The eutectic in the Sn–Cd binary system is 177°C, close to ideal as a substitute for Sn–Pb from the point of view of melting point. However, Cd is highly toxic.) The NCMS Pb-Free Project member companies selected solders with liquidus temperatures less than 225°C and with an equilibrium pasty range (the difference between liquidus and solidus temperatures) less than 30°C. The IDEALS and iNEMI projects limited its candidate solders to eutectic and near-eutectic, Sn-rich solders.

For Sn-rich solders, the composition dependence of the liquidus temperature can be estimated from a simple linear equation for additions of Ag, Bi, Cu, Ga, In, Pb, Sb, and Zn to Sn [1]. For Ag, Bi, Cu, and Pb, the coefficients were derived from the

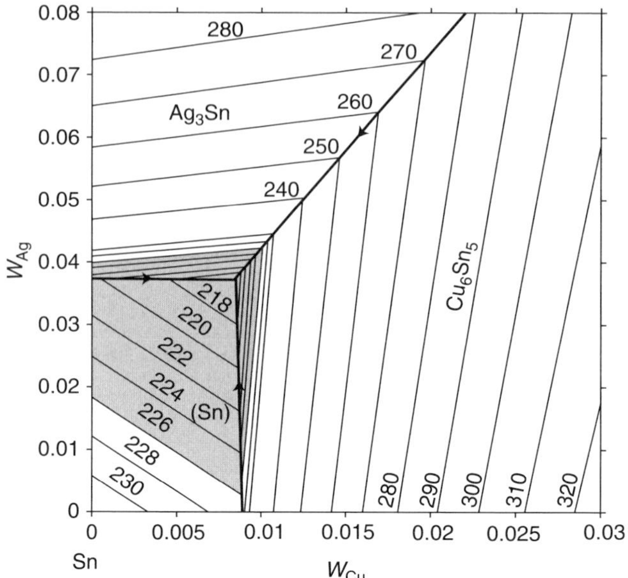

Figure 1.5. Sn–Ag–Cu phase diagram [21].

slopes of the Sn–X (X = Ag, Bi, Cu, Pb) binary phase diagram liquidus line

$$T_\ell = 232^\circ\mathrm{C} - 3.1W_{\mathrm{Ag}} - 1.6W_{\mathrm{Bi}} - 7.9W_{\mathrm{Cu}} - 3.5W_{\mathrm{Ga}} - 1.9W_{\mathrm{In}} - 1.3W_{\mathrm{Pb}} + 2.7W_{\mathrm{Sb}} - 5.5W_{\mathrm{Zn}} \quad (1.1)$$

where the coefficients are in units of °C, and W_X is the amount of element X in mass fraction · 100. This equation is valid for the following alloy additions to Sn (expressed in mass fraction · 100): Ag < 3.5, Bi < 43, Cu < 0.7, Ga < 20, In < 25, Pb < 38, Sb < 6.7, and Zn < 6.

Using this equation, the maximum decrease from the melting point of pure Sn with additions of Ag and Cu is 15–16°C, in agreement with the measured ternary eutectic temperature in the Sn–Ag–Cu system of 217°C, as seen in Figure 1.5 [21]. From Eq. (1.1), many alloy compositions with Bi, In, and Zn additions can be identified with liquidus temperatures of 183°C, the eutectic temperature of Sn–Pb eutectic solder. The problem with most of these alloys is that their solidus temperatures are significantly lower than 183°C. This issue of limiting the pasty range is particularly serious for through-hole joints: as noted above, alloys with a large pasty range may exhibit fillet lifting.

The melting behavior of three compositions in the Sn–Ag–Cu system that have been used commercially as replacements for Sn–Pb eutectic solders is illustrated in Figure 1.6; two additional compositions are shown for comparison. These three commercially available solder compositions are Sn–3.0Ag–0.5Cu, Sn–3.5Ag–0.9Cu, and Sn–3.9Ag–0.6Cu. A comparison of the calculated fraction solid as a

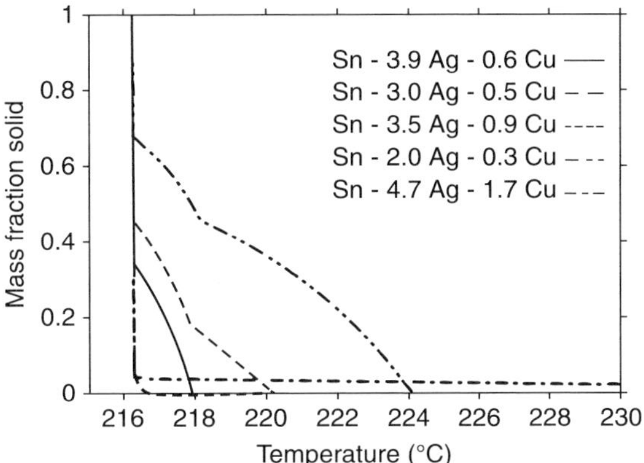

Figure 1.6. Comparison of calculated fraction solid as a function of temperature for five different Sn–Ag–Cu alloys.

function of temperature for these five SAC alloys illustrates an important point regarding the sensitivity of the melting behavior to changes in composition. For near-eutectic alloys and compositions higher in Ag and Cu than the eutectic composition, the total fraction of intermetallic phases over wide composition ranges is small and is difficult to detect using standard DTA measurement systems. The "effective" liquidus temperatures measured will, therefore, be 217°C for a wide range of compositions. In reflow soldering, it is likely that this small fraction of intermetallic phase will have a correspondingly small effect on solder flow and wetting, even if the solder in the joint never becomes completely liquid. Figures 1.7*a* and 1.7*b* show the equilibrium fractions of different phases, Sn, Ag_3Sn, and Cu_6Sn_5, as a function of temperature during heating for Sn–3.0Ag–0.5Cu and Sn–3.9Ag–0.6Cu, respectively.

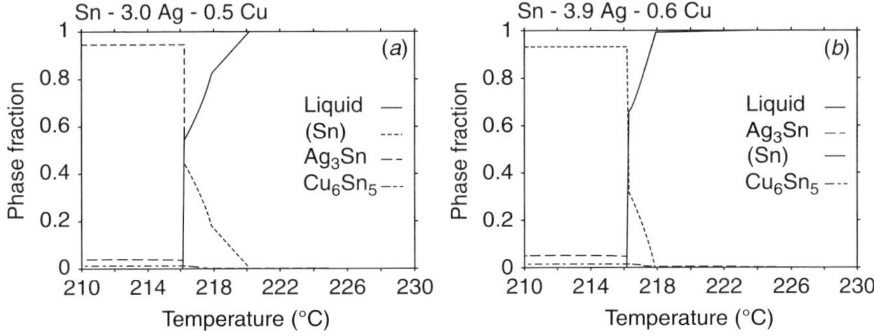

Figure 1.7. (*a*) Calculated melting path for Sn–3.0Ag–0.5Cu. (*b*) Calculated melting path for Sn–3.9Ag–0.6Cu.

Another useful representation of the melting behavior of SAC alloys as a function of temperature and composition is an isothermal section through the Sn–Ag–Cu phase diagram as presented in Figures 1.8*a–d*. The experimentally determined eutectic composition of Sn–3.5 (±0.2)–Ag–0.9(±0.2)Cu is indicated by the black square in Figure 1.8, where the numbers in the parentheses in the alloy formula indicate the laboratory measurement uncertainty. When we consider the typical tolerance ranges of alloy compositions in solder pastes (±0.2), the melting range for the eutectic composition Sn–3.5Ag–0.9Cu becomes approximately 13°C. Likewise, when the typical tolerance of (±0.2) in alloy composition is included, the iNEMI alloy Sn–3.9Ag–0.6Cu, shown by the medium gray square, has a melting range of 12°C. The third tin–silver–copper alloy Sn–3.0 Ag–0.5Cu, shown by the light gray square, has a melting range of 5°C. In spite of these seeming

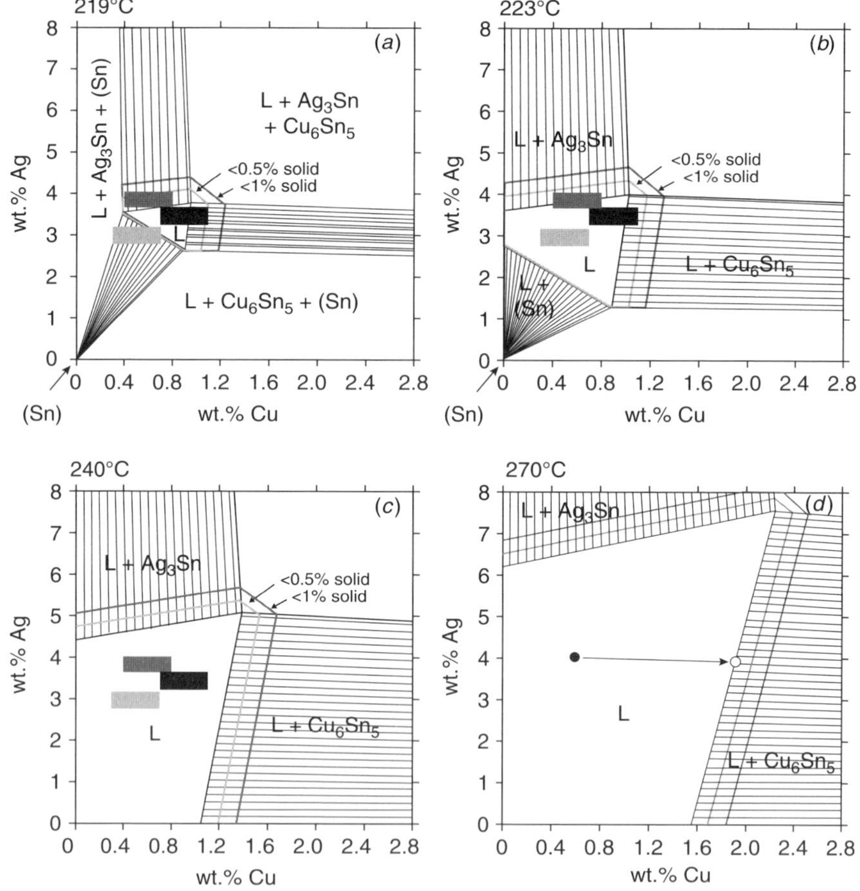

Figure 1.8. Isothermal sections through the Sn–Ag–Cu phase diagram: (*a*) 219°C, (*b*) 223°C (*c*) 240°C, (*d*) 270°C. Copper dissolution in the solder can change the composition from the initial (filled circle) to the final composition (open circle).

differences, in practice these alloys all melt in a remarkably similar way, making a wide range of alloy compositions acceptable in terms of their melting behavior.

Figures 1.8*a–d* show the compositions over which there is <0.5% and <1% solid as the temperature increases from 217°C, the eutectic temperature, to 219°C to 223°C to 240°C to 270°C. In Figure 1.8*a* the region marked "L" and bounded by the black triangle is the range of compositions that are completely liquid at 219°C. The regions outlined in green and red are compositions with less than 0.5% and 1% solid, respectively, at temperatures higher than 219°C. The values of 0.5% and 1% were chosen since the presence of less than 1% solid is expected to have no effect on the reflow behavior of solder pastes. The remaining solid-phase particles at this fraction are significantly smaller than the solder alloy powder particles from which they formed and will have a negligible effect on melting and coalescence of the alloy powders as they melt. As you can see from this plot, both the NEMI and the eutectic alloys have less than 1% solid remaining at 219°C. Beyond these two alloys, a wide range of alloys meets this criterion of having less than 1% solid remaining at 219°C. At 223°C (Figure 1.8*b*), the range of compositions broadens further, with all three alloys having less than 0.5% solid remaining. At 240°C (Figure 1.8*c*), the range of compositions with 0%, less than 0.5%, and less than 1% solid remaining is extremely broad. For practical reflow purposes, the effective liquidus temperatures measured will, therefore, be 217°C for a wide range of compositions as result of the small volume fraction of solid remaining above 217°C and includes all three alloys indicated. The isothermal phase diagrams in Figure 1.8 can be used to estimate the change in solder composition of a Sn–Ag–Cu alloy held at 219°C, 223°C, 240°C, or 270°C in contact with copper from the leads or pads, as well as in contact with silver from a board surface finish, and, upon solidification, estimate the amount of intermetallic in the solder joint. Figure 1.8*d* shows the initial alloy composition of Sn–3.9Ag–0.6Cu and the final composition as determined by the solubility limit of copper in the alloy at 270°C. Chada et al. [27] performed a comprehensive experimental study of the solubility limit of Cu in molten Sn–Ag–Cu solder alloys; their experimental results were in agreement with these calculations.

In terms of solidification of the solder joints as the assemblies are cooled, SAC alloys show similar behavior. As the joint cools, intermetallics form in the solder joint, both at the interfaces with the board and component and in the solder itself. The amount and types of intermetallic will be determined by the starting composition of the alloy and how much copper and other metals from the board and component have dissolved into the molten solder. Tin–silver–copper solder alloys actually cool with a significant amount of liquid to about 190°C because solid tin has difficulty forming. At about 190°C, all of the alloys quickly solidify to 100% solid.

Based on these analyses, the tin–silver–copper system is quite forgiving in terms of its insensitivity of melting and solidification behavior to composition over a wide composition range. Therefore, a minimal effect of solder composition on assembly processing should occur for compositions within this range. The same holds true for wave soldering. The temperatures for wave soldering are much higher than for reflow soldering and are determined by many factors, including the activity of the flux and the board design. The solder alloy composition will affect how much copper and other metals will dissolve in the bath, so one might conclude that the base solder should contain high amounts of copper. A tradeoff in copper concentration actually occurs: Low initial copper concentrations encourage fast dissolution from the boards and the components, while high initial copper concentrations encourage intermetallic formation in colder sections of the bath. This tradeoff has led us to suggest a copper concentration limit in the alloy of 0.5% to 0.6%.

1.3.1.2. Solidification Behavior

1.3.1.2.1. Nonequilibrium Effects. The pasty ranges based on equilibrium phase diagrams are the minimum pasty ranges that will occur during solidification. Nonequilibrium segregation and metastable phase formation may extend these ranges. In systems that exhibit substantial changes in the solubility of solid Sn during cooling, the amount of liquid present during cooling can be greater than predicted from the equilibrium phase diagram. Tin-based solder systems that exhibit this effect include Sn–Bi, Sn–In, and Sn–Pb. For example, as a Sn–rich Sn–Bi solder alloy is cooled from its liquidus temperature, the first solid that forms is Sn containing significantly less Bi than the Sn (Bi) solid solution at the eutectic temperature. If there is sufficient solid-state diffusion to maintain the equilibrium solid composition as the alloy cools, the final liquid transforms to solid at the equilibrium temperature and composition. If diffusion in the solid does not establish the equilibrium solid composition at each temperature as the alloy cools, the remaining liquid becomes increasingly Bi-rich and will solidify at the eutectic temperature. For a Sn–6Bi solder, the liquidus temperature is approximately 224°C and the equilibrium pasty range is approximately 26°C; in the limit of no diffusion in the solid, the pasty range can be as large as 85°C. In the NCMS Pb-Free Solder Project, DTA measurements of Sn–6Bi detected a measurable fraction of eutectic liquid that solidified at 139°C and, therefore, a nonequilibrium pasty range of 85°C.

This effect is illustrated in Figures 1.9 and 1.10 for the ternary Sn–Ag–Bi system with calculations of the solid fraction as a function of temperature and composition based on the phase diagram "lever rule" and nonequilibrium solidification, as represented by the Scheil equation [1–4, 20]. The liquidus projection of the ternary phase diagram is shown in Figure 1.9, where the lines correspond to compositions with the same liquidus temperatures. Considering the composition Sn–15Ag–7.5Bi, the last liquid solidifies at 185°C based on the equilibrium phase diagram; however, as a result of segregation during solidification, some liquid is predicted to still be present until the ternary eutectic temperature is attained at 138°C. The amount of nonequilibrium liquid present depends on the cooling conditions and will be between the limits defined by the two curves for Sn–3.5Ag–7.5Bi in Figures 1.10*a* and 1.10*b*. Since there is little solubility of Ag and Cu in Sn, nonequilibrium solidification due to interdiffusion in the solid plays little role in the behavior of SAC alloys.

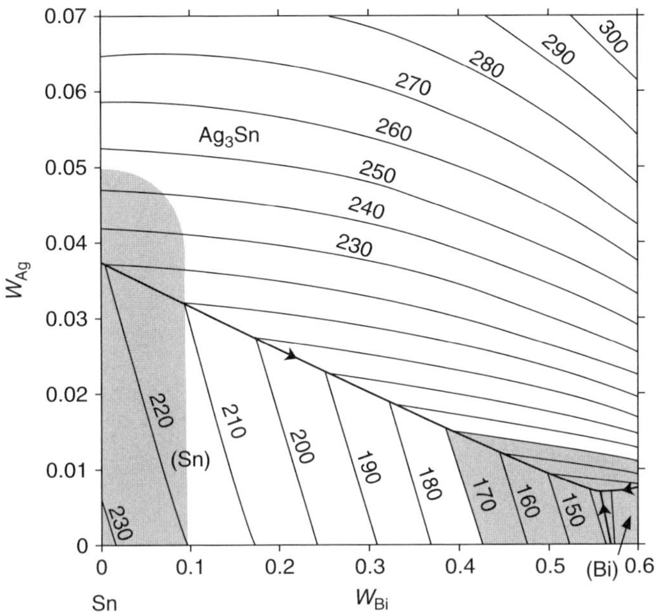

Figure 1.9. The liquidus projection of the Sn–Ag–Bi phase diagram. The hatched regions represent compositions with <30°C pasty range [1].

1.3.1.2.2. Metastable Phase Formation. The other characteristic of most Pb-free solder systems is the formation of nonequilibrium phases during cooling due to the difficulty in nucleating one or more phases, as illustrated using the Sn–Ag–Cu system [21]. The Sn–Ag–Cu phase diagram, the calculated solidification path, and DTA results for the Sn–4.7Ag–1.7Cu are shown in Figures 1.5, 1.11*a*, and 1.11*b*, respectively. At equilibrium, solidification begins with the formation of Cu_6Sn_5 at 265°C; at 238°C, formation of Ag_3Sn begins and the remaining liquid should transform to a mixture of Sn, Ag_3Sn, and Cu_6Sn_5 at the ternary eutectic of 217.5°C. However, during cooling in a DTA experiment (Figure 1.11*b*) from the liquid state, the first phases to form are Ag_3Sn and Cu_6Sn_5 at 244°C; and since solid Sn is difficult to nucleate, the liquid supercools by approximately 20°C while Ag_3Sn and Cu_6Sn_5 continue to form until the remaining liquid solidifies at 198.5°C. The latent heat (or heat of fusion) is released, leading to the solder self-heating to 217°C. This phenomenon is known as recalescence and can also be exhibited in the Sn–Pb system, but typically with a supercooling of 5°C. For the Sn–Ag, Sn–Cu, and Sn–Ag–Cu systems in particular, the existence of liquid below the eutectic temperature in the Sn–Ag, Sn–Cu, and Sn–Ag–Cu means that intermetallic phases form and coarsen in the liquid for significantly longer than expected from equilibrium behavior and the liquid becomes Sn-rich by the continued formation of the intermetallics. When the solid Sn phase finally nucleates in the Sn-rich

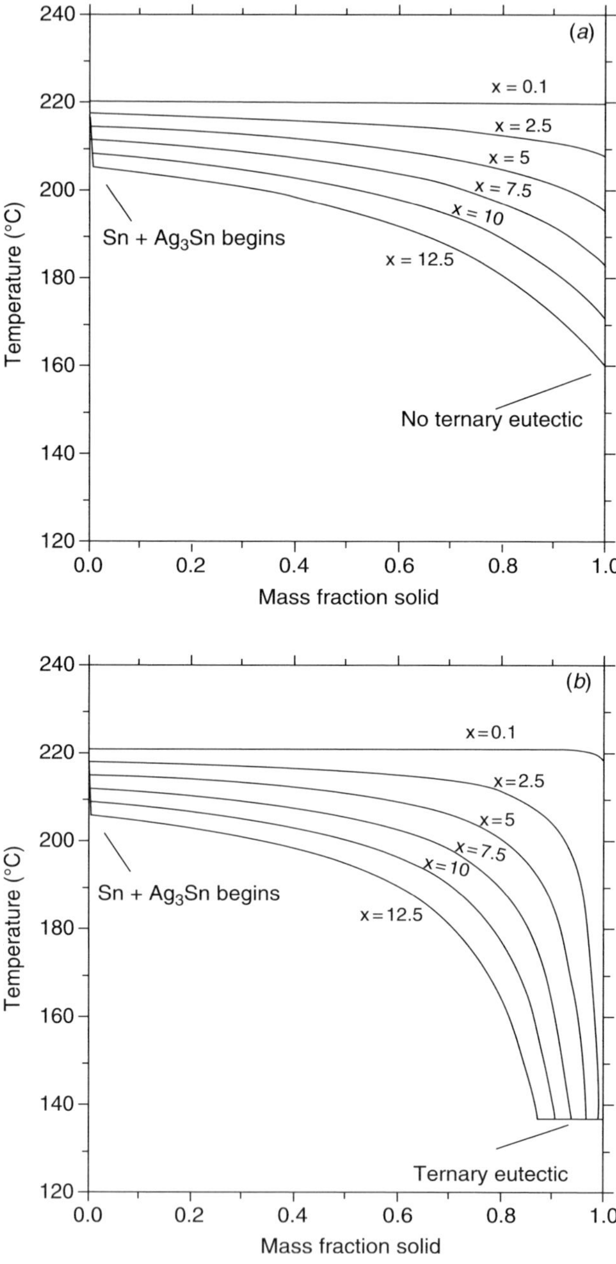

Figure 1.10. (*a*) Lever solidification calculation for Sn–Ag–Bi [1]. (*b*) Scheil solidification calculation for Sn–Ag–Bi [1].

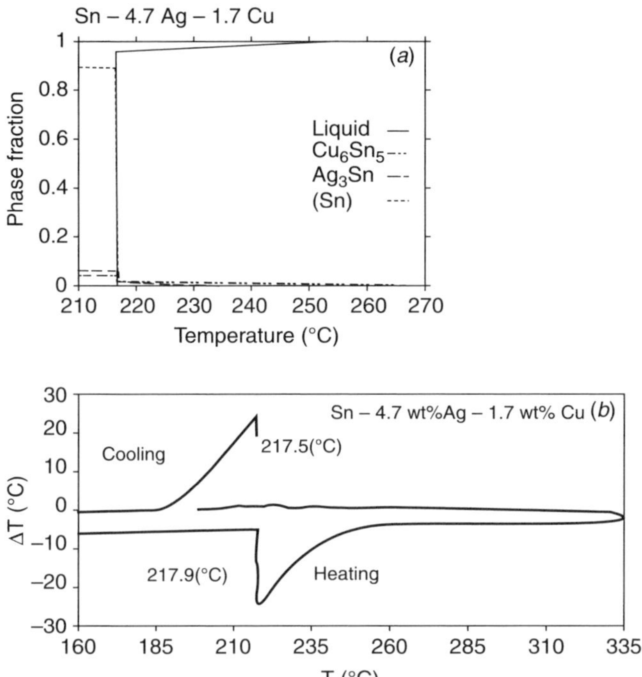

Figure 1.11. (*a*) Calculated solidification path for Sn–4.7Ag–1.7Cu. (*b*) DTA heating and cooling curves for Sn–4.7Ag–1.7Cu [21].

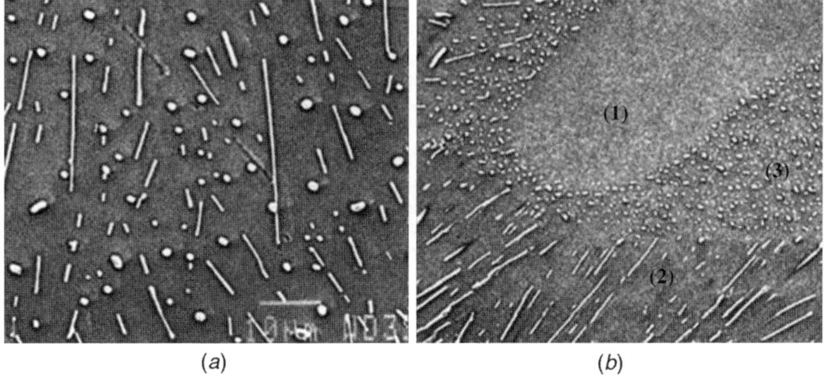

Figure 1.12. SEM micrographs of eutectic structures: (*a*) ternary eutectic structure (matrix, Sn: needle shape, Ag_3Sn; and disk shape, Cu_6Sn_5). (*b*) Region with coexisting Sn + Cu_6Sn_5 and Sn + Ag_3Sn fine two-phase regions near an Sn dendrite arm. Region labels: **1**, Sn; **2**, Sn + Ag_3Sn; **3**, Sn + Cu_6Sn_5; **4**, Sn + Ag_3Sn + Cu_6Sn_5 [21].

liquid phase, the solid Sn phase grows as large Sn dendrites. This is the origin of the commonly observed multiphase, heterogeneous microstructures characteristic of SAC alloys (shown in Figure 1.12) which contain tin dendrites, rather than a classic "eutectic" microstructure characteristic of Sn–Pb. A comprehensive experimental and theoretical study by Moon et al. [21] on the Sn–Ag–Cu system provides more detailed discussion of the microstructures, melting behavior, and solidification behavior, which applies to Sn–Ag, Sn–Cu, and alloys of Sn–Ag–Cu with other alloy additions.

1.3.1.2.3. Fillet Lifting. A failure phenomenon for through-hole joints that occurs for some Pb-free solders during solidification that does not occur for eutectic Sn–Pb is "fillet lifting." Fillet lifting, as shown in the micrograph in Figure 1.13, is characterized by the complete or partial separation of a solder joint fillet from the intermetallic compound on the land to the shoulder of the through hole. This phenomenon was first identified in 1993 by Vincent and co-workers in the DTI-sponsored Pb-free solder project in which fillet lifting was attributed to the presence of the Sn–Bi–Pb ternary eutectic (98°C) resulting from Pb contamination of Bi-containing solders from the Sn–Pb HASL board finish [34–36]. This effect is now known to occur without Pb contamination for some Pb-free solder alloys, particularly for high-Sn Bi-containing alloys.

From work by Suganuma [37], Boettinger et al. [38], and Takao and Hasegawa [39], fillet lifting has been found to be a result of "hot tearing," a mechanism that leads to relief of thermally induced stresses *when the solder is between 90% and 100% solid.* The differential shrinkage due to CTE mismatch between the board and the solder generates the stresses; at lower solid fractions, fluid flow occurs relieving the stresses. As the volume fraction of liquid decreases, the stresses are carried by the dendritic matrix and failures occur at the weakest point, the location

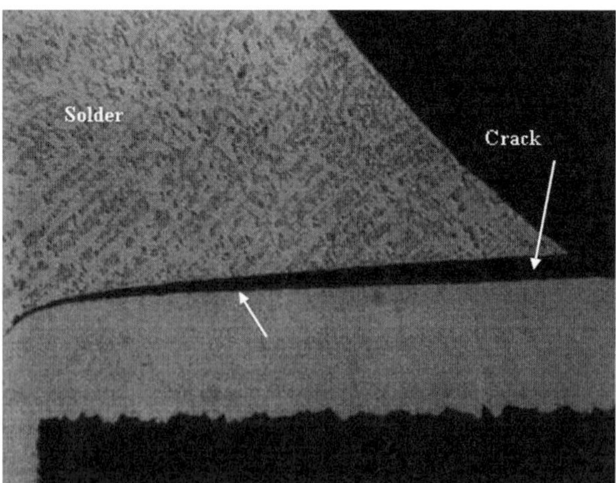

Figure 1.13. Optical microscope cross section of fillet lifting in a through hole solder joint with Sn–3.5Ag–5Bi alloy.

with the highest remaining liquid fraction: the board-side intermetallic compound/solder interface. The tendency for hot tearing increases as the pasty range increases and the temperature difference between 90% and 100% solid ($\Delta T_{90\% \rightarrow 100\%}$) increases and is typically worse for alloys with a large nonequilibrium pasty range, like Sn–Bi or Sn–Ag–Bi. Takao and Hasegawa [39] have quantified the tendency to fillet lifting as a function of alloy composition in terms of the enthalpy change as a function of composition and temperature during cooling which corresponds directly to $\Delta T_{90\% \rightarrow 100\%}$.

In the NCMS Pb-free solder project, the "hot tearing" hypothesis was tested by taking Sn–3.5Ag, an alloy that showed minimal fillet lifting, and transforming it into an alloy showing close to 100% cracked joints with the addition of 2.5 wt% Pb [1]. The addition of 2.5% Pb increased the pasty range from 0°C to 34°C. These results predicted that Pb contamination from Sn–Pb surface finishes would lead to fillet lifting in alloys that in their uncontaminated state showed little or no fillet lifting. Subsequent wave soldering experiments by Multicore [40], Nortel [41], and others exhibited fillet lifting in through-hole joints with Sn–Ag, Sn–Cu, or Sn–Ag–Cu solders and Sn–Pb surface finished components and/or boards. It should also be remembered that Sn–Ag, Sn–Cu, and Sn–Ag–Cu fillets may also show fillet lifting for thick boards without Pb contamination, as indicated in the NCMS Pb-Free Solder Project with Sn–Ag eutectic [1].

1.3.2. Solidification and Surface Porosity

As noted in Moon et al. [21] and in the discussion of Figure 1.11*b* above, solidification of Sn–Ag–Cu alloys occurs with the formation of Sn dendrites as seen in the as-solidified structure in Figure 1.14*a*. The formation of Sn dendrites is accompanied by the redistribution of the interdendritic liquid and, ultimately, by a retraction of liquid as it solidifies. (The volume of the liquid is larger than the volume of the solid it becomes by solidification.) This retraction of the interdendritic liquid leads to a rough surface as indicated by the arrows in Figure 1.14*a* and,

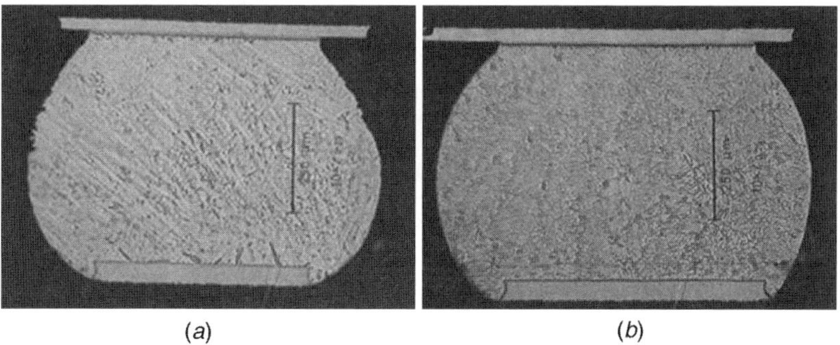

(*a*) (*b*)

Figure 1.14. (*a*) Cross section showing surface roughness of Sn–Ag–Cu alloy as solidified. (*b*) Comparison of surface roughness of Sn–Pb and Sn–Ag–Cu solder joints.

correspondingly, to an overall greater surface roughness than Pb–Sn eutectic, as seen in the SEM micrographs in Figure 1.14*b*. These micrographs indicate why the surfaces of properly soldered Pb-free solder joints appear significantly rougher than correspondingly well-soldered Pb-Sn solder joints and why the visual inspection criteria must be changed for Pb-free solder joints: the dendritic microstructure causing the roughness is an intrinsic characteristic of the Pb-free alloys. The scale of the surface roughness depends on several factors, including the final joint composition and the cooling rate of the joint. The IDEALS project examined the effect of these surface "pores" caused by retraction of the solidifying interdendritic liquid on where the solder joint fails during thermal cycling. They determined that the surface "pores" between the dendrites were not preferential sites for solder joint failure [5–8].

1.3.3. Contamination of Pb-Free Solders

The use of a solder-based board or component surface finish with a different composition than the solder paste or wave soldering alloy may result in different properties of the solder joints than expected from solder joints made from the paste or wave composition alloy alone. For Pb-contamination in particular, the liquidus temperature decreases by 1.3°C (per mass fraction Pb · 100), which can be calculated using Eq. (1.1). The next question to be answered is how the Pb contamination affects the solidus temperature, the lowest temperature where liquid exists. When Pb-free solder alloys are contaminated by Pb from the pre-tinned layer, the last liquid that solidifies may form a low melting eutectic. This case was studied in detail for Pb-contaminated Sn–Bi solders by Moon et al. [19] using DTA methods in conjunction with calculations of the equilibrium phase diagram and Scheil solidification. They found that contamination of Sn–Bi eutectic, Sn–5Bi, and Sn–10Bi alloys by 6% Pb results in the formation of a measurable fraction of low melting Sn–Bi–Pb eutectic at 95°C.

Since the freezing ranges of other Pb-free solder alloys may be similarly susceptible to Pb contamination, the freezing behavior of four additional solder alloys was studied by Kattner and Handwerker [22] using lever rule and Scheil freezing path calculations of the original solder alloy and the contaminated solder. The level of contamination was chosen to be 6% Pb from Moon et al.'s estimate of a Pb concentration of 6% (mass fraction) in the solder from contamination by the component

TABLE 1.3. Modified Solder Compositions as a Result from Contamination with 16% of Sn–37% Pb Solder[a]

Original Solder Composition	Contaminated Solder Composition
Sn–3.5% Ag	Sn–2.9% Ag–6% Pb
Sn–4% Ag–1% Cu	Sn–3.4% Ag–0.8% Cu–6% Pb
Sn–3.5% Ag–4.8% Bi	Sn–2.9% Ag–4% Bi–6% Pb
Sn–3.4% Ag–1% Cu–3.3% Bi	Sn–2.8% Ag–0.8% Cu–2.8% Bi–6% Pb

[a]Compositions are in percentage of mass fraction.

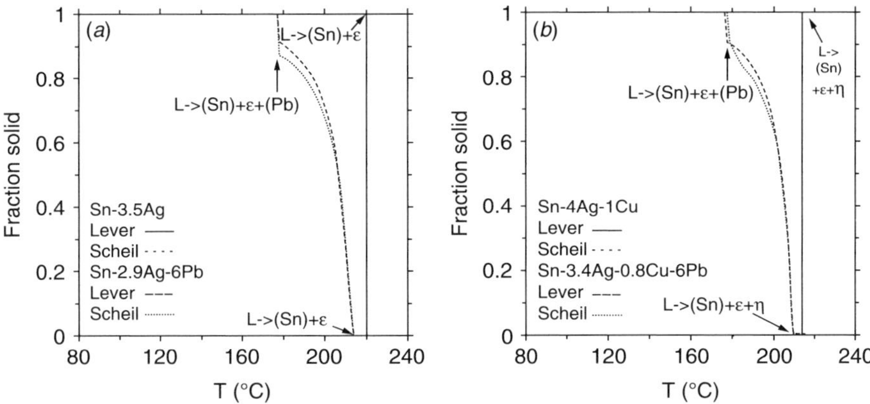

Figure 1.15. Lever and Scheil calculations for fraction solid as a function of temperature for Pb-free solders without and with 6% Pb contamination: (*a*) Sn–3.5Ag and (*b*) Sn–Ag–Cu [22].

lead and board pre-tinning. The original solder compositions and those resulting from contamination are listed in Table 1.3.

The calculations were carried out using the NIST solder database [42], the Thermo-Calc software package [43] and the Scheil and Lever programs [44]. Figures 1.15*a–b* show the calculated fraction solid as a function of temperature for Sn–3.5 Ag and Sn–4Ag–1 Cu and for the corresponding alloys contaminated by 6% Pb. Contamination of the binary eutectic alloy Sn–3.5Ag with 6% Pb lowers the liquidus temperature from 221°C to 213°C and lowers the solidus from 221°C to 177°C, creating an alloy with an equilibrium pasty range of 44°C. Likewise, Pb contamination of Sn–4Ag–1 Cu alloy leads to a 2°C increase in the liquidus temperature from 225°C to 227°C and a decrease in solidus temperature from 215°C to 177°C.

1.3.4. Wetting and Solderability

Wetting of a liquid on a solid is determined by the relative energies of the liquid–vapor surface tension, the solid–liquid interfacial energy, and the solid–vapor interfacial energy. The thermodynamics of an alloy plays a central role in determining its intrinsic surface tension. It is well known that the surface tension of pure Sn is significantly higher than Sn–Pb eutectic, as measured by White as a function of temperature from pure Sn to pure Pb [45]. Ohnuma et al. [46, 47] have used thermodynamic parameters to predict the surface tension and viscosity of the Sn-based liquid solder as a function of composition, as shown in Figure 1.16. The difference in surface tension between Sn–Pb alloys and Pb-free Sn-based alloys translates into generally higher contact angles for Pb-free alloys [48].

Evaluation of solderability in manufacturing has considerably greater complexity than wetting of molten solder on a substrate in a controlled laboratory environment

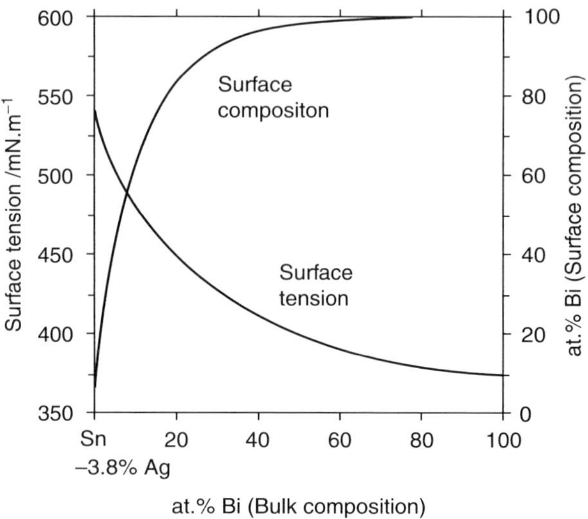

Figure 1.16. Calculated dependence of surface tension as a function of composition for Sn-based liquids.

[48], but simple wetting balance and area-of-spread measurements are useful for (a) separating the effects of some of these factors and (b) as screening tools when comparing different solder alloys. (Manufacturing issues are discussed in detail in Chapter 8.) Through numerous national and international Pb-free solder R&D projects using wetting balance measurements, solderability was found to be a serious issue only for Zn-containing alloys, and then only for concentrations greater than 1% Zn [1].

For Pb-free alloys not containing Zn, their wetting characteristics on a specific metal substrate depend on the compositions of the solder and the substrate, the temperature of the solder and the substrate, the size and thermal conductivity of the substrate, the liquidus temperature of the solder, the surface condition of the substrate, the gaseous experimental environment (oxygen, air, nitrogen), and, last but not least, the flux. A comparison of wetting balance data for various Pb-free solder alloys on copper from the IDEALS and NCMS projects indicates that (1) in general, the temperature for similar wetting balance performance to eutectic Sn–Pb scales with the liquidus temperature of the Pb-free solder and (2) the effects of the variables listed above are separable. Figure 1.17 from the IDEALS Pb-Free Project [5–8] shows the time to 2/3 wetting force for five Pb-free solder alloys compared with Sn–40Pb at three temperatures per alloy, $T_\ell + 25°C$, $T_\ell + 35°C$, and $T_\ell + 50°C$. With the exception of Sn–0.7Cu, the characteristic wetting times are virtually indistinguishable using ACTIEC 5 flux (Figure 1.17*a*). When the flux is changed to pure Rosin flux, four of the five Pb-free solders are again virtually identical to Sn–40Pb (Figure 1.17*b*). Only Sn–0.7Cu–0.5Bi shows significantly poorer wetting than the other five solders. Figure 1.18 shows similar results from the NCMS Pb-Free Solder Project [1–4].

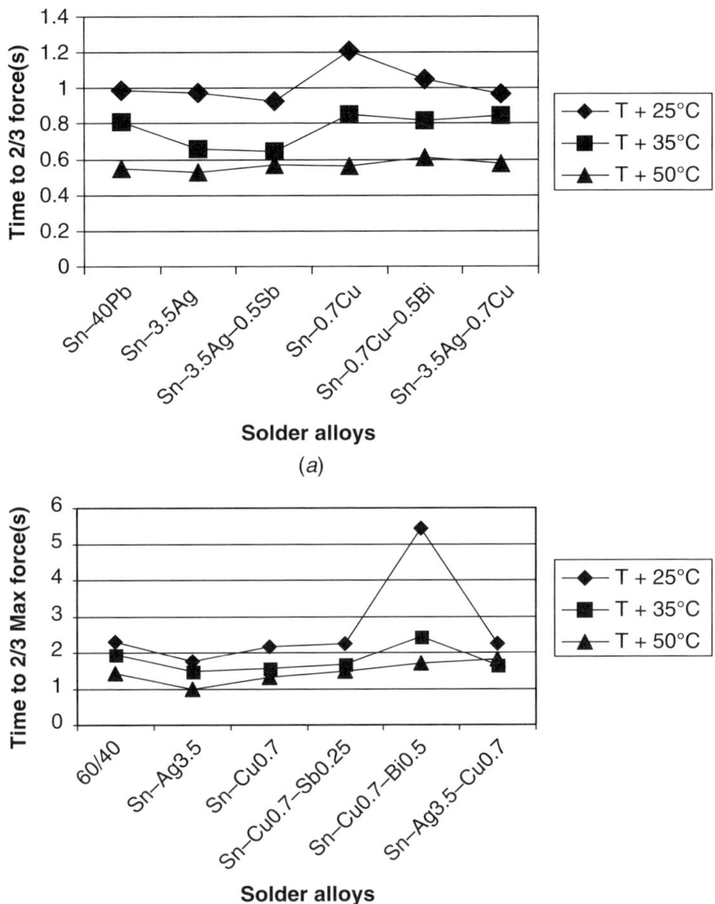

Figure 1.17. (*a*) IDEALS project wetting data: wetting balance parameter, time to 2/3 force as a function of alloy composition for three different temperatures relative to the liquidus temperature for Actiec 5 flux [7]. (*b*) IDEALS project wetting data: wetting balance parameter, time to 2/3 force as a function of alloy composition for three different temperatures relative to the liquidus temperature for rosin flux [7].

1.3.4.1. Effect of Surface Finish on Wetting. The range of possible wetting behavior quickly broadens as manufacturing variables are included. In the NCMS Pb-Free Solder Project, the solderability of component leads was characterized using a semiquantitative "wetting figure-of-merit" as a function of solder composition, solder reflow temperature profile, and surface finish. The wetting performance of each alloy was evaluated during SMT assembly with pastes containing a conventional no-clean RMA (rosin mildly activated) flux. The wetting performance

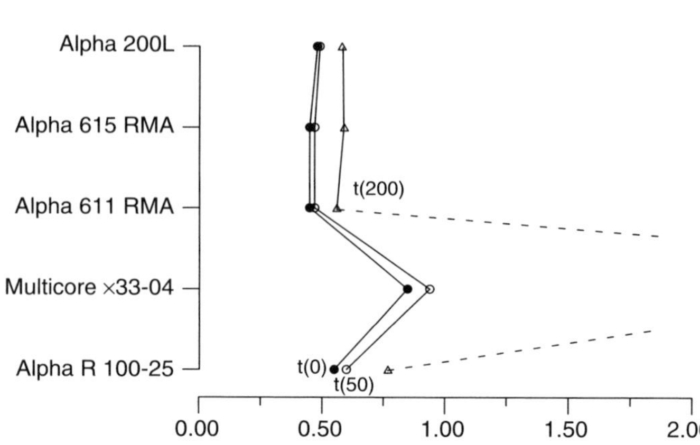

Figure 1.18. NCMS project wetting data: three wetting balance time parameters as a function of flux type and temperature for Sn–3.5Ag [1].

of the Pb-free solders was almost as good as the eutectic Sn–Pb control, except when soldering to the imidazole OSP finish. At least one Pb-free alloy matched the wetting performance of the eutectic Sn–Pb control alloy for each metal finish tested, other than the imidazole OSP-coated Cu. Most metallic surface finishes improved the spreading of the Pb-free solders. In the case of the Ni/Au finish, all Pb-free solders exhibited wetting scores indicating the best performance possible. Immersion Sn finish also enhanced the spreading of the Pb-free solders, most significantly in the case of the Sn–58Bi eutectic. On both the Ni/Pd and Pd-over-Cu finishes, the Sn-rich solders exhibited adequate wetting and spreading (equivalent to Sn or Ni/Au surface finishes), whereas Sn–58Bi and Sn–2.8Ag–20 in exhibited considerably reduced spreading. The IDEALS and NCMS results demonstrate that Pb-free alloys can be differentiated based on these experiments, even ranked relative to each other, and to Sn–Pb eutectic.

Bradley and Hranisavljevic [49] have provided a thermodynamic, that is, liquidus temperature, basis for understanding the effect of alloy composition and surface finish/lead metal on wetting behavior. They determined the temperatures at the start of solder paste coalescence and at full wetting, when all the solder particles are completely coalesced into a molten solder mass, for the alloys Sn–3.5Ag, Sn–3.8Ag–0.7Cu, Sn–1Ag–3Bi, Sn–1Ag–4.8Bi, and Sn–1Ag–7.5Bi and for four surface finishes. Figure 1.19 shows that the measured temperatures for all the solder alloy pastes, except for Sn–3.8Ag–0.7Cu, decrease with dissolution of the surface finish metal into the solder, with copper surface finish producing the greatest effect, followed by gold. This decrease in solder coalescence temperatures corresponds to a decrease in the liquidus temperatures produced by dissolution of the surface finish into the solder alloy. The alloy Sn–3.8Ag–0.7Cu is close to the ternary eutectic composition and the solidus temperature is, thus, unaffected by Cu dissolution.

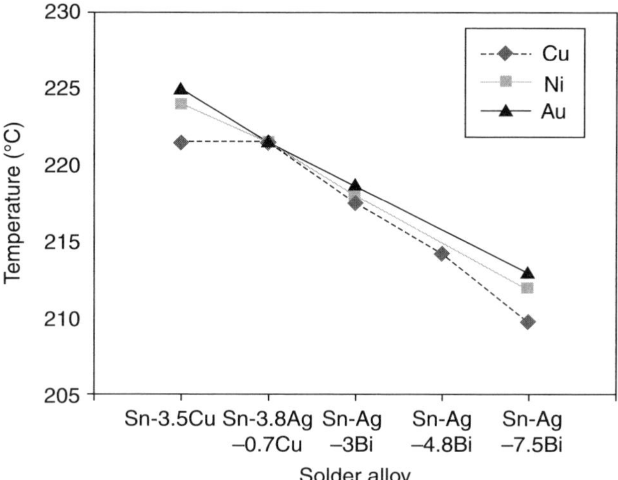

Figure 1.19. Temperature at which solder paste begins to coalesce as a function of PCB substrate finish. Note the effect of Cu on depressing the coalescence temperature and therefore the effective wetting temperature of alloys without Cu [49].

1.3.5. Reliability Concerns

Solder alloys can be easily ranked based on mechanical property values obtained in a particular test or for a particular application. For solder alloys, however, one of the most important mechanical properties is resistance to thermomechanical fatigue (TMF). Unfortunately, the most widely accepted method to determine TMF resistance is using accelerated thermal cycling tests on PWBs, tests that are both time- and labor-intensive. A complicating factor is that for many materials, using methods and conditions that accelerate fatigue produces failure modes that are not relevant for the real product application.

For Sn–Pb solders, there is general acceptance of accelerated thermal cycling of PWB test vehicles under specific conditions as a reasonable method to assess TMF resistance. This acceptance comes from decades of industrial experience relating specific thermal cycling conditions to wear-out failure in specific classes of product. From the NCMS Pb-Free Project, the Pb-free solder alloys were able to withstand different amounts, types, and rates of loading which are dependent upon the different coefficients of thermal expansion (CTE) and mechanical properties of the board, components, and alloys, solder joint geometry, solder microstructure, and residual stresses. Taken together for a given alloy, these properties can produce solder joint performance better for some components than eutectic Sn–Pb and worse for other components on the same board, and they may be different for different thermal cycling conditions [1–4].

In the NCMS Pb-Free Solder Project [1–4], the only surface mount components with obvious fatigue failures after more than 6700 cycles of 0°C to 100°C, or 5000 cycles of −55°C to +125°C, were leadless ceramic chip carriers (LCCC) and 1206 chip resistors. No leaded surface mount devices exhibited failures. There were no unexpectedly early or catastrophic chip carrier or passive component failures. Those failures that occurred followed the same component order as observed for eutectic Sn–Pb. The ranking of alloys relative to eutectic Sn–Pb varied with thermal cycling conditions and component type. (This is discussed in greater detail below.) The effects of thermal cycling conditions and component type on the relative performance of Pb-free solders can also be seen by the thermal cycling results on fleXBGA and PBGA packages (Tables 1.4 and 1.5) from the NCMS High Temperature, Fatigue Resistant Solder Project [13, 14].

The NCMS alloy ranking results, which change with component type, demonstrate the dangers of using a single component, a small subset of typical solder joint configurations, or a set of laboratory experiments, such as creep tests, to predict general behavior. The open questions, however, are what performance is necessary for Pb-free solder alloys to be acceptable in most product applications and how well laboratory measurements and accelerated thermal cycling of test vehicles predict the performance of a given solder alloy relative to Sn–Pb eutectic. In this section, the thermal, compositional, and microstructure origins of the mechanical properties of Pb-free alloys are examined to illustrate what mechanical behavior might be expected relative to Sn–Pb eutectic.

The temperature and composition dependence of alloy mechanical properties is illustrated from the NCMS and IDEALS Project results using laboratory test methods. In the IDEALS Project, a range of physical properties of the Pb-free solder alloys was measured, including coefficient of thermal expansion (CTE), elasticity, yield stress, and plastic behavior. Figure 1.20*a* illustrates a significant point about lead-free solders as compared with Sn–Pb. The yield stress of eutectic Sn–40Pb is lower than for Sn–3.5Ag, Sn–0.7Cu–0.5Sb, and Sn–3.5Ag–0.7Cu for all temperatures. The solder Sn–0.7Cu exhibited the lowest yield stress at low temperatures, but becomes virtually the same as the other Pb-free alloys above 125°C. If the data are replotted using temperature normalized to their liquidus temperatures (homologous temperature) as seen in Figure 1.20*b*, additional information can be obtained about their relative behavior. The yield stresses of Sn–40Pb, Sn–3.5Ag, Sn–0.7Cu–0.5Sb, and Sn–3.5Ag–0.7Cu are similar at low homologous temperature. At higher homologous temperatures, the yield stress of Sn–40Pb continues to decrease with increasing temperature, approaching zero at the eutectic temperature. For the Pb-free solders, the yield stress shows a lower dependence on temperature as the homologous temperature increases.

Maintaining strength with increasing temperature is characteristic of precipitation hardened materials. In the case of Pb-free solders, the precipitation hardening is provided by the presence of the intermetallic phases, dispersed in and between the Sn dendrites. Creep results for Pb-free solder alloys display similar transitions in behavior, leading to changing alloy rankings of creep resistance as the temperature and strain rate change.

TABLE 1.4. Comparison of Pb-Free Alloys Reliability for fleXBGA Package (TC2 Cycling, 0°C to +100°C)

Alloy Code	Alloy Elements	Number on Test	Number Failed	First Failure (Cycle)	Mean Life (Cycle)	Rank by First Failure	Rank by Mean Life
A1	Sn−3.5Ag	15	6	6288	10300	5	3
A11	Sn−4Ag−1Cu	15	6	6967	9456	4	4
A14	Sn−4Ag−0.5Cu	15	11	6073	8861	6	6
A21	Sn−2.5Ag−0.8Cu−0.5Sb	14	6	8089	9238	3	5
A32	Sn−4.6Ag−1.6Cu−1Sb−1Bi	14	0	N/A	N/A	1	1
A62	Sn−3.4Ag−1Cu−3.3Bi	15	0	N/A	N/A	1	1
A66	Sn−3.5Ag−1.5ln	15	13	5630	6448	7	7
B63	Sn/Pb Control	14	14	3418	4465	8	8

TABLE 1.5. Comparison of Pb-Free Alloys Reliability for PBGA Package (TC1 Cycling, −40°C to 125°C)

Alloy Code	Alloy Elements	Number on Test	Number Failed	First Failure (Cycle)	Second Failure (Cycle)	Mean Life (Cycle)	Rank by First Failure	Rank by Mean Life
A11	Sn−4Ag−1Cu	14	11	4476	4686	5428	5	6
A14	Sn−4Ag−0.5Cu	14	3	5195	6054	N/A	3	3
A21	Sn−2.5Ag−0.8Cu−0.5Sb	14	6	3450	4621	6734	6	4
A32	Sn−4.6Ag−1.6Cu−1Sb−1Bi	15	0	N/A	N/A	N/A	1	1
A62	Sn−3.4Ag−1Cu−3.3Bi	14	1	5875	N/A	N/A	2	2
A66	Sn−3.5Ag−1.5In	14	9	5102	5207	5784	4	5
B63	Sn/Pb Control	14	14	3395	3462	3710	7	7

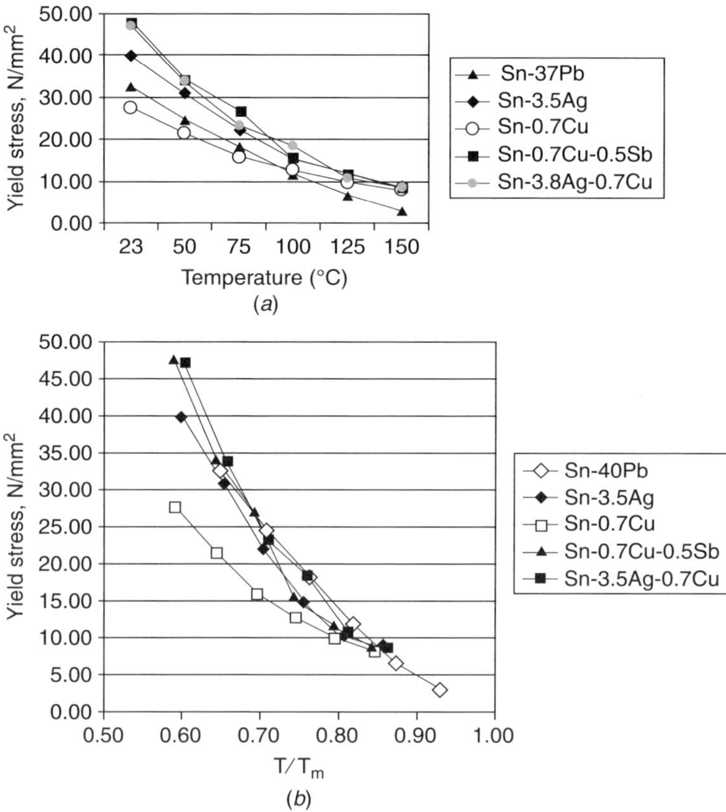

Figure 1.20. (*a*) Yield stress as a function of composition and temperature [3, 7]. (*b*) Yield stress as a function of composition and fraction of liquidus temperature in K [3, 7].

1.4. R&D ISSUES REMAINING IN LEAD-FREE SOLDER IMPLEMENTATION

Following these analyses, NEMI chose a single lead-free (Sn–3.9Ag–0.6Cu) alloy in 1999 to replace eutectic Sn–Pb solder for high-volume surface mount applications. Once the solder selection was made, the NEMI task force then focused on the open questions of component survivability during assembly, process optimization, and assembly reliability. (1) Components, board materials, and fluxes, for example, were all designed for peak temperatures around 220°C, not 240–260°C, needed for SAC alloys. If these materials are not stable, new materials may be needed immediately to allow the products to survive the manufacturing process. (2) The assembly processes are far from optimum. What is the processing window that leads to acceptable joints? Does the reduced wetting relative to Sn–Pb eutectic lead to a reduction in reliability? How does this change with various board and component surface finishes? Certainly the cosmetics are changed. Fluxes will need reformulation to provide adequate performance at higher process temperatures. Cleaning agents, if used, may require modification to effectively clean residues of modified fluxes from printer wiring board assemblies processed at higher temperatures. Inspection procedures and joint specifications will have to be revised from current practice. (3) In terms of reliability, more extensive testing is required for hybrid applications, a wide range of passive components, small outline integrated circuits (SOIC), quad flat packages (QFP), ball grid arrays (BGA), chip scale packages (CSP), and direct chip attach (DCA) devices. The reliability of through-hole assemblies was a completely open question. Results in these areas are discussed in the chapters that follow.

A major remaining question is on the acceleration factors for Pb-free solders, which quantify the relationship between solder composition, thermal cycling conditions, component and board materials and geometries, and useful assembly life. Based on the viewpoints of different individuals, the current "common wisdom" on the ability of ATC tests to predict Pb-free solder joint lifetime spans the extremes: ATC tests are believed to either underestimate or overestimate the reliability of lead-free solders compared with Sn–Pb eutectic. The complexity of the situation was enunciated by Bartello et al. [50] in their examination of the relative performance of Sn–Pb eutectic and Pb-free solders as a function of ATC cycling conditions for a single component type, ceramic ball grid arrays (CBGA). For 0–100°C ATC testing with cycles times of 30–240 minutes, the ATC performances of CBGAs assembled with Sn–3.8Ag–0.7Cu and Sn–3.5Ag–3.0Bi (SAB) alloys were superior to those assembled with eutectic Sn–Pb CBGAs. When the thermal cycling condition was changed to −40°C to 125°C at cycle times from 42 minutes to 240 minutes, the ATC performance of CBGAs assembled with the SAC alloy was inferior to eutectic Sn–Pb CBGAs. Using the same temperature difference as 0°C to 100°C, the performance of the SAC alloy was superior to those assembled with eutectic Sn–Pb CBGAs. Similar results were obtained for the Sn–3.5Ag–3Bi alloy, with the added complication that for −40°C to 125°C ATC test conditions, the CBGAs assembled with SAB performed better than the Sn–Pb eutectic controls for short cycle times (42 minutes), but performed worse at longer ATC cycle times (240 minutes). These results are similar to those obtained by Woosley and co-workers [51] for a wider variety of components, but a smaller set of cycling conditions. Additional data on the lifetimes of commercial products coupled with further ATC testing are needed before these issues can be resolved.

1.5. SUMMARY

The behavior of solder alloys in manufacturing and in use can be understood in terms of their thermodynamic properties, the kinetics of reactions, including wetting, and their temperature, stress, and strain-rate-dependent mechanical properties. While the performance of a specific solder alloy cannot be quantitatively predicted in manufacturing or in product applications based on laboratory experiments, the metallurgical concepts outlined in this chapter combined with laboratory measurements have been used to identify alloys that have potential for commercial use. The thermodynamics of alloy melting, solidification, and wetting forms the basis for understanding solder joint formation in reflow and wave soldering applications. Once a printed circuit board is successfully assembled, the reliability of its solder joints in use depends not only on solder's thermomechanical properties, including thermal expansion coefficient and the response of the alloy microstructure to the applied stress, but also on the properties and the response of the components and the circuit board in the system. By examining the dependence of the mechanical properties of Pb-free and Sn–Pb eutectic solder as a function of temperature, strain rate, and stress, the underlying mechanisms responsible for solder behavior changing as a function of alloy composition are beginning to be revealed. As the microelectronics community becomes more experienced with SAC alloys, the remaining issues regarding assembly and reliability will undoubtedly be addressed. In terms of a "standard" alloy, there will remain, at least for several years, a tension between (a) component manufacturers who want to push peak assembly temperatures down to those used for Sn–Pb eutectic and (b) board assemblers who want as large a process window as possible to allow high yield surface mount assembly of large, complex boards.

Disclaimer: Commercial equipment and materials are identified in order to adequately specify certain procedures. In no case does such identification imply recommendation or endorsement by the National Bureau of Standards, nor does it imply that the materials or equipment identified are necessarily the best available for the purpose.

REFERENCES

1. NCMS Lead-Free Solder Project Final Report, NCMS, National Center for Manufacturing Sciences, 3025 Boardwalk, Ann Arbor, Michigan 48108-3266, Report 0401RE96, August 1997, and CD-ROM database of complete dataset, including micrographs and raw data, August 1999. Information on how to order these can be obtained from http://www.ncms.org/
2. I. Artaki, D. Noctor, C. Desantis, et al., Research trends in lead-free soldering in the US: NCMS Lead-Free Solder Project (Keynote), pp. 602–605, 1999. IEEE Computer Society. Proceedings—EcoDesign '99: First International Symposium on Environmentally Conscious Design and Inverse Manufacturing, February 1–3, 1999, Tokyo, Japan.
3. C. A. Handwerker, E. E. de Kluizenaar, K. Suganuma, and F. W. Gayle, Major international lead-free solder studies, in K. J. Puttlitz and K. A. Stalter, Eds., *Issues and Implementation of Pb-free Technology in Microelectronics*, McGraw-Hill, New York, 2004.
4. C. A. Handwerker, NCMS lead-free solder project: A summary of results, conclusions and recommendations, IPC Work '99: An International Summit on Lead-Free Electronics Assemblies, Proceedings, October 23–28, 1999; Minneapolis, MN.
5. M. Harrison and J. H. Vincent, Improved design life and environmentally aware manufacturing of electronic assemblies by lead-free soldering, <http://www.lead-free.org/research/index.html>

6. M. R. Harrison and J. Vincent, IDEALS: Improved design life and environmentally aware manufacturing of electronics assemblies by lead-free soldering, in *Proceedings, IMAPS Europe '99* (Harrogate, GB), June 1999.
7. The Synthesis Report for the IDEALS project can be downloaded from.: <http://www.alphametals.com/products/lead_free/PDF/synthesis.pdf>, <http://www.marconicaswell.com/tech/emtec.htm>, and <http://www.cordis.lu>
8. M. H. Biglari, M. Oddy, M. A. Oud, et al., Pb-free solders based on SnAgCu, SnAgBi, SnAg, and SnCu, for wave soldering of electronic assemblies, in *Proceedings, Electronics Goes Green 2000+* (Berlin, Germany), September 2000.
9. Second European Lead-Free Soldering Technology Roadmap, February 2003 and Framework for an International Lead-Free Soldering Roadmap, December 2002, Soldertec, available at http://www.lead-free.org
10. E. Bradley, NEMI Pb-free interconnect task group report, IPC Work '99: An International Summit on Lead-Free Electronics Assemblies, Proceedings, October 23–28, 1999, Minneapolis, MN.
11. J. Bath, C. Handwerker, and E. Bradley, Research Update: Lead-Free Solder Alternatives, Circuits Assembly, May 2000, pp. 31–40.
12. A. Rae and C. A. Handwerker, Circuits Assembly, April 2004.
13. F. Gayle, G. Becka, J. Badgett, et al., High temperature lead-free solder for microelectronics, *J. Miner. Metals Mater. Soc.* **53**(6), 17–21, 2001.
14. Frank W. Gayle, Fatigue-resistance, high temperature solder, *Adv. Mater. Processes* **159**(4), 43–44, 2004.
15. K. Suganuma, Research and development for lead-free soldering in Japan, IPC Work '99: An International Summit on Lead-Free Electronics Assemblies, Proceedings, October 23–28, 1999, Minneapolis, MN.
16. JEITA Lead-Free Roadmap 2002 for Commercialization of Lead-Free Solder, September 2002, Lead-Free Soldering Roadmap Committee, Technical Standardization Committee on Electronics Assembly Technology, JEITA (Japan Electronics and Information Technology Industries Association).
17. NEDO Research and Development on Lead-Free Soldering, Report No.00-ki-17, JEIDA, Tokyo, Japan, 2000.
18. *Lead-Free Soldering—An Analysis of the Current Status of Lead-Free Soldering*, Report from the UK Department of Trade and Industry. Copies can be obtained from the ITRI website: http://www.lead-free.org/
19. K. W. Moon, W. J. Boettinger, U. R. Kattner, C. A. Handwerker, and D. J. Lee, The effect of Pb contamination on the solidification behavior of Sn–Bi solders, *J. Electron. Mater.* **30**(1), 45–52, 2001.
20. U. R. Kattner and W. J. Boettinger, On the Sn–Bi–Ag ternary phase-diagram, *J. Electron. Mater.* **23**, 603–610, 1994.
21. K. W. Moon, W. J. Boettinger, U. R. Kattner, F. S. Biancaniello, and C. A. Handwerker, Experimental and thermodynamic assessment of Sn–Ag–Cu solder alloys, *J. Electron. Mater.* **29**(10), 1122–1136, 2000.
22. U. R. Kattner and C. A. Handwerker, Calculation of phase equilibria in candidate solder alloys, *Z. Metallkunde* **92**(7), 740–746, 2001.
23. U. R. Kattner, Phase diagrams for lead-free solder alloys, *JOM—J. MINER. Metals Mater. Soc.* **54**(12), 45–51, 2002.

24. Z. Mei, F. Hua, and J. Glazer, Thermal reliability of 58Bi–42Sn solder joints on Pb-containing surfaces, in *Proceedings, Design and Reliability of Solders and Solder Interc.*, TMS, 1997, pp. 229–239.
25. Paul Vianco, *Proceedings of IPC Works '99, An International Summit on Lead-Free Electronics Assemblies*, October 1999, Paper S-03-3 (31 pages) Other Vianco papers can be obtained at the IPC website: http://www.leadfree.org.
26. K. Suetsugu, *Development and Application of Lead-Free Solder Bonding Technology*, Matsushita report, 1999.
27. S. Chada, W. Laub, R. A. Fournelle, and D. Shangguan, *J. Electron. Mater.* **28**, 1194, 1999.
28. M. E. Loomans and M. E. Fine, *Metall. Mater.Trans.* A 31A 1155, 2000.
29. E. Gebhardt and G. Petzow, *Z. Metallkunde* 50, 597, 1959.
30. D. Shangguan and A. Achari, Lead-free solder development for automotive electronics packaging applications, *SMI*, 423–428. 1995.
31. M. Arra, D. Xie, and D. Shangguan, Performance of lead-free solder joints under dynamic mechanical loading, in *52nd Electronic Components and Technology Conference*, 2002, pp. 1256–1280.
32. T. B. Massalski, H. Okamoto, P. R. Subramanian, and L. Kacprzak, Eds., *ASM Binary Alloy Phase Diagrams*, ASM International, 1990.
33. Y. Kariya, N. Williams, C. Gagg, and W. Plumbridge, Tin pest in Sn–0.5 wt.% Cu lead-free solder, JOM- *J. Miner. Metals Mater. Soc.* **53**, 39–41, 2001.
34. J. H. Vincent, B. P. Richards, D. R. Wallis, et al., Alternative solders for electronics assemblies, Part 2: UK progress and preliminary trials, *Circuit World*, 19, 32–34, 1993.
35. Alternative Solders for Electronic Assemblies—Final Report of DTI Project 1991–1993, GEC Marconi, ITRI, BNR Europe, and Multicore Solders. DTI Report MS/20073, issued 10.26.93.
36. J. H. Vincent and G. Humpston, Lead-free solders for electronic assembly, *GEC J. Res.* **11**, 76–89, 1994
37. K. Suganuma, Microstructural features of lift-off phenomenon in through hole circuit soldered by Sn–Bi, *Scripta Mater.* **38**(9), 1333–1340, 1998.
38. W. J. Boettinger, C. A. Handwerker, B. Newbury, T. Y. Pan, and J. M. Nicholson, Mechanism of fillet lifting in Sn–Bi alloys, *J. Electron. Mater.* **31**(5), 545–550, 2002.
39. H. Takao and H. Hasegawa, Influence of alloy composition on fillet-lifting phenomenon in Sn–Ag–Bi alloys, *J. Electron. Mater.* **30**, 513–520, 2001.
40. Peter Biocca, Solder Paste: What are the process requirements to achieve reliable lead-free wave soldering?, http://www.leadfreemagazine.com/pages/papers/Q_A_Kester.pdf
41. Peter Biocca, Reliable lead-free wave soldering and SMT processes, http://ap.pennnet.com/Articles/Article_Display.cfm?Section=Articles&Subsection=Display&ARTICLE _ID = 216210
42. NIST Thermodynamic Database for Solder Systems, http://www.metallurgy.nist.gov/phase/solder/solder.html
43. J.-O. Andersson, T. Helander, L. Höglund, P. Shi, and B. Sundman, *Calphad* **26**, 273–312, 2002.

44. W. J. Boettinger, U. R. Kattner, S. R. Coriell, Y. A. Chang, and B. A. Mueller, Development of multicomponent solidification micromodels using a thermodynamic phase diagram data base, in *Modeling of Casting, Welding and Advanced Solidification Processes*, VII, M. Cross and J. Campbell, Eds., TMS, Warrendale, PA, 1995, pp. 649–656.

45. D. W. G. White, Surface tensions of Pb, Sn, and Pb–Sn alloys, *Metall. Trans.* **2**, 3067–3071, 1971.

46. I. Ohnuma, X. J. Liu, H. Ohtani, K. Anzai, R. Kainuma, and K. Ishida, Development of thermodynamic database for micro-soldering alloys, Electronics Packaging Technology Conference, 2000. (EPTC 2000), Proceedings of 3rd Conference, 2000, pp. 91–96.

47. I. Ohnuma, M. Miyashita, K. Anzai, et al., Phase equilibria and the related properties of Sn–Ag–Cu based Pb-free solder alloys, *J. Electron. Mater.* **29**(10), 1137–1144, 2000.

48. N.-C. Lee, Prospect of lead-free alternatives for reflow soldering, IPC Work '99: An International Summit on Lead-Free Electronics Assemblies, Proceedings, October 23–28, 1999, Minneapolis, MN.

49. E. Bradley and J. Hranisavljevic, ECTC, 50th Electronic Components & Technology Conference, *IEEE Transa. Electron. Packaging Manuf.* **24**(4), 255–260, 2001.

50. J. Bartelo, S. R. Cain, D. Caletka et al., Thermomechanical fatigue behavior of selected lead-free solders, in *Proceedings, IPC SMEMA* Council APEX 2001, Paper # LF2-2.

51. G. Swan, A. Woosley, K. Simmons, T. Koschmieder, T. T. Chong, and L. Matsushita, Development of lead-(PB) and halogen free peripheral leaded and PBGA components to meet MSL3 at 260°C peak reflow profile, 1, 9-11-2000, in *Proceedings, Electronics Goes Green 2000+*, September 11–13, 2000, Berlin, Germany, VDE Verlag, pp. 121–126.

CHAPTER 2

Review and Analysis of Lead-Free Solder Material Properties

JEAN-PAUL CLECH

2.1. INTRODUCTION

The life of Sn–Pb or lead-free solder joints is limited by creep-fatigue damage that accumulates cyclically in solder materials. Insight into the thermomechanical response of solder joints is thus critical to the design and deployment of reliable electronic circuit board assemblies. Understanding the mechanics of lead-free soldered assemblies is also essential to the development of accelerated test plans and predictive reliability models, as well as essential to their use as effective tools for product reliability assessment.

Accelerated testing provides distributions of failure times whose relevance to service life is determined by extrapolation of the test data to use conditions based on the appropriate acceleration factors (AFs). Since AFs are defined as the ratio of life under test and use conditions, their determination requires up-front predictions of solder joint lives under both sets of conditions. Such models are also of use for reliability analysis of circuit boards at the design stage. While a variety of life prediction models have been developed for Sn–Pb assemblies, to this author's knowledge, such models are not currently available for lead-free soldered assemblies. The development of life prediction models requires a detailed understanding of failure modes, a constitutive model that captures the thermomechanical response of lead-free solders in electronic assemblies and a reliability database that is needed for the empirical correlation of failure times under test and field conditions.

Some reliability data have been acquired or are in the process of being gathered by several lead-free consortia. Efforts are also underway to characterize the mechanical behavior of lead-free solders, and numerous studies have been published with emphasis on secondary creep of solder. The derived creep rate equations are an

Lead-Free Electronics. Edited by Bradley, Handwerker, Bath, Parker, and Gedney

important part of constitutive models; however, their application to circuit board assemblies still is the subject of validation studies.

This report provides a quantitative review of the thermomechanical properties of lead-free solders, with emphasis on Sn–Ag–Cu (SAC) and Sn–Ag alloys. The Sn–Ag–Cu alloys of near-eutectic composition are labeled "SAC" throughout the memo, including the iNEMI-selected Sn–3.9Ag–0.6Cu alloy. SAC solders are the alloys of choice for solder reflow assemblies. Eutectic Sn–3.5Ag (as well as Sn–0.7Cu) is also recommended for wave-soldering applications. The microstructure of both alloys consists of an Sn matrix with finely dispersed intermetallic precipitates. We thus expect similarities in the constitutive response of SAC and eutectic Sn–Ag solders. Given the more abundant literature on Sn–Ag, its properties are reviewed in detail in an attempt to better understand the qualitative behavior of precipitate-strengthened solder alloys.

To put things in perspective, this report starts with a review of Sn–Pb properties and lessons learned from the development of constitutive and life prediction models for near-eutectic Sn–Pb assemblies. This understanding of Sn–Pb behavior, although not fundamentally complete, has proven valuable to industry practitioners. It is also worthwhile observing that the characterization of Sn–Pb solder and the development of reliability models for Sn–Pb assemblies spans over *three decades* of research in industry and academia. For example, one of the earliest and well-known solder joint reliability models is the Norris-Landzberg (1969) model for non-underfilled, flip-chip assemblies using Sn–Pb solder with high-lead contents [1]. While some of the lessons learned with Sn–Pb will transfer to lead-free applications, a significant amount of research is needed for lead-free reliability models to come up to par with their Sn–Pb equivalents.

The report then proceeds with an extensive, although not exhaustive, review of material properties for Sn–Ag and SAC alloys. Gaps in the material property database are identified and suggestions are offered for additional testing and analysis that are required to develop constitutive models for engineering use.

2.2. TIN–LEAD PROPERTIES AND MODELS

2.2.1. Complexity of Problem

Solder joints of electronic assemblies are complex interconnection elements that cannot be studied using the traditional techniques of structural analysis or fatigue of engineering metals. The acknowledged complexity of the mechanics of solder joints arises from the following:

- The problem is three-dimensional (3D) with solder joints subjected to a system of distributed, multiaxial forces and moments exerted by the interconnected parts. Even when taking advantage of symmetries, all joints are not equal because of varying distances to the neutral axis of an assembly and variability in joint geometry and metallurgy.

- A solder joint is a multilayered, nonhomogeneous structure. Reflowed solder is sandwiched between thin layers of intermetallic compounds. In the case of Sn–Pb, solder itself is made up of lead- and tin-rich phases with variations in composition near the intermetallic layers—for example, tin-depleted regions on the board side due to the formation of Cu–Sn intermetallic compounds during reflow. Moreover, the Sn–Pb microstructure evolves in service. The microstructure coarsens due to thermally activated grain growth, a phenomenon that takes place under stress or at constant temperature.
- The mechanical behavior of solder is highly nonlinear and temperature-dependent. Soft solder creeps readily at ambient temperature (and below) and creep rates increase drastically with temperature.
- Failure of Sn–Pb solder joints is a complex sequence of events involving microstructure coarsening, matrix creep, grain boundary sliding, micro-void formation and linking, crack initiation, and crack growth. In the case of SAC and Sn–Ag solder joints, the damage accumulation process leads to finer, less visible coarsening of the microstructure.
- Most often, electrical opens resulting from solder joint failures are intermittent and may be difficult to detect accurately. Electrical continuity may still be maintained when a solder joint is fully cracked because of contacts between asperities on the opposite surfaces of the crack. This may result in hard-to-detect failures and "no trouble found" (NTF) diagnosis during troubleshooting.

In spite of all this, significant progress has been made in the understanding of Sn–Pb solder joint mechanics, fatigue, and failure. Although the study of Sn–Pb assembly reliability is a semiempirical science, a vast body of knowledge, engineering models, test data, and experimental findings has accumulated over the last 30–50 years, providing useful insight into the mechanical behavior of solder joints of real assemblies.

2.2.2. Creep and Constitutive Models for Near-Eutectic Sn–Pb

2.2.2.1. Overview. One lesson learned from Sn–Pb studies is that, because of the spread in material properties, there is no unique constitutive model for SMT solder joints. In the end, a reasonable agreement between life predictions and test results determines whether a constitutive model is of practical use.

The mechanical behavior of solder depends on the joint microstructure and is affected by many parameters such as intermetallics, joint or specimen size, cooling rate of the assembly after soldering, aging in service, and so on. Test factors such as specimen or load eccentricity, temperature variations, and measurement errors also contribute to the scatter in the mechanical properties of solder. Nevertheless, simplified constitutive models have been developed to help characterize the mechanical behavior of Sn–Pb solder and enable first-order stress/strain analysis of solder joints using methods of classical mechanics or numerical techniques such as the Finite Element Method (FEM).

2.2.2.2. Solder Creep. Under constant load or stress, solder undergoes progressive inelastic deformations over time. This time-dependent deformation is called creep and the associated strains that develop over time are creep strains. When the test specimen is subjected to a constant load, the initial, instantaneous response includes elastic and time-independent plastic flow. Creep then proceeds in three stages of primary, secondary, and tertiary creep (Figure 2.1). During primary creep, metals strain-harden. The strain rate decreases over time, as hardening of the metal becomes more difficult. Specimen deformations keep increasing with secondary creep proceeding at a steady strain rate. Note also that the initial deformation that occurs upon loading of the test specimen includes both elastic and plastic strains. Often, these initial deformations are not reported on in the context of creep studies. However, they cannot be neglected, a priori. Moreover, these initial deformations, which depend on the loading rate, may become important under service conditions with intermediate to rapid temperature ramps. For most metals, secondary creep is the dominant deformation mode at homologous temperatures above half the melting point, T_M, in degrees Kelvin. For eutectic Sn–Pb, with a melting point of 183°C, $T_M = 456°\text{K}$ and $1/2\ T_M = 228°\text{K} = -45°\text{C}$. That is, standard Sn–Pb solder readily creeps well below room temperature. The last stage of deformation is tertiary creep where strain rates increase rapidly until the test specimen ruptures. In Sn–Pb solder, tertiary creep proceeds by void formation and growth along grain boundaries, micro-cracking, and necking of tensile specimens.

Numerous researchers have investigated the mechanical and creep behavior of near-eutectic Sn–Pb solder (e.g., Refs. 2–15). Most investigations have focused on steady-state creep, with the secondary creep rate $\mathring{\varepsilon}_{SS}$ often given as a function of stress, σ, and the absolute temperature, T:

$$\mathring{\varepsilon}_{SS} = A\left(\frac{b}{g}\right)^p \frac{E(T)}{kT}\left(\frac{\sigma}{E(T)}\right)^n \exp\left(-\frac{\mathring{Q}}{kT}\right) \tag{2.1}$$

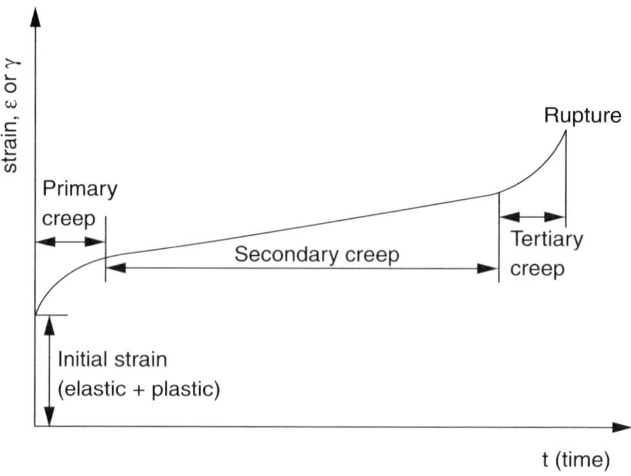

Figure 2.1. Creep curve: Strain versus time under constant stress (or load) and temperature.

where A is a material constant, b is a dislocation characteristic length or magnitude of slip or Burgers vector, g is the material grain size, $E(T)$ is the temperature-dependent Young's modulus, k is Boltzmann's constant, the exponents p and n are constants, and Q is the activation energy of the rate-controlling diffusion mechanism. Equation (2.1) is known as Dorn's equation [16] and is often simplified as

$$\dot{\varepsilon}_{SS} = A_1 g^{-p} \sigma^n \exp\left(-\frac{Q_a}{kT}\right) \tag{2.2}$$

where A_1 is another material constant and Q_a is an apparent activation energy. Equation (2.2) shows the strong dependence of creep rates on stress and temperature as well as grain size (in the case of Sn–Pb).

Since creep properties, as well as strength and other mechanical properties, vary with specimen size, the mechanical response of tiny solder joints differs from that of bulk solder test specimens. Solder deformations, including creep, have been measured on solder joints of actual electronic assemblies (e.g., Refs. 9 and 11) and for several solder alloys: 60Sn–40Pb, 62Sn–36Pb–2Ag, 96.5Sn–3.5Ag, 97.5Pb–2.5Sn, 100 In, and 50In–50Pb [4, 17]. These models are presented, briefly, hereafter because they have been found to be of use to practicing engineers. The reader is referred to the original publications for additional details as well as for relevant information on the experimental techniques that were used and that could be applied to the study of lead-free solders.

2.2.2.3. Motorola/Darveaux Constitutive Model. Darveaux and his co-workers at Motorola [4] conducted extensive mechanical testing of flip-chip and Ball Grid Array (BGA) solder joints and characterized the time-independent plastic flow and creep deformations of several solder alloys. Their constitutive model is described below for several alloys of electronic solder. Robert Darveaux implemented this model into two commercial finite element codes, ANSYS™ and ABAQUS™. His original publication [4] includes detailed recommendations on how to input material constants in the pre-processor of those two programs. One important feature of Darveaux's creep model is that it was found to apply *consistently* to several solder alloys: 60Sn–40Pb, 62Sn–36Pb–2Ag, 96.5Sn–3.5Ag, 97.5Pb–2.5Sn, 100In, and 50In–50Pb, and over a wide range of temperatures and several orders of magnitude in strain rates.

The initial, instantaneous strain that develops at the start of a creep test includes an elastic strain and an inelastic strain that represents time-independent plastic flow. The plastic strain, γ_P, is described by a plastic flow or strain hardening law of the form

$$\gamma_P = C_6 \left(\frac{\tau}{G}\right)^m \tag{2.3}$$

where τ is the applied shear stress and $G = G(T)$ is the temperature-dependent shear modulus:

$$G(T) = G_0 - G_1\big(T(^\circ K) - 273\big) \tag{2.4}$$

TABLE 2.1. Solder Material Constants for Shear Modulus and Plastic Flow Rule

	Alloy			
	60Sn–40Pb	62Sn–36Pb–2Ag	96.5Sn–3.5Ag	97.5Pb–2.5Sn
		Elastic Constants		
G_0 (Mpsi)	1.9	1.9	2.8	1.3
G_1 (kpsi/°K)	8.1	8.1	10.0	1.5
		Strain-Hardening		
C_6	2.34E13	1.21E13	2.04E11	6.36E7
m	5.58	5.53	4.39	3.10

After Darveaux et al. [4].

with $G_0 = 1.9$ Mpsi (shear modulus at 0°C) and $G_1 = 8.1$ kpsi/°K for both alloys of 60Sn–40Pb and 62Sn–36Pb–2Ag. C_6 and m are material constants. The elastic constants and the plastic flow parameters for several solder alloys, including Sn–Ag eutectic, are given in Table 2.1. Note that the elastic constants and the power-law exponent m are about the same for 60Sn–40Pb and 62Sn–36Pb–2Ag. However, the constant C_6 is about twice as low for 62Sn–36Pb–2Ag. This implies that, under equal loads, tin–lead with 2% silver will see half as much initial plastic strain than 60Sn–40Pb.

During primary or transient creep, the creep strain is given by the equation:

$$\gamma_C^1 = \frac{d\gamma_S}{dt}t + \gamma_T\left[1 - \exp\left(-B\frac{d\gamma_S}{dt}t\right)\right] \tag{2.5}$$

where the creep strain superscript "1" refers to primary creep, γ_T is the transient creep strain, B is the transient creep coefficient, and $d\gamma_S/dt$ is the steady-state creep rate. The primary creep constants for several alloys are given in Table 2.2.

The primary creep rate is

$$\frac{d\gamma_C^1}{dt} = \frac{d\gamma_S}{dt}\left[1 + \gamma_T B\exp\left(-B\frac{d\gamma_S}{dt}t\right)\right] \tag{2.6}$$

Initially, at time $t = 0$, the primary creep rate is a factor $(1 + \gamma_T B)$ times greater than the steady-state creep rate. For 60Sn–40Pb, this factor is $(1 + \gamma_T B) = 1 + 0.026 \times 403 = 11.48$; that is, the initial transient creep rate is over an order of magnitude higher than the steady creep rate. For Sn–3.5Ag, the rate factor is even larger: $(1 + \gamma_T B) = 1 + 0.167 \times 131 = 21.88$. Thus, *primary creep may not be negligible in applications with high-temperature ramp rate or under thermal cycling conditions with short dwell times.*

TABLE 2.2. Primary Creep Constants for Common Solder Alloys

	Alloy			
	60Sn–40Pb	62Sn–36Pb–2Ag	96.5Sn–3.5Ag	97.5Pb–2.5Sn
γ_T	0.026	0.040	0.167	0.115
B	403	152	131	137

After Darveaux et al. [4].

TABLE 2.3. Steady-State Creep Parameters for Common Solders

	Alloy			
	60Sn–40Pb	62Sn–36Pb–2Ag	96.5Sn–3.5Ag	97.5Pb–2.5Sn
C_4(°K/s/psi)	0.198	0.0989	3.13E-3	1.62E7
α	1300	1300	1500	1000
n	3.3	3.3	5.5	7.0
Q(eV)	0.548	0.548	0.50	1.10
C_5(1/s)	2.778E5	1.39E5	2.46E5	4.60E11
α_1	8.0E-4	8.0E-4	6.3E-4	8.0E-4
Q_a(eV)	0.70	0.70	0.75	1.15

After Darveaux et al. [4].

More general relationships were found to apply to steady-state creep of solder in shear:

$$\overset{\circ}{\gamma}_S = \frac{d\gamma_S}{dt} = C_4 \frac{G(T)}{T}\left[\sinh\left(\alpha \frac{\tau}{G}\right)\right]^n \exp\left(\frac{-Q}{kT}\right) \tag{2.7}$$

or in a simplified form:

$$\overset{\circ}{\gamma}_S = C_5[\sinh(\alpha_1 \tau)]^n \exp\left(\frac{-Q_a}{kT}\right) \tag{2.8}$$

$\overset{\circ}{\gamma}_S$ is the steady-state strain rate, $G(T)$ is the temperature-dependent shear modulus, T is the absolute temperature (in degrees Kelvin), τ is the applied stress, n is a constant exponent that depends on the controlling creep mechanism, k is Boltzmann's constant ($k = 8.620 \times 10^{-5}$ eV/°K), Q is the creep activation energy, Q_a is the apparent activation energy, and α, C_4, and C_5 are constants. The above constants and activation energies are given for several solder alloys in Table 2.3.

2.2.2.4. DEC'S Model. Knecht, Fox, and Shine of the Digital Equipment Corporation (DEC) conducted isothermal mechanical and fatigue testing of 63Sn–37Pb solder joints on a 7.6-mm × 7.6-mm, 16 I/O Leadless Ceramic Chip Carrier (LCCC) test vehicle [9, 11]. The LCCCs were mounted on stainless steel or G-10 organic substrates with copper pads matching the component terminations (0.64-mm × 1.0-mm pads). Solder joint thickness varied from 4 to 12 mils. Testing was conducted at temperatures of 25°C, 60°C, and 100°C. The solder joints were subjected to shear in a torsion apparatus powered by an electric motor. Shear deformations were determined from the reflection of a laser beam on the mirrored edges of LCCC components. Creep data were obtained from curves of isothermal creep strain versus time under constant torque. Stable stress–strain hysteresis loops were obtained from the first cycles of fatigue tests conducted under constant plastic strain range. From these experiments and associated data reduction, failure mode analysis, and modeling, Shine and Fox [11] concluded the following:

- The creep rate versus stress curves are bilinear, with creep regions associated with two different mechanisms:
  - Grain boundary creep at low stresses and creep rates
  - Matrix creep at high stresses and creep rates
- Solder joints undergo creep-fatigue failures associated with steady-state creep. Creep damage occurs by a combination of grain-boundary and matrix creep, with scanning electron microscope (SEM) photos of fatigued solder joints showing intercrystalline voids and cracks. Isothermal fatigue life cycles have an inverse relationship to integrated matrix creep.
- The grain size of thick joints is larger than that of thin joints and solder grain size increases during fatigue testing. Under identical loads, thin joints with initially smaller grain sizes are expected to have a longer fatigue life than thick joints with larger grains.

Using the hysteresis loops and creep data from the above experiments, Knecht and Fox [9] developed a simple constitutive model for eutectic (63Sn–37Pb) solder in shear. The constitutive equations from that study are summarized hereafter with some minor modifications. Knecht and Fox used their model to conduct finite element analysis of solder joints in SMT assemblies and to correlate fatigue life data to integrated matrix creep strains.

The average shear strain γ is given as the sum of an elastic strain γ_E, a time-independent plastic strain γ_P and a secondary creep strain γ_C, that is,

$$\gamma = \gamma_E + \gamma_P + \gamma_C \tag{2.9}$$

The elastic strain component is

$$\gamma_E = \frac{\tau}{G(T)} \tag{2.10}$$

where the temperature-dependent shear modulus is

$$G(T) = \frac{E(T)}{2(1 + \upsilon)} \tag{2.11}$$

Poisson's ratio for near-eutectic solder is $\nu = 0.4$ and the temperature-dependent Young's modulus is

$$E(T) \text{ (in GPa)} = 32 - 0.088T(^{\circ}C) \tag{2.12a}$$

or

$$E(T) \text{ (in psi)} = 1.45 \times 10^{5}(32 - 0.088T(^{\circ}C)) \tag{2.12b}$$

The time-independent plastic strain is given by the following plastic flow rule:

$$\gamma_P = \left(\frac{\tau}{\tau_P(T)}\right)^2 \tag{2.13}$$

where the temperature-dependent plasticity parameter τ_P is obtained by curve-fitting the τ_P versus temperature data in the original publication by Knecht and Fox [9], that is,

$$\tau_P(\text{in psi}) = 49{,}367 - 299 \times T(^{\circ}C) \tag{2.14}$$

Finally, creep strains are obtained by integration of the steady-state creep rate equation:

$$\mathring{\gamma}_C(\text{s}^{-1}) = 8.31 \cdot \exp\left(-\frac{0.5\,\text{eV}}{kT}\right) \cdot \tau^2 + 1.12 \cdot \exp\left(-\frac{0.84\,\text{eV}}{kT}\right) \cdot \tau^{7.1} \quad \text{(with } \tau \text{ in MPa)} \tag{2.15}$$

or

$$\mathring{\gamma}_C(\text{s}^{-1}) = 3.952 \times 10^{-4} \cdot \exp\left(-\frac{0.5\,\text{eV}}{kT}\right) \cdot \tau^2 + 5.327 \times 10^{-5} \cdot \exp\left(-\frac{0.84\,\text{eV}}{kT}\right) \cdot \tau^{7.1} \quad \text{(with } \tau \text{ in psi)} \tag{2.16}$$

The first term on the right-hand-side of Eqs. (2.15) and (2.16) is for grain boundary creep with a stress exponent: $n_{\text{GB}} = 2$ and an activation energy: $\Delta H_{\text{GB}} = 0.5$ eV. The second term on the right-hand-side of Eqs. (2.15) and (2.16) is for matrix creep with a stress exponent $n_{\text{MC}} = 7.1$ and an activation energy $\Delta H_{\text{GB}} = 0.84$ eV. As shown by Knecht and Fox, these values of stress exponents and activation energies are consistent with steady state creep parameters reported in the literature (see Table 2.4).

TABLE 2.4. Comparison of Near-Eutectic Sn–Pb Creep Parameters

Reference	ΔH_{GB} (eV)	n_{GB}	ΔH_{MC} (eV)	n_{MC}
Grivas et al. [5]	0.5	2.0	0.84	7.1
Kashyap and Murty [8]	0.46	1.7	1.0	11.1
Solomon [19, 20]	—	3.1	—	7.3
Baker [21]	0.69	3.4	—	—
Lam et al. [22]	0.45	2.4	0.9	5.7
Wong et al. [14]	0.47	3.0	0.47	7.0

After Knecht et al. [18].

Last, note that, even though the experimental results of Shine and Fox [11] suggest a grain size effect, the constitutive model developed by Knecht and Fox does not include any grain size parameter. The most likely reason for this is that those effects are difficult to quantify when the grain size is not a control parameter of an experiment. However, Knecht and Fox stated that "*using 'real' joints will generate practical results*" [9] and their constitutive model is thought to be useful for the engineering analysis of electronic solder joints.

Boon Wong and his colleagues at Hughes Aircraft Company [14] compiled Sn–Pb eutectic steady-state creep data from seven sources. The data, which covered several orders of magnitude in stress and creep rates, was found to fit in a correlation band as shown in Figure 2.2. The raw data fell within a band of height about *one order of magnitude* in the vertical direction. The width and height of the band are typical of the scatter usually associated with creep data. Available data points from Shine et al. [23] are found to fit within or close to Wong's correlation band. In Figure 2.2, the normal stress, σ_{SS}, is scaled with a temperature-dependent

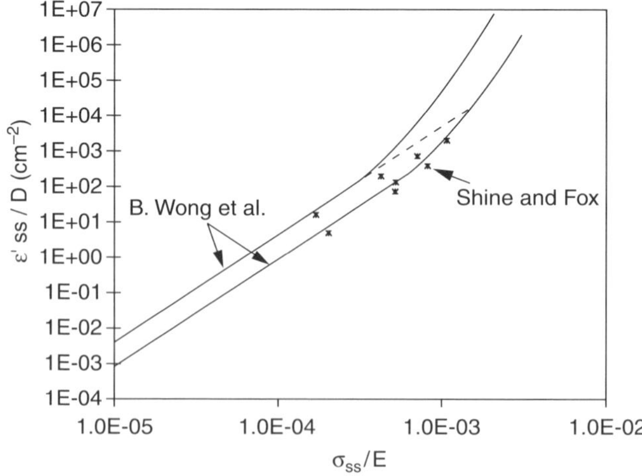

Figure 2.2. Creep rate plot for Sn-Pb eutectic solder.

Young's modulus, $E = E(T)$, on the horizontal axis. Young's modulus is given as in Eq. (2.12a). The steady-state creep rate on the vertical axis is normalized with a Arrhenius-type, temperature-dependent diffusion parameter, $D = D(T)$:

$$D(T) = \exp\left(-\frac{Q_C}{RT(^\circ K)}\right) = \exp\left(-\frac{5473}{T(^\circ K)}\right) \tag{2.17}$$

where Q_C is an average activation energy for creep of eutectic Sn–Pb solder: $Q_C = 45$ kJ/mole $= 0.47$ eV and R is the universal gas constant: $R = 8.314$ J/$^\circ$K $\cdot$ mole. The correlation band in Figure 2.2 has two distinct slopes that reflect different creep mechanisms:

- A dislocation glide regime with a stress exponent of 3 in the low-stress region
- A dislocation climb regime with a stress exponent of 7 in the higher-stress region

Based on the above correlation of creep data, Wong gave the following expression for average creep rates:

$$\frac{\mathring{\varepsilon}_{SS}}{D} = 1.7 \times 10^{12}\left(\frac{\sigma_{SS}}{E}\right)^3 + 8.9 \times 10^{24}\left(\frac{\sigma_{SS}}{E}\right)^7 \tag{2.18}$$

The equations of the lower and upper bounds of the correlation band are [24]

$$\frac{\mathring{\varepsilon}_{SS}}{D} = 8.4 \times 10^{11}\left(\frac{\sigma_{SS}}{E}\right)^3 + 7.8 \times 10^{23}\left(\frac{\sigma_{SS}}{E}\right)^7 \quad \text{(lower bound)} \tag{2.19}$$

$$\frac{\mathring{\varepsilon}_{SS}}{D} = 4.2 \times 10^{12}\left(\frac{\sigma_{SS}}{E}\right)^3 + 4.7 \times 10^{25}\left(\frac{\sigma_{SS}}{E}\right)^7 \quad \text{(upper bound)} \tag{2.20}$$

The above stress exponents are consistent with the steady-state creep equation in DEC's steady-state creep model. Note also that Wong's creep rate equations do not include grain size dependence. Wong argued that the experimental data were inconclusive at the time [14], and this did not warrant any attempt at including grain size effects. Wong also stated that grain size dependence is not expected in either the climb or glide-controlled creep regimes. One last important aspect of Wong's steady-state creep model is that it was derived based on the correlation of creep data from several independent sources, with the correlation holding over a wide range of temperatures (-60°C to 150°C), stresses, and creep rates.

The above compilation of creep data was used successfully in a solder joint life prediction model developed by Wong et al. [14]. The upper bound of the correlation band was also used as the creep rate equation in the Solder Reliability Solutions (SRS) life prediction model [25, 26]. The upper bound was selected in order to maximize strain rates, thus building in some conservatism in the model. The SRS model has since been validated by over 60 experiments [27, 28]. Last, based on the Sn–Pb experience, we can expect creep rates from compiled test results for lead-free solders to spread over *one order of magnitude*. Such spread in the data did not impede the development of first-order life prediction models for Sn–Pb assemblies.

2.2.2.5. Hall's Stress–Strain Hysteresis Loop. One significant contribution to the field of solder joint mechanics, and also most enjoyable reading, is the shear strain and hysteresis loop measurements and theory developed by Peter Hall at AT&T Bell Laboratories [29–31]. Using strain gauge measurements and a simplified analysis of shear forces exerted on the solder joints of LCCC assemblies, Hall showed that the stress–strain response of solder joints during temperature cycling is a boomerang-shaped hysteresis loop (Figure 2.3). The shape of the loop reflects the temperature-dependent inelastic deformations of solder and elastic deformations of the entire assembly. The thermal expansion mismatch between board and component is accommodated by shear of the solder joints and simultaneous stretching and bending of the board and component. These elastic deformations of the interconnected parts provide compliance to the assembly, suggesting practical ways to reduce solder joint stresses by designing boards and components that are more compliant.

The hysteresis loop in Figure 2.3 illustrates the complexity of the solder joint stress–strain response during a large temperature cycle between −25°C and 125°C. The dwell times at the temperature extremes were two hours. The ramp times were five hours with a rather slow ramp rate of 0.5°C/min. The data points, shown as circles, are derived from Hall's strain measurements on noncastellated, 25-mil-pitch, 84 I/O LCCCs mounted on FR-4. The solder joints were 11 mils high. A complete description of test vehicles and details of the experimental setup, measurement techniques, and data reduction procedures are given in Hall's publications [29–31]. Pao conducted similar measurements on a double-beam, alumina-on-aluminum test vehicle [10, 32].

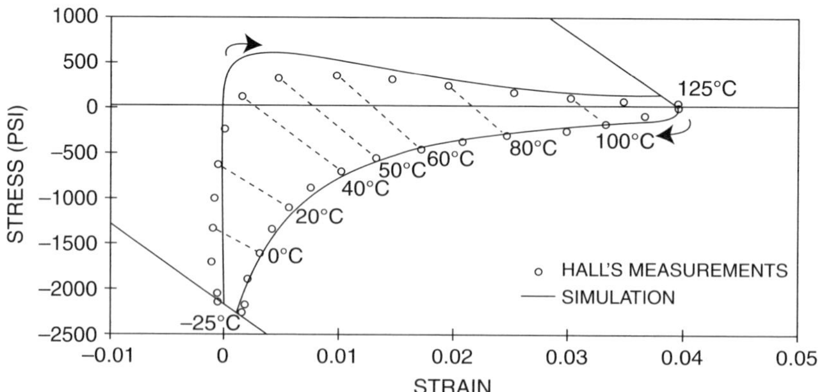

Figure 2.3. Solder joint hysteresis loop during thermal cycling between −25°C and 125°C (1 MPa = 145 psi).

The shape of the hysteresis loop in Figure 2.3 reflects the nonlinear, temperature-dependent constitutive behavior of solder. Important features of the loop are described hereafter:

- The loop is described clock-wise. Isothermal stress reduction lines are drawn as dashed lines between the data points corresponding to equal temperatures during the ramp-up and ramp-down phases of the thermal cycle. The stress reduction lines are shown for every 10 or 20°C temperature increment. The stress reduction lines are almost parallel to each other, with an average slope in very good agreement with the slope predicted by Hall's assembly stiffness model.
- During the dwell periods at the temperature extremes, stresses are reduced along the stress-reduction lines (shown as solid lines) for those temperatures:
  - At 125°C, where solder is very soft, shear strains are large and initial stresses are relatively low, less than 200 psi. Creep rates are very high and stress reduction is rapid. The intersect of the stress reduction line with the strain axis is the maximum available strain due the thermal expansion mismatch between the board and the LCCC component.
  - At −25°C, initial stresses are much higher, of the order of −2300 psi. In spite of high stresses, stress reduction is limited during the two-hour dwell because creep rates are rather small at cold temperatures.
- During the ramp-up phase of the thermal cycle:
  - As temperature goes up, starting at −25°C, solder is relatively strong and the shear strain remains about constant. Shear forces are unloaded almost elastically.
  - Past 35°C, when the shear force is close to zero, stresses build up due to plastic flow of solder in the opposite direction and strains start increasing with the added thermal expansion mismatch between board and component.
  - Creep accelerates as temperature keeps going up. Slightly past 50°C, the creep rates are so high that stress reduction prevents any further build-up of stress. Creep strains develop at a faster rate, contributing to rapid increments in the total shear strain.
  - When temperature approaches 125°C, strains keep increasing with the thermal expansion mismatch between board and component and stresses relax at a rather fast rate.
- During the ramp-down phase of the thermal cycle:
  - As temperature goes down, starting at 125°C, the cycle is reversed and shear strains decrease.
  - Initially, and down to about 50–60°C, shear strains decrease at a high rate. Stresses start building up in the opposite direction.
  - From 50–60°C to −25°C, stresses become larger and built up at faster rate since solder becomes stronger at lower temperatures.

The loop in Figure 2.3 was obtained by using the plastic flow rule and creep constitutive model from DEC. Loop simulations have been conducted by several investigators using slightly different approaches and constitutive models (e.g., Refs. 31–42).

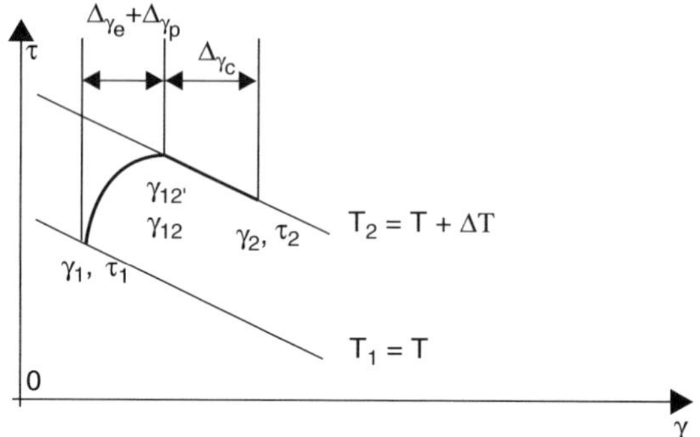

Figure 2.4. Solder joint stress–strain simulation when temperature increases from T to $T+\Delta T$.

The basic algorithm that is used to generate hysteresis loops during thermal cycling follows the stress/strain curves shown in Figure 2.4. During a small time-step from t to $t+\delta t$, temperature increases from T to $T+\Delta T$. At time t, the stress/strain curve intersects the stress reduction line for temperature $T_1 = T$. Due to the increase in temperature, stresses build up instantaneously from τ_1 to τ_{12} with a change in strain from γ_1 to γ_{12} that includes an elastic strain increment $\Delta\gamma_e$ and plastic flow with a strain increment $\Delta\gamma_p$. Stresses then relax from τ_{12} to τ_2, and the shear strain increases by creep (increment $\Delta\gamma_c$) along the stress reduction line at $T_2 = T+\Delta T$. Knowing the stress–strain state (γ_1, τ_1) at time t, the stress/strain state (γ_2, τ_2) at time $t+\delta t$ is obtained by solving the following system of four equations with four unknowns γ_{12}, τ_{12}, γ_2, and τ_2:

- From the stress reduction line at $T_2 = T+\Delta T$:

$$\gamma_{12} + \frac{\tau_{12}}{\kappa} = \frac{L\Delta\alpha(T_2 - T_0)}{h_S} \tag{2.21}$$

$$\gamma_2 + \frac{\tau_2}{\kappa} = \frac{L\Delta\alpha(T_2 - T_0)}{h_S} \tag{2.22}$$

- For the strain increment due to elastic deformation and plastic flow as per, for example, DEC's plastic flow rule:

$$\gamma_{12} - \gamma_1 = \frac{\tau_{12} - \tau_1}{G(T_1)} + \frac{2\tau_1 \times \Delta\tau}{\tau_P^2(T_1)} \qquad \text{with } \Delta\tau = \tau_{12} - \tau_1 \tag{2.23}$$

- For the creep strain developing during the short dwell of duration δt at T_2:

$$\gamma_2 - \gamma_{12} = \dot{\gamma}_{SS}(T_2, \tau_{12}) \times \delta t \tag{2.24}$$

where $\dot{\gamma}_{SS}(T_2, \tau_{12})$ is the steady-state creep rate under an applied stress τ_{12} at temperature T_2.

The above algorithm can be implemented in a computer program or even in a spreadsheet. Usually, a zero stress–strain state is used, somewhat arbitrarily, to initialize the algorithm. Small enough time steps are required to follow the prescribed temperature profile closely and to generate the stress–strain response with the desired accuracy. A few cycles are required to obtain a closed and stable hysteresis loops for stiff systems like leadless assemblies. Many more iterations are needed for compliant systems like leaded assemblies with compliant leads. Hysteresis loops can also be obtained using finite element models that include material options for a constitutive model with elastic, plastic flow, and creep (see, e.g., Refs. 4, 35, 42, and 43).

2.2.3. Fatigue Life Correlations

2.2.3.1. Coffin–Manson and Morrow's Fatigue Laws. Hysteresis loops provide useful information for engineering evaluation of solder joint reliability. For example, the width of the loop gives an estimate of the inelastic strain range experienced by solder joints. The inelastic strain range is used in Coffin–Manson type of fatigue laws. Another, more general approach consists in using the hysteresis loop area which is a measure of the amount of cyclic strain energy density that is imparted to solder joints. Strain energy density is used in Morrow's type of fatigue laws [44, 45]. Where cycles to failure are given as a function of the cyclic inelastic strain energy density, ΔW_{in}:

$$N = \frac{C}{\Delta W_{\mathrm{in}}^{n}} \tag{2.25}$$

where C is a material constant and the exponent n has been found to be in the range 0.7 to 1.6 for several engineering metals, including soft solders.

2.2.3.2. Sn–Pb Solder Joint Reliability Models. Several engineering models have been developed to predict solder joint reliability based on Coffin–Manson [e.g., Ref. 1 and the standard IPC-SM785, Guidelines for Accelerated Reliability Testing of Surface Mount Solder Attachments (November 1992)], and Morrow's fatigue laws (e.g., Refs. 4, 25, and 46). For example, the Darveaux model uses inelastic strain energy obtained from finite element analysis to correlate fatigue lives from over 100 experiments. The Solder Reliability Solutions (SRS) model [25] derives strain energy densities from simplified, one-dimensional structural models and correlates failure data from over 60 experiments.

Figure 2.5 shows the SRS correlation of fatigue life data from 19 accelerated tests [25]. The correlation gives joint characteristic lives scaled for the solder crack area, $\alpha_{\mathrm{joint}}/A$, versus cyclic inelastic strain energy. The life prediction model based on the best-fit line going through the data (centerline of correlation band in Figure 2.5) was frozen based on 19 experiments. Figure 2.6 is similar to Figure 2.5, with 35 data points added in for model validation [28]. Lessons

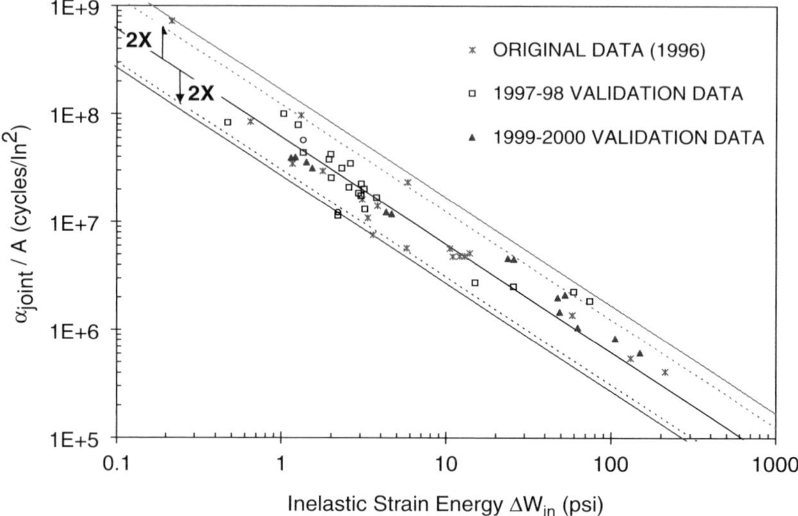

Figure 2.5. SRS correlation of accelerated test data [25].

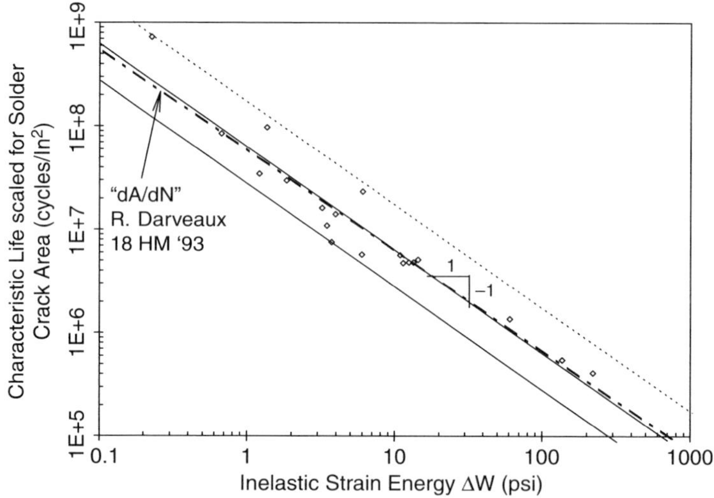

Figure 2.6. Fit of validation data to initial correlation of solder joint fatigue lives [28].

learned from developing this type of life prediction model, as well as from Darveaux's model, are:

- Given the semianalytical, semiempirical nature of the models, it is important that they be validated over time when data becomes available for new types of packages or assemblies.
- For these models to be more reliable, the empirical correlation of life data should hold over *several orders of magnitude on both axis*—that is, for inelastic

strain energy (or another damage parameter) and the crack propagation rate or other life parameter. Note, for example, that the data plotted in Figures 2.5 and 2.6 extend over three orders of magnitude on the horizontal and vertical axis.

2.2.3.3. SAC Versus Sn–Pb Fatigue Data. Figure 2.7 shows correlations of *isothermal, mechanical* fatigue lives versus inelastic strain energy for eutectic Sn–Pb and Sn–3.5Ag–0.75Cu lap-joint specimens tested at 25°C. This suggests that SAC thermomechanical failure data may correlate to strain energy as in the case of near-eutectic Sn–Pb electronic assemblies. In Figure 2.7, the slope of the best-fit line through the 63Sn–37Pb data is close to −1, similar to the corresponding slopes in the Darveaux and SRS models for near-eutectic Sn–Pb assemblies. However, the slope of the line through the Sn–3.5Ag–0.75Cu data points in Figure 2.7 appears to be slightly less than −1 [i.e., the exponent n in Eq. (2.25) would be greater than 1]. This suggests that existing methodologies for predicting Sn–Pb solder joint lives may apply to SAC electronic assemblies. However, both the intercept and the slope of life correlations such as shown in Figures 2.5 and 2.6 would have to be adjusted using extensive empirical data.

2.2.4. Conclusions on Sn–Pb Properties

The material that was summarized in this section illustrates the intricacies involved in the development of engineering constitutive and life prediction models for Sn–Pb solders and assemblies. As of yet, the corresponding knowledge base for lead-free solders is not as extensive and much work remains to be done to develop adequate constitutive and life prediction models for Sn–Ag and SAC electronic solder joints. The following sections are strictly focused on reviewing material properties and mechanical behavior of those two alloy families. While covering a wide range of data, this review is by no means exhaustive.

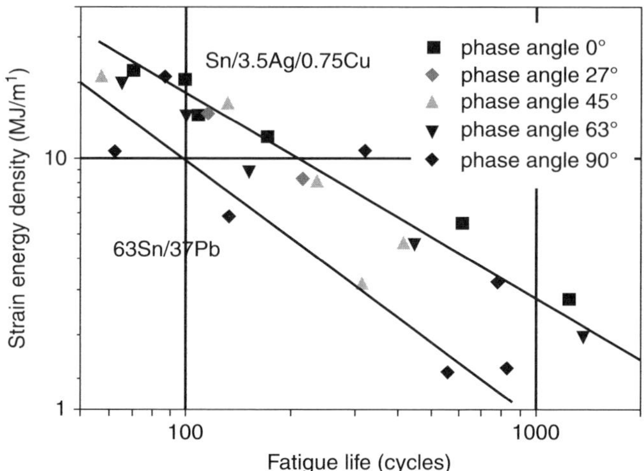

Figure 2.7. Sn–Pb and SAC fatigue life versus strain energy correlations. (After Ref. 47.)

2.3. TIN–SILVER PROPERTIES AND CREEP DATA

2.3.1. Overview and Conclusions

This section summarizes the review and analysis of isothermal tensile, compression, and shear creep data for Sn–3.5Ag eutectic solder with a melting point of 221°C. First-order creep models are developed for possible use in stress–strain analysis [e.g., Finite Element Analysis (FEA)], and solder joint life prediction models. The raw creep data is tabulated in Appendix A for possible further analysis by others. Other properties such as the CTE and Young's modulus of Sn–3.5Ag were also collected. The intent of this review is to pull together hard data from the existing literature on Sn–3.5Ag solder and develop simple, first-order creep models in an attempt to bridge datasets obtained from independent sources. It is also hoped that the gathered data will be of use for others to develop more sophisticated constitutive models. The analysis of the Sn–3.5Ag data leads to the following conclusions:

- Tensile creep data for bulk Sn–3.5Ag solder shows significant scatter.
- Compression data is limited. Nevertheless, the data suggest an *uneven* behavior of Sn–3.5Ag solder with higher strength in compression than in tension.
- Shear creep data from lap joint or plug and ring specimens also shows significant scatter.
- Creep data for Sn–3.65Ag and Sn–4Ag appear to follow similar trends as for Sn–3.5Ag, suggesting a small effect of silver contents in the range 3.5–4% Ag.
- Shear creep data from ceramic chip carrier (CCC) assemblies correlates well with flip-chip solder joint data.
- Regression of the data also shows some sensitivity to the regression procedure.
- The review also points to lesser data available at stresses less than 10 MPa. Future work should consider these lower stress levels since they are representative of stress conditions experienced by solder joints of electronic assemblies in use.
- Most studies of Sn–3.5Ag solder focus on ultimate strength and secondary or steady-state creep with very few investigating initial deformations or primary creep. Complete stress–strain curves are rarely published. These curves, obtained from constant strain rate tests, as well as thermal cycling hysteresis loops would be of much use for the validation of material constitutive models.

Since Sn–3.5Ag is a precipitate-strengthened Sn-based alloy, the findings of this review may also be of use in the development of guidelines or test procedures for the characterization of Sn–Ag–Cu alloys.

2.3.2. Bulk Sn–3.5Ag Tensile Creep

2.3.2.1. Source and Plot of Data. Isothermal tensile creep data from five independent sources are plotted as strain rate (/second) versus tensile stress (MPa) in

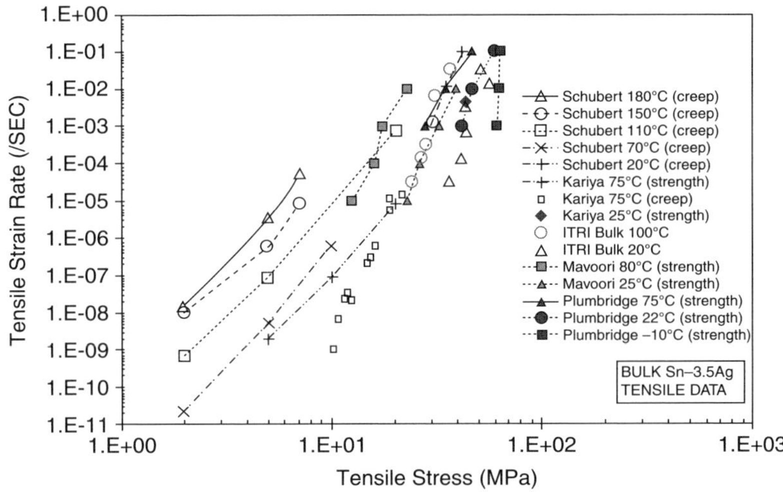

Figure 2.8. Log–log plot of isothermal tensile creep data for bulk Sn–3.5Ag solder.

Figure 2.8. The data covers about two orders of magnitude on the stress axis (from 2 to 64 MPa) and 10 orders of magnitude on the strain rate axis (from 2.2×10^{-11} to 10^{-1}/s). Test temperatures are in the range −10°C to 180°C. The raw data (59 points total) are listed in Table A.1 in Appendix A. The data were obtained from the figures and tables in their respective publications, listed in Table 2.5. Most tests were tensile "strength" tests carried out at a constant strain rate. Schubert et al. [48] conducted constant load creep tests. Kariya et al. [49, 50] ran both constant load and constant strain rate tests. Note that only the Schubert data cover the low stress region below 10 MPa. This is the region of most interest for solder joints of electronic assemblies under a wide range of use conditions.

TABLE 2.5. Bulk Sn–3.5%Ag Tensile Data: Source of Data and Test Parameters

Publication	Figure/Table in Publication	Strain Rate (/second)	Stress (MPa)	Test Temperatures
Schubert et al. [48]	Figure 7	2.2×10^{-11} to 7.6×10^{-4}	1.97–20.25	20°C, 70°C, 110°C, 150°C, 180°C
Kariya and Plumbridge [49]	Figures 3, 4	10^{-9} to 10^{-1}	10.1–42.5	75°C
Kariya and Otsuka [50]	Figure 4	5×10^{-3}	44	25°C
ITRI No. 656 [51]	Tables, p. 24	3.33×10^{-5} to 3.33×10^{-2}	24.2–56.8	20°C, 100°C
Mavoori et al. [52]	Table I	10^{-5} to 10^{-2}	12.5–39.4	25°C, 80°C
Plumbridge and Gagg [53]	Figure 2b	10^{-3} to 10^{-1}	28–64	−10°C, 20°C, 75°C
Plumbridge [54]	Figure 2b			

TABLE 2.6. Bulk Sn–3.5%Ag Tensile Specimen Geometries

Publication	Specimen Shape	Specimen and/or Gauge Length	Cross Section
Schubert et al. [48]	Dog-bone	60 mm (30-mm gauge)	3 mm × 3 mm
Kariya et al. [49]	Cylinder	60 mm gauge	11.28 mm
Kariya et al. [50]	Cylinder	64 mm (20 mm)	7 mm
ITRI No. 656 [51]	Cylinder	25-mm gauge	12.7 mm
Mavoori et al. [52]	Dog-bone	12 × 12.5 × 6.5-mm gauge section	Not specified
Plumbridge [53] Plumbridge [54]	Cylinder	60-mm bar, 50-mm gauge length	11.28 mm

Test specimens were dog-bone or cylinder-shaped tensile specimens with specimen and/or gauge length and cross-sectional dimensions given in Table 2.6. Melting and cooling conditions as well as thermal treatment of specimens, when available, are summarized in Table 2.7. The reader is referred to the original publications for additional details on specimen preparation. Note that:

- The specimens by Schubert et al. [48] have a smaller cross section: 3 mm × 3 mm compared to 7 mm, 11.28 mm, and 13 mm in diameter for the specimens by Kariya et al. [49, 50] and Plumbridge [53, 54].
  - The above dimensions are larger than typical sizes of 0.1–0.2 mm (4–8 mils) for flip-chip solder joints or 0.5 mm (20 mils) for ball grid array (BGA) solder joints. That is, bulk specimen cross sections are one to two orders of magnitude larger than actual solder joints.

TABLE 2.7. Bulk Sn–3.5%Ag Tensile Specimen Treatments

Publication	Melting Conditions	Specimen Treatment
Schubert et al. [48]	NA[a]	NA[a]
Kariya et al. [49]	20°K above melting point	Water-quenched
Kariya et al. [50]	Chill-cast	Cooling: unspecified Heat-treated for 1 hour at 100°C = 373°K (=80% of melting point)
ITRI No. 656 [51]	50°C above liquidus	Water-cooled from the base for uniaxial solidification
Mavoori et al. [52]	NA	Cooling: unspecified Heat-treated at 150°C for 24 hours. Aged at room temperature for 6–10 days
Plumbridge [53] Plumbridge [54]	20°C above meting point	Rapid cooling at ~ 40°C/s (water-quenched)

[a]NA, not available.

- The Kariya et al. [49, 50] specimens were rapidly cooled by water-quenching while cooling conditions for others were not reported in detail. Some specimens were heat-treated while no heat-treatment is reported for others. Heat-treatment conditions also differ among laboratories—for example, 24 hours at 150°C (Mavoori et al. [52]) versus 1 hour at 100°C (Kariya et al. [49, 50]).

2.3.2.2. Microstructures. The microstructure of the as-cast Sn–3.5Ag solder consists of a β-Sn matrix with dispersed Ag_3Sn intermetallic phases. Specific microstructural features for the alloys and specimens under study are summarized in Table 2.8, when available.

2.3.2.3. Data Analysis. In an attempt to consolidate the data in Figure 2.8, the creep rates are fitted to a hyperbolic sine model:

$$\mathring{\varepsilon}(/\text{second}) = A \cdot [\sinh(B \cdot \sigma(\text{MPa}))]^n \cdot \exp\left(-\frac{Q(\text{J/mole})}{RT(^\circ\text{K})}\right) \qquad (2.26)$$

which gives the steady-state strain rate: $\mathring{\varepsilon} = d\varepsilon_{\text{creep}}/dt$ (/second) as a function of stress σ (MPa) and absolute temperature T (°K). $R = 8.314$ J/(K · mole) ≈ 2 cal/(K · mole) is the universal gas constant. The model constants A, B, n, and the activation energy Q are obtained by regression of the data in Table A.1. The regression was done using the nonlinear, multiple variable curve-fitting program "Datafit" by Oakdale Engineering (http://www.curvefitting.com). Because of the wide range of low strain rates, the analysis was conducted on $Y = \ln(\mathring{\varepsilon})$ and the regression function was specified as

$$Y = \ln(\mathring{\varepsilon}) = \text{LNA} - \frac{Q_a}{T(\text{K})} + n \cdot \ln[\sinh(B \cdot \sigma(\text{MPa}))] \qquad (2.27)$$

TABLE 2.8. Microstructure of Bulk Sn–3.5Ag Tensile Specimens (Prior to Testing)

Publication	Micrographs (in Publication)	Features/Description
Schubert et al. [48]	Figure 6	Ag_3–Sn intermetallic precipitates sparsely dispersed in tin matrix
Kariya et al. [49]	NA	NA
Kariya et al. [50]	NA	NA
ITRI No. 656 [51]	NA	NA
Mavoori et al. [52]	Figure 1a	β-Sn dendrites of random orientation
Plumbridge [53]	NA	Microstructure associated with rapid cooling
Plumbridge [54]	NA	NA

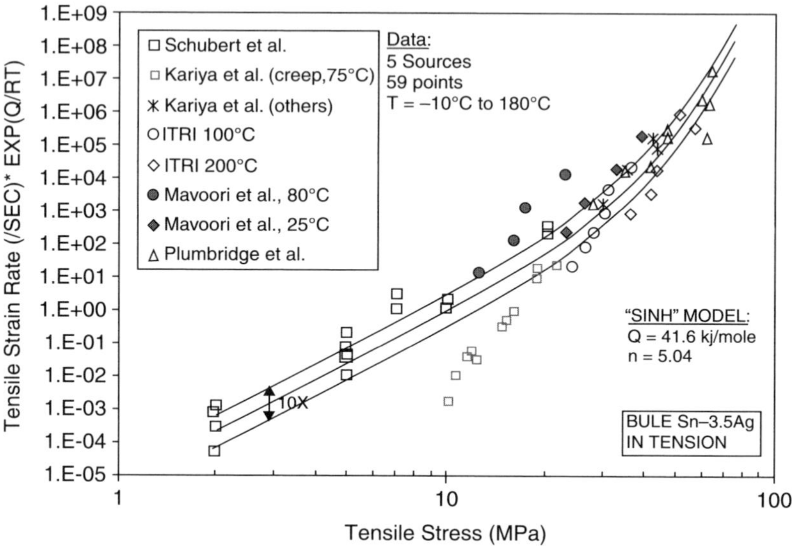

Figure 2.9. Curve-fitting of bulk Sn–3.5Ag tensile creep data to hyperbolic sine model. Centerline is a plot of $\overset{\circ}{\varepsilon} \exp(Q/RT)$ versus σ, where $\overset{\circ}{\varepsilon}$ is given by Eq. (2.28).

where T (°K) and σ (MPa) are entered as independent variables and the regression constants LNA and Q_a are defined as LNA $= \ln(A)$ and $Q_a = Q/R$. The regression results are given as central values with standard deviations:

- LNA $= 6.951 \pm 3.005$ (from which the central value of A is $A = 1044$)
- $B = 0.0615 \pm 0.012$
- $n = 4.894 \pm 0.579$ (range: $n = 4.315$–5.473)
- $Q_a = 6855 \pm 1029$ (from which the central value of Q is $Q = 57$ kJ/mole ~ 0.59 eV)

The constant LNA has the largest standard deviation error. The error on the activation energy is $\pm 15\%$ and the range of activation energies is: 48.4 to 65.5 kJ/mole.

The fit of the data to the "sinh" model is shown in Figure 2.9. The equation of the centerline in Figure 2.9 is

$$\overset{\circ}{\varepsilon}\ (\text{/second}) = 1044 \cdot [\sinh(0.0615 \cdot \sigma(\text{MPa}))]^{4.89} \cdot \exp\left(-\frac{6855}{T(^{\circ}K)}\right) \tag{2.28}$$

The dashed lines in Figure 2.9 are an arbitrary factor $\sqrt{10} \approx 3.16$ times above and below the best-fit line.

2.3.2.4. Discussion. The significant scatter of the data around the centerline in Figure 2.9 highlights the difficulties encountered when trying to consolidate data from multiple sources or when using a simplified, first order model such as the hyperbolic sine model. Often, these discrepancies are not shown since many publications do not include comparisons of raw data to independent test results. We conclude that:

- Simplified creep models such as the one in Eq. (2.26) or other models that are fit to data from a single source should be used with caution. Creep models derived from a single source of data may not be as encompassing as it appears and these models should be bounced against independent experiments.
- Much work remains to be done to resolve discrepancies among the Sn–3.5Ag tensile datasets and before a reliable constitutive model can be developed for small solder joints of electronic assemblies.
- Similar challenges hold for other lead-free solder alloys.

2.3.2.5. Comparison to Sn–4Ag Tensile Creep Data. Figure 2.10 shows Sn–4Ag creep data on the same plot as the Sn–3.5Ag tensile master curve from Figure 2.9. The raw data, digitized from Figure 3 in Jones et al. [55] and Figure 10 in Neu et al. [56], are listed in Table A.2 in Appendix. Basic specimen information was given as follows:

- For the Jones et al. experiment: cast, water-chilled tensile specimens, 40 mm long (15-mm gauge length), 2 mm thick. Isothermal tensile tests were strain-controlled.
- For the Neu et al. data: cylinders 127 mm long by 12 mm in diameter (alloy was heated 50°C above liquidus, mold was air-cooled), gauge section machined to 15.3 mm in length and 6.35 mm in diameter. The data are from isothermal strain-rate jump tests.

Sn–4Ag, with a melting range from 221°C to 228°C, has a composition near that of the eutectic Sn–3.5Ag alloy. Although the Sn–3.5Ag master curve is an

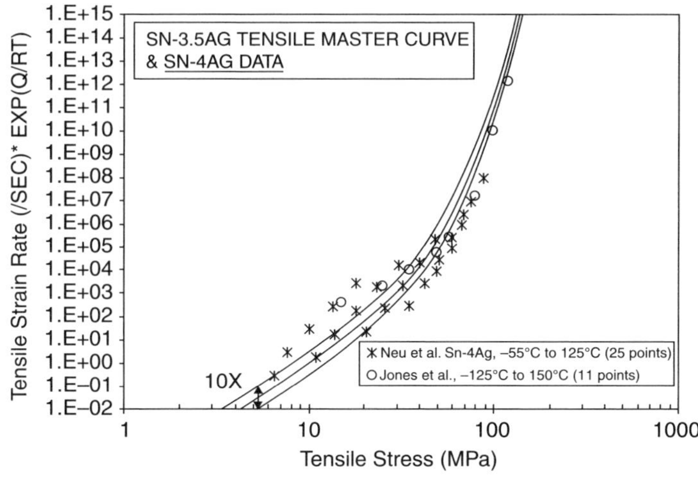

Figure 2.10. Sn–4Ag tensile creep data compared to master curve of Sn–3.5Ag tensile creep data.

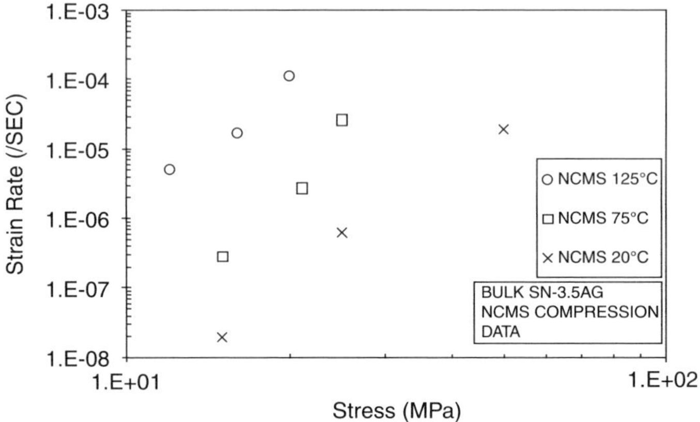

Figure 2.11. Bulk Sn–3.5Ag compression creep data.

approximate fit through the Sn–3.5Ag data, Figure 2.10 shows that, to a first order, the Sn–4Ag data points fit reasonably well to the correlation band for Sn–3.5Ag. That is, the creep resistance and tensile strength of the two alloys with close silver contents (4% versus 3.5%) appear to follow similar trends.

2.3.3. Sn–3.5Ag Compression Creep

2.3.3.1. Raw Data. NCMS conducted isothermal load-controlled, compression creep tests on cylindrical specimens 0.4 in. (10.16 mm) in diameter by 0.8 in. (20.32 mm) in length. Prior to chill-casting the specimens, Sn–3.5Ag solder was heated up 40°C above its melting point (221°C). Tests were conducted at 20°C, 75°C, and 125°C. The raw data are listed in Table A.3 in Appendix A and is plotted in Figure 2.11.

2.3.3.2. Data Analysis. We were not able to fit the compression data to a hyperbolic sine model, thus the data was fit to a simpler power law model:

$$\mathring{\varepsilon}(/\text{second}) = A \cdot [\sigma(\text{MPa})]^n \cdot \exp\left(-\frac{Q(\text{J/mole})}{RT(^\circ\text{K})}\right) \tag{2.29}$$

with the constants A, n, and Q as previously defined. The analysis was done on $Y = \ln(\mathring{\varepsilon})$ with the regression function specified as

$$\text{Y} = \ln(\mathring{\varepsilon}) = \text{LNA} - \frac{Q_a}{T(\text{K})} + n \cdot \ln[\sigma(\text{MPa}))] \tag{2.30}$$

where LNA = $\ln(A)$ and $Q_a = Q/R$. The regression results from the "Datafit" program are given as central values with standard deviations:

- LNA = -9.268 ± 2.081 (from which the central value of A is $A = 9.44\text{e-}5$).
- $n = 6.05 \pm 0.70$ (range: $n = 5.35$–6.75).

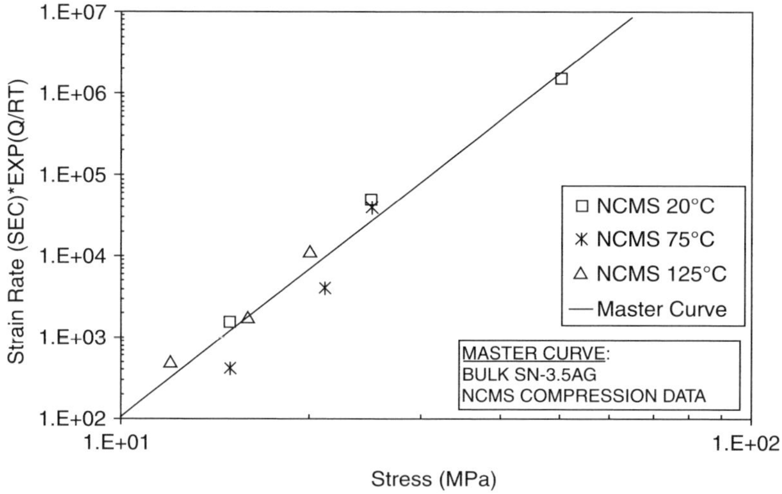

Figure 2.12. Power-law fit to bulk Sn–3.5Ag compression creep data (on the vertical axis: $Q = 61.1$ kJ/mole).

- $Q_a = 7349 \pm 753$ (from which the central value of Q is $Q = 61.1$ kJ/mole ~ 0.63 eV). Note that the activation energy in compression, $Q = 61.1$ kJ/mole is within 7.2% of the previous value: $Q = 57.0$ kJ/mole for tensile creep of bulk Sn–3.5Ag specimens.

The best fit line in Figure 2.12 is plotted from the master curve equation:

$$\overset{\circ}{\varepsilon}\,(\text{/second}) = 9.44 \times 10^{-5} \cdot [\sigma(\text{MPa})]^{6.05} \cdot \exp\left(-\frac{7349}{T(^{\circ}\text{K})}\right) \tag{2.31}$$

2.3.3.3. Comparison of Tensile and Compressive Creep. Figure 2.13 shows the NCMS compression creep data compared to the plot of the master curve for the tensile creep data (Figure 2.9). Although the master curve is an approximate fit through the tensile data, Figure 2.13 clearly shows that Sn–3.5Ag is stronger in compression than in tension. The red curve in Figure 2.13 is a manual fit through the compressive data and is 1.4 times to the right of the centerline for the tensile data.

Thus, Sn–3.5Ag behaves like an *uneven* material with higher strength or more creep resistance in compression than in tension. It is important to keep this uneven behavior in mind, especially when conducting finite element analysis. Finite element modeling of Sn–3.5Ag solder joints should be done with a finite element code that can handle compressive and tensile creep constitutive equations separately.

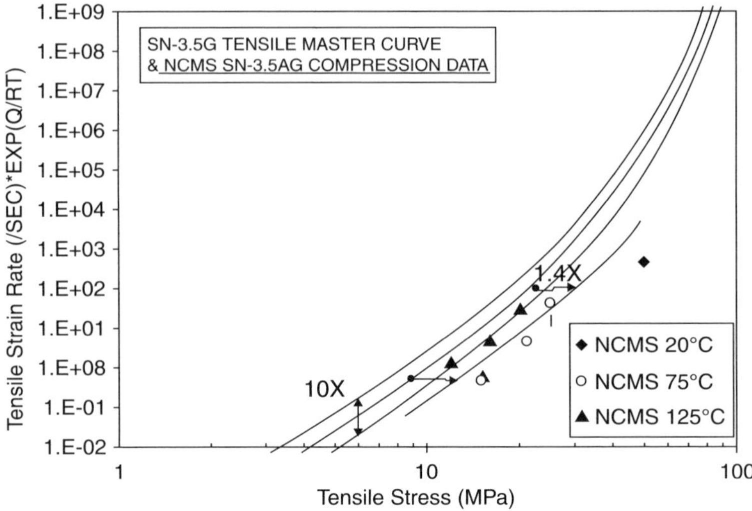

Figure 2.13. Sn–3.5Ag compression creep data compared to master curve of tensile creep data ($Q = 57$ kJ/mole on vertical axis).

2.3.4. Creep of Sn–3.5Ag LAP Shear and Plug-and-Ring Joints

2.3.4.1. Shear Joint Specimen Sizes. Isothermal shear testing data was collected from several sources (e.g., first authors quoted in Figure 2.14). Sizes of the various joints are shown in Figure 2.14 with joint dimensions given as thickness and length (or diameter) in shear. The joint depth is not shown and the arrows indicate the direction of the applied shear forces. Joint thickness is from 0.1 mm to 0.43 mm and joint length is from 0.2 mm to 16 mm.

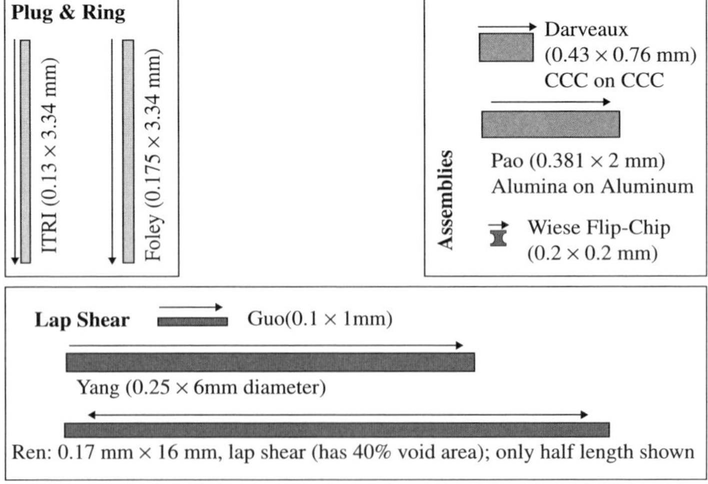

Figure 2.14. Sn–3.5Ag shear joint sizes. Arrows indicate direction of shear forces.

Creep properties are given as a function of an average shear stress τ that is defined as the total shear force, F, divided by the minimum load bearing area A in shear: $\tau = F/A$. This is an approximate description of the state of stress in the solder joints since lap joint or plug and ring shear stresses are not distributed uniformly in the shear direction. The shear stress distribution in the length direction is expected to have maximum values at the joint ends (where cracks initiate) and follow some type of hyperbolic sine function. Thus, we expect differences in shear strength values derived, for example, from the small flip-chip joints and the longer ring and plug or the 16-mm-long lap shear joints shown in Figure 2.14. Moreover, long joints increase the risk of void formation during soldering of lap shear specimens. This was reported in the Ren et al. [57] experiment where the void contents was 40% of the load bearing area in shear.

2.3.4.2. Sn–3.5Ag Lap Joint and Plug-and-Ring Shear Data. We first look at data for lap shear and plug-and-ring joints such as shown schematically in Figure 2.14. The corresponding creep data was collected from publications by ITRI (No. 656 [51]), Ren et al. [57], Hernandez et al. [58], Yang et al. [59], Guo et al. [60], Igoshev and Kleiman [61] and Glazer [62, 63]. Hernandez used copper ring-and-plug specimens, and Ren, Yang, and Guo used copper strips for their lap shear specimens. The raw data, plotted in Figure 2.15, is given as temperature, average shear stress, and shear strain rate in Table A.4 in Appendix A.

The various datasets in Figure 2.15 show some continuity as well as discrepancies. For example, the 25°C Glazer data, the 23°C Hernandez data, and the ITRI 20°C datasets appear to follow the same trendline. However, the 158°C Yang data follows a trendline that would be below most of the 100°C ITRI data or seems to

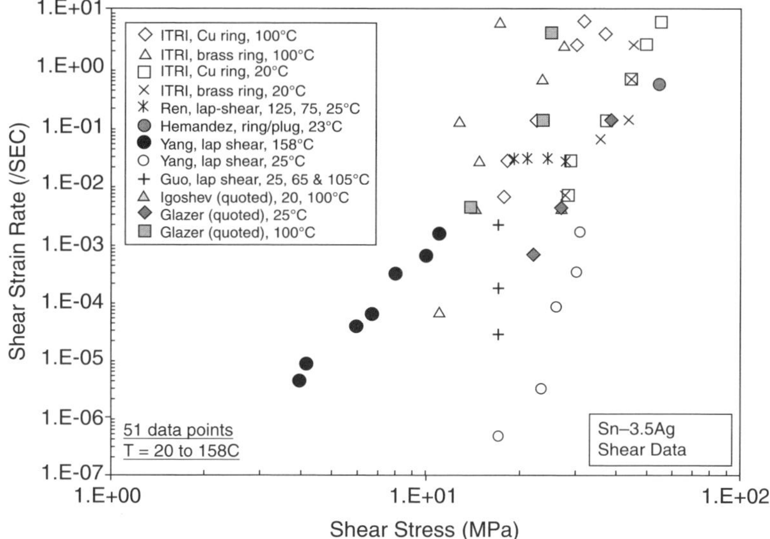

Figure 2.15. Creep data for Sn–3.5Ag lap shear and plug-and-ring joints.

line up with the 100°C Glazer data. Note also that, except for the 158°C data by Yang et al., most of the data are in the stress range above 10 MPa. As for the tensile data, the shear data do not cover the less than 10-MPa stress range that is of interest under service conditions.

2.3.4.3. Analysis of Sn–3.5Ag Lap Joint and Plug-and-Ring Shear Data.

The shear creep data in Figure 2.15 are fitted to a hyperbolic sine model:

$$\mathring{\gamma}(/\text{second}) = A \cdot [\sinh(B \cdot \tau(\text{MPa}))]^n \cdot \exp\left(-\frac{Q(\text{J/mole})}{RT(^\circ\text{K})}\right) \tag{2.32}$$

where τ is the average shear stress and $\mathring{\gamma}$ is the average shear strain-rate. The regression constants are:

- LNA = $\ln(A) = 27.43 \pm 6.38$ (from which the central value of A is $A = 8.179\text{e}11$)
- $B = 0.0266 \pm 0.0118$
- $n = 8.67 \pm 1.09$ (range: $n = 7.58$ to 9.76)
- $Q_a = Q/R = 9310 \pm 1185$ (from which the central value is $Q = 77.4$ kJ/mole ~ 0.80 eV)

Note that the exponent $n = 8.67$ for the shear data is significantly higher than the average value $n = 4.89$ that was obtained previously for the bulk specimen tensile creep data. The activation energy: $Q = 77.4$ kJ/mole is 36% larger than the average value $Q = 57$ kJ/mole for the tensile creep data.

The data in Figure 2.15 are replotted in Figure 2.16 where the best-fit line [plot of $\mathring{\gamma}\exp(Q/RT)$ versus τ] is obtained from the master curve equation:

$$\mathring{\gamma}(/\text{second}) = 8.179 \times 10^{11} \cdot [\sinh(0.0266 \cdot \tau(\text{MPa}))]^{8.67} \cdot \exp\left(-\frac{9310}{T(^\circ K)}\right) \tag{2.33}$$

2.3.4.4. Comparison to Sn–3.65Ag and Sn–4Ag Data.

Figure 2.17 shows shear creep data for Sn–3.65Ag and Sn–4Ag alloys on the plot of the Sn–3.5Ag lap shear and plug-and-ring master curve (Figure 2.16). The raw data—after Foley et al. [64] and GE/DeVore, and Westerman [65]—is listed in Table A.5 in Appendix A. Basic specimen information was given as follows:

- The Sn–3.65Ag data are from ring-and-plug joint strength measurements by Foley et al. [64]. Test temperatures were 23°C and 125°C.
- The Sn–4Ag data are from bulk alloy shear strength data in the DeVore/GE handbook of solder properties. Test temperatures were −130°C, 25°C, and 150°C.

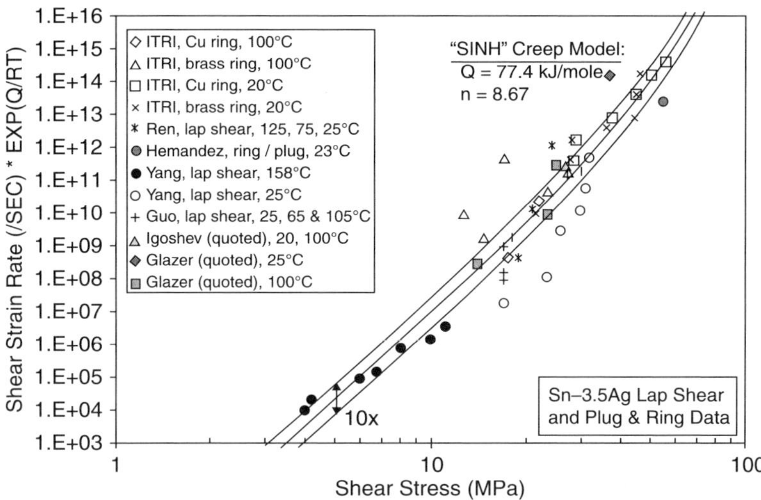

Figure 2.16. Fit of "sinh" model to Sn–3.5Ag lap shear and plug-and-ring data.

Except for the GE/−130°C data point at the top of the chart, the other data points are within or close to the correlation band for the Sn–3.5Ag shear data. Again, this suggests a possibly low sensitivity of creep behavior to silver contents for Sn–Ag alloys near the eutectic composition.

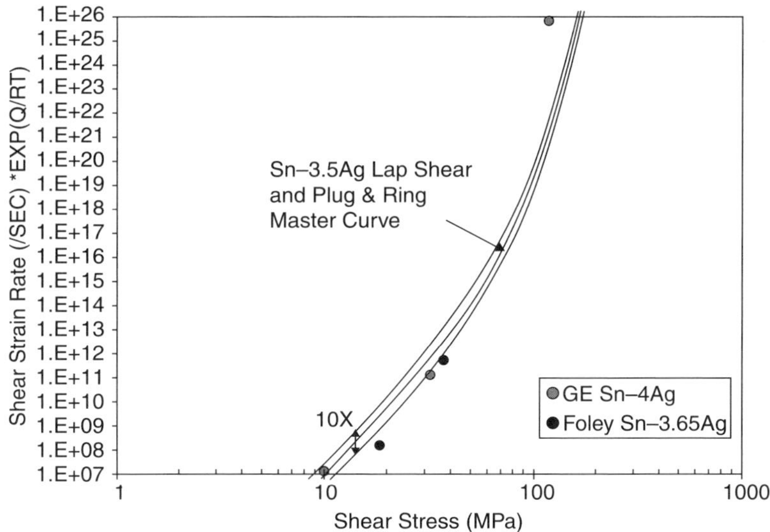

Figure 2.17. Fit of Sn–3.65Ag and Sn–4Ag data to Sn–3.5Ag lap shear and plug-and-ring correlation band ($Q = 77.4$ kJ/mole on the vertical axis).

2.3.5. Creep of Sn–3.5Ag CCC Solder Joints in Shear

2.3.5.1. Darveaux's Sn–3.5Ag Data. Table A.6 in Appendix A lists shear creep data for actual solder joints of area-array ceramic chip carriers (CCCs) mounted on identical CCCs. The data were obtained and digitized from Figure 17 in Darveaux and Banerji [17]. CCC on CCC assemblies had been aged for several months at room temperature or 100 hours at 100°C. The test setup was a double-lap shear configuration. Tests were run at 27°C, 80°C, and 132°C.

2.3.5.2. Fit of Darveaux Data to Lap Shear and Plug-and-Ring Correlation Band. Figure 2.18 shows Darveaux's solder joint data on the plot of the Sn–3.5Ag lap shear and plug-and-ring master curve (Figure 2.16). The Darveaux's data are within or fairly close to the 10× correlation band. However, the Darveaux's dataset also show some curvature that is more pronounced than that of the master curve for the lap shear and plug-and-ring data. This suggests careful consideration of lap or plug and ring joint data before it can be applied to actual solder joints of electronic assemblies.

2.3.5.3. Regression of Darveaux's CCC Shear Data. In Darveaux et al. [4], the CCC solder joint data was fitted to a hyperbolic sine model such as given by Eq. (2.32). The corresponding equation in Darveaux et al. [4] is Eq. (13.3), and the regression constants are given in Table 13.1 of the same publication. After conversion of stresses from psi to MPa, Darveaux's creep rate equation is written as

$$\mathring{\gamma}\,(/\text{second}) = 2.46 \times 10^{5} \cdot [\sinh(0.0913 \cdot \tau(\text{MPa}))]^{5.5} \cdot \exp\left(-\frac{8721}{T(^{\circ}\text{K})}\right) \tag{2.34}$$

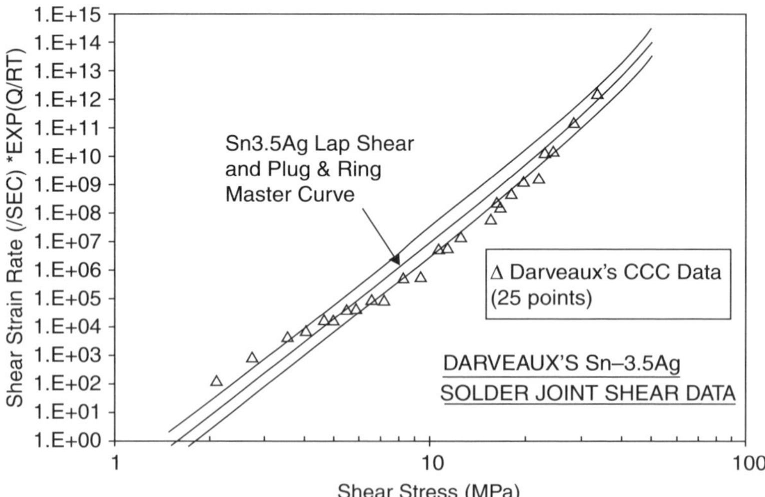

Figure 2.18. Fit of Sn–3.5Ag solder joint data to Sn–3.5Ag lap shear and plug-and-ring correlation band ($Q = 77.4$ kJ/mole on the vertical axis).

Analysis of the CCC data with the Datafit program gives the following regression constants:

- LNA = $\ln(A) = 12.84 \pm 1.158$ (from which the central value of A is $A = 3.77e5$)
- $B = 0.0983 \pm 0.0072$
- $n = 5.548 \pm 0.270$
- $Q_a = 8957 \pm 371$ (from which the central value of Q is $Q = 74.5$ kJ/mole ~ 0.77 eV)

Note that the above value of $Q = 74.5$ kJ/mole is within 4% of the activation energy $Q = 77.4$ kJ/mole obtained from the plug and ring and the lap joint shear data. The equation of the best-fit line through the CCC data are obtained as

$$\mathring{\gamma}\,(/\text{second}) = 3.77 \times 10^5 \cdot [\sinh(0.0983 \cdot \tau(\text{MPa}))]^{5.55} \cdot \exp\left(-\frac{8957}{T(\text{K})}\right) \qquad (2.35)$$

The ratios of constants in the creep rate equations (2.34) and (2.35) are:

- For the creep rate constant, A: 3.77e5/2.46e5 = 1.53.
- For the stress multiplicative factor, B: 0.0983/0.0913 = 1.0766.
- For Q_a, or for the activation energy: 8957/8721 = 1.027.
- For the exponent, n, of the sinh() functions: 5.548/5.5 = 1.009.

The largest difference in the regression parameters is by a factor 1.53× for the creep rate constants, A. This may not show on plots of strain rates versus stress on log–log scales because the vertical axis covers several orders of magnitude. In terms of graphical display, either model appears to work well as seen in Figure 2.19 where we plotted the data and master curves for the two regression equations. The regression in Darveaux's paper was reportedly done by "trial and error." The Datafit software uses a nonlinear solver. However, the 1.53× difference in the constants A is large enough to have an impact on solder joint stress/strain analysis results.

2.3.6. Creep of Sn–3.5Ag Flip-Chip Solder Joints in Shear

2.3.6.1. Flip-Chip Sn–3.5Ag Shear Data. Shear data for Sn–3.5Ag solder joints were obtained from two sources:

- Shear creep tests conducted by Wiese et al. [66] on hourglass-shaped, flip-chip solder joints of silicon-on-silicon Sn–3.5Ag assemblies.
  - Solder bumps were deposited using paste and processes from two manufacturers. Pads were 0.1 mm × 0.1 mm or 0.2 mm × 0.2 mm, and the hour-glass shaped solder joints were 0.15 mm or 0.2 mm in height. The average shear

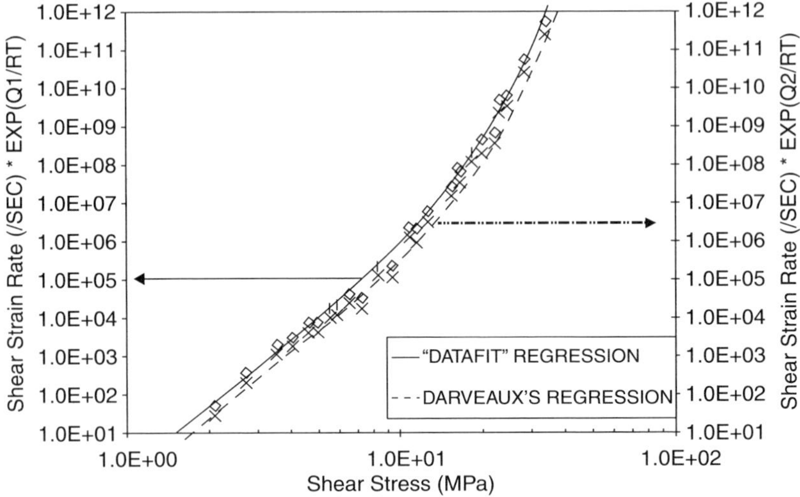

Figure 2.19. Fit of Darveaux's Sn–3.5Ag data to regression models using Eqs. (2.34) and (2.35). The primary *Y*-axis (to the left) uses activation energy $Q1 = 74.5$ kJ/mole from the Datafit regression. The secondary *Y*-axis (to the right) uses activation energy $Q2 = 72.5$ kJ/mole from Darveaux's regression.

stress was defined as the applied force divided by the narrow area or minimum cross-section area of the hourglass-shaped solder joints.

- Test results given in Figure 19 in Wiese et al. [66] were converted from shear to tensile data. The digitized raw data is given in Table A.7 in Appendix A. The creep tests were run at 5°C, 10°C, and 50°C. Bump paste was from two manufacturers identified by the labels DG and PT. In Table A.7, we re-converted the tensile data to shear using the same Von Mises transformation as was used by Wiese et al. [Eqs. (9) and (10) in Ref. 66].

- Shear creep tests by Yang et al. [67]:
  - Flip-chip joints from silicon-on-silicon assemblies (33 × 33 I/O) using standard reflow were apparently barrel-shaped with a minimum load bearing area at the joint to chip interface.
  - Limited isothermal test data were available at 25°C and 80°C. The digitized raw data, from Figure 12 in Yang et al. [67], is listed in Table A.8 in Appendix A.

2.3.6.2. Comparison of Flip-Chip and CCC Solder Joint Shear Data. The raw shear creep data for Sn–3.5Ag flip-chip and CCC solder joints is plotted in Figure 2.20. Some datasets show the expected continuity while others do not. For example:

- The 25°C flip-chip data by Yang et al. [67] shows continuity with the 27°C CCC data by Darveaux et al. [4]. However, the same data falls below the 10°C data points by Wiese et al. [66].

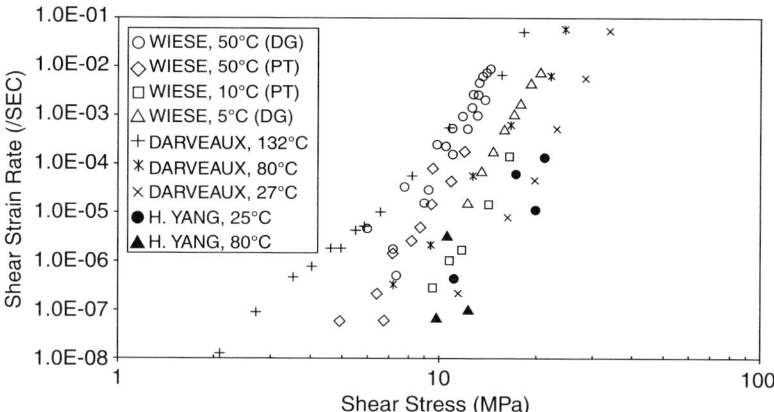

Figure 2.20. Plot of Sn–3.5Ag flip-chip and CCC joint shear creep data (DG and PT are labels for paste manufacturers).

- Besides differences in the average stress and strain definitions, which by themselves may lead to discrepancy in the data, differences in the high stress locations and the solder joint shapes may be in part responsible for the differences between the Wiese et al. [66] data and the other two datasets by Yang et al. [67] and Darveaux et al. [4]. Also, from finite element analysis [42], stress concentration patterns are expected to be quite different for barrel-shaped and hour-glass-shaped flip-chip solder joints. Last, differences in the amount of terminal metals dissolved in solder may contribute to the higher creep resistance of the Darveaux's CCC and Yang's flip-chip solder joints.
- Of the three 80°C flip-chip data points by Yang et al. [67], one of them lines up with the Darveaux CCC data at 80°C but the two other data points are over one order of magnitude below. Strangely, these two data points also fall below the Yang's data points at 25°C.

The Wiese et al. and Yang's flip-chip data is shown on the plot of the Darveaux's CCC data correlation in Figure 2.19. The master curve of the CCC data is as given by Eq. (2.35) in the previous subsection. The 10× correlation band around the CCC data in Figure 2.21 is also defined by upper- and a lower-bound lines that are an arbitrary factor $\sqrt{10} = 3.16$ times above and below the centerline. The flip-chip data of Wiese et al. [66] and Yang et al. [67] is also shown on the same plot:

- As discussed above, the 25°C flip-chip data by Yang et al. [67] fits rather well within or close to the CCC master curve and correlation band.
- One of the 80°C data points from Yang et al. [67] is close to the correlation band (near its lower bound) whereas the two other 80°C data points are outside and below the band.

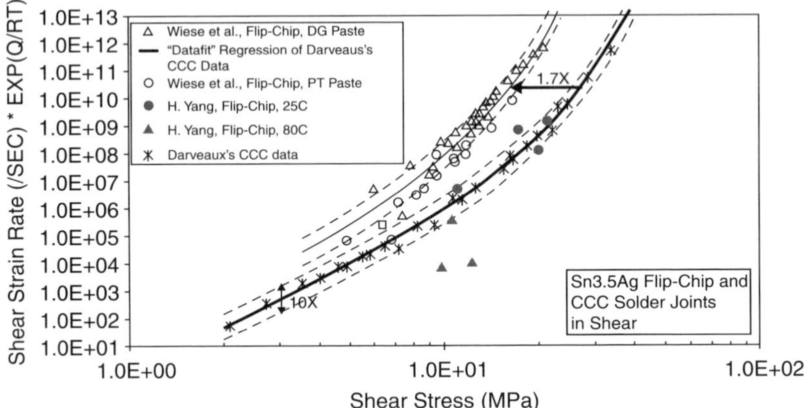

Figure 2.21. Fit of Sn–3.5Ag flip-chip data to correlation of Darveaux's CCC shear data (DG and PT are labels for paste manufacturers). $Q = 74.5$ kJ/mole on the vertical axis.

- The flip-chip data from Wiese et al. [66] is outside the CCC data correlation band and falls within a band that is offset from the CCC master curve by a factor 1.7× along the stress axis. This indicates that the flip-chip data correlates to the CCC data (within a 1.7× factor in stress) and follow similar trends in terms of the temperature effect. However, the CCC joints appear to be more creep resistant and to have higher strength. This is possibly due to more strengthening of the CCC joints close to the solder–component interface, differences in stress concentration factors for a barrel-shaped versus an hour glass shaped joint or a combination of those two effects.
    - The Wiese et al. [66] data in Figure 2.21 also show a potential "paste/manufacturing" effect. Looking at the shifted master curve and the 10× correlation band around it, the majority of the "PT" paste data points are below the centerline, whereas most of the "DG" data points are close to or above the centerline.

While all of this remains to be better understood, the above correlations of data for flip-chip and BGA-type solder joints is encouraging for future reliability analysis of solder joints of electronic assemblies. This also suggests that, since the ultimate goal of solder testing and modeling is to apply constitutive models to real electronic solder joints, it is worthwhile determining solder properties from measurements on test vehicles that closely resemble production assemblies.

2.3.7. Other Properties of Sn–3.5Ag

2.3.7.1. Young's Modulus Versus Temperature. Young's modulus versus temperature is plotted in Figure 2.22, similar to Figure 9 in Wiese et al. [66]. The original source of the data was publications by Darveaux et al. [4] and Lau and Pao [42]. The data were digitized and, for each dataset, we added a linear trendline

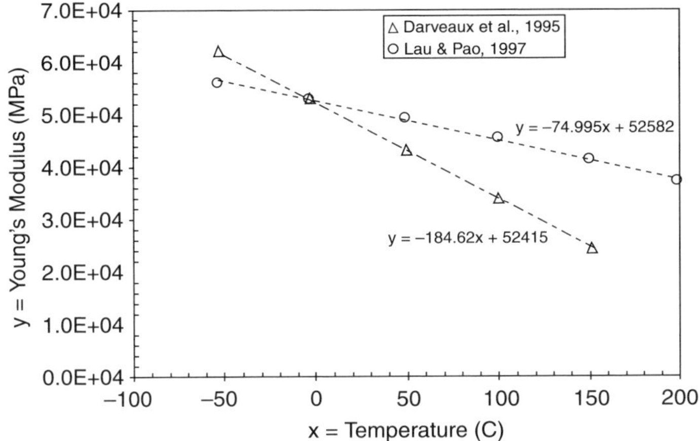

Figure 2.22. Plot of Sn–3.5Ag Young's modulus E (MPa) versus temperature T (°C).

and its equation so the temperature-dependence of Young's modulus is readily available. The temperature-dependent Young's modulus is entered as material properties in stress–strain analysis programs such as FEA codes.

2.3.7.2. Poisson's Ratio. Another elastic property that is of use in stress–strain analysis programs or to convert Young's modulus E to a shear modulus, G, is Poisson's ratio. Lau et al. [68] used a Poisson's ratio $\nu = 0.4$ in the finite element modeling of Sn–3.5Ag assemblies. We were not able to locate any report with direct measurements or experimental determination of Poisson's ratios for Sn–3.5Ag alloy.

2.3.7.3. Coefficients of Thermal Expansion (CTE). Measured or quoted coefficients of thermal expansion (CTEs) for Sn–3.5Ag solder are given in Table 2.9. As expected, the CTE results show a slight temperature dependence.

TABLE 2.9. Sn–3.5Ag CTE Values

CTE (ppm/°C)	Measurement/Specimen/Temperature Range	Reference
22.1 (Vendor A) 23.2 (Vendor B)	• Thermomechanical analysis (TMA) • Specimen: 6.4 mm ± 0.2 mm diameter × 1.6 ± 0.1 mm thick • Temperature range: 50–160°C	Lau et al. [69]
	• Thermomechanical analysis (TMA) • Specimen: 3 mm × 3 mm × 1.6 mm thick • Temperature range:	Schubert et al. [48] Wiese et al. [66]
20.2–21.7 →	20–150°C	
19.4–20.4 →	20–100°C	
21.5–22.9 →	100–150°C	
$21.85 + 0.02039\,T$	• NA (temperature T in degrees C)	Lau et al. [68]

From the data in Table 2.9, an overall average value for the CTE of Sn–3.5Ag is about 21.5 ppm/°C.

For comparison purposes, a value that is often quoted for the CTE of eutectic Sn–Pb is 24 ppm/°C. The CTE of Sn–3.5Ag appears to be slightly lower than that of eutectic Sn–Pb. This is beneficial to attachment reliability since a lower CTE of the solder alloy reduces local CTE mismatches between the solder joint and the interconnected parts.

2.3.7.4. Other Physical Properties of Sn–3.5Ag. Other properties of interest for predictive modeling of solder joint geometry using, for example, a computer program such as Surface Evolver are:

- Solder density, ρ in lb/in.^{3} (or g/cm^3).
  - ρ (g/cm^3) = 7.5 (NIST Boulder database).
    - Note that an error seems to have propagated through the literature since both the ITRI publication (No. 656) and the GE/DeVore handbook quote $\rho = 10.38$ g/cm^3 for Sn–3.5Ag.
    - The NIST-quoted value is consistent with the rule of mixture calculation: $\rho = 96.5\%\ \rho_{Sn} + 3.5\%\ \rho_{Ag}$. The density of pure Sn and Ag are $\rho_{Sn} = 7.31$ g/cm^3 and $\rho_{Ag} = 10.5$ g/cm^3, which gives: $\rho_{Sn3.5Ag} = 7.42$ g/cm^3.
- Surface tension, γ, in units of mNm^{-1}, to be specified in terms of the soldering atmosphere (e.g., air, nitrogen...):
  - $\gamma = 431$ mNm^{-1} in air, $\gamma = 493$ mNm^{-1} in nitrogen (at 50°C above liquidus), after Glazer [62], also quoted in NIST-Boulder database.

2.3.8. Discussion

2.3.8.1. Data Scatter. The scatter of the Sn–3.5Ag data is better described as erratic rather than random since none of the reported studies was statistical in nature. There are many possible reasons for this, including differences in specimen size and geometry, specimen preparation and treatment, interfacial metallurgical effects in the case of shear specimens, test and loading conditions (e.g., tensile test at constant strain rate versus load-controlled creep test), possible specimen misalignment and other experimental differences and errors. Also, load-controlled creep tests are not exactly constant-stress tests although the data is often treated as if they were. Last, the initial loading rate may have an impact on the entire deformation history but, often, the initial loading rate is not reported.

The various tests that were investigated did not follow a unique standard and the reports that were examined had variable levels of completeness. For example, the microstructural features of solder joints or specimens were not always available and, when they were, representative dimensions (such as the length and distribution density of intermetallic precipitates) were not given and were difficult to evaluate on micrograph reproductions.

2.3.8.2. Multiaxial Conditions. The Sn–3.5Ag compression, bulk solder data suggests higher strength or creep resistance in compression than in tension. However, we were not able to find a correlation between shear and tensile test results.

The often-used Von Mises yield criterion leads to a simple (and useful) transformation between tensile (σ, ε) and shear (τ, γ) stresses and strains: $\sigma = \tau\sqrt{3}$, and $\varepsilon = \gamma/\sqrt{3}$. This criterion has been validated at strain rates above 10^{-4}/s for near-eutectic Sn–Pb based on mechanical testing of bulk tensile and torsion specimens [70]. The Von Mises criterion and the resulting stress/strain transformations require further investigation before they can be applied to Sn–3.5Ag.

2.3.8.3. Constitutive Modeling. The Sn–3.5Ag creep data was fitted to simplified hyperbolic sine creep models. These do not constitute a full-fledge constitutive model but provide a simple equation that allows for some consolidation of secondary creep data. The hyperbolic sine models have been found to work well for metals that exhibit significant creep deformations and they are easily implemented in commercial finite element codes.

The reviewed data do not allow for the development of a complete constitutive model since mostly secondary or steady-state creep was reported on. However, it is noteworthy that the study by Darveaux et al. [4] provides an example of a constitutive model for Sn–3.5Ag solder joints. The Darveaux model is of the additive type whereby strains are broken up in their elastic, plastic, primary and secondary creep components. The reader is referred to the original work by Darveaux et al. [4] for pertinent details of this model. Yang et al. [67] also stressed the need to investigate primary creep for Sn–3.5Ag solder. Given the parallelism between Darveaux's CCC and Wiese et al.'s flip chip data, it is worthwhile applying the Darveaux model to the stress/strain analysis of Sn–3.5Ag solder assemblies. As always, such an analysis would have further merit if it were validated against independent experimental data.

More advanced constitutive models have been proposed for Sn–3.5Ag and Sn–4Ag, such as Unified Creep Plasticity (UCP) models (see Wen [71] and Neu et al. [56], for example). These models treat plastic and creep strains as a single visco-plastic or inelastic strain. In the formulation of inelastic strain rates, applied stresses are reduced by an internal back-stress that reflects the resistance of intermetallic precipitates, or other obstacles, to dislocation motions. Back-stresses are state variables that follow their own evolution equations. The latter are semiempirical and may call for a large number of fitting constants depending on the complexity of the back-stress model. To this author's knowledge, the user implementation of UCP models in commercial finite element codes is not readily available.

2.3.8.4. Recommendations. Based on our review of Sn–3.5Ag properties, further analysis is warranted before the full range of deformations can be characterized and before a workable constitutive model can be used in engineering applications. Whatever constitutive model is developed, it is important that the model be validated against independent data. Three types of stress/strain measurements are useful for model validation:

- Stress–strain curves that are acquired in isothermal, constant strain rate tests. Many reports only publish the Ultimate Tensile Strength, or UTS, from a tensile test at a given strain rate and temperature. The entire stress–strain curve would be of use as the portion of it prior to reaching the ultimate strength captures a more complete history of inelastic deformations. An appropriate constitutive model should be capable of simulating this initial portion of the stress–strain curve.
- Similarly, few creep test reports include plots of deformations versus time. These would be useful for the investigation of initial deformations (plastic flow, primary creep) as well as for the validation of constitutive models. See Villain et al. [72], for such examples of Sn–3.5Ag deformation curves.
- Last, since one major goal of alloy testing is to develop models for the analysis of solder joints of electronic assemblies under thermal cycling conditions, hysteresis loops of average shear stresses and strains in actual solder joints are needed. Such hysteresis loops will provide for the ultimate validation of solder alloy constitutive models. Guidelines and test procedures for acquiring such hysteresis loops are available in Hall [29–31], Lau and Pao [42], and Pao et al. [10, 32].

2.4. TIN–SILVER–COPPER PROPERTIES AND CREEP DATA

2.4.1. Overview

This next main section summarizes our review and analysis of isothermal tensile creep data for bulk lead-free solders of composition close to that of the NEMI-selected Sn–3.9Ag–0.6Cu alloy. As with the review of Sn–3.5Ag properties, the ultimate goal of this work is to develop preliminary constitutive models for use in stress–strain analysis [e.g., Finite Element Analysis (FEA)] and solder joint life prediction models. The raw creep data for SAC alloys are tabulated in Appendix B. The data were acquired for comparative analysis and to identify gaps and differences among existing datasets:

- Isothermal bulk solder creep data from publications by Kariya and Plumbridge [49], Schubert et al. [48], Wiese et al. [66], and Neu et al. [70] was tabulated and plotted. Specimen geometry and preparation, as well as a description of the microstructures, are summarized. The data cover two orders of magnitude on the stress axis and eight orders of magnitude on the strain rate axis. Test temperature is in the range −55°C to 150°C.
- The raw data are presented in a tabular format for inclusion in a materials property database and further analysis by others.

The intent of the analysis is to develop simple, first-order creep models in an attempt to bridge data from several sources. Sub-datasets that do not fit in are also highlighted and possible reasons are given for differences in trends. Main findings of this review and analysis are:

- Although there appears to be a few possible outliers, and the SAC type alloys have slightly different Ag and Cu weight percentages (including the Sn–2.5Ag–0.8Cu–0.5Sb Castin^{TM} alloy), the creep datasets from three independent sources show first order consistency.
- Most of the data fit a simple hyperbolic sine creep equation with a "stress" exponent close to 5 and an activation energy of 59 kJ/mole.
- Additional SAC data by Kim et al. [73] for specimens that were rapidly cooled also fit the first-order creep model while data for slowly cooled specimens is offset by two orders of magnitude in terms of creep rates. As expected, cooling rates have a significant effect on mechanical properties (because of differences in microstructures). Those effects are larger than the effect of alloy composition for the alloys that were investigated and will have to be considered in the development of more advanced constitutive models.
- The review of SAC data also points to lesser data available at strain rates less then 10^{-6}/s, stresses less than 10 MPa and temperatures above 75°C or combinations thereof. Future work should consider gathering creep data under those conditions since they are representative of stress conditions experienced by solder joints of electronic assemblies in use.

The results of the creep data analysis are for bulk solder in tension and do not include creep data for small size solder joints as found in actual electronic assemblies or solder joints in shear.

2.4.2. "SAC" Creep Data

2.4.2.1. Source and Plot of Data. Isothermal creep data from three independent sources are plotted as tensile strain rate (/second) versus tensile stress (MPa) in Figure 2.23. The data cover two orders of magnitude on the stress axis and eight orders of magnitude on the strain rate axis. Test temperatures are in the range −55°C to 150°C. The original data, given in Table B.1 in Appendix B, were digitized by Kil-won Moon of NIST. The raw data were obtained from the figures listed in Table 2.10, in their respective publications:

These investigations of SAC properties are fairly recent, when compared to the Sn–Pb or Sn–Ag literature, since it is only recently that SAC alloys have received increased attention from industry.

2.4.2.2. Specimens. Test specimens were dog-bone- or cylinder-shaped tensile specimens with specimen and/or gauge length and cross-sectional dimensions given in Table 2.11. Melting and cooling conditions as well as thermal treatment

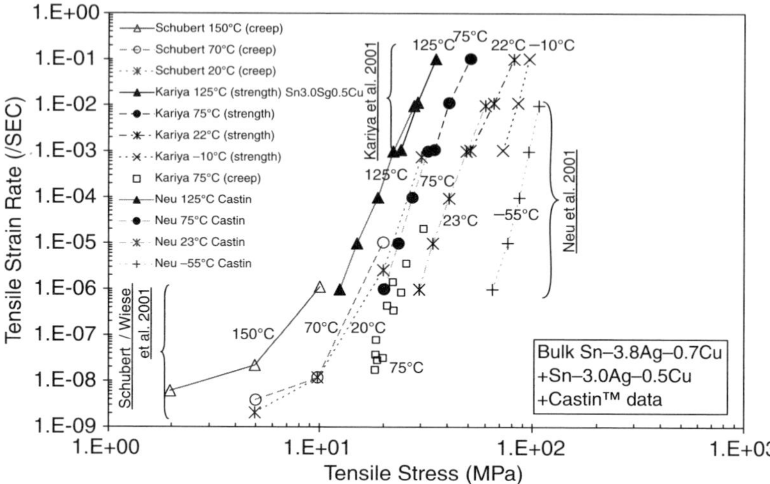

Figure 2.23. Log–log plot of isothermal creep data for Sn–Ag–Cu and Castin™ (Sn–2.5Ag–0.8Cu–0.5Sb) bulk solders.

TABLE 2.10. Source of Data, Alloy Composition and Test Temperatures

Publication	Figure(s) in Publication	Alloy (s)	Test Temperatures
Schubert et al. [48]	Figure 10	95.5Sn–3.8Ag–0.7Cu	20°C, 70°C, 150°C
Wiese et al. [66]	Figure 15		
Neu et al. [56]	Figure 11	96.2Sn–2.5Ag–0.8Cu–0.5Sb	−55°C, 23°C, 75°C, 125°C
Kariya and Plumbridge [49]	Figures 2, 3, and 4	95.5Sn–3.8Ag–0.7Cu 96.5Sn–3.0Ag–0.5Cu	−10°C, 22°C, 75°C, 125°C

TABLE 2.11. Specimen Geometries

Publication	Alloys and Melting Points	Specimen Shape	Specimen and/ or Gauge Length	Cross Section
Schubert et al. [48] Wiese et al. [66]	95.5Sn–3.8Ag–0.7Cu (MP[a] 217°C)	Dog-bone	60 mm (30-mm gauge)	3 mm × 3 mm
Neu et al. [56]	96.2Sn–2.5Ag–0.8Cu–0.5Sb (MP 210 to 215°C)	Cylinder	127 mm (12.7-mm gauge)	13 mm
Kariya and Plumbridge [49]	95.5Sn–3.8Ag–0.7Cu (MP 217°C) 96.5Sn–3.0Ag–0.5Cu (MP ∼217°C)	Dog-bone	60-mm gauge	11.28 mm

[a]MP, melting point of alloy.

TABLE 2.12. Melting Conditions and Specimen Treatments

Publication	Alloys (nominal composition) and Melting Points	Melting Conditions	Specimen Treatment
Schubert et al. [48] Wiese et al. [66]	95.5Sn–3.8Ag–0.7Cu (MP 217°C)	NA[a]	NA
Neu et al. [56]	96.2Sn–2.5Ag–0.8Cu–0.5Sb (MP 210–215°C)	50°C above liquidus	Air-cooled Aged at room temperature for 21 days Stored at −20°C
Kariya and Plumbridge [49]	95.5Sn–3.8Ag–0.7Cu (MP 217°C) 96.5Sn–3.0Ag–0.5Cu (MP ∼217°C)	20°K above meting point	Rapid cooling (water-quenched)

[a]NA, not available.

of specimens, when available, are summarized in Table 2.12. The reader is referred to the original publications for additional details on the specimen preparation.

The alloys have nominal compositions with elemental contents in the following ranges:

- 95.5 to 96.5 wt% for Sn; 2.5 to 3.9 wt% for Ag; and 0.5 to 0.8 wt% for Cu.

Chemical analysis of the 95.5Sn–3.8Ag–0.7Cu (nominal) alloy studied by Schubert et al. gave a composition of 95.4 wt% Sn, 3.91 wt% Ag and 0.71 wt% Cu (from Table 1 in Schubert et al. [48]).

Comparing the specimen information from the three sources, note that:

- The specimens by Schubert et al. [48] have a significantly smaller cross section: 3 mm × 3 mm versus 11.28 mm and 13 mm in diameter for the specimens by Kariya and Plumbridge [49], and Neu et al. [56], respectively.
  - The above dimensions compare to typical sizes of 0.1–0.2 mm (4–8 mils) for flip-chip solder joints, or 0.5 mm (20 mils) for ball grid array (BGA) solder joints.
- The Kariya and Plumbridge [49] specimens were rapidly cooled by water-quenching while the Neu et al. [56] specimens were air-cooled although the cooling rate was not reported. The cooling conditions for the Schubert et al. [47] specimens were not reported.

2.4.2.3. Microstructures. The microstructure of the as-cast SAC solders consists of the Sn matrix with precipitate-free β-Sn regions and eutectic (or near-eutectic) regions having Ag_3Sn or Cu_6Sn_5 intermetallic phases dispersed in the Sn matrix. Specific microstructural features for the alloys and specimens under study are given in Table 2.13.

TABLE 2.13. Microstructure of Bulk "SAC" Specimens

Publication	Alloys	Micrographs	Features
Schubert et al. [48] Wiese et al. [66]	95.5Sn–3.8Ag–0.7Cu	See Figure 9 in Schubert et al. [47] and Figure 14 in Wiese et al. [63]	Elongated rod-like Ag_3Sn precipitates. Tiny dispersed Cu_6Sn_5 precipitates of size 50 nm dispersed in Sn matrix.
Neu et al. [56]	96.2Sn–2.5Ag–0.8Cu–0.5Sb	See Figures 2 and 3 in Neu et al. [70]	Sn-rich dendritic phase with Ag_3Sn particles in interdentritic phase. Cu phase sparsely distributed through microstucture. No. Sb detected in EDX analysis.
Kariya and Plumbridge [49]	95.5Sn–3.8Ag–0.7Cu 96.5Sn–3.0Ag–0.5Cu	See Figure 1 in Kariya and Plumbridge [49]	Eutectic regions with β-Sn grain size of ∼1 μm: • Eutectic regions of fine dispersed Ag_3Sn spherical intermetallics (size: submicrons). • Eutectic regions of tiny dispersed Cu_6Sn_5 intermetallics. Large precipitate-free regions with β-Sn grain size of ∼5 μm.

2.4.2.4. Test Procedures. The test conditions and procedures are summarized in Table 2.14. The isothermal tensile tests are of three types:

- Conventional "creep" tests (with constant load rates) where the linear portion of the strain-versus-time response gives the "minimum" or steady strain rate (Figure 2.24*a*).
- Conventional "strength" tests (stress–strain tests ran at constant strain rates) where the maximum stress gives the ultimate tensile strength (UTS) of the specimen at a given strain rate (Figure 2.24*b*). Once the UTS has been reached, the typical response of solder is further elongation of the specimen at constant stress or with a slow drop in load.
- Strain rate jump tests which are similar to the above stress–strain tests except that the strain rate was increased by a factor of 10 after each increment of 1.5% strain.

TABLE 2.14. Test Procedures and Conditions

Publication	Alloys	Test Type	Range of Strain Rates (/second)	Temperatures (°C)
Schubert et al. [48] Wiese et al. [66]	95.5Sn–3.8Ag–0.7Cu	Constant load creep test	2×10^{-9} to 7.5×10^{-4}	20, 70, and 150
Neu et al. [56]	96.2Sn–2.5Ag–0.8Cu–0.5Sb	Constant strain rate jump test	10^{-6} to 10^{-2}	−55, 23, 75, and 125
Kariya and Plumbridge [49]	95.5Sn–3.8Ag–0.7Cu 96.5Sn–3.0Ag–0.5Cu	Constant strain rate tensile ("strength") test Constant load creep test	10^{-3} to 10^{-1}	−10, 22, 75, and 125

Noteworthy observations about these tests are:

- In general, the first two types of tests give similar strain-rate-vesus-stress equations since the *strain rate is constant* during both the strength test and the steady-state phase of the creep test. However, one of the limitations of the creep tests is that they are not run at constant stress since it is easier to control loads rather than stress. The reported stress is the initial stress defined as the applied load divided by the initial cross section of the specimen.
- As pointed out by Neu et al. [56], the strain rate jump tests allow for the acquisition of tensile strengths at different strain rates from a single specimen. However, Neu et al. mentionned that the strength results may be affected by the history of prior deformations of a given specimen when jumping from one strain rate to the next.

The reviewed literature focuses on *steady-state creep*. To our knowledge, the *other deformation modes* (i.e., the initial, instantaneous plastic flow and primary

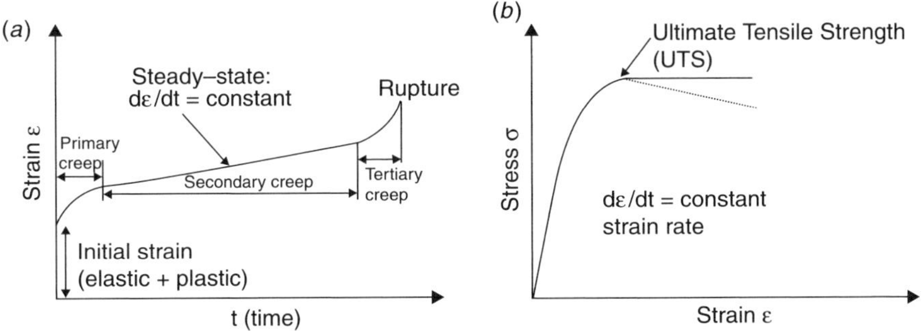

Figure 2.24. Output of (*a*) creep test and (*b*) "strength" test.

creep) have not been reported on for SAC-type alloys. Based on studies by Darveaux et al. [4, 17], these deformation modes are not negligible for near-eutectic Sn–Pb and Sn–Ag alloys and are worthwhile investigating further for SAC solders.

2.4.3. "SAC" Creep Data Analysis and Modeling

2.4.3.1. Fit of Kariya and Schubert Models. Isothermal lines representing the Kariya and Schubert creep models are plotted in Figures 2.25 and 2.26, respectively. The original test data are also shown in Figure 2.23. The Kariya model is a power-law model that did fit the Kariya and Plumbridge's [49], Sn–3.0Ag–0.5Cu and Sn–3.8Ag–0.7Cu data equally well. The Kariya model gives steady-state strain rates $\overset{\circ}{\varepsilon} = d\varepsilon_{\text{creep}}/dt$ as a function of stress σ and absolute temperature T ($R = 8.314$ J/mole):

$$\overset{\circ}{\varepsilon}(/\text{second}) = 1.37 \times 10^{46} \cdot \left(\frac{\sigma\,(\text{MPa})}{E\,(\text{MPa})}\right)^{13.2} \cdot \exp\left(-\frac{61\ \text{kJ/mole}}{RT\,(\text{K})}\right) \tag{2.36}$$

with E (MPa) = 76087 − 109 × T(K))

Schubert et al. [48]/Wiese et al. [66], proposed the following power-law break-down/hyperbolic sine model with model constants obtained by regression of their Sn–3.8Ag–0.7Cu bulk solder creep data:

$$\overset{\circ}{\varepsilon}(/\text{second}) = 0.0026 \cdot [\sinh(0.185\sigma\,(\text{MPa}))] \cdot \exp\left(-\frac{38.7\ \text{kJ/mole}}{RT(\text{K})}\right) \tag{2.37}$$

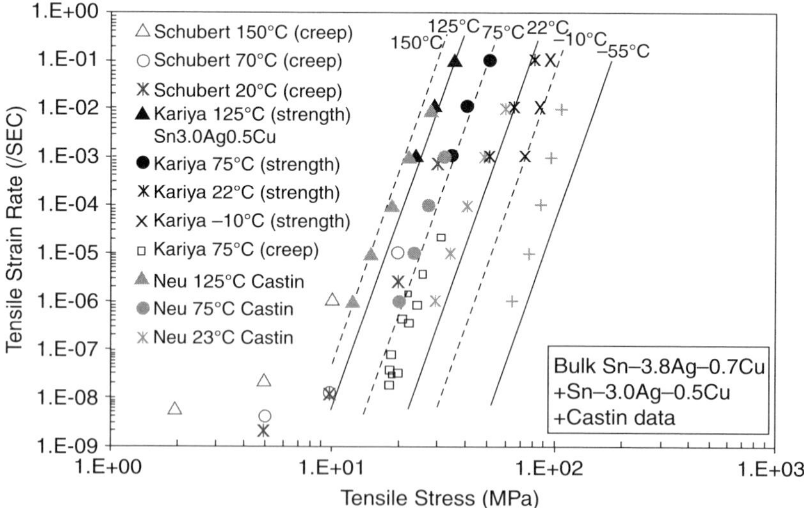

Figure 2.25. Plot of SAC creep data and isothermal lines of Kariya and Plumbridge's [49] model (Castin^{TM} = Sn–2.5Ag–0.8Cu–0.5Sb).

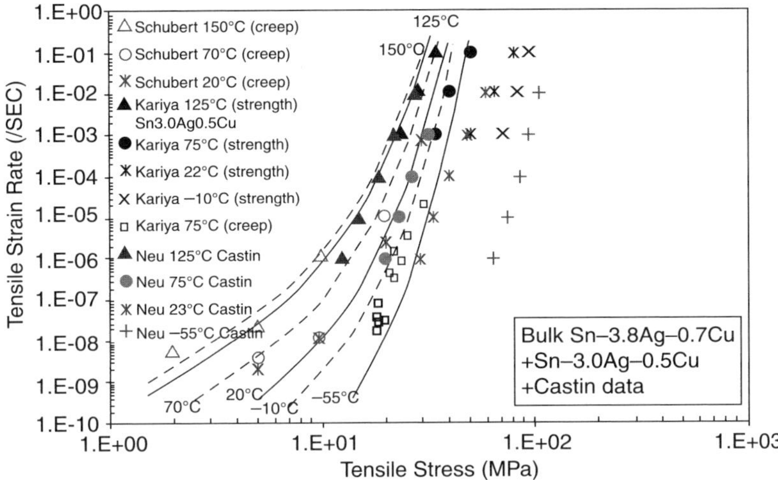

Figure 2.26. Plot of SAC creep data and isothermal lines of Schubert et al.'s [48]/Wiese et al.'s [66] model (Castin™ = Sn–2.5Ag–0.8Cu–0.5Sb).

From Figure 2.25:

- The Kariya model fits the Kariya and Neu datasets nicely except for a slight departure from the data at −55°C and strain rates above 10^{-3}/s. This shows consistency between the SAC and Castin™ (Sn–2.5Ag–0.8Cu–0.5Sb) data.
- The Kariya model does not capture the Schubert data.

From Figure 2.26:

- As expected, the Schubert model fits the Schubert et al. [48]/Wiese et al. [66] Sn–3.8Ag–0.7Cu data well.
- The Schubert model also fits the 125°C Kariya and Neu data well.
- The Schubert model is at a significant departure from the Kariya and Neu datasets at temperatures of 70–75°C and below.

Assuming that the models apply to all SAC alloys under study, one of the main differences between the Kariya and Schubert models is that the models attempt to fit data in different stress regions, mostly under 10–20 MPa for the Schubert data and above 10 MPa for the Kariya's data. Based on Sn–Pb experience, solder joints stresses in Sn–Pb assemblies are typically under in the range 1–10 MPa. This suggests the need to gather SAC creep data in the lower stress range (<10 MPa).

In the next subsections, the above data are reviewed further and we attempt to fit a simple hyperbolic sine model to the Kariya and Plumbridge [49], Neu et al. [56], and Schubert et al. [48] datasets. The intent of this exercise is to bridge the datasets with points below and above the 10-MPa stress level. Because the alloys under consideration have slightly different compositions, and because some of the data points may show as outliers (as discussed below), the proposed empirical model needs be tested against additional, independent creep data.

2.4.3.2. Review of SAC Data. The publications by Kariya and Plumbridge [49], Neu et al. [56], and Schubert et al. [48]/Wiese et al. [66] provide a wealth of data that, taken together, covers a wide range of stress, strain rate and temperature conditions. Looking at the data plotted in Figure 2.23, the following empirical observations are made:

- The 150°C Schubert et al. [48] dataset shows continuity with the Neu et al. [56] and Kariya and Plumbridge [49] datasets at 125°C.
- To a first order, the 70°C Schubert al. [48] dataset shows approximate continuity with the Neu et al. [56] data and Kariya and Plumbridge [48] "strength" data at 75°C.
- However, the 75°C Kariya and Plumbridge [49] creep data do not seem to show continuity with the 70°C Schubert et al. [63] data.
- The 20°C Schubert et al. [48] data do not show continuity with the other datasets at 23°C (Neu et al. [56]) and 22°C (Kariya and Plumbridge [48]). It is also not understood why the 20°C Schubert et al. [48] data points at 5, 10, and even 20 MPa give creep rates fairly close to those at 70°C.
- The last two datasets (i.e., the −10°C Kariya and Plumbridge [49] data and the −55°C Neu et al. [56] data) seem to fit the general temperature trends of the other Kariya and Plumbridge [49] and Neu et al. [56] datasets.

In summary:

- Out of the 12 isothermal datasets shown in Figure 2.22, two of them (the 20°C Schubert et al. [48] data and the 75°C Kariya and Plumbridge [49] "creep" data) appear as potential outliers. They are nevertheless included in the regression analysis since, after correspondence with the authors, there was no apparent reason to treat them separately.
- The other 10 datasets show first-order consistency and will have more weight in the regression analysis. The corresponding data covers temperatures in the range −55°C to 150°C, stresses in the range 2–100 MPa, and strain rates in the range 3.8×10^{-9}/s to 1×10^{-3}/s.

2.4.3.3. Regression Analysis. Nonlinear regression of the SAC data in Table B.1 gives the following creep rate equation:

$$\mathring{\varepsilon}\,(/\text{second}) = 2631 \cdot [\sinh(0.04525 \cdot \sigma\,(\text{MPa}))]^{4.96} \cdot \exp\left(-\frac{59\ \text{kJ/mole}}{RT\,(\text{K})}\right) \quad (2.38)$$

The regression results from the "Datafit" program—as per Eqs. (2.26) and (2.27)—are given as central values with standard deviations:

- LNA $= \ln(A) = -7.875 \pm 2.590$ (from which the central value of A is $A = 2631$).
- $n = 4.96 \pm 0.76$ (range: $n = 4.20$ to 5.72).
- $Qa = 7097 \pm 972$ (from which the central value of Q is $Q = 59.0$ kJ/mole).

The model and the data are shown in Figure 2.27 where the master curve (solid line) or centerline of the correlation band is plotted as

$$Y \equiv \overset{\circ}{\varepsilon}(/\text{second}) \cdot \exp\left(\frac{59\,\text{kJ/mole}}{RT(\text{K})}\right) = 2631 \cdot [\sinh(0.04525 \cdot \sigma\,(\text{MPa}))]^{4.96} \quad (2.39)$$

The dashed lines are "lower" and "upper" bounds of the correlation band that are an arbitrary factor $\sqrt{10} \approx 3.16$ times above and $\sqrt{10} \approx 3.16$ times below the master curve. The two datasets that were discussed as being potential outliers are further away from the correlation band. The other datasets fall within or close to the bounds of the correlation band. The hyperbolic sine model gives an activation energy $Q = 59$ kJ/mole which compares to 61 kJ/mole in the Kariya et al. power-law model. The exponent of the hyperbolic sine function is $n \sim 5$, higher than the exponent $n = 3$ in the Schubert et al. [48]/Wiese et al. [66] power-law breakdown model.

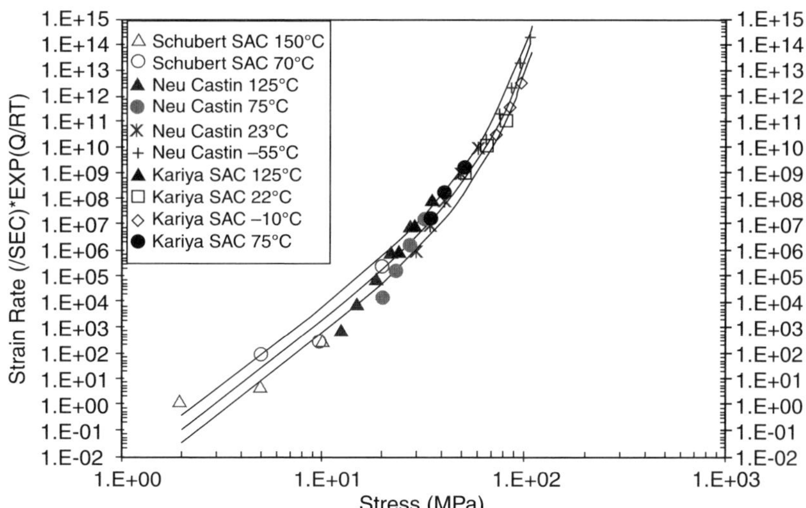

Figure 2.27. Fit of "sinh" creep model to the SAC data ($Q = 59$ kJ/mole on the vertical axis; Castin™ = Sn–2.5Ag–0.8Cu–0.5Sb).

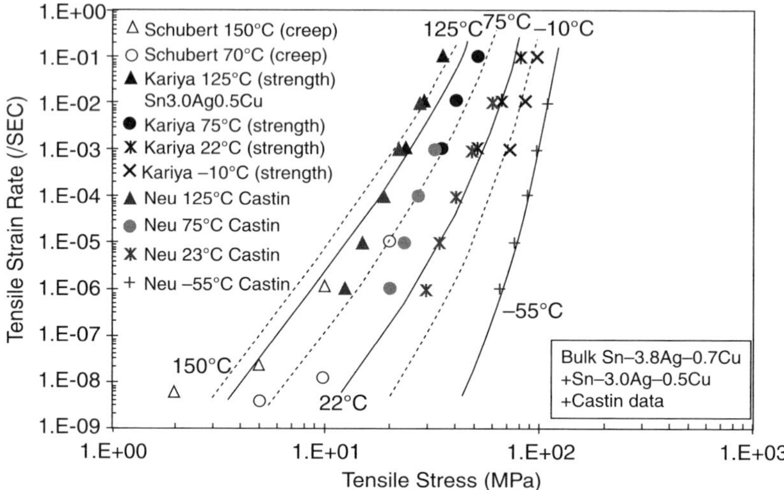

Figure 2.28. Plot of SAC creep data and isothermal lines of first-order SAC creep model.

The raw SAC data—except for the two datasets labeled as potential outliers—and isothermal lines of the hyperbolic sine model are plotted in Figure 2.28. Figures 2.27 and 2.28 suggest that, to a first order, the simplified SAC creep model provides for a reasonable fit of the data somewhat independent of the SAC alloy composition.

2.4.4. Fit of Additional Data to First-Order Creep Model

2.4.4.1. Kim et al.'s Data and Effect of Cooling Rate. Since the first-order SAC creep model given by Eq. (2.38) is essentially empirical, it is important to bounce the model against additional, independent test data with the goal of identifying limitations of the model. Figures 2.29 and 2.30 show how the model compares to data gathered by Kim et al. [73] for Sn–3.5Ag–0.7Cu, Sn–3.0Ag–0.5Cu and Sn–3.9Ag–0.6Cu tensile specimens. The specimens were either slowly cooled (SC) or rapidly cooled (RC) with cooling rates of 0.012°C/s and 8.3°C/s, respectively. The cooling rates were quoted as average values in the temperature range 230–180°C. Kim et al. [73] stated that "*RC is equivalent to the cooling speed for soldering in practical conditions in industry.*" Actually, maximum cooling rates on reflowed surface-mount assemblies are typically in the range 1–3°C/s, 4–6°C/s maximum. Technically, these cooling rates should be pondered for the mass and/or volume or surface areas of test specimens and electronic solder joints.

The raw data, obtained from constant rate stress–strain tests conducted at room temperature, is given in Table B.2 in Appendix. The reader is referred to Kim et al.'s paper [72] for further details on the experimental conditions and the authors' discussion on creep strength versus alloy composition and microstructure. Note also that specimens that were moderately cooled (MC)—at a rate of 0.43°C/s from 230°C to 180°C—showed similar microstructures and UTS results as for the RC specimens.

Figures 2.29 and 2.30 show that the data for RC specimens fit the SAC creep model well. The solid line in Figure 2.30 was obtained by back-solving Eq. (2.38)

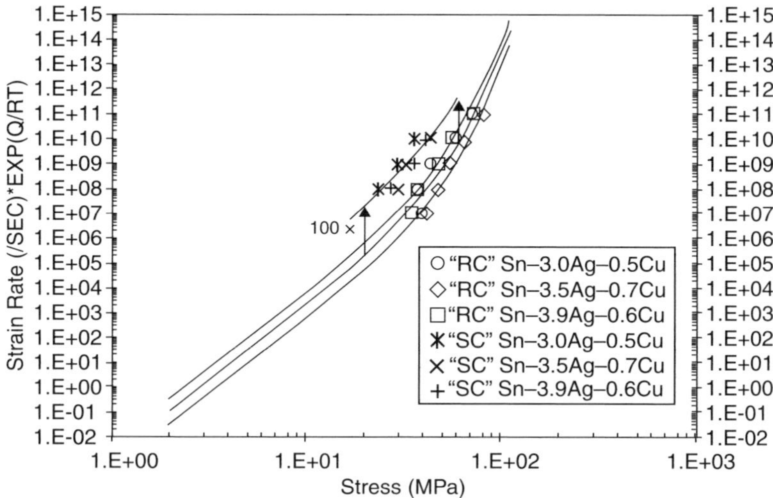

Figure 2.29. Fit of Kim et al.'s [73] tensile data to the first-order SAC creep model.

for stress at a given strain rate. In Figure 2.29, the RC data falls within or very close to the model correlation band. Figure 2.30 shows that the RC data fall on either side of the centerline of the SAC model (shown as a solid line) and that the effect of alloy composition is rather small (compared to other effects), in the range 12–25%. This

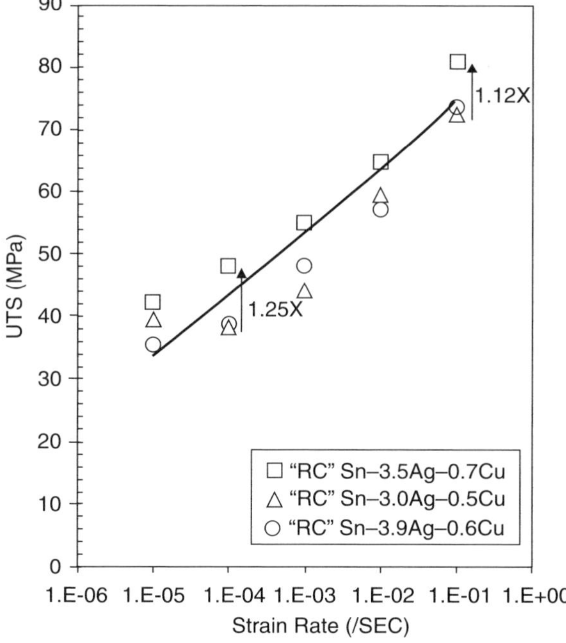

Figure 2.30. Fit of first-order SAC model (solid line) to Kim et al.'s [73] data for "rapidly cooled" specimens.

spread in the data is typical of standard deviations in creep strength measurements from a single experiment.

Figure 2.29 also shows that the data for the SC specimens falls outside the correlation band of the SAC model. The offset in terms of creep rate is a factor of about 200 times. This is shown by the solid line going through the SC data points which is set a factor of 200 times above the centerline of the SAC model. In terms of stress, and as pointed out by Kim et al. [72], the strength of SC specimens is about 50% that of RC specimens.

Clearly, there is a *strong effect of cooling conditions* which is explained by Kim et al. [73] in terms of the microstructure. Cooling rates were not reported quantitatively in the Kariya and Plumbridge [49], Neu et al. [56], Schubert et al. [48]/Wiese et al. [66] publications. However, Kariya and Plumbridge [49] indicated that their specimens were cooled rapidly ("water-quenched"). Thus, it seems appropriate that the SAC first-order creep model agrees with the RC data but does not agree with the SC data of Kim et al. [73].

2.4.4.2. NCMS Compression Creep Data.

Results of creep compression tests at 20°C, 75°C, and 125°C for the NCMS-studied Sn–4.7Ag–1.7Cu alloy (NCMS [74]) are plotted in Figure 2.31. Test specimens were short cylinders of dimensions: 0.4″ in diameter by 0.8 in. in height. Interestingly, the specimens were *cooled rapidly* since, according to the NCMS report, molten solder was poured into a casting "mold that was chilled." The compression creep tests were run in load control mode.

The NCMS data fall within or close to the lower bound of the correlation band of the SAC creep model. Although the NCMS alloy has higher Ag and Cu contents and the SAC creep model is based on tensile creep data, the model seems to apply, at

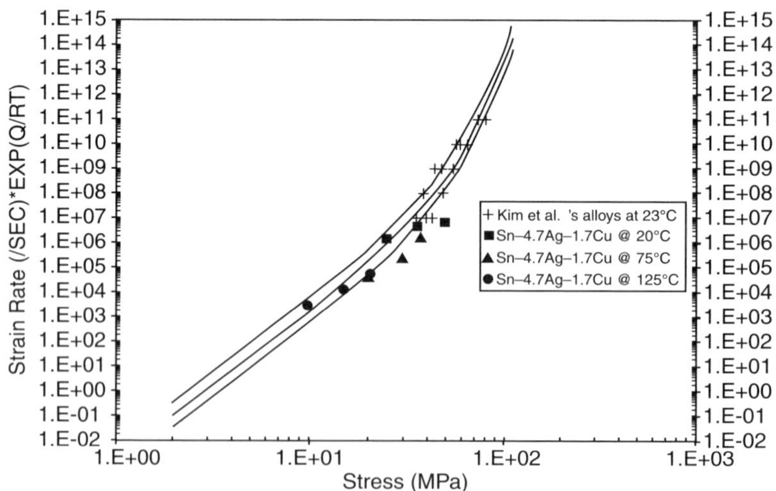

Figure 2.31. Fit of NCMS Sn–4.7Ag–1.7Cu compression creep data to the first-order SAC creep model.

least to a first-order, to compression creep. However, the fact that the data is slightly to the right of the tensile master curve and seems to follow a trendline with different slopes, suggests that the SAC alloy may also be an uneven material (similar to Sn–3.5Ag solder). Additional compression data would be needed to better characterize the compressive versus tensile response of SAC solders.

The Kim et al. [73] creep data for the SAC specimens that were rapidly cooled is also shown on the same plot (Figure 2.31) to illustrate that the two datasets provide for a first-order validation of the creep model at different stress levels: above 35 MPa for the Kim et al. [73] "RC" data, below 35–50 MPa for the NCMS data.

2.4.4.3. Flip-Chip Solder Joint Shear Data.

Wiese et al. [66], tested flip-chip solder joints of composition Sn–3.8Ag–0.7Cu in shear at 5°C, 25°C, and 50°C. After conversion from shear to tension, their data (shown in Figures 2.19 and 2.22 in Wiese et al. [66]) was fit to the following power-law model:

$$\overset{\circ}{\varepsilon}\,(/\text{second}) = 2 \times 10^{-21} \cdot (\sigma\,(\text{MPa}))^{18} \cdot \exp\left(-\frac{83.1\ \text{kJ/mole}}{RT(\text{K})}\right) \tag{2.40}$$

The above model is plotted on Figure 2.32 as

$$Y \equiv \overset{\circ}{\varepsilon}\,(/\text{second}) \cdot \exp\left(\frac{Q}{RT(\text{K})}\right) = 2 \times 10^{-21} \cdot (\sigma\,(\text{MPa}))^{18} \times \exp\left(-\frac{(83.1 - 59.0)\ \text{kJ/mole}}{RT(\text{K})}\right) \tag{2.41}$$

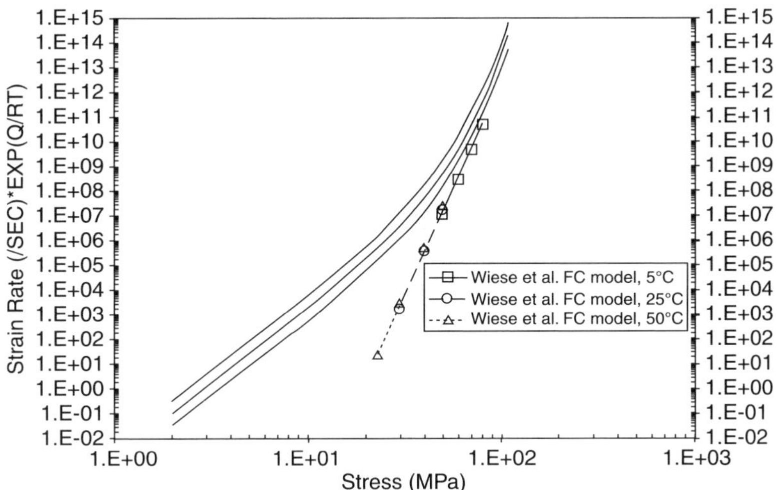

Figure 2.32. Flip-chip solder joint creep model versus master curve of bulk SAC tensile creep model.

where $Q = 59.0$ kJ/mole is the activation energy of the SAC tensile creep model. Data points that represent Eq. (2.41) are calculated at temperatures of 5°C, 25°C, and 50°C and are plotted in the stress range of the original test data obtained by Wiese et al. [66].

Clearly, the flip-chip creep model and the master curve of the tensile SAC creep model follow different trends. A similar difference between creep in shear and in tension was pointed out by Wiese et al. [66]. There are several possible reasons for this, although it is not clear how these differences can be resolved:

- Significant differences in microstructures and dispersion of chip metallization elements or intermetallics in the flip-chip joints, the effects of which are not captured in the SAC creep model.
- Equation (2.40) was obtained by fitting shear creep data that were converted to tensile data using the uniaxial stress and strain transformations: $\sigma = \tau\sqrt{3}$ and $\varepsilon = \gamma/\sqrt{3}$. The latter are based on the application of a multiaxial Von Mises yield criterion. This widely used criterion is for materials with time-independent plastic flow and, in general, has not been verified under creep conditions. The applicability of the Von Mises criterion to SAC solders has not been demonstrated either and may be questionable as in the case of Sn–3.5Ag.
- The shear creep data derived by Wiese et al. [66] converts shear forces and displacement rates into average shear stresses and shear strain rates assuming a uniform shear distribution in the minimum section of hourglass-shaped flip-chip joints. While this is a first-order engineering approach at handling the force-displacement rate data, it may be an oversimplifying assumption because of the complexity of shear strain distributions in solder joints.

A similar discrepancy between creep in shear and in tension has been reported by Darveaux et al. [4] for Sn–Pb and Sn–3.5Ag solders.

2.4.5. Microstructure and Cooling Rate Effects

In addition to Kim et al.'s [73] data, the Joo and Yu [75] Sn–3.5Ag–0.7Cu data also confirms the strong effect of cooling rate and microstructure on creep properties. Joo and Yu [75] conducted creep tests at 100°C on two types of dog-bone shaped tensile specimens:

- *"TS" Specimens*: 5-mm-thick, 8-mm-wide × 32-mm-long tensile section; cast alloys were cold-rolled at about 50% and heat-treated or thermally stabilized (TS) for 12 hours at 393 K (120°C). The resulting size of the β-Sn dendritic globules (prior to testing) was 50–100 μm.
- *"FC" Specimens*: 1-mm-thick, 8-mm-wide × 19-mm-long tensile section; specimens that were cast in a thin aluminum mold were water quenched from the melting point at a cooling rate of about 145 K/s (FC, fast-cooled). The resulting size of the β-Sn dendritic globules (prior to testing) was 5–10 μm.

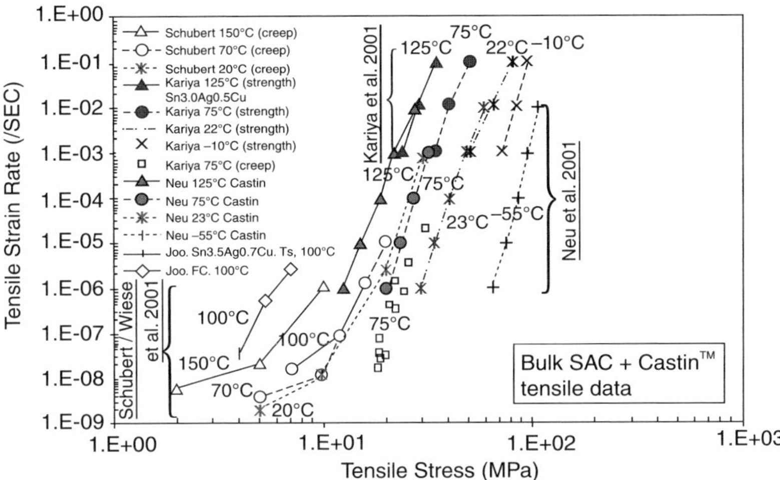

Figure 2.33. Fit of Joo and Yu's [75] Sn–3.5Ag–0.7Cu, 100°C data to plot of raw SAC tensile data. TS, thermally stabilized; FC, fast cooled (at 145 K/s).

The secondary creep results are added to our first plot of raw SAC creep data (Figure 2.23) as shown in Figure 2.33. The raw data, digitized from Figure 5 in Joo and Yu [75], is listed in Table B.4 in Appendix B. To a first order, and in spite of slight differences in Ag and Cu contents, the Joo and Yu [75] FC data seem to fit in with the other datasets:

- The 100°C FC data falls between the 70°C and 150°C Schubert et al. [48] data. We cannot be more conclusive on this because the Schubert et al. [48] paper did not report on cooling rates and the size of the β-Sn globules.
- By extrapolation to stress levels in the range 10–20 MPa, the 100°C FC data also shows continuity with the Kariya and Plumbridge [49] data and would fall somewhere between the Kariya and Plumbridge [49] datasets at 75°C and 125°C. This is encouraging since the Kariya and Plumbridge [49] specimens were also rapidly cooled (water-quenched) with a grain size of the β-Sn globules given at approximately 5 μm, similar to the 5 to 10-μm globules in Joo and Yu's [75] FC specimens.

On the other end, the 100°C TS data from Joo and Yu [75] is offset from the rest and lies about two orders of magnitude above the FC data in the direction of the strain rate axis. As discussed in Joo and Yu [75], and from elementary material science, this offset between the FC and TS data is mostly due to differences in the microstructures (5–10 μm β-Sn globules for FC versus 50–100 μm globules for TS). The two order-of-magnitude effect from the Joo and Yu [75] data is also consistent with the 200× difference in strain rates for the SC and RC specimens of Kim et al. [73].

2.4.6. Other “SAC” Properties

2.4.6.1. Young’s Modulus Versus Temperature. Young’s modulus is given versus temperature in Table 2.15 and the data is plotted in Figure 2.34. For the two Sn–Ag–Cu alloys, the data was digitized from Figure 2 in Schubert et al. [48]. For Sn and Castin, the data were obtained from equations given in Kariya and Plumbridge [49] and Neu et al. [56], respectively. For each dataset, a linear trendline and its equation are shown in Figure 2.32.

The temperature-dependent Young’s modulus is used as material properties in stress–strain analysis programs such as FEA codes. Young’s modulus is also used to scale steady-state stresses in some creep rate models.

2.4.6.2. Poisson’s Ratio. Another elastic property that is also used in stress–strain analysis programs or to convert Young’s modulus to a shear modulus, G, is Poisson’s ratio. Neu et al. [56] reported a Poisson’s ratio $\nu = 0.4$ for Castin (after Whitelaw et al. [76]), however, we did not find literature values of Poisson’s ratios for the other SAC alloys.

2.4.6.3. Coefficients of Thermal Expansion (CTE). Measured or quoted CTEs are given in Table 2.16. The CTE values given by Schubert et al. [48] for

TABLE 2.15. Young’s Modulus Versus Temperature Data

Alloy/Source of Data	T(°C)	T(K)	E (MPa)
95.8Sn–3.5Ag–0.7Cu (data in Schubert et al. [48])	18.1	291.1	4.657E + 04
	48.7	321.7	4.456E + 04
	74.1	347.1	4.341E + 04
	99.2	372.2	4.208E + 04
95.5Sn–3.8Ag–0.7Cu (data in Schubert et al. [48])	22.2	295.2	4.595E + 04
	49.6	322.6	4.395E + 04
	100.0	373.0	3.525E + 04
Sn (equation in Kariya and Plumbridge [49])	−55.0	218.0	5.23E + 04
	0.0	273.0	4.63E + 04
	17.0	290.0	4.45E + 04
	52.0	375.0	4.07E + 04
	102.0	375.0	3.52E + 04
Castin (equations in Neu et al. [56])	−55.0	218.0	4.78E + 04
	0.0	273.0	4.72E + 04
	102.0	375.0	4.61E + 04
	17.0	290.0	4.69E + 04
	52.0	325.0	4.65E + 04
	102.0	375.0	4.61E + 04

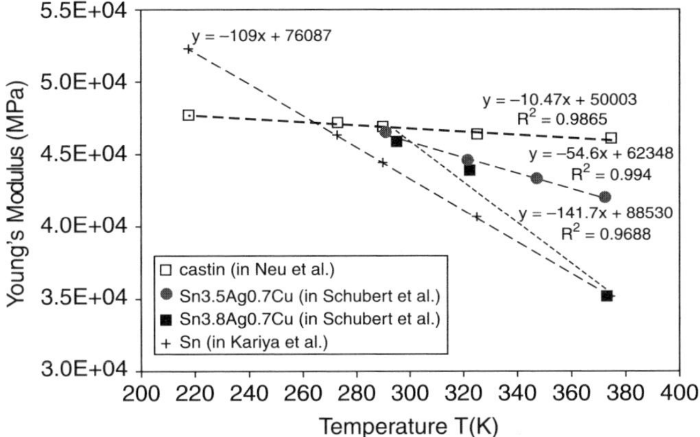

Figure 2.34. Plot of SAC Young's Modulus E (MPa) versus temperature T (K).

the SAC alloys are lower than the CTE of 23.5 ppm/°C that is quoted for pure tin. The CTE of the Castin alloy is higher at 26.9 ppm/°C. A typical value quoted for the CTE of eutectic Sn–Pb is 24 ppm/°C.

2.4.6.4. Other Physical Properties. Other properties of interest for predictive modeling of solder joint geometry using, for example, a computer program such as Surface Evolver [78], are:

- The solder density, ρ in $lb/in.^3$ (or g/cm^3). Values quoted in the NIST-Boulder database [79] are:
  - ρ (g/cm^3) = 7.39 for Castin; 7.39, 7.44 for Sn–4Ag–0.5Cu; 7.5 for Sn–3.8Ag–0.7Cu (Multicore solder).

TABLE 2.16. CTE Literature Data

Alloy	Reference	Temperature Range	CTE (ppm/°C)	Measurement Technique/Notes
95.5Sn–3.8Ag–0.7Cu	Schubert et al. [48]	20°C to 100°C 20°C to 100°C 100°C to 150°C	17.6 16.7 18.8	TMA, 5 K/min heating rate from 20°C to 170°C 3-mm × 3-mm × 12-mm-high specimens
	Warwick [77]	NA	23.5	NA (quoted data for pure tin)
96.2Sn–2.5Ag–0.8Cu–0.5Sb	Neu et al. [56]	−55°C to 125°C	26.9	Thermal strain during temperature cycling under zero load. CTE was found to be temperature independent.

- The surface tension, γ, in units of mNm^{-1}, to be specified in terms of the soldering atmosphere (e.g., air, nitrogen, etc.). We were not able to find measured values of γ for SAC alloys.

The thermal conductivity, k in units of W/mK, is also used for heat transfer analysis:

$k = 57.26\,\mathrm{W/mK}$, quoted for Castin in the NIST-Boulder database [79].

Missing values of the above physical properties need be added to material property databases when the data are available.

2.4.7. Conclusions on SAC Properties

- Most tensile creep data from five independent tests show consistency over the temperature range, -55°C to 150°C:
  - The results for SAC alloys with 2.5–3.9% Ag and 0.5–0.8% Cu contents were used to develop a first-order hyperbolic sine tensile creep model.
  - To a first order, the creep model applies to bulk SAC solders of slightly different compositions in the range: 95.5 to 96.5%Sn with 2.5–3.9% Ag and 0.5–0.8% Cu contents, including Castin™ (96.2Sn–2.5Ag–0.8Cu–0.5Sb).
  - Most of the data were at stress levels greater than 10 MPa. More data are needed to test the model (or any other constitutive model) below 10 MPa since solder joints of electronic assemblies are likely to experience such stress levels in use.
- Comparison of the model to creep rate data for *slowly cooled* specimens showed that, under those conditions, the model is offset from the data by a large factor ($\sim 200\times$). Thus, the effect of cooling rates appears to be an important parameter that needs to be investigated further.
  - This strong effect of slow versus rapid cooling rates was observed in two independent experiments (Kim et al. [73]; and Joo and Yu [75]). Fast cooling rates result in a more creep resistant, finer microstructure with β-Sn globules of approximate size 5–10 μm.
- The comparison of the tensile-creep, SAC bulk solder model to a shear-creep, SAC flip-chip solder joint model showed that the two models follow different trends. Such a discrepancy is not new and has been observed with Sn–Pb creep data in the past. Nevertheless, this is thought to be a significant issue that will have to be addressed in future investigations.
- As for Sn–3.5Ag, most of the SAC mechanical properties are provided for stresses above 10 MPa. By the same token, only secondary (or steady-state) creep data are reported on since secondary creep is the overwhelmingly dominant deformation mode under high-stress conditions.
  - Deformation modes at lower stress levels may include significant, or at least non-negligible primary creep as was found by Darveaux et al. [4] and Yang et al. [67] for Sn–3.5Ag.

2.5. ALLOY COMPARISONS

2.5.1. Lead-Free Creep Parameters

Values of creep parameters from the above analysis of SAC and Sn–3.5Ag creep data are summarized in Table 2.17:

- Interestingly, the activation energies for SAC and Sn–3.5Ag bulk tensile or compressive specimens are in the same narrow range (57.0–61.1 kJ/mole for central values) and the ranges of Q values overlap.
- The exponents n of the hyperbolic sine models for SAC and Sn–3.5Ag bulk tensile specimens are also very close ($n = 4.96$ and 4.89, respectively, based on central values).
- Similarly, the activation energies for Sn–3.5Ag lap and plug-and-ring joints or chip carrier joints in shear also fall in a narrow range (74.5–77.4 kJ/mole for central values). However, these values of Q are about 35% higher than activation energies for the tensile data.

2.5.2. Creep Rate Comparisons

Figure 2.35 shows plots of tensile creep rates versus stress for eutectic Sn–Pb and lead-free alloys at temperatures of $-55°C$, 25°C, and 100°C. These temperatures cover most service and common test conditions. For each alloy, creep rates are plotted as a centerline and an error band. For Sn–Pb, we used Hughes'model and its lower and upper bounds [Eqs. (2.18) to (2.20)]. For the lead-free alloys, we used the centerlines from the above regression analysis [Eq. (2.28) for Sn–3.5Ag, and Eq. (2.38) for the SAC alloy] as well as upper and lower bounds that are a factor $\sqrt{10} \approx 3.16$ times above and below the centerline. Figure 2.35 shows that:

- In general, SAC is slightly more creep resistant (i.e., strain rates are lower) than Sn–3.5Ag.

TABLE 2.17. Creep Parameters: Activation Energy Q and "Stress" or "sinh" Exponent for Different Specimen Types and Loading Mode[a]

Alloy	Specimen	Load	Q (min, max), kJ/mole	n (min, max)
SAC	Bulk	Tension	59.0 (50.9, 67.1)	4.96 (4.20, 5.72)
Sn–3.5Ag	Bulk	Tension	57.0 (48.4, 65.5)	4.89 (4.31, 5.47)
		Compression (power law)	61.1 (54.8, 67.4)	6.05 (5.35, 6.75)
	Lap joint/plug and ring	Shear	77.4 (67.5, 87.2)	8.67 (7.58, 9.76)
	Chip-carrier joints	Shear	74.5 (71.3, 77.6)	5.54 (5.27, 5.82)

[a]Central values, and minimum and maximum values in parentheses are obtained from standard deviations given by "Datafit" regression software.

- At low stress (say, less than 20 MPa), creep rates for both SAC and Sn–3.5Ag are much lower than for Sn–Pb.
- However, at greater stresses, the lead-free alloys have higher strain rates than eutectic Sn–Pb. This inversion of creep resistance occurs at stresses of about 20–30 MPa for Sn–3.5Ag and 40–50 MPa for the SAC alloy.

The latter observation suggests that under high-stress conditions such as those encountered in some accelerated thermal cycles, the lead-free alloys may accumulate more creep damage than may Sn–Pb assemblies. This is supported by thermal cycling results for chip resistor assemblies where the lead-free assemblies had shorter test lives than Sn–Pb assemblies (Swan et al. [80], Woodrow [81]). Whether this leads to reduced reliability under field conditions is a question that cannot be answered as of yet, since, to this author's knowledge, no life prediction or acceleration model is available for lead-free assemblies.

2.6. GENERAL CONCLUSIONS/RECOMMENDATIONS

A review of material properties of Sn–3.5Ag and SAC solder alloys was conducted. The review is not exhaustive or complete and is limited to publications in the English language. A lot of scatter was observed in the data although some trends are clearly visible, as discussed in the "Conclusions" sections of the Sn–Ag and SAC text above. More general conclusions and recommendations are as follows:

- Discrepancies were noticed between shear and tensile data.
  - The applicability of the Von Mises yield criterion has not been demonstrated for the alloys of interest. For example, the Von Mises criterion does not seem to apply to Sn–3.5Ag in the low stress regime.
- Discrepancies were also noticed between bulk solder results and data obtained from solder joints of electronic assemblies.
- Since mechanical properties are strongly influenced by the microstructure, it is important that the latter be described in a quantitative manner for proper interpretation and use of the data.
- The microstructure itself, for example the size of the β-Sn globules in Sn–Ag and SAC alloys, depends very much on the specimen cooling rates. While some studies report on both, others do not mention either one, which makes it difficult, if not impossible, to put the data in perspective with the results from other experiments.
  - For reference purposes, standards (e.g., JEDEC standards) for the assembly of electronic circuit boards recommend maximum cooling rates of 6°C/s. This is relatively slow cooling when compared to rates of 50–150°C/s when water quenching cast alloy laboratory specimens.
  - For the purpose of developing accurate constitutive models for solder joints of electronic assemblies, the initial microstructure of test specimens should have features (e.g., β-Sn dendrite globules) of similar size as identical

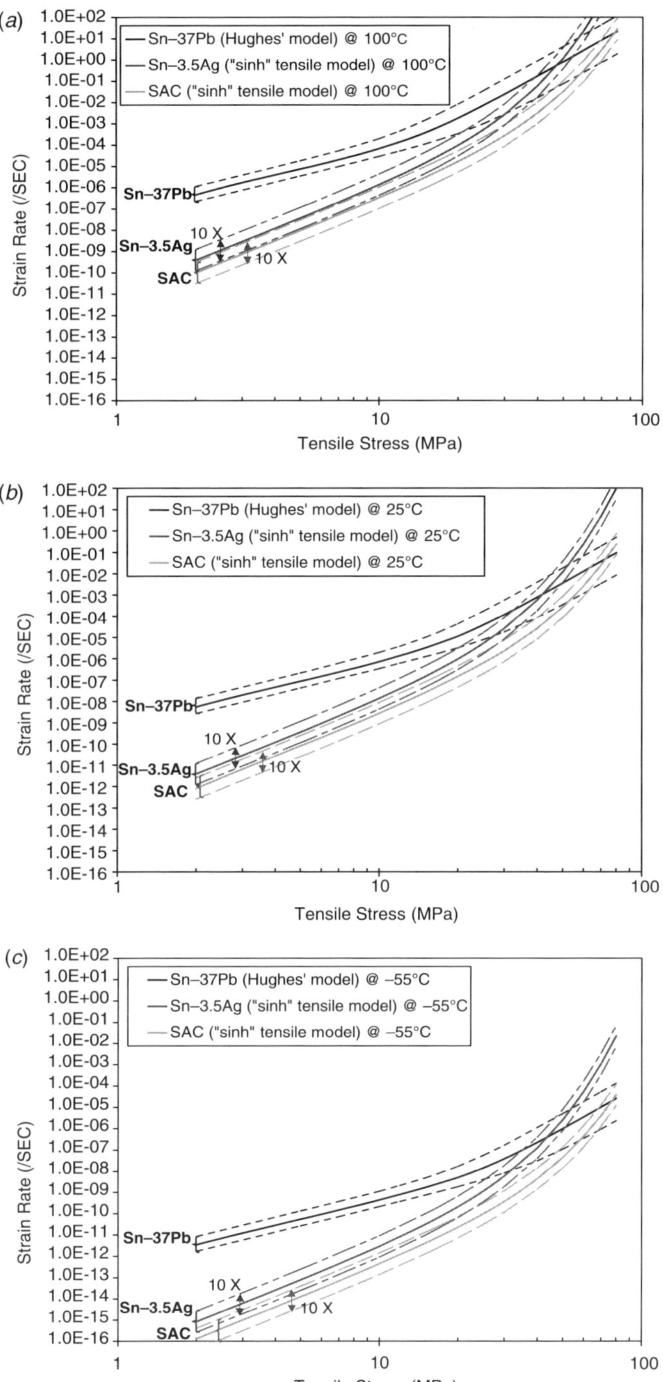

Figure 2.35. Comparison of tensile creep rates versus stress for eutectic Sn–Pb, Sn–3.5Ag, and SAC solder at (*a*) 100°C, (*b*) 25°C, and (*c*) −55°C.

features measured in real solder joints after assembly. This is not only a function of cooling rates but also a function of test specimen and solder joint sizes (for example, in terms of volume, a flip-chip solder joint is about 200 times smaller than a conventional BGA solder joint).

- Very little data are available at stress levels below 10 MPa, the stress range of interest under many service conditions. While creep tests conducted at low stress can be time-consuming, testing at higher temperatures can accelerate this process.
- Simplified creep models that were developed in this study show that, under low-to-moderate stress conditions, the lead-free alloys are more creep resistant than eutectic Sn–Pb. However, under higher-stress conditions, that trend is reversed and the lead-free alloys creep faster than Sn–Pb. The main implication is that accelerated thermal cycling profiles that were developed for Sn–Pb reliability studies may have to be modified and optimized for lead-free assemblies. Along the same lines, life prediction or acceleration models are needed to interpret test results and enable a meaningful extrapolation of lead-free test data to service conditions.
- The results of most mechanical tests emphasize secondary or steady-state creep deformations or strength. Other deformation modes, such as initial deformations, rapid plastic flow and primary creep, are rarely reported on because they appear less significant under high-stress conditions. However, these deformation modes need to be investigated further since they may become more significant under rapid thermal cycling conditions with high ramp rates and/or short dwell times.
- Data from a given publication are rarely bounced against data from other sources. Their applicability to solder joints of electronic assemblies requires further analysis and validation.

APPENDIX A: TIN–SILVER CREEP DATA

TABLE A.1. Bulk Sn–3.5Ag Tensile Specimens: Isothermal Stress–Strain Rate Data (59 Data Points Plotted in Figure 2.1)

Source	*T* (°C)	Stress (MPa)	Strain Rate (1/s)
Schubert et al. [48]	180	1.968	1.387E-08
Sn–3.5Ag (creep)	180	4.991	3.537E-06
	180	7.067	5.352E-05
	150	2.003	1.002E-08
	150	4.966	5.879E-07
	150	7.059	8.600E-06
	110	1.996	6.790E-10
	110	4.973	8.888E-08
	110	20.229	7.645E-04
	70	1.978	2.254E-11
	70	5.039	5.086E-09
	70	9.956	5.937E-07
	20	5.027	1.873E-09
	20	10.029	8.857E-08
	20	20.257	8.272E-06
ITRI [51] Sn–3.5Ag	100	36.700	3.333E-02
	100	31.400	6.667E-03
	100	30.400	1.333E-03
	100	28.000	3.333E-04
	100	26.600	1.333E-04
	100	24.200	3.333E-05
	20	51.200	3.333E-02
	20	56.800	1.333E-02
	20	43.700	3.333E-03
	20	44.100	6.667E-04
	20	41.900	1.333E-04
	20	36.300	3.333E-05
Mavoori et al. [52]	25	23	1.00E-05
Sn–3.5Ag (strength)	25	26.3	1.00E-04
	25	32.8	1.00E-03
	25	39.4	1.00E-02
	80	12.5	1.00E-05
	80	16	1.00E-04
	80	17.5	1.00E-03
	80	23	1.00E-02

(Continued)

TABLE A.1. *Continued*

Source	T (°C)	Stress (MPa)	Strain Rate (1/s)
Plumbridge [54] Sn–3.5Ag	−10	64	1.000E-01
		63	1.000E-02
		62	1.000E-03
	22	59.8	1.000E-01
		47	1.000E-02
		42	1.000E-03
	75	47	1.000E-01
		35	1.000E-02
		28	1.000E-03
Kariya and Plumbridge [49] Sn–3.5Ag (creep)	75	10.113	1.053E-09
	75	10.661	6.615E-09
	75	11.536	2.335E-08
	75	11.965	3.513E-08
	75	12.362	2.056E-08
	75	14.795	2.012E-07
	75	15.216	3.057E-07
	75	16.058	5.923E-07
	75	18.857	5.746E-06
	75	18.837	1.154E-05
	75	21.675	1.467E-05
Kariya and Plumbridge [48] Sn–3.5Ag (strength)	75	29.950	1.032E-03
	75	35.490	1.085E-02
	75	42.496	1.015E-01
	25	44.000	5.000E-03

TABLE A.2. Sn–4Ag Bulk Tensile Creep Data (34 Data Points)

Source	T (°C)	Stress (MPa)	Strain Rate (1/s)
Neu et al. [56] Sn–4Ag	125	6.521	1.014E − 06
	125	7.607	9.712E − 06
	125	10.014	9.965E − 05
	125	13.447	1.017E − 03
	125	17.936	9.786E − 03
	75	10.851	1.009E − 06
	75	13.718	9.880E − 06
	75	17.938	1.025E − 04
	75	23.162	9.921E − 04
	75	30.691	9.786E − 03
	23	20.376	1.002E − 06
	23	25.843	1.004E − 05
	23	32.197	1.005E − 04
	23	39.793	9.921 E − 04
	23	48.436	9.515E − 03
	−15	3.51E + 01	1.01E − 06
	−15	4.24E + 01	9.76E − 06
	−15	5.05E + 01	9.90E − 05
	−15	5.88E + 01	9.95E − 04
	−15	6.87E + 01	9.70E − 03
	−55	4.91 E + 01	9.84E − 07
	−55	5.94E + 01	9.77E − 06
	−55	6.75E + 01	9.87E − 05
	−55	7.57E + 01	9.84E − 04
	−55	8.75E + 01	9.78E − 03
Jones et al. [55] Sn–4Ag	−125	119	3.00E − 03
	−100	98	3.00E − 03
	−50	80	3.00E − 03
	−25	73	3.00E − 03
	0	58	3.00E − 03
	25	49	3.00E − 03
	60	35	3.00E − 03
	100	25	3.00E − 03
	150	15	3.00E − 03

TABLE A.3. NCMS Bulk Sn–3.5Ag Compression Data (9 Data Points)

Source	T (°C)	Stress (MPa)	Strap Rate (1/s)
NCMS [74] Sn–3.5Ag (compression)	20	15.0	2.0E − 08
	20	25.0	6.2E − 07
	20	50.0	1.9E − 05
	75	15.0	2.8E − 07
	75	21.1	2.8E − 06
	75	25.0	2.6E − 05
	125	12.0	5.0E − 06
	125	16.0	1.7E − 05
	125	20.0	1.1E − 04

TABLE A.4. Sn–3.5Ag Lap Shear/Plug-and-Ring Data (51 Data Points)

Source	T (°C)	Shear Stress (MPa)	Shear Strain Rate (/second)
ITRI [51] Sn–3.5Ag (Cu ring)	100	31.80	6.410E + 00
	100	29.90	2.564E + 00
	100	22.50	1.282E − 01
	100	18.10	2.564E − 02
	100	17.60	6.410E − 03
	20	55.90	6.410E + 00
	20	50.00	2.564E + 00
	20	44.60	6.410E − 01
	20	37.70	1.282E − 01
	20	28.90	2.564E − 02
	20	28.40	6.410E − 03
ITRI [51] Sn–3.5Ag (brass ring)	100	17.10	6.410E + 00
	100	27.40	2.564E + 00
	100	23.50	6.410E − 01
	100	12.70	1.282E − 01
	100	14.70	2.564E − 02
	100	11.00	6.410E − 05
	20	46.10	2.564E+00
	20	45.10	6.410E − 01
	20	44.60	1.282E − 01
	20	36.30	6.410E − 02
	20	27.90	2.564E − 02
	20	27.90	6.410E − 03

(*Continued*)

TABLE A.4. *Continued*

Source	T (°C)	Shear Stress (MPa)	Shear Strain Rate (/second)
Ren et al. [57] lap shear specimen	25	24.28	2.941 E − 02
	75	20.93	2.941 E − 02
	125	18.93	2.941 E − 02
Hernandez et al. [58]	23	55.00	5.250E − 01
Igoshev and Kleiman [61] (quoted)	20	26.8	4.000E − 03
	100	14.2	4.000E − 03
Guo et al. [60]	25	17	2.500E − 05
	65	17	1.600E − 04
	105	17	2.000E − 03
Yang et al. [67]	158	4	4.00E − 06
	158	4.2	8.00E − 06
	158	6	3.50E − 05
	158	6.7	6.00E − 05
	158	8	3.00E − 04
	158	10	6.00E − 04
	158	11	1.40E − 03
	25	17	4.50E − 07
	25	23.4	3.00E − 06
	25	25.9	8.00E − 05
	25	30	3.20E − 04
	25	31	1.50E − 03
Glazer's review [62]	25	22	6.20E − 04
	25	27	4.00E − 03
	25	39	1.30E − 01
	25	37	4
	100	14	4.00E − 03
	100	23.5	1.30E − 01
	100	25	4

TABLE A.5. Sn–4Ag and Sn–3.65Ag Lap Shear Data (5 Data Points)

Source	T (°C)	Stress (MPa)	Strain Rate (1/s)
DeVore and Westerman [65] 96Sn–Ag	−130	118.00	3.333E − 03
	25	32.00	3.333E − 03
	150	10.00	3.333E − 03
Foley et al. [64] Sn–3.65Ag	23	37.2	9.524E − 03
	125	18.4	9.524E − 03

TABLE A.6. Darveaux's Sn–3.5Ag Solder Joint Shear Data (25 Points)

Source	T (°C)	Stress (MPa)	Strain Rate (1/s)
Darveaux et al. [4] Sn–3.5Ag solder joint shear data	132	2.10	1.266E − 08
	132	2.73	8.812E − 08
	132	3.53	4.887E − 07
	132	4.04	7.507E − 07
	132	4.62	1.805E − 06
	132	4.96	1.832E − 06
	132	5.50	4.224E − 06
	132	5.86	5.058E − 06
	132	6.53	1.002E − 05
	132	8.26	5.405E − 05
	132	10.75	5.536E − 04
	132	15.54	6.429E − 03
	132	18.21	4.955E − 02
	80	7.20	3.200E − 07
	80	9.37	2.123E − 06
	80	12.61	5.592E − 05
	80	16.65	6.136E − 04
	80	22.17	6.201E − 03
	80	24.59	5.873E − 02
	27	11.50	2.161E − 07
	27	16.28	8.207E − 06
	27	19.84	4.515E − 05
	27	23.11	5.120E − 04
	27	28.49	5.730E − 03
	27	33.89	5.396E − 02

TABLE A.7. Wiese et al.'s Sn–3.5Ag Flip-Chip Data[a]

		Normai		Shear	
Temperature (°C)/(Paste)	T (°C)	Stress (MPa)	Strain Rate (/second)	Stress (MPa)	Strain Rate (/second)
50 (PT)	50	1.184E + 01	3.684E − 08	6.834E + 00	6.382E − 08
	50	1.421E + 01	1.566E − 06	8.202E + 00	2.712E − 06
	50	1.649E + 01	4.610E − 05	9.522E + 00	7.985E − 05
50 (PT)	50	8.495E + 00	3.396E − 08	4.905E + 00	5.882E − 08
	50	1.108E + 01	1.303E − 07	6.395E + 00	2.257E − 07
	50	1.244E + 01	8.496E − 07	7.183E + 00	1.472E − 06
	50	1.501E + 01	2.770E − 06	8.668E + 00	4.798E − 06
	50	1.649E + 01	8.324E − 06	9.522E + 00	1.442E − 05
	50	1.883E + 01	2.306E − 05	1.087E + 01	3.994E − 05
	50	2.057E + 01	1.085E − 04	1.188E + 01	1.879E − 04
10 (PT)	10	1.649E + 01	1.664E − 07	9.522E + 00	2.883E − 07
	10	1.873E + 01	6.132E − 07	1.081E + 01	1.062E − 06
	10	2.035E + 01	1.000E − 06	1.175E + 01	1.732E − 06
	10	2.456E + 01	8.324E − 06	1.418E + 01	1.442E − 05
	10	2.867E + 01	8.157E − 05	1.655E + 01	1.413E − 04
50 (DG)	50	1.279E + 01	2.945E − 07	7.384E + 00	5.100E − 07
	50	1.604E + 01	1.734E − 05	9.262E + 00	3.003E − 05
	50	1.812E + 01	1.330E − 04	1.046E + 01	2.304E − 04
	50	2.035E + 01	5.105E − 04	1.175E + 01	8.841E − 04
	50	2.199E + 01	1.534E − 03	1.269E + 01	2.657E − 03
	50	2.298E + 01	2.714E − 03	1.327E + 01	4.701E − 03
	50	2.362E + 01	3.466E − 03	1.364E + 01	6.003E − 03
	50	2.429E + 01	4.249E − 03	1.402E + 01	7.360E − 03
	50	2.497E + 01	5.210E − 03	1.441E + 01	9.023E − 03
05 (DG)	5	2.115E + 01	9.031E − 06	1.221E + 01	1.564E − 05
	5	2.336E + 01	4.080E − 05	1.349E + 01	7.066E − 05
	5	2.538E + 01	1.042E − 04	1.466E + 01	1.804E − 04
	5	2.758E + 01	2.885E − 04	1.592E + 01	4.997E − 04
	5	2.947E − 01	6.008E − 04	1.702E + 01	1.041E − 03
	5	3.097E + 01	1.021E − 03	1.788E + 01	1.768E − 03
	5	3.347E + 01	2.606E − 03	1.932E + 01	4.513E − 03
	5	3.576E + 01	4.249E − 03	2.065E + 01	7.360E − 03
50 (DG)	50	1.244E + 01	1.000E − 06	7.183E + 00	1.732E − 06
	50	1.561E + 01	9.407E − 06	9.010E + 00	1.629E − 05
	50	1.915E + 01	8.849E − 05	1.106E + 01	1.533E − 04
	50	2.103E + 01	3.130E − 04	1.214E + 01	5.422E − 04
	50	2.260E + 01	5.769E − 04	1.305E + 01	9.991E − 04
	50	2.389E + 01	1.153E − 03	1.379E + 01	1.998E − 03

(*Continued*)

TABLE A.7. *Continued*

		Normai		Shear	
Temperature (°C)/(Paste)	T (°C)	Stress (MPa)	Strain Rate (/second)	Stress (MPa)	Strain Rate (/second)
50 (DG)	50	1.037E + 01	2.659E − 06	5.985E + 00	4.606E − 06
	50	1.352E + 01	1.959E − 05	7.803E + 00	3.393E − 05
	50	1.705E + 01	1.443E − 04	9.843E + 00	2.500E − 04
	50	1.904E + 01	3.005E − 04	1.099E + 01	5.205E − 04
	50	2.057E + 01	5.538E − 04	1.188E + 01	9.592E − 04
	50	2.186E + 01	8.324E − 04	1.262E + 01	1.442E − 03
	50	2.260E + 01	1.473E − 03	1.305E + 01	2.551E − 03

[a]Original data in shear were plotted as normal (tensile) strain rate versus tensile stress (3rd and 4th columns) in Wiese et al. [66]. Tensile data were reconverted to shear strain rate and shear stress (in 5th and 6th columns).

TABLE A.8. Yang et al. [67] Flip-Chip Sn–3.5Ag Shear Creep Data at 25°C and 80°C (7 Data Points)

Source	T (°C)	Stress (MPa)	Strain Rate (1/second)
Yang et al. [67] Sn–3.5Ag flip-chip solder joint shear data	25	11.1	4.30E − 07
	25	20	1.10E − 05
	25	17.3	6.10E − 05
	25	21.3	1.30E − 04
	80	12.3	1.00E − 07
	80	10.6	3.40E − 06
	80	9.8	7.00E − 08

APPENDIX B: TIN–SILVER–COPPER CREEP DATA

TABLE B.1. SAC Isothermal Stress–Strain Rate Data Points (52 Data Points)

Source/Alloy	T (°C)	Stress (MPa)	Strain Rate (1/s)
Schubert et al. [48]	150	1.97	6.040E − 09
Sn–3.8Ag–0.7Cu (creep)	150	4.95	2.190E − 08
	150	10.05	1.128E − 06
	70	5.00	3.829E − 09
	70	9.81	1.193E − 08
	70	19.97	1.031E − 05
	20	4.96	2.037E − 09
	20	9.81	1.193E − 08
	20	19.97	2.582E − 06
	20	30.12	7.488E − 04
Neu et al. [56]	125	12.49	9.821E − 07
*Castin 96.2Sn–2.5Ag–0.8Cu–0.5Sb	125	15.02	9.868E − 06
	125	18.83	9.781E − 05
	125	22.15	9.963E − 04
	125	27.66	9.642E − 03
	75	20.14	1.001E − 06
	75	23.47	9.851E − 06
	75	27.33	9.771E − 05
	75	32.41	9.926E − 04
	23	29.45	9.843E − 07
	23	34.20	9.815E − 06
	23	40.69	9.379E − 05
	23	48.84	9.884E − 04
	23	59.96	9.797E − 03
	−55	64.85	9.846E − 07
	−55	76.44	9.921E − 06
	−55	86.73	9.843E − 05
	−55	95.82	9.821E − 04
	−55	106.85	9.642E − 03
Kariya and Plumbridge [49]	125	24.05	1.077E − 03
Sn–3.0Ag–0.5Cu (strength)	125	28.88	1.140E − 02
	125	35.16	1.020E − 01

(Continued)

TABLE B.1. *Continued*

Source/Alloy	T (°C)	Stress (MPa)	Strain Rate (1/s)
Kariya and Plumbridge [49]	75	34.75	1.070E − 03
Sn–3.8Ag–0.7Cu (strength)	75	40.56	1.119E − 02
	75	51.09	1.027E − 01
	22	50.94	1.048E − 03
	22	65.48	1.102E − 02
	22	81.55	1.000E − 01
	−10	72.24	1.041E − 03
	−10	85.16	1.100E − 02
	−10	96.31	1.032E − 01
Kariya and Plumbridge [49]	75	18.30	1.749E − 08
Sn–3.8Ag–0.7Cu (creep)	75	18.37	3.727E − 08
	75	18.72	2.837E − 08
	75	18.47	7.870E − 08
	75	19.87	3.162E − 08
	75	20.78	4.367E − 07
	75	22.26	3.435E − 07
	75	22.05	1.431E − 06
	75	24.18	8.515E − 07
	75	25.51	3.595E − 06
	75	30.85	2.054E − 05

TABLE B.2. Room Temperature (~23°C) SAC Creep Data by Kim et al. [73]

Alloy/Condition	Strain Rate (1/s)	Stress (MPa)
Sn–3.5Ag–0.7Cu RC[a]	1.00E − 05	42.3
	1.00E − 04	48.3
	1.00E − 03	54.9
	1.00E − 02	64.7
	1.00E − 01	81
Sn–3.0Ag–0.5Cu RC	1.00E − 05	39.6
	1.00E − 04	38.5
	1.00E − 03	44.2
	1.00E − 02	59.3
	1.00E − 01	72.3
Sn–3.9Ag–0.6Cu RC	1.00E − 05	35.3
	1.00E − 04	38.5
	1.00E − 03	48.1
	1.00E − 02	57.1
	1.00E − 01	73.4
Sn–3.5Ag–0.7Cu SC[a]	1.00E − 04	30
	1.00E − 03	34.4
	1.00E − 02	44.1
Sn–3.0Ag–0.5Cu SC	1.00E − 04	23.8
	1.00E − 03	30
	1.00E − 02	36.2
Sn–3.9Ag–0.6Cu SC	1.00E − 04	27.4
	1.00E − 03	36
	1.00E − 02	41.9

[a]RC, rapidly cooled, 15 data points; SC, slowly cooled, 9 data points.

TABLE B.3. NCMS Compression Creep Data for Bulk Sn–4.7Ag–1.7Cu Alloy (9 Data Points; After NCMS Report, [74])

T (°C)	Strain Rate (/second)	Stress (MPa)
20	1.16E − 06	20
	3.55E − 06	25
	5.25E − 06	35.6
75	2.80E − 06	20
	1.63E − 05	30
	1.10E − 04	37
125	3.50E − 06	10
	1.60E − 05	15
	7.12E − 05	20.84

TABLE B.4. 100°C SAC Tensile Creep Data by Joo and Yu [75]

Source/Alloy	T (°C)	Stress (MPa)	Strain Rate (1/s)
Joo and Yu [75], ECTC 2002			
Sn–3.5Ag–0.7Cu			
TS[a]	100	4.00	3.79E − 08
	100	5.26	5.26E − 07
	100	6.99	2.66E − 06
FC[a]	100	7.02	1.55E − 08
	100	12.00	8.78E − 08
	100	15.88	1.25E − 06

[a]TS, thermally stabilized, 3 data points; FC, fast cooled, 3 data points.

ACKNOWLEDGMENTS

Support has been provided by the U.S. National Institute of Standards and Technology (NIST). The author is grateful to Dr. Carol Handwerker of NIST for support and review of this work as well as to Mr. Kil-won Moon of NIST for digital acquisition of most of the raw data. The author also thanks Dr. Alek Zubelewicz, when at Motorola, Jasbir Bath, Edwin Bradley, and other members of the NEMI lead-free project (alloy properties group) for insightful discussions on the data and on the topic of constitutive modeling. Many thanks also to Professor William Plumbridge of The Open University (UK) and the late Dr. Andreas Schubert of the Fraunhofer Institute (Berlin, Germany) for discussions on their Sn–Ag–Cu data.

REFERENCES

1. K. C. Norris and A. H. Landzberg, Reliability of controlled collapse interconnections, *IBM J. Res. Dev.*, 266–271, May 1969.
2. E. Baker and T. J. Kessler, The influence of temperature on stress relaxation in a chill-cast, tin–lead solder, *IEEE Trans. Parts, Hybrids and Packaging* **PHP-9**, 243–246, 1973.
3. P. E. Chen, R. Motluck, and M. T. Skubiak, Time and temperature effects on the mechanical behavior of tin–lead solders, in *Proceedings of the International Conference on Mechanical Behavior of Materials*, Kyoto, August 15–20, 1971, *Mechanical Behavior of Materials*, published by the Society of Materials Science, Japan, 1972, Vol. III, pp. 243–254.
4. R. Darveaux, K. Banerji, A. Mawer, and G. Dody, Reliability of plastic ball grid array assemblies, in J. H. Lau, Ed., *Ball Grid Array Technology*, McGraw-Hill, New York, 1995, pp. 379–442.
5. D. Grivas, K. L. Murty, and J. W. Morris, Deformation of Pb/Sn eutectic alloys at relatively high strain rates, *Acta Metall.* **27**, 731–737, 1979.
6. P. Hacke, A. F. Sprecher, and H. Conrad, Computer simulation of thermo-mechanical fatigue of solder joints including microstructure coarsening, *ASME Trans. J. Electron. Packaging* **115**, 153–158, 1993.

7. P. L. Hacke, A. F. Sprecher, and H. Conrad, Thermomechanical fatigue of 63Sn–37Pb solder joints, in J. H. Lau, Ed., *Thermal Stress and Strain in Microelectronics Packaging*, Van Nostrand Reinhold, New York, 1993, pp. 467–499.
8. B. P. Kashyap and G. S. Murty, Experimental constitutive relations for the high temperature deformation of a Pb–Sn eutectic alloy, *Mater. Sci. Eng.* **50**, 205–213, 1981.
9. S. Knecht and L. Fox, Integrated matrix creep: Application to accelerated testing and lifetime prediction, in J. H. Lau, Ed., *Solder Joint Reliability: Theory and Applications*, Van Nostrand Reinhold, New York, 1991, pp. 508–544.
10. Y.-H. Pao, S. Badgley, L. Baumgartner, R. Allor, and R. Cooper, Measurement of mechanical behavior of high lead lead–tin solder joints subjected to thermal cycling, *ASME Trans. J. Electron. Packaging* **114**, 135–144, 1992.
11. M. C. Shine and L. R. Fox, Fatigue of solder joints in surface mount devices, *Low Cycle Fatigue*, ASTM Special Technical Publication STP 942, 1988, pp. 588–610.
12. D. Stone, S. Hannula, and C.-Y. Li, The effects of service and material variables on the fatigue behavior of solder joints during the thermal cycle, in *Proceedings, 35th IEEE Electronic Components Conference*, 1985, pp. 46–51.
13. D. S. Stone and M. M. Rashid, Constitutive models, in D. Frear, H. Morgan, S. Burchett and J. Lau, (Eds.) *The Mechanics of Solder Alloy Interconnects*, Van Nostrand Reinhold, 1994, pp. 87–157.
14. B. Wong, D. E. Helling, and R. W. Clark, A creep rupture model for two-phase eutectic solders, *IEEE Trans. on Components, Hybrids and Manuf. Technol.* **II**(3), 284–290, 1988.
15. B. Wong and D. E. Helling, A mechanistic model for solder joint failure prediction under thermal cycling, *ASME Trans. J. Electron. Packaging* **112**, 104–112, 1990.
16. J. E. Bird, A. K. Mukherjee, and J. E. Dorn, *Quantitative Relation Between Properties and Microstructure*, Israel University Press, Jerusalem, Israel, 1969, pp. 255–342.
17. R. Darveaux and K. Banerji, Constitutive relations for tin-based solder joints, *IEEE Trans. on Components, Hybrids, and Manuf. Technol.* **15**(6), 1013–1024, 1992.
18. S. Knecht and L. Fox, Integrated matrix creep: application to accelerated testing and lifetime prediction, Chapter 16, *Solder Joint Reliability: Theory and Applications*, ed. J. H. Lau, Van Nostrand Reinhold, New York, 1991, pp. 508–544.
19. H. D. Solomon, Creep, strain rate sensitivity and low cycle fatigue of 60/40 solder, *Brazing and Soldering*, No. 11, Autumn 1986, pp. 68–75.
20. H. D. Solomon, Low-frequency, high temperature, low-cycle fatigue of 60Sn–40Pb, *in Low Cycle Fatigue*, ASTM Special Technical Publication STP 942, 1988, pp. 342–370.
21. E. Baker and T. J. Kessler, The influence of temperature on stress relaxation in a chill-cast, tin-lead solder, *IEEE Transactions on Parts, Hybrids and Packaging*, **PHP-9**, 243–246, 1973.
22. S. T. Lam, A. Arieli, and A. K. Mukherjee, Superplastic behavior of Pb–Sn eutectic alloy, *Mater. Sci. Engi.* **40**, pp. 73–79, 1979.
23. M. C. Shine and L. R. Fox, Fatigue of solder joints in surface mount devices, *Low Cycle Fatigue*, ASTM Special Technical Publications STP 942, 588–610, 1987.
24. J-P. Clech and J. A. Augis, Temperature cycling, structural response, and attachment reliability of surface mount leaded packages, in *Proceedings, 8th Annual International Electronic Packaging Society Conference*, Dallas, November 1988, pp. 305–325.

25. J.-P. Clech, Solder reliability solutions: A PC-based design-for-reliability tool, in *Proceedings, Surface Mount International Conference*, Sept. 8–12, 1996, San Jose, CA, Vol. I, pp. 136–151. Also in *Soldering and Surface Mount Technology*, Wela Publications, British Isles, **9**(2), July 1997, pp. 45–54.
26. J.-P. Clech, Flip-chip/CSP assembly reliability and solder volume effects, in *Proceedings, Surface Mount International Conference*, San Jose, CA, August 25–27, 1998, pp. 315–324.
27. J.-P. Clech, BGA, Flip-chip and CSP solder joint reliability, in *Proceedings, MicroMat2000 Conference*, Berlin, Germany, April 17–19, 2000, pp. 153–158.
28. J.-P. Clech, Solder joint reliability of CSP versus BGA assemblies, in *Proceedings, System Integration in Micro Electronics*, SMT ESS & Hybrids Conference, Nuremberg, Germany, June 27–29, 2000, pp. 19–28.
29. P. M. Hall, Strain measurements during thermal chamber cycling on leadless ceramic chip carriers soldered to printed boards, in *Proceedings, 34th Electronic Components Conference*, New Orleans, LA, May 14–16, 1984, pp. 107–116.
30. P. M. Hall, Forces, moments, and displacements during thermal chamber cycling of leadless ceramic chip carriers soldered to printed boards, *IEEE Trans. on Components, Hybrids and Manuf. Technol.* **7**(4), 314–327, 1984.
31. P. M. Hall, Creep and stress relaxation in solder joints, in J. H. Lau, Ed., *Solder Joint Reliability: Theory and Applications*, Van Nostrand Reinhold, New York, 1991, pp. 306–332.
32. Y.-H. Pao, S. Badgley, R. K. Govila, and E. Jih, Thermomechanical and fatigue behavior of four lead and lead-free solder joints, ASME EEP, Vol. 4–2, *Advances in Electronic Packaging, Proceedings of the 1993 ASME International Electronic Packaging Conference*, Binghamton, NY, September 29–October 2, 1993, Vol. 2, pp. 937–949.
33. W. M. Sherry and P. M. Hall, Materials, structures, and mechanics of solder joints for surface mount microelectronics technology, in *Proceedings, Interconnection Technology in Electronics*, Fellbach, Germany, February 1986, pp. 47–61.
34. J.-P. Clech and J. A. Augis, Engineering analysis of thermal cycling accelerated tests for surface mount attachment reliability evaluation, in *Proceedings, 7th Annual International Electronic Packaging Society Conference*, Boston, November 1987, pp. 385–410.
35. F. Lovasco, Solder joint modeling using finite element analysis, in *Proceedings, 8th IEPS Conference*, Dallas, TX, November 7–10, 1988, pp. 410–415.
36. R. Subrahmanyan, J. R. Wilcox, and C.-Y. Li, A damage integral approach to thermal fatigue of solder joints, in *Proceedings, Electronic Components Conference, IEEE*, 1989, pp. 240–252.
37. J. R. Wilcox, R. Subrahmanyan, and C.-Y. Li, Thermal stress and inelastic deformations in solder joints, in *Proceedings, ASM Electronic Packaging Conference—Materials and Processes*, ASM International, 1989, pp. 203–211.
38. C.-Y. Li, R. Subrahmanyan, J. R. Wilcox, and D. Stone, A damage integral methodology for thermal and mechanical fatigue of solder joints, in J. H. Lau, Ed., *Solder Joint Reliability: Theory and Applications*, Van Nostrand Reinhold, New York, 1991, pp. 361–383.
39. D. C. Whalley and D. J. Chambers, Thermal and mechanical modeling of electronic interconnects for the automotive environment, in *Proceedings, IMA Conference on Mathematics in the Automotive Industry*, Coventry, UK, September 1989, pp. 283–297.

40. D. C. Whalley and D. J. Chambers, Evaluation of SMD solder joints using finite element analysis, in *Proceedings, 7th European Hybrid Microelectronics Conference (ISHM)*, Hamburg, May 1989.
41. D. C. Whalley, Reliability prediction tools for SMD solder joints, *Soldering and Surf. Mount Technol.*, No. 9, 8, 9 and 15, 1991.
42. J. H. Lau and. Y.-H. Pao, *Solder Joint Reliability of BGA, CSP, Flip Chip. and Fine Pitch SMT Assemblies*, McGraw-Hill, New York, 1997, ISBN 0-07-036648-9.
43. R. Darveaux, Solder joint fatigue model, in *Proceedings, Design & Reliability of Solders and Solder Interconnections*, Symposium held during the TMS Annual Meeting, Orlando, FL, February 10–13, 1997, pp. 213–218.
44. J. Morrow, Cyclic plastic strain energy and fatigue of metals, in *Internal Friction, Damping and Cyclic Plasticity*, 67th Annual Meeting, American Society for Testing and Materials, Chicago, June 22, 1964, ASTM Special Technical Publication No. 378, ASTM, Philadelphia, pp. 45–86.
45. R. Ohtani, T. Kitamura, A. Nitta and K. Kuwabura, High-temperature low cycle fatigue crack propagation and life laws of smooth specimens derived from the crack propagation laws, in *Low Cycle Fatigue*, eds. Solomon, Halford, Kaisand, and Leis, ASTM STP 942, ASTM Philadelphia, 1985, pp. 1163–1180.
46. J.-P. Clech, J. C. Manock, D. M. Noctor, F. E. Bader, and J. A. Augis, A Comprehensive Surface Mount Reliability model covering several generations of packaging and assembly technology, *IEEE Trans. Components, Hybrids and Manuf. Technol.* **16**(8), 949–960, 1993.
47. T. S. Park and S.-B. Lee, Isothermal low cycle fatigue tests of Sn/3.5Ag/0.75Cu and 63Sn/37Pb solder joints under mixed-mode loading cases, on CD-ROM, *Proceedings, 52nd Electronic Components & Technology Conference*, San Diego, CA, May 28–31, 2002.
48. A. Schubert, H. Walter, R. Dudek, B. Michel, G. Lefranc, J. Otto, and G. Mitic, Thermo-mechanical properties and creep deformation of lead-containing and lead-free solders, in *Proceedings, 2001 International Symposium on Advanced Packaging Materials*, pp. 129–134.
49. Y. Kariya and W. J. Plumbridge, Mechanical properties of Sn–3.0mass%Ag–0.5mass%Cu alloy, in *Proceedings, 7th Symposium on Microjoining and Assembly Technology in Electronics*, Feb. 1–2, 2001, Yokohama, Japan, pp. 383–388.
50. Y. Kariya and M. Otsuka, Mechanical fatigue characteristics of Sn–3.5Ag–X (X = Bi, Cu, Zn and In) solder alloys, *J. Electron. Mater.* **27**(11), 1229–1235, 1998.
51. ITRI, Mechanical properties of solders and solder joints, International Tin Research Institute, Publication No. 656.
52. H. Mavoori, J. Chin, S. Vaynman, B. Moran, L. Keer, and M. Fine, Creep, stress relaxation, and plastic deformation in Sn–Ag and Sn–Zn eutectic solders, *J. Electron. Mater.* **26**(7), 783–790, 1997.
53. W. J. Plumbridge and C. R. Gagg, Effects of strain rate and temperature on the stress-strain response of solder alloys, *J. Mater. Sci.: Mater. Electron.*, No. 10, 461–468, 1999.
54. W. J. Plumbridge, Solders as high temperature engineering materials, *Mater. High Temp. Sci. Rev.* **17**(3), 381–387, 2000.
55. W. K. Jones, Y. Liu, and M. Shah, Mechanical properties of Pb–Sn, Pb–In, and Sn–In, *Soldering and Surface Mount Technol.* **10**(1), 37–41, 1998.

56. R. W. Neu, D. T. Scott, and M. W. Woodmansee, Thermomechanical behavior of 96Sn–4Ag and Castin Alloy, *ASME Trans. J. Electron. Packaging* **123**(3), 238–246, 2001.
57. W. Ren, M. Lu, S. Liu, and D. Shangguan, Themal mechanical property testing of new lead-free solder joints, *Soldering and Surf. Mount Technol.* **9**(3), 37–40, 1997.
58. C. L. Hernandez, P. T. Vianco, and J. A. Rejent, Effect of interface microstructure on the mechanical on the mechanical properties of Pb-free hybrid microcircuit solder joints, in *Proceedings, IPC/SMTA Electronics Assembly Expo*, October 27–29, 1998, Providence, RI, pp. S19-1-1 to S19-1-8.
59. W. Yang, L. E. Felton, and R. W. Messler, Jr., The effect of soldering process variables on the microstructure and mechanical properties of eutectic Sn–Ag/Cu solder joints, *J. Electron. Mater.* **24**(10), 1465–1472, 1995.
60. F. Guo, J. P. Lucas, and K. N. Subramanian, Creep behavior in Cu and Ag particle-reinforced composite and eutectic Sn–3.5Ag and Sn–4.0Ag–0.5Cu non-composite solder joints, *J. Mater. Sci.: Mater. Electron.* **12**, 27–35, 2001.
61. V. I. Igoshev and J. I. Kleiman, Creep phenomena in lead-free solders, *J. Electron. Mater.* **29**(2), 244–250, 2000.
62. J. Glazer, Microstructure and mechanical properties of Pb-free solder alloys for low-cost electronic assembly: A review, *J. Electron. Mater.* **23**(8), 693–700, 1994.
63. J. Glazer, Metallurgy of low temperature Pb-free solders for electronic assembly, IMR/273, The Institute of Materials and ASM International, 1995, pp. 65–93.
64. J. C. Foley, A. Gickler, F. H. Leprevost, and D. Brown, Analysis of ring and plug shear strengths for comparison of lead-free solders, *J. Electron. Mater.* **29**(10), 1258–1263, 2000.
65. J. A. DeVore and L. Westerman, *Solder and Solder Joint Properties Handbook*, GE, Electronics Laboratory, Syracuse, NY, 1980.
66. S. Wiese, A. Schubert, H. Walter, R. Dudek, F. Feustel, E. Meusel, and B. Michel, Constitutive behavior of lead-free solders vs. lead-containing solders—Experiments on bulk specimens and flip-chip joints, in *Proceedings, 51st Electronic Components and Technology Conference*, Orlando, FL, pp. 890–902, 2001.
67. H. Yang, P. Deane, P. Magill, and K. L. Murty, Creep deformation of 96.5Sn–3.5Ag solder joints in a flip-chip package, in *Proceedings, 1996 Electronic Components and Technology Conference*, pp. 1136–1142.
68. J. H. Lau, S. H. Pan and C. Chang, Creep analysis of water level chip scale package (WLCSP) with 96.5Sn-3.5Ag and 100In lead-free solder joints and microvia build-up printed circuit board, *ASME Transactions, Journal of Electronic Packaging*, June 2002, Volume 124, Issue 2, pp. 69–76.
69. J. H. Lau and C. Chang, TMA, DMA, DSC, and TGA of lead free solders, *Soldering and Surface Mount Technology* **11**(2), 17–24, 1999.
70. S. R. Low and R. J. Fields, Multiaxial mechanical behavior of 63SSn–37Pb solder, in *Proceedings, 41st Electronic Components and Technology Conference*, 1991, Atlanta, GA, pp. 292–298.
71. S. Wen, Thermomechanical fatigue life prediction for several solders, Ph.D. dissertation, Department of Mechanical Engineering, Northwestern University, Evanston, IL, 2001.
72. J. Villain, O. Bruller, and T. Qasim, Creep deformation of the lead-free solder alloy Sn–3.5Ag at high homologues temperatures using laser extensometry with miniprobes, in

Proceedings, SMT ES&S Hybrid Conference, Nuremberg, Germany, June 27–29, 2000, pp. 143–147.

73. K. S. Kim, S. H. Huh, and K. Suganuma, Effects of cooling speed on microstructure and tensile properties of Sn–Ag–Cu alloys, *Mater. Sci. Engi. A* **333**(1–2), 106–114, 2002.
74. NCMS Lead Free Solder Project, NCMS Report 0401RE96 (on CD-ROM), June 1998.
75. D. K. Joo and J. Yu, Effects of microstructure on the creep properties of the lead-free Sn–3.5Ag–Cu solders, on CD-ROM, *Proceedings, 52nd Electronic Components & Technology Conference*, San Diego, CA, May 28–31, 2002.
76. R. S. Whitelaw, R. W. Neu, and D. T. Scott, Deformation behavior of two lead-free solders: Indalloy 227 and Castin alloy, *ASME Trans. J. Electron. Packaging* **121**, pp. 99–107.
77. M. Warwick, Implementing lead free soldering—European consortium research, SMTA, *J. SMT* **12**(4), 1–12, 1999.
78. Surface Evolver program, available at: http://www.geom.umn.edu/software/download/evolver.html, with additional information under the web site of the NIST solder joint design program at: http://www.ctcms.nist.gov/programs/solder
79. NIST-Boulder database, Database for solder properties with emphasis on new lead-free solders, *Properties of Lead-Free Solder*, Release 3.0, May 31, 2001, by T. Siewert, S. Liu, D. R. Smith and J. C. Madeni, National Institute of Standards and Technology & Colorado School of Mines, available on the web at: http://www.boulder.nist.gov/div853/
80. G. Swan, A. Woosley, N. Vo, and T. Koschmieder, Development of lead-free peripheral leaded and PBGA components to meet MSL3 at 260°C peak reflow profile, in *Proceedings, IPC SMEMA Council APEX 2001*, Paper # LF2–6.
81. T. A. Woodrow, Reliability and leachate testing of lead-free solder joints, in *Proceedings, IPC and JEDEC International Conference on Lead-Free: Electronic Components and Assemblies*, San Jose, CA, May 1–2, 2002 (paper in conference CD-ROM).

CHAPTER 3

Lead-Free Solder Paste Technology

NING-CHENG LEE

3.1. INTRODUCTION

Solder paste is the primary solder material used for high-speed, high-volume, and precision-soldering applications [1]. The creamy characteristics of this material allows it to be manipulated with automated precision-deposition technology, such as printing, dispensing, or pin-transferring. The tacky nature of solder paste enables it to be used as a temporary glue during component placement and soldering steps. The mass reflow soldering technologies such as forced air convection, infrared, conduction, or vapor phase reflow generally deliver a consistent heating and a well-controlled soldering process. The solder paste technology using lead-containing solder alloy has matured after continuous development and evolution for three decades. The design rules are well established, the processes are well fine-tuned, and the reliability and failure mode are well studied. However, the global move toward lead-free soldering [2–7] casts a curve ball to this well understood solder paste technology. The primary challenges are poor solder wetting and poor components or substrates performance caused by the high soldering temperature. Other challenges are also of concern, such as insufficient Pb-free solder alloys reliability data, lack of alternatives for high-Pb solder alloys, and problems caused by mixed solder alloys or Pb-contamination during the transition period. This chapter will introduce the state of the art of lead-free solder paste technology, including the materials, the process, the reliability, and the challenges encountered.

3.2. MATERIALS

Solder paste is made of mixture of creamy flux and solder powder. The chemistry of flux, type of alloy, amount of metal load, particle size distribution, and solder paste rheology are application-specific and are briefly introduced below.

Lead-Free Electronics. Edited by Bradley, Handwerker, Bath, Parker, and Gedney

3.2.1. Solder Alloy

A number of lead-free alloys have been investigated in the past decade [5–7]. Although many appear to be promising, alloys adequate for solder paste applications are much limited in variety. In a survey conducted in Japan in 2001, out of 32 lead-free products manufactured with lead-free soldering, 66% use Sn–Ag–Cu, 19% use Sn–Ag–Cu–Bi, 9% use Sn–Ag–BiIn, and 6% use Sn–Zn–Bi [3]. The Japanese Electronic Industry Association (JEIDA) provides recommendations about alloys based on applications [4, 7, 8]. For reflow soldering, the alloys recommended for medium and high temperature include: (1) Sn–Ag family: 96.5Sn–3.5Ag, Sn–(2–4)Ag–(0.5–1)Cu, or Sn(2–4)Ag–(1–6)Bi, including those with 1–2% In. The process window for Sn–Ag(Cu) is narrow due to high temperature concern. In addition, the compatibility of Bi bearing alloy with Sn–Pb surface finish and 42 alloy is in question. (2) Sn–8Zn–(0–3)Bi, particularly Sn–8Zn–3Bi. There is specific corrosion concern with Sn–Zn alloy. Protection–coating such as NiAu is required for copper electrodes for high-temperature use. For low temperature: Sn–(57–58)–Bi, particularly 42Sn–57Bi–1Ag, is recommended. The compatibility with Sn–Pb surface finish and with 42 alloy is questionable. The first choice per JEIDA is 96.5Sn–3.0Ag–0.5Cu. National Electronic Manufacturing Initiative (NEMI) recommend 95.5Sn–3.9Ag–0.6Cu and 96.5Sn–3.5Ag for reflow soldering [7]. This recommendation is fairly similar to that favored in Europe. The status of implementation and performance of those alloys will be discussed in more details later.

3.2.2. Solder Powder

3.2.2.1. Powder Manufacturing. Atomization is a process converting metal into very fine particles. After atomization process, the powder is typically sized by methods such as sieving process or air classification process for desired particle size range. The categories of solder powder size are shown in Table 3.1. The atomization technologies for lead-free solder powder are virtually the same as those for lead-containing solder powder. Table 3.2 shows the methods of

TABLE 3.1. Classification of Solder Powder Size, Expressed as % of Sample by Weight—Nominal Sizes [9]

Category	None Larger than	Less than 1% Larger than	80% Minimum between	10% Maximum Less than
Type 1	160 μm	150 μm	150 and 75 μm	20 μm
Type 2	80 μm	75 μm	75 and 45 μm	20 μm
Type 3	50 μm	45 μm	45 and 25 μm	20 μm
Type 4	40 μm	38 μm	38 and 20 μm	20 μm
Type 5	30 μm	25 μm	25 and 15 μm	15 μm
Type 6	20 μm	15 μm	15 and 5 μm	5 μm

Note: Type 7 has been referred to by some manufacturers as size ranges from 2 μm to 11 μm.

TABLE 3.2. Methods of Atomization [10]

Commercial Methods	Near-Commercial Methods	Other Methods
Water atomization	Ultrasonic gas atomization	Centrifugal shot casting process
Oil atomization	Rotating disk atomization	Spinning cup atomization
Gas atomization	Electron beam rotating disk process	Centrifugal impact atomization
Vacuum atomization	Roller atomization	Laser spin atomization
Rotating electrode atomization		Durarc process
		Vibrating electrode atomization

atomization. Among those, gas atomization, centrifugal atomization, and ultrasonic atomization are more commonly used for type 3 or type 4 solder powder typically used for SMT assembly applications, while atomization in oil appears to be favorable for ultra-fine solder powder such as type 5 or type 6 powder. At this stage, solder powder finer than type 6 is being attempted for ultra-fine pitch wafer solder bumping applications. Besides the methods listed in Table 3.2, some other methods, such as solder jetting process, are also investigated for manufacturing of solder powder.

Features important for solder powder such as spherical shape, low oxide content, and size distribution are discussed in details by Lee [1]. The solder powder not only has to be consistent in particle size distribution, but also has to be highly spherical in order to facilitate a good flow of paste during deposition stage. Spherical powder with smooth surface also reflects the surface of solder powder being very low in oxide during atomization process. This allows the surface tension of molten solder serve as dominant force which converts the solder droplet into a spherical ball. Figure 3.1 shows examples of back–scattering electron microscope (BSE) pictures of Pb-containing solder powder, including 63Sn–37Pb, 62Sn–36Pb–2Ag, and 10Sn–88Pb–2Ag. The surface of type 3 powder of 63Sn–37Pb and 62Sn–36Pb–2Ag is fairly smooth, with distinct tin-rich phase (dark phase) and Pb-rich phase (light phase). 63Sn–37Pb type 7 powder exhibits similar dual phases morphology, with the surface wrinkle being more noticeable due to a higher magnification. For 10Sn–88Pb–2Ag, the surface is slightly rougher than eutectic Sn–Pb system, presumably due to the Pb crystal structure.

Lead-free solder powder often exhibits a rougher surface texture than eutectic tin-lead solder powder. Figure 3.2 shows BSE pictures of type 3 and type 6 solder powder of 95.5Sn–3.8Ag–0.7Cu and 96.5Sn–3.5Ag. In the case of type 3 powder, a relatively regular, orange peel-like surface texture is easily recognizable for both alloys. This is mainly attributable to the dendrite formation of beta-tin in the high-tin alloys. The wrinkle formation is also more noticeable for type 6 powder than 63Sn–37Pb, particularly in the case of 96.5Sn–3.5Ag. Similar wrinkle formation can also be observed for type 7 99.3Sn–0.7Cu solder powder, as shown

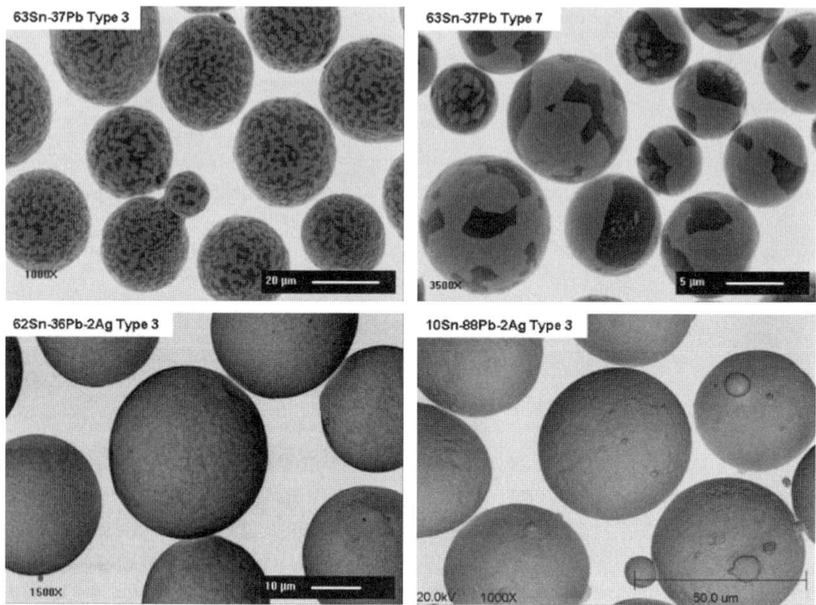

Figure 3.1. Back-scattering electron microscope pictures of solder powder: 63Sn–37Pb type 3 and type 7, 62Sn–36Pb–2Ag type 3, and 10Sn–88Pb–2Ag type 3.

Figure 3.2. BSE pictures of type 3 and type 6 solder powder of both 95.5Sn–3.8Ag–0.7Cu and 96.5Sn–3.5Ag.

by the 3500× picture in Figure 3.3. The roughness appears to be more severe, apparently due to the high tin content. It should be pointed that for very fine powder, such as type 6 or type 7, many extremely fine particles coexist with the powder with nominal size, possibly due to the difficulty in sizing, as demonstrated by the 750× picture of 99.3Sn–0.7Cu solder powder shown in Figure 3.3. Those extremely fine particles often pose challenges on solder paste performance, including both paste stability and soldering performance [1].

In the case of 58Bi–42Sn, a distinct Sn-rich (dark phase) and Bi-rich (light phase) two-phase morphology is also observed, with Bi phase as the slightly dominant phase. The crystalline texture of Bi-rich phase results in a bulging formation, as indicated by the 3500× picture of type 5 powder in Figure 3.3.

3.2.2.2. Solder Oxidation. Low oxide content of solder powder is very critical for satisfactory solder paste soldering performance. As a result, understanding and controlling of solder oxidation are essential for solder paste manufacturing process.

For Sn–Pb solders, the oxide film is a mixture of tin oxide and lead oxide with the tin oxide being the predominant constituent. Auger electron spectroscopy study conducted by de Kluizenaar [11] indicates formation of three regions on the 60Sn–40Pb solder surface: (1) a thin outer layer of SnO_2 about 2 nm in thickness, (2) under the SnO_2 layer is a 6-nm layer of SnO with finely dispersed metallic lead, and (3) further below is a 6-nm layer of SnO and metallic tin and lead, with solder metal underlying below this layer. However, the composition of oxide is affected by the oxygen partial

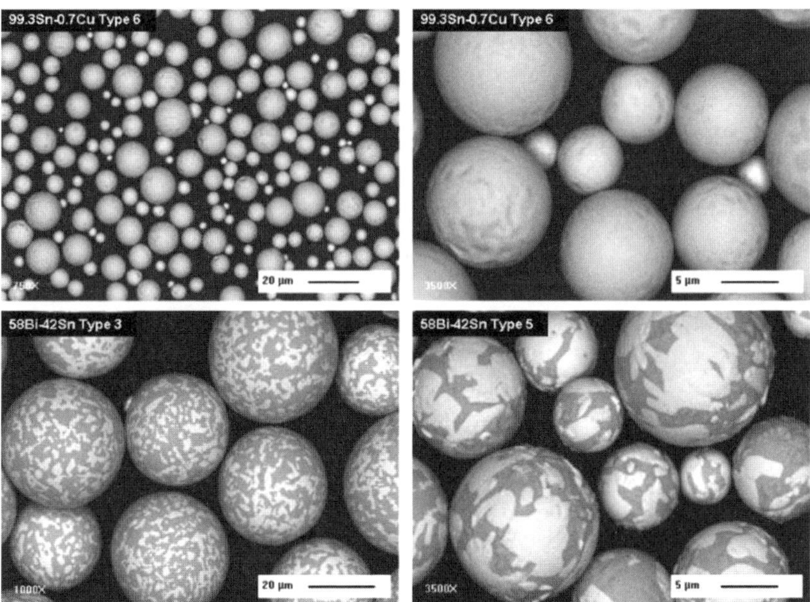

Figure 3.3. BSE pictures of type 7 99.3Sn–0.7Cu and type 3 and 5 for 58Bi–42Sn solder powder.

pressure. At the higher pressures, the rate of oxidation is sufficiently fast that the available tin is rapidly consumed, and oxidation of lead follows. At sufficiently low partial pressures of oxygen, the rate of oxidation is slower than the rate of tin diffusion from the bulk of the alloy to its surface. Thus, there is always a sufficiently supply of tin for the oxidation reaction, and the lead remains in the metallic state. As a result, only a tin oxide forms on the surface [12].

For lead-free solders, such as Sn–Ag–Cu, Sn–Ag, Sn–Cu, Sn–Ag–Bi, Sn–Ag–Bi–In, and Sn–Zn, the oxidation rate and oxides formed are different from that of Sn–Pb solders. Table 3.3 shows the free energies of oxide formation for elements utilized in those lead-free solders. Except for Ag and Cu, thermodynamically all of the rest elements commonly encountered in Pb-free solders have a greater tendency to form oxides than Pb.

Kinetically, the oxidation rate of metal, expressed as amount of metal oxide formed per unit area, depends on the material type and can be expressed by the following equation:

$$\frac{\Delta m}{A} = k\sqrt{t} \tag{3.1}$$

where Δm is mass increase in kilograms, A is surface area in square meters, t is time in seconds, and k (growth coefficient) $= k_0 \exp(-B/T)$, where here T is absolute temperature and, k_0 and B are constants.

For 60Sn–40Pb at 240°C, the k value is approximately $10^{-6}\ \mathrm{kg \cdot m^{-2} \cdot s^{-1}}$ [15]. Consistent with the thermodynamic free energy of oxide formation data, the value of k for 100% tin is higher than that for 60Sn–40Pb and is reported as twice that for 60Sn–40Pb [15, 16].

TABLE 3.3. Thermodynamic Properties of Elements of Common Solder Alloys: Standard Free Energies of Oxide Formation [13]

Oxide	Free Energy $\Delta_f G^\circ$ (kJ/mol)
Ag_2O	−11.2
Bi_2O_3	−493.7
Cu_2O	−146
CuO	−157.3
In_2O_3	−830.7
SnO	−251.9 (tetragonal)
SnO_2	−515.8 (tetragonal)
PbO	−187.9 (yellow), −188.9 (red)
PbO_2	−217.3
Sb_2O_4	−796.3 [14]
Sb_2O_5	−829.8 [14]
ZnO	−320.5

The oxide species of 100% tin has been studied since more than 30 years ago. Recently, Hillman and Chumbley [17] studied the tin oxidation products of rolled tin foil using sequential electrochemical reduction analysis (SERA). Consistent with previous reports [18–20], their results also indicate that SnO formed first, followed by the formation of SnO_2 outer crust. Furthermore, SERA data suggest that the growth of SnO crystal presumably initiates at multiple locations. While the SnO crystal film continues to grow and cover a larger surface area of tin, SnO_2 crystal also starts to grow on top of the SnO film. The growth of SnO_2 crystal film may be such that it fully covers the SnO crystal film, including the boundary between those multiple SnO film domains. For 100% tin foil aged at ambient condition for 2 years, the total oxide thickness measured by SERA was 58.6Å.

For lead-free solders, the oxidation is more complicate than the 100% tin. The oxidation of many solders may differ considerably from what suggested by free energy data. Miric and Grusd [21] studied the oxide thickness of solder preforms of several lead-free solder alloys when oxidized at a temperature 140°C above the melting point of alloys. Results, as shown in Table 3.4, indicate that oxide formation rate of liquid solder may change significantly with increasing time, such as that of 63Sn–37Pb increasing rapidly after 10 min, while that of 99.3Sn–0.7Cu, 96.5Sn–3.5Ag, and 91Sn–9Zn slowing down after 10 min. Although some alloys may exhibit an oxidation rate in line with the free energy data, such as Bi-, Sb-, In-, and Zn-containing alloys showing higher oxide thickness at 10 min than Pb-containing alloy, some may show the opposite results, such as Sn–Cu or Sn–Ag versus 63Sn–37Pb. Presumably besides the solder atom diffusion rate as stated above, the oxide packing condition may also impart a significant impact on oxygen permeation rate, and consequently on the oxide thickness formed.

TABLE 3.4. Initial and Final Oxide Thickness of Solder Preforms when Oxidized at a Temperature 140°C Above the Melting Point of Alloys [21]

Alloy	Oxidation Temperature (°C)	Oxide Thickness (angstroms)			Dominant Oxide Type
		Initial	After 10 min	After 50 min	
Sn99.3–Cu0.7	367	20	50	50	Sn oxide
Sn96.5–Ag3.5	361	30	50	50	Sn oxide
Sn63–Pb37	323	30	50	500	Sn oxide
Bi58–Sn42	278		350	800	Sn oxide
Sn95–Sb5	380	20	875	1425	Sn oxide
Sn91–Zn9	339	70	200	325	Zn oxide
52In–48Sn	257	20	175	600	In oxide

Note: Before oxidizing the preform in air, the initial oxides were removed by heating the preform in nitrogen to 500°C and then holding it for 10 minutes in a flow of hydrogen (hydrogen reduces oxides); afterwards, the preform was cooled in nitrogen to a temperature that was 140°C above the solder's melting point. Then, a nitrogen flow was switched to air flow, to start oxidation. Finally, the preform was cooled to room temperature in nitrogen, and the oxide thickness was measured using auger electron spectroscopy.

The solder powder oxidation rate at ambient temperature differs from that of liquid state reported above. The oxidation rate of lead-free solder powder was studied for 95.5Sn–3.8Ag–0.7Cu (type 3) [22]. Here the metal oxide was monitored by determining the overall powder weight difference before and after powder coalescence in the presence of flux. The weight reduced at coalescence process is attributed to the removal of solder oxides, presumably mainly tin oxides, by fluxing reaction. The oxide was analyzed for fresh powder and powder stored for some time in vacuum-sealed aluminum bags at room temperature. 63Sn–37Pb powder of various powder size distribution was also studied as control. Figure 3.4 shows the oxide content of 95.5Sn–3.8Ag–0.7Cu type 3 (25–45 μm) powder and 63Sn–37Pb powder with size range 20–25 μm, 20–38 μm, and 45–75 μm, as a function of age after manufacturing. The oxide content of fresh 95.5Sn–3.8Ag–0.7Cu solder powder ranges from 0.06 to 0.09% (w/w), slightly lower than the typical value of 0.11% for that of fresh 63Sn–37Pb type 3 powder. For 63Sn–37Pb, the oxide content of finer powder is higher than that of coarser powder. Upon aging, all powders seem to show a similar oxidation behavior, with the oxide increasing rate decreases with increasing time, suggesting the oxide increase rate may follow Eq. (3.1) as shown above. Finer powder seems to exhibit a more rapid oxidation rate than coarser powder. Another study indicates that freezer storage would slow down the oxide formation rate of 95.5Sn–3.8Ag–0.7Cu, and oxide increases from 0.07% to 0.08% after aging for 2 months. For 95.5Sn–3.8Ag–0.7Cu (type 3), in general an oxide level greater than 0.10% tends to result in a compromised solder balling performance, suggesting freezer storage condition may be desirable for this powder.

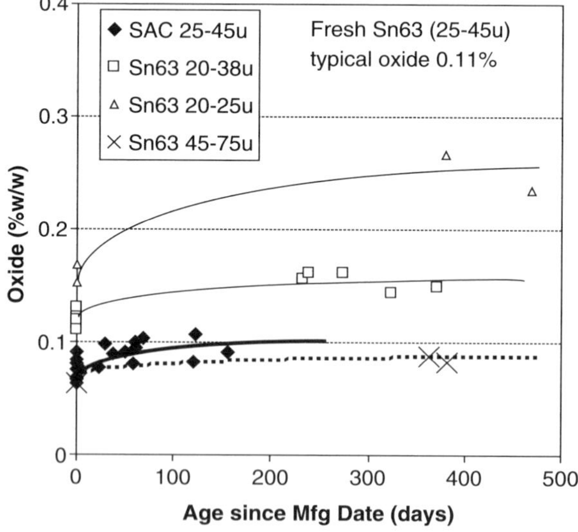

Figure 3.4. Oxide content of 95.5Sn–3.8Ag–0.7Cu (SAC) type 3 (25–45 μm) powder and 63Sn–37Pb (Sn63) powder with size range 20–25 μm, 20–38 μm, and 45–75 μm of various lots stored in aluminum bags at room temperature.

The oxidation rate of solder powder may also be affected by the microstructure. For instance, 58Bi–42Sn is composed of Bi precipitates in Sn phase, with a volume ratio of Sn phase/Bi phase being 49:51 [23, 24]. Presence of Bi disrupts the protective tin oxide layer that forms on Sn and Sn–Pb alloys and results in higher oxide content.

3.2.3. Flux

A solder flux needs to perform a number of important functions at the same time. It must promote thermal transfer to the area of the solder joint, enhance wetting of the solder on the base metal, and prevent oxidation of the metal surfaces at soldering temperatures. Among those, the primary task is to remove the tarnish layer from the metal joint that is about to be soldered. For most of the fluxes used, the flux reactions can be simulated with the interactions at the metal/metal oxide/electrolyte solution interface. The fluxing reactions that can occur at the oxide–solution interface include acid–base reactions and reduction reactions. Variables such as the structure of the metal oxide, temperature, pH, concentration of the electrolyte, and the chemical nature of the solute and solvent all affect the reaction rates and mechanisms [1].

3.2.3.1. Flux Reaction. The most common type of flux reaction is acid–base reaction. In general, this can be accomplished with the use of organic acids, such as carboxylic acids, or inorganic acids, such as halogen acids, as fluxes. The reactions between flux and metal oxides can be exemplified by the simplified equations as shown below:

$$\mathrm{MO}_n + 2n\mathrm{RCOOH} \rightarrow \mathrm{M(RCOO)}_n + n\mathrm{H_2O}$$
$$\mathrm{MO}_n + 2n\mathrm{HX} \rightarrow \mathrm{MX}_n + n\mathrm{H_2O}$$

where M stands for metal, O represents oxygen, RCOOH represents carboxylic acids, and X stands for halides, such as F, Cl, or Br.

Reduction reaction often utilizes reducing atmosphere such as hydrogen and a soldering temperature higher than 300°C. The reaction can be exemplified by the simplified equation as shown below:

$$\mathrm{MO}_n + n\mathrm{H_2} \rightarrow \mathrm{M} + n\mathrm{H_2O}$$

For solder paste applications, acid–base reaction is the most common approach, and reduction reaction is rarely used. As discussed by Lee [1], in general, the chemicals to be used as flux for solder paste have to be nonreactive enough toward metals at room temperature so that proper shelf life of the solder paste can be granted. In addition, the chemicals also have to be retainable in the solder paste during handling of materials. For those sakes, chemicals either too reactive or too volatile are not pertinent as ingredient for the fluxes used in solder pastes. The most commonly used fluxes for solder pastes include organic acids, organic bases, organic halogen compounds, and organic halide salts, as will be discussed in the next session.

3.2.3.2. Flux for Lead-Free Solders. Although the above is true for all solder pastes, there are still considerably differences between flux chemistries of eutectic Sn–Pb solder pastes and lead-free solder pastes, as listed below.

1. Flux Activity. Early attempts of industry of lead-free solder pastes typically result in very poor wetting, or even incomplete reflow [25, 26]. This is attributed to the poor solder wetting ability of lead-free alloys [25]. The flux used in lead-free solder pastes needs to be more aggressive than conventional fluxes used for 63Sn–37Pb solder pastes in order to compensate for the inferior wetting of lead-free alloys.

2. Alloy Compatibility. For solder paste applications, good compatibility between flux and solder powder is critical for achieving good shelf life, good stencil life, and good soldering performance. For instance, excessive corrosion reaction between flux and solder powder at storage temperature can result in pitting on the powder and hardening of solder paste [1]. The compatibility between fluxes and lead-free solder alloys is case-dependent. For instance, some fluxes react more aggressively with Pb-containing alloys than with Pb-free alloys, while some other fluxes show an opposite behavior. Table 3.5 shows the solder weight loss (%) due to corrosion when solder ribbons of various alloys of dimension 0.2 in. (width) $\times$ 2.0 in. (length) $\times$ 0.01 in. (thickness) were immersed in various fluxes at 60°C for 2 weeks.

Although flux D consistently shows a higher corrosivity than other fluxes toward all alloys, the relative corrosivity of the rest fluxes is alloy dependent. For instance, compared with flux A, flux C is less corrosive toward 63Sn–37Pb and 95.5Sn–3.8Ag–0.7Cu, but more corrosive toward 96.5Sn–3.5Ag.

On the other hand, Ag in the lead-free solder powder has been reported to cause degradation of viscosity stability due to the catalytic cleavage of the organic halide activator across its double bonds. This results in the formation of low-molecular-weight acids for some flux chemistries [27], and is verified by a decrease in pH and increase in acid number and paste viscosity with increasing time of paste.

Several elements are more challenging in the compatibility. For instance, indium is reactive toward many fluxes. The pastes often thicken up after 1–2 weeks of storage at ambient condition. Zinc is particularly notorious for its high reactivity

TABLE 3.5. Solder Weight Loss (%) Due to Corrosion for Solder Ribbons with Dimension 0.2 in. (width) × 2.0 in. (length) × 0.01 in. (thickness) Immersed in Fluxes at 60°C for 2 Weeks

Alloy	Flux A	Flux B	Flux C	Flux D
63Sn–37Pb	1.02%	0.90%	0.28%	5.98%
95.5Sn–3.8Ag–0.7Cu	0.38%	1.20%	0.08%	2.26%
96.5Sn–3.5Ag	0.97%	3.06%	1.49%	10.78%

toward fluxes. For most of flux chemistries available in industry, except few special cases [28], this reaction results in significant outgassing, and Zn-containing solder pastes can be converted into foamed but thickened pastes after overnight storage at room temperature.

Huang and Lee [25] studied the shelf life behavior of 110 solder pastes, including 10 fluxes (F1 to F10), 10 lead-free alloys, and 63Sn–37Pb as control. The shelf life of solder pastes was determined by monitoring the viscosity stability of solder pastes at 25°C over a period of one month. A changing viscosity, typically increasing with time, is considered undesirable. For each solder paste sample, the viscosity is determined at 1 day, 7 days, and 30 days after paste manufacturing. The percentage change of viscosity for each sample is calculated for the period from 1 day to 7 days (change rate A) and for the period from 7 days to 30 days (change rate B). The overall instability is calculated with the equation shown below:

$$\text{Overall instability} = 0.3 \times (\text{change rate } A) + 0.7 \times (\text{change rate } B)$$

The shelf life is expressed as shelf life index (SLI) and is defined in Table 3.6. A higher value in SLI represents a longer shelf life.

Shelf life of solder pastes is dictated by the reaction between flux and solder powder, and is accordingly a function of flux chemistry, as reflected by the average of shelf life index determined on 11 solder pastes for each flux, as shown in Table 3.7. Fluxes F5, F6, and F10 are most reactive, while flux F9 is the least reactive at room temperature toward solder alloys.

Compared with flux chemistry, the alloy type can have a more drastic effect on shelf life, as shown in Table 3.8. Therefore, 89Sn–8Zn–3Bi exhibits virtually no shelf life due to the high reactivity of Zn, while the rest alloys display a shelf life roughly comparable or longer than that of 63Sn–37Pb.

In view of the above phenomena, it is clear that the flux chemistries have to be carefully tailored in order to minimize the undesirable reaction with alloys. Lead-free alloys may or may not be more prone to corrosion caused by fluxes than 63Sn–37Pb. The difference among various lead-free alloys and 63Sn–37Pb is not significant, except for those alloys containing reactive elements such as Zn or In.

3. Thermal Stability. Most of the lead-free solders are high-tin alloys, with melting temperature higher than 210°C, therefore require a higher reflow

TABLE 3.6. Definition of Shelf Life Index (SLI)

SLI	Description
0	Overall instability > 25%
2	Overall instability = 20–25%
4	Overall instability = 15–20%
6	Overall instability = 10–15%
8	Overall instability = 5–10%
10	Overall instability = 0–5%

TABLE 3.7. Summary of Shelf Life Index (SLI) of Flux Systems with Variety of Alloys[a]

Flux	SLI
F1 (NC, air, no X, probe)	8.4
F2 (NC, air, no X, probe)	7.6
F3 (NC, air, X)	6.9
F4 (NC, air, X)	6.9
F5 (RMA, air, X)	5.3
F6 (NC, air, no X)	4.5
F7 (NC, N2, no X, low R)	8.5
F8 (NC, N2, no X, ultra-low R)	8.5
F9 (WS, air, no X, medium temperature)	9.1
F10 (WS, air, X, high temperature)	4.4

[a]The SLI value is determined by the average of shelf life index determined on 11 solder pastes for each flux. NC, no clean; WS, water-soluble; X, halogen; R, residue [25].

TABLE 3.8. Summary of Shelf Life Index (SLI) of Solder Alloy Systems with Variety of Fluxes[a]

Alloy	SLI
63Sn–37Pb	7.2
96.5Sn–3.5Ag	6.6
99.3Sn–0.7Cu	6.8
95.5Sn–3.8Ag–0.7Cu	9.4
93.6Sn–4.7Ag–1.7Cu	6.8
96.2Sn–2.5Ag–0.8Cu–0.5Sb	9.2
91.7Sn–3.5Ag–4.8Bi	9.0
90.5Sn–7.5Bi–2Ag	8.4
58Bi–42Sn	6.4
95Sn–5Sb	6.6
89Sn–8Zn–3Bi	0.8

[a]The SLI value is determined by the average of shelf life index determined on 10 solder pastes for each alloy.

temperature than that of eutectic Sn–Pb solder. Since thermal decomposition often results in loss of fluxing ability, or crosslinking or hardening of flux residue, this inevitably places a greater demand on the flux thermal stability in order to achieve satisfactory soldering, cleaning, or probe testability.

3.3. RHEOLOGY

One of the primary advantages of solder paste over other solder materials is its ability to be deposited with precision and automation. Another major advantage of solder paste is its ability to serve as a temporary glue during components placement and reflow process. However, to realize both advantages the rheology of solder paste has to be properly engineered.

3.3.1. Relation Between Rheology and Printing

The relation between the rheology of solder paste and printing performance can be summarized below.

1. Materials with lower compliance and higher meta-rigidity will have less tendency to ooze out underneath the stencil during printing, therefore have less tendency to be smeared. Higher elastic properties will help the material to pull together during stencil release, hence will reduce the chance of clogging [1, 28].
2. Higher elastic properties, higher solid characteristics, and higher rigidity will help reducing slump. Higher solid characteristics and higher rigidity will provide slump resistance via both high storage modulus and high loss modulus. Similar to the case of elastic properties, the high storage modulus contribute to slump resistance via elastic nature. The high loss modulus will contribute to slump resistance via kinetic mechanism—that is by slowing down the slumping process via high viscosity [1, 28].
3. High elastic property, low compliance, and low solid characteristics are required in order to achieve high tack value. In general, tack is considered to be a function of both cohesion and adhesion. A high cohesion of material is required in order to prevent tack failure due to rupture occurred through the material itself, while a high adhesion is needed in order to avoid interfacial failure. Both high elastic properties and low compliance will contribute to a high cohesion properties. A low solid characteristics could enhance the wetting between the solder paste and the devices and accordingly improve the adhesion [1, 28].
4. By plotting the thixotropic index (TI) value versus viscosity value determined at 6 s^{-1} for a series of solder pastes, Harada observed that there is a "window" for good printing performance [29]. The desired TI value ranges from about -0.5 to -0.7, while the desired viscosity ranges from about 1300–2700 poise. In general, too high a TI value result in low tack and paste sliding during printing, while too low a TI value result in slump. On the other hand, too high a viscosity result in clogging and squeegee hanging, while too low a viscosity results in smearing and slumping. Since the optimum window is highly empirical and may vary with flux chemistry, solder powder size and content, stencil aperture design, as well as printing parameters, care should be taken before adopting any criteria for the purpose of solder paste selection.

The rheology of solder paste, particularly the viscosity, the thixotropic property and the tackiness, is mainly governed by the flux material. However, the rheology can also be affected by solder powder [1, 30]. The viscosity of solder paste maintained the same as flux vehicle viscosity initially with increasing metal content, then increases exponentially with further increase in metal content at metal volume percentage greater than 40% (v/v) of solder paste. The thixotropic property drops initially with increasing metal content, then increases again at metal volume percentage greater than 50% (v/v) with further increase in metal content.

The relation between metal content and tack is fairly complicated, as shown in Figure 3.5 [30]. At low metal content, the data are interfered by gasketing effect. At metal loads beyond 40%, the tack shows an increase followed by a decrease with further increase in the metal load. The increase in tack can be attributed to the increasing cohesive force due to increasing filler reinforcement effect. The decrease in the tack at metal volume content beyond 53% presumably can be related to the gradually increasing insufficiency of the flux/vehicle binder for the powders. The local tack maximum observed indicates that 53% (v/v), or 90.5% (w/w) for 63Sn–37Pb, may be the optimum metal content for achieving a high tack value. The tack time decreases almost linearly with increasing metal content, apparently due to a decreasing amount of tacky flux material.

Solder powder size also affects rheology. With decreasing powder size, the viscosity and tack value increase, while the TI value and paste slump decrease. The tack time shows an initial increase, then a decrease with further decrease in powder size [30]. The initial increase can be explained by the increasing diffusion path length for the solvent to reach the paste surface before it dries out. In addition, the increase in powder surface area also helps to retain the solvent longer due to increasing powder-solvent surface adsorption. The declining trend beyond the peak can probably be attributed to the skin formation effect caused by excessive chemical reaction between flux and the fine powder. This dry skin formed consequently reduces the adhesive force of the paste, and result in a short tack time.

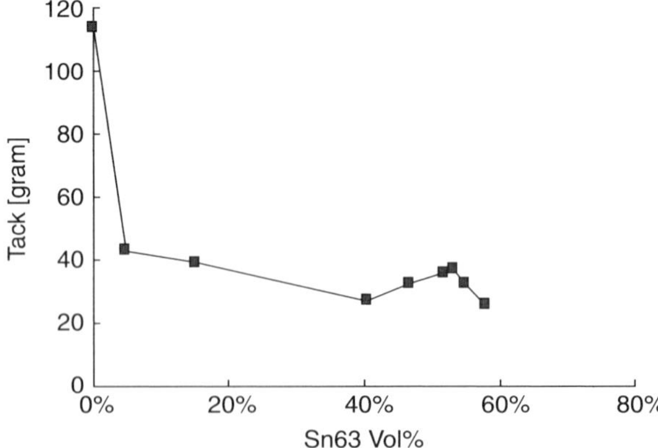

Figure 3.5. Relation between tack and metal content of Sn63 solder paste (−325/+500 mesh powder) [29].

3.3.2. Lead-Free Solder Paste Rheology and Relevant Properties

Changing from eutectic Sn–Pb to lead-free solders does not necessarily pose a change in behavior of solder rheological properties. The rheology and relevant properties, such as viscosity, thixotropic property, and tack value, are mainly dictated by flux chemistry and are insensitive to alloy type, except for the alloys containing reactive elements. The lead-free solder pastes have similar Brookfield viscosity, Malcom viscosity, and tack value as eutectic Sn–Pb pastes. The printability often is slightly poorer than typical 63Sn–37Pb pastes for many first-generation lead-free solder pastes, presumably due to the focused effort on meeting the challenge of soldering performance.

Table 3.9 shows the print volume standard deviation for a selected print pattern for six commercial solder pastes, including a leading lead-free solder paste and five 63Sn–37Pb solder pastes [27]. This leading lead-free solder paste (Paste D) displays a print volume consistency within the range of 63Sn–37Pb solder pastes. Although it does not print as consistent as the leading 63Sn–37Pb solder pastes such as A and B, it still outperforms commercially acceptable 63Sn–37Pb solder pastes E and F considerably and easily meets the print consistency requirement.

Table 3.9 also shows the elasticity of solder pastes, or the extent of recovery after a retardation/relaxation cycle in creep test, with elasticity being defined as (J1–J2)/J1. Here J1 = Creep Equilibrium (maximum) value of compliance for retardation step. J2 = Creep Equilibrium (minimum) value of compliance for relaxation step. Compliance J is strain divided by corresponding stress. The experiment was run with 5 minutes of retardation at 400 dynes/cm^2, followed by 5 minutes of relaxation at 0 dyne/cm^2. The print volume standard deviation roughly increases with decreasing elasticity. This is consistent with Bao and Lee's observation that the print defect is inversely proportional to the elastic properties (structure recovery) and meta-rigidity (yield stress) [28].

The elasticity of solder paste can also be reflected by the shape of the thixotropy loop, as exemplified by that of paste A, D, E, and F, as shown in Figure 3.6. The thixotropy loop is obtained from the equilibrium flow curve. To determine the

TABLE 3.9. Standard Deviation of Print Volume for a Selected Print Pattern and Elasticity of Solder Pastes [31]

Solder Paste	Standard Deviation of Print Volume (cubic mils)	Elasticity of Solder Paste, (J1 – J2)/J1
Paste A (Sn–Pb, no-clean)	279	41.50%
Paste B (Sn–Pb, no-clean)	485	29.94%
Paste C (Sn–Pb, water soluble)	611	45.23%
Paste D (Pb-free, no-clean)	656	35.87%
Paste E (Sn–Pb, no-clean)	1047	31.77%
Paste F (Sn–Pb, water soluble)	1255	19.36%

equilibrium flow curve, the strain/shear rate is monitored under each applied stress. When equilibrium is obtained, the shear stress is increased. This sequence of steps is repeated until the required stress range has been achieved. Then the whole process is repeated except the shear stress being decreasing gradually. The thixotropy loop is the equilibrium flow curve resulted from increasing stress from a starting value to a preset maximum value, then followed by decreasing stress back to the starting value. A smaller area enclosed by the loop reflects a faster recovery, or a higher elasticity. In this study, the stress is varied between 100 and 25,000 dyne/cm^2. In Figure 3.6, the elasticity, or the reciprocal of the area enclosed by the thixotropy loop, exhibits an order of D > A > E > F. This is slightly different from but close to the elasticity order obtained from creep retardation/relaxation test, A > D > E > F, as shown in Table 3.9. Again, the elasticity behavior of lead-free solder sample D falls well within the range of Sn–Pb solders, indicating that

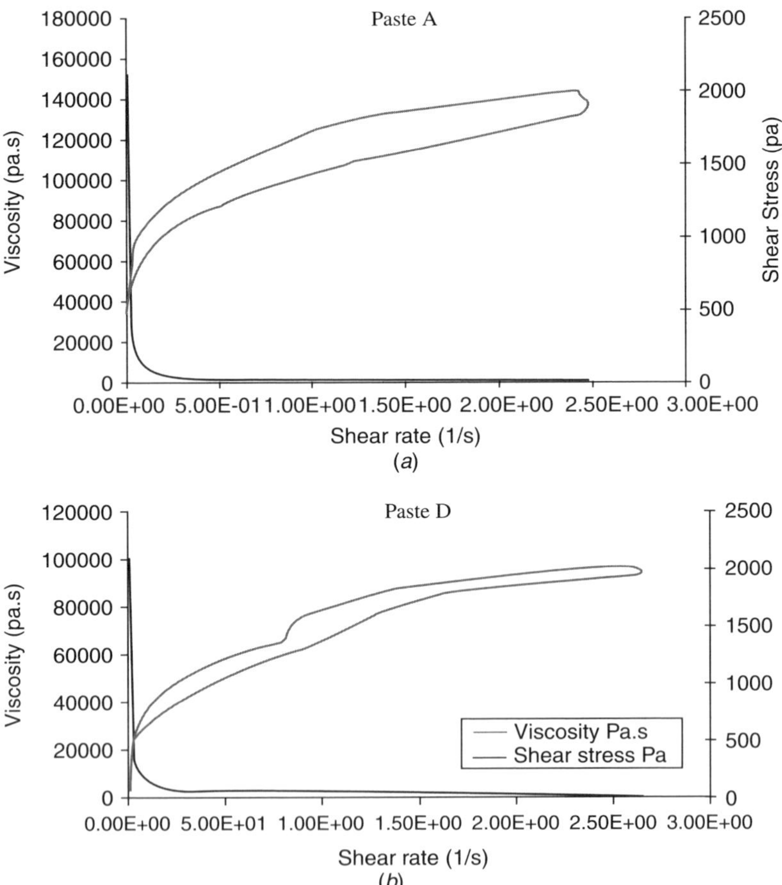

Figure 3.6. Thixotropy loop of solder paste samples A, D, E, and F obtained from equilibrium flow curve [31].

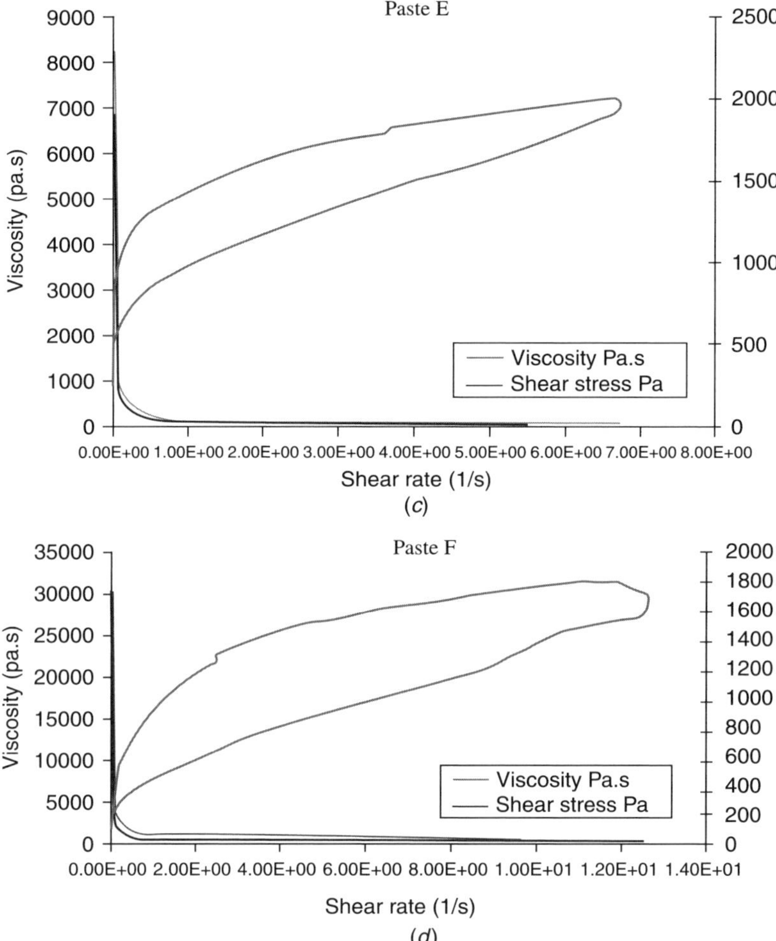

Figure 3.6. (*Continued*).

lead-free solder paste poses no issue in meeting the desirable high elasticity rheological behavior.

In some cases, the lead-free no-clean paste may have extremely long print life and tack time [26, 32]. This is reasonable, since print life and tack time are mainly dictated by solvent volatility instead of alloy type, unless some reactive elements are incorporated in the alloy.

Figure 3.7 shows the print volume and volume range of a series of lead-free solder pastes A1, B1, B3, C1, and C2, when freshly printed onto 12-mil SMD pads [26]. In general, the performance of most lead-free solder pastes studied is fairly comparable with the control Sn63 solder paste when freshly printed. Bath also studied the print volume [33], and reported that the relative print volume of Sn–Ag–Cu paste versus 63Sn–37Pb paste is print pattern dependent. Therefore,

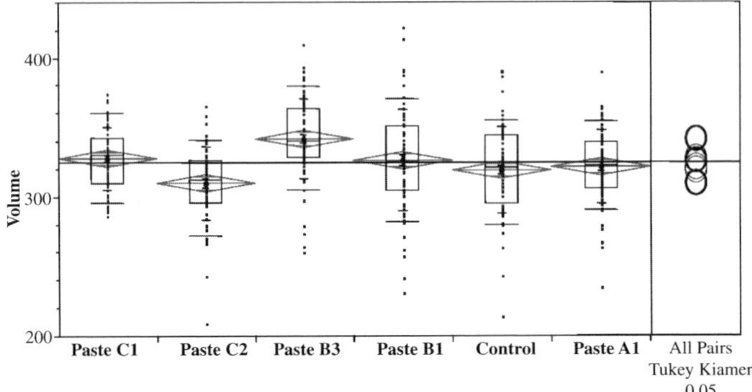

Figure 3.7. Solder paste volumetric measurement for 12-mil SMD pads at abandon time = 0 hr. All samples are Sn–Ag–Cu solder pastes except the control which is Sn–Pb solder paste [26].

Sn–Ag–Cu paste volume (2500 cubic mils) is lower than that of 63Sn–37Pb (3000 cubic mils) for 16-mil pitch pad paste volume (theoretical 3840 cubic mils) and standard deviation of paste volume for TSOP48. However, lead-free volumes are equal or greater than 63Sn–37Pb paste volumes for 256PBGA, 169CSP, 208CSP, 256CBGA, R2512. In general, printing of Sn–Ag–Cu paste is slightly poorer than 63Sn–37Pb paste.

In another work, Ashmore and Goldsmith studied three lead-free (one Sn95.5–Ag3.8–Cu0.7, two Sn96.5–Ag3–Cu0.5) and one Sn–Pb (63Sn–37Pb) solder pastes [34] using enclosed head technology. They concluded that all lead-free materials are compatible with the enclosed head technology, although on average the volume deposited for lead-free is reduced by 15%. A high print speed and low pressure is beneficial to printing process.

Besides Sn–Ag–Cu solder pastes, 58Bi–42Sn and 57Bi–42Sn–1Ag solder pastes have also been reported to exhibit good printability and stencil life [35, 36]. On the other hands, water-soluble lead-free Sn–Ag–Cu solder pastes appear to lag behind their no-clean counterpart, and often exhibit poorer print behavior [37].

3.3.3. Tack Time

Tack time is a function of cohesive force, adhesive force, solvent drying rate, and flux-solder reaction rate at ambient temperature. Huang and Lee [25]. Studied the tack time behavior of a series of lead-free solder pastes. The tack time of solder paste is determined by the following procedure. (1) Print solder paste onto ceramic coupons, as prescribed by J-STD-006 procedure. (2) Condition the specimen under 76% relative humidity. (3) Measure the tack value, per J-STD-006 procedure, of the conditioned specimen at fresh, 8 hours, 24 hours, 48 hours, and 72 hours. The specimen is discarded after each tack measurement. The tack data are expressed as tack time index (TTI), which is defined in Table 3.10. A higher value in TTI represents a longer tack time.

TABLE 3.10. Definition of Tack Time Index (TTI)

TTI	Description
0	A decreasing tack curve reaching a value at <10 g at the third day
2	A decreasing curve reaching 10–20 g at the third day
4	A decreasing curve reaching 25–20 g at the third day
6	The tack initially increases, reaching the maximum, and continuously decreases
8	A continuously increasing curve
10	Constant over 3 days

TABLE 3.11. Summary of Tack Time Index (TTI) of Flux Systems with Variety of Alloys[a]

Flux	TTI
F1 (NC, air, no X, probe)	7.6
F2 (NC, air, no X, probe)	6.5
F3 (NC, air, X)	7.6
F4 (NC, air, X)	7.6
F5 (RMA, air, X)	1.1
F6 (NC, air, no X)	4.4
F7 (NC, N2, no X, low R)	0.0
F8 (NC, N2, no X, ultra-low R)	6.2
F9 (WS, air, no X, medium temperature)	1.1
F10 (WS, air, X, high temperature)	5.6

[a]The TTI value is determined by the average of shelf life index determined on 11 solder pastes for each flux [25].

TABLE 3.12. Summary of Tack Time Index (TTI) of Solder Alloy Systems with Variety of Fluxes[a]

Alloy	TTI
63Sn–37Pb	4.8
96.5Sn–3.5Ag	5.0
99.3Sn–0.7Cu	4.6
95.5Sn–3.8Ag–0.7Cu	5.0
93.6Sn–4.7Ag–1.7Cu	5.6
96.2Sn–2.5Ag–0.8Cu–0.5Sb	5.2
91.7Sn–3.5Ag–4.8Bi	5.2
90.5Sn–7.5Bi–2Ag	5.0
58Bi–42Sn	5.8
95Sn–5Sb	5.2
89Sn–8Zn–3Bi	1.2

[a]The TTI value is determined by the average of shelf life index determined on 10 solder pastes for each alloy [25].

Tack time of solder pastes is a strong function of flux chemistry, as reflected by the average of tack time determined on 11 solder pastes for each flux, as shown in Table 3.11. Fluxes F5, F7, and F9 are considerably shorter in tack time than the rest fluxes, mainly due to the high volatility of solvents used in fluxes.

Except for Zn-containing alloy, alloy appears to have no impact on the tack time, as shown in Table 3.12. Zn is particularly reactive toward most fluxes. The resultant metal salts typically are fairly high in viscosity and thus present a dry and hard appearance. Other than that, all of the rest alloys are comparable in TTI value, suggesting that tack time should be a non-factor in the conversion of Sn–Pb into Pb-free soldering.

3.4. APPLICATIONS

Similar to lead-containing solder pastes [1], lead-free solder pastes are also used for solder bumping [38–42] or assembly soldering applications.

3.4.1. Solder Bumping

Solder paste has been used for solder bumping for wafer, CSP, and BGA, mainly due to its low cost advantage. This is particularly true for wafer bumping. One common wafer bumping process involves printing solder paste into the cavity of photoresist on wafer without using a stencil, with each cavity registered for a pad. The paste is then reflowed to form a solder bump on the pad, followed by cleaning the flux residue and stripping off the photoresist. Although most of the wafer bumping utilizes eutectic Sn–Pb solder, with some using high Pb solders, some lead-free wafer solder bumping has been implemented or attempted, including Au–Sn and Sn–Ag–Cu solders. The solder powder size used is typically type 4 or type 5 for coarser pitch, and type 6 for finer pitch. In one effort, bumping was performed using fine grain 96.2Sn–3.5Ag–0.3Cu solder paste for 100-μm-pitch bumping process shown in Figure 3.8 [39]. Standard under bump metallurgy (UBM) was used on the pads.

Other bumping approaches include the low cost print-detach-reflow process, mainly for CSP or BGA solder bumping. The transfer efficiency is reported to increases with increasing area ratio, decreasing pitch, decreasing stencil thickness, decreasing challenge, and is not sensitive to aspect ratio of aperture to solder particle size. Successful implementation of this paste bumping process also relies on stencil manufacturing technology capable of providing an aperture pattern with spacing considerably smaller than the stencil thickness. Slow print speed is also essential for adequate printing. A non-shiny non-smooth stencil surface is considered beneficial for aiding paste rolling [42]. The challenge level of this process is greater than the wafer bumping process using photoresist as a temporary stencil described above in terms of solder volume and volume consistency for solder bumps formed. At this stage, most of the paste bumping performed for CSP or BGA is for peripheral array pattern, due to an easy aperture designing for overprint

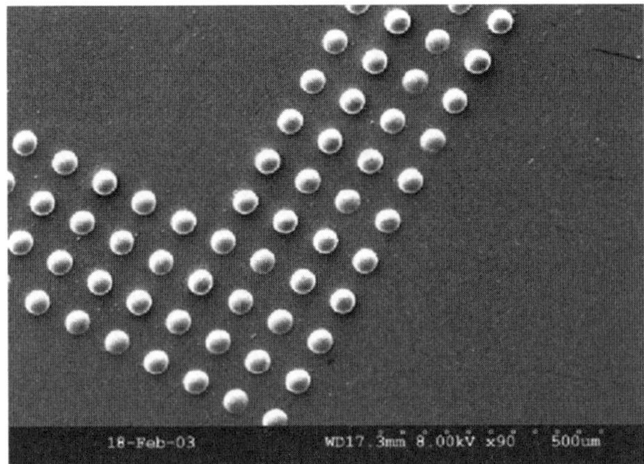

Figure 3.8. 96.2Sn–3.5Ag–0.3Cu solder bumps on patterns representing a corner of a 100-μm pitch peripheral flip chip [39].

purpose. While Sn–Pb paste bumping has been adopted in industry for more than 10 years, lead-free paste bumping has only been attempted recently.

3.4.2. SMT Assembly

The primary use of solder paste in SMT industry is for assembly process, with stencil printing as the most dominant deposition method. Other deposition methods include dispensing and pin-transfer. In this chapter, the major emphasis will be on assembly using printing process.

3.5. REFLOW SOLDERING

3.5.1. Reflow Technology

In general, to reflow lead-free solder pastes, all existing reflow technologies used for eutectic Sn–Pb are still applicable, such as forced gas convection, infrared, conduction, vapor phase, laser, soft beam, and hot bar reflow technologies [1]. Among those, forced gas convection mass reflow is still the most favorable choice. Due to the higher process temperature required, some of the reflow ovens need to be redesigned. Furthermore, more heating zones may be desired in order to have better control for board temperature. Inerting ability is also important for new model design.

Vapor-phase reflow was one of the main stream reflow technologies in 1980s and early 1990s. Since then, it has been replaced by forced gas convection reflow, mainly due to the high cost and high process defect rate of vapor phase, such as tombstoning and wicking. However, with the emergence of lead-free soldering, the vapor phase has been receiving new attention, mainly due to its ability to maintain a controlled maximum temperature. This feature is particularly attractive when thermal damage is of a major concern for both components and boards. Perfluorinated reflow fluids of various boiling temperature have been developed. Examples include perfluoropolyether, such as oxidized polymerized 1,1,2,3,3,3-hexafluoro-propene, with boiling points ranging from 155°C up to 260°C, and molecular weight ranging from 700 up to 1210, respectively [43]. For lead-free soldering, the boiling points utilized include 230°C, 240°C, and 260°C. For instance, NEC was noted to use 240°C vapor fluid. In Europe, the vapor phase reflow is estimated to be less than 10% of overall reflow process at this stage [44].

Choice of reflow technology may also have an impact on solder joint strength. Herzog et al. [45] studied reflow of two lead-free alloys 95.5Sn–3.8Ag–0.7Cu and 96.5Sn–3.5Ag + 99Sn–1Cu on electroless tin, OSP, and Ni–Au surface finishes. The differences of the shear strengths of solder joints are very small between forced convection or vapor phase soldering for the two alloys. There is a higher strength loss after vapor phase between room temperature and homologous temperature $T_h = 0.75$ or 0.81 than for forced convection soldering, as shown in Table 3.13. Presumably, this could be attributed to the loss of long-range order of solder joints. For vapor-phase reflow, the device cools down slower, thus is expected to form solder joints with larger dendrite formation, or a higher long range order. The softening and disintegration of this higher long-range order consequently may result in a higher sensitivity toward increasing temperature.

Miyazaki et al. investigated ΔT, the difference of temperature between the bottom of BGA (the hardest to rise in temperature) and the PCB surface (the easiest to rise in temperature), as a function of reflow technology [46]. The results shown in Table 3.14 indicate that forced convection, or heated gas blowing furnace, can reduce the ΔT to 10°C with high preheat temperature, while ΔT is 29°C when using infrared furnace with low preheat temperature.

3.5.2. Profile

3.5.2.1. Historical Background. Reflow solder paste is the primary method of forming solder joints at board level assembly in SMT industries. The reflow profile historical background and optimal profile for Sn–Pb soldering have been discussed in details by Lee [1]. In general, when properly done, the reflow process provides a high yield, high reliability, and low cost advantages. Among all of the process considerations, reflow profile is one of the most important factors in determining soldering defect rate. The types of defects affected by reflow profile include component cracking, tombstoning, wicking, solder balling, bridging, solder beading, cold joints, excessive intermetallics formation, poor wetting, voiding, skewing, charring, delamination, leaching, dewetting, solder or pad detachment, and so on. Therefore, it is extremely important to have the reflow profile engineered properly in order to achieve both high yield and high reliability.

A reflow profile can be roughly divided into three major elements: the peak temperature, the heating stage, and the cooling stage. Each element has its impact on the reflow results. Typically, a slow ramp-up rate is desired in order to minimize hot slump, bridging, tombstoning, skewing, wicking, opens, solder beading, solder

TABLE 3.13. Mean Shear Strength of CR Components after Reflow Soldering [45]

			Mean Shear Strength in N								
			Room Temperature			$T_h = 0.75$			$T_h = 0.81$		
Reflow Method	PCB Finish		Electr. Tin	OSP	Ni–Au	Electr. Tin	OSP	Ni–Au	Electr. Tin	OSP	Ni–Au
Forced convection	95.5Sn–3.8Ag–0.7Cu	0603	31.87	30.42	32.11	29.40	30.42	27.98	23.36	24.24	25.98
	95.5Sn–3.8Ag–0.7Cu	0805	56.74	53.80	50.39	46.02	47.82	40.53	34.69	35.49	36.53
	99Sn–1Cu + 96.5Sn–3.5Ag	0603	24.37	25.40	23.99	21.19	20.30	21.02	16.52	16.49	19.86
	99Sn–1Cu + 96.5Sn–3.5Ag	0805	37.17	41.90	36.25	31.98	31.69	30.71	26.33	23.94	26.61
Vapor phase	95.5Sn–3.8Ag–0.7Cu	0603	30.40	30.72	28.03	21.70	22.62	23.47	18.59	17.16	20.40
	95.5Sn–3.8Ag–0.7Cu	0805	48.97	52.99	51.87	34.01	38.01	36.61	29.85	30.40	34.34
	99Sn–1Cu + 96.5Sn–3.5Ag	0603	23.53	23.75	23.37	17.93	17.38	17.66	14.90	15.30	16.57
	99Sn–1Cu + 96.5Sn–3.5Ag	0805	41.27	41.68	41.38	26.84	27.39	27.80	22.19	22.90	26.81

TABLE 3.14. Difference of Temperature (ΔT) as a Function of Reflow Technology and Preheat Temperature [46]

Furnace:	Infrared		Forced Convection	
Preheat temperature:	High	Low	High	Low
ΔT (°C):	22	29	10	15

balling, and components cracking. A minimized soaking zone reduces poor wetting, solder balling, and opens. Use of low peak temperature lessens charring, delamination, intermetallics, leaching, dewetting, and voiding. A rapid cooling rate helps reducing intermetallics, charring, leaching, dewetting, and grain size. However, a slow cooling rate reduces solder or pad detachment. The optimized profile favors that the temperature ramps up slowly until nearly reaching the melting temperature. The temperature is then gradually raised further up to above the melting temperature within about 30 seconds, then raised rapidly until reaching the maximum temperature about 30–40°C above the melting temperature. Alternatively, the temperature can be raised linearly but slowly, approximately 0.5–1°C/s, from room temperature up to the peak temperature. After that, the temperature is brought down with a rapid cooling rate, approximately 4°C/s if possible. The conventional profile was developed due to the limitation of past reflow technologies. Implementation of the optimized profile requires the support of a heating-efficient reflow technology with a controllable heating rate. Vapor phase reflow can provide a rapid heating, but has difficulty to control the heating rate. Infrared reflow can regulate the heating rate, but is sensitive to variation in parts features. Emergence of the forced air convection reflow provides controllable heating rate. In addition, it is not sensitive to variation in parts features, thus allows the realization of the optimized profile.

3.5.2.2. Minimal Peak Temperature. For lead-free soldering, the promising alloy options, such as Sn–Ag–Cu, Sn–Ag–Bi, Sn–Ag, Sn–Cu, are mostly high in tin content, with melting temperature ranging from 210°C to nearly 230°C. Among those, the most prominent one is ternary eutectic Sn–Ag–Cu system, with a melting temperature of 217°C. The discussion on reflow profile will focus on that for Sn–Ag–Cu system.

The minimal peak temperature of a reflow profile is a function of solder melting temperature and wetting ability. For 63Sn–37Pb, the minimal peak temperature used usually is about 30°C above the melting temperature. If a similar practice is applied to Sn–Ag–Cu system, a minimal peak temperature of 247°C would be resulted. Apparently, this will be a significant challenge to the thermal stability of both components and boards, and a truly minimal, and hopefully lower, peak temperature acceptable for both process and reliability is desperately needed [26, 46–50].

Johnson et al. reported the wetting balance performance of 95.5Sn–3.8Ag–0.7Cu on five PWB surface finishes, Ni–Au, Sn, Pd, OSP, and Ag, at 230–255°C [50]. Among the five finishes studied, OSP exhibits the lowest wetting force, or spreading, at low temperature end, and the F_{max} for OSP offers marginal spreading

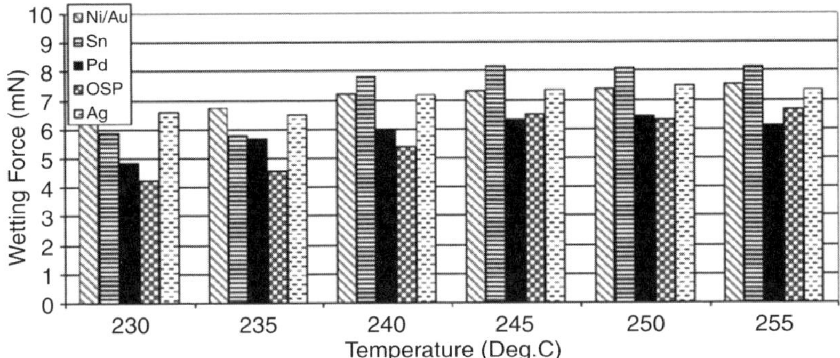

Figure 3.9. Wetting force of 95.5Sn–3.8Ag–0.7Cu on five PWB surface finishes at 230–255°C [50].

at 230°C, as indicated by Figure 3.9. This spreading suggests that 230°C can be the minimal peak temperature of reflow profile. Optimal wetting is almost achieved at temperatures of 235°C, and temperatures over 240°C only slightly increase wetting force.

Compared with wetting force, the wetting time varies significantly more with temperature, and ranges from 0.2 seconds to 4 seconds, as indicated by Figure 3.10. However, since reflow process often takes several minutes, with dwell time above liquidus temperature generally longer than 30 seconds, the wetting time appears to be more meaningful for wave soldering process, and has less significance in the reflow profile assessment.

The stipulation of 230°C being the minimal peak temperature at reflow is supported by several other works [26, 46–48]. Miyazaki et al. [46] reported that reflow soldering is possible even if peak temperature is 230°C and nitrogen atmosphere with oxygen content less than 100 ppm is used for 96.5Sn–3.0Ag–0.5Cu. In their work, the solder paste contains 12% flux, using rosin activated flux with

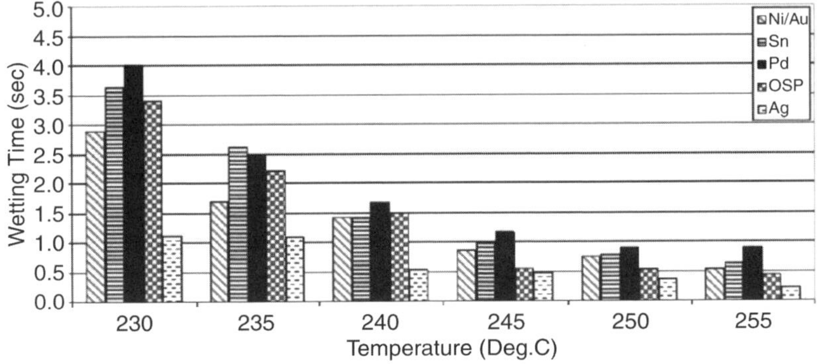

Figure 3.10. Wetting time of 95.5Sn–3.8Ag–0.7Cu on five PWB surface finishes at 230–255°C [50].

TABLE 3.15. Shear Test Results for 0201 Components Based on the Solder Material and Peak Temperature [47]

	63Sn–37Pb	Sn–Ag–Cu			
Peak temperature (°C)	215	224	228	242	257
Shear force average (gm)	649	702	669	781	758
Standard deviation (gm)	101	159	138	96	96
Maximum shear force (gm)	920	1040	1080	1040	1050
Minimum shear force (gm)	370	350	350	460	500

0.06% halogen content in flux. Albrecht et al. [48] reported that peak temperatures between 230°C (nitrogen) and 240°C (normal atmosphere) can be applied. Houston et al. also reported that temperatures below 240°C can be used for Sn–Ag–Cu reflow although temperatures below 232°C would not be recommended [47]. Although the maximum shear strength of 0201 component solder joints may still be comparable, the standard deviation, the average shear strength, and the minimum shear strength all decrease at peak temperature lower than 232°C, as indicated in Table 3.15. Figure 3.11 shows the cross-sectional SEM images of solder joints of 0201 components when reflowed at 232°C or below. At peak temperature below 232°C, either the joint fillet shape is not a smooth curve or the joint underneath the component is irregular in shape. Only at 232°C, the joint shape appears to be normal, supporting a minimal peak temperature of 232°C is required.

However, the wetting ability of solder paste is not only determined by solder alloy type, but also by the flux used. The discussions above regarding nitrogen being required at 230°C are not applicable to all fluxes. Butterfield et al. [26] studied 19 no-clean solder pastes of 95.5Sn–3.8Ag–0.7Cu (melting temperature 217°C), and concluded that one solder paste, besides meeting all other performance

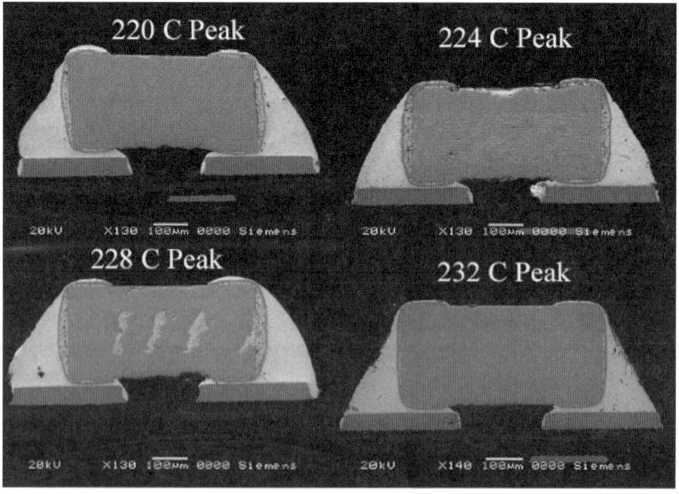

Figure 3.11. SEM images of 0201s reflowed below the recommended peak temperature [47].

Figure 3.12. Cross-sectional view of 95.5Sn–3.8Ag–0.7Cu solder joints reflowed under air at peak temperature 229°C, 237°C, and 245°C and dwell time 60, 70, and 80 seconds, respectively. Also shown is that of 62Sn–36Pb–2Ag reflowed at 210°C [26]. All joints are formed on copper pads with Entek OSP surface finish.

requirements, wets satisfactorily on Entek OSP finish at temperature as low as 229°C when reflowed under air. Figure 3.12 shows satisfactory 95.5Sn–3.8Ag–0.7Cu solder joints are formed when reflowed under air at peak temperature 229°C, 237°C, and 245°C, with dwell time above liquidus temperature being 60, 70, and 80 seconds, respectively. The wetting performance is very comparable with that of 62Sn–36Pb–2Ag solder joint reflowed at 210°C.

Therefore, it can be concluded that the minimal peak temperature can be pushed down to at least as low as 12°C above the melting temperature of Sn–Ag–Cu and still achieve satisfactory wetting when reflowed under air. At this stage, this can be achieved only with the use of the leading flux/solder paste. It is expected that, with time, more fluxes can be developed from more flux suppliers to match this performance.

3.5.2.3. Maximal Peak Temperature. The interest in the maximal peak temperature tolerable is mainly driven by the concern about the thermal stability of components and boards, particularly for plastic components. Due to the higher reflow temperature employed for lead-free soldering, as discussed above, the parts may not survive the reflow process. As a result, it is crucial to identify the maximal peak temperature which may be experienced by the lead-free reflow process. This peak temperature will then serve as the target temperature of which the thermal stability of parts should be met.

For eutectic Sn–Pb reflow, IPC/JEDEC J-STD-020 currently classifies components as either large ($\geq$2.5 mm thick or has a volume $\geq$350 mm^3) or small ($<$2.5 mm thick or has a volume $<$350 mm^3). Large components must be evaluated to a peak reflow temperature of 225 +0/−5°C and small components to a peak reflow temperature of 240 +0/−5°C. Kelly et al. [51] investigated the component

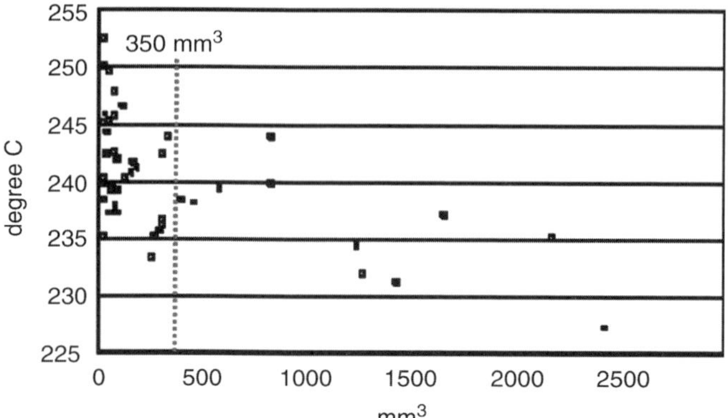

Figure 3.13. Plot of body temperature versus volume with large volume (CCGA) packages removed [51].

body temperature as a function of component volume for lead-free reflow process, with results shown in Figure 3.13. It appears that the component body temperature categorization based on breakdown of component volume is still applicable for lead-free reflow. The results are adopted by J-STD-020B [52, 53], with the same breakdown in component volume, and large components must be evaluated to a peak reflow temperature of 245 +0/−5°C, and small components to a peak reflow temperature of 250 +0/−5°C.

However, it should be noted that although the target maximum peak temperature used for qualifying components has been assessed by NEMI as 245–250°C and specified in J-STD-020B, component manufacturers, such as Texas Instruments, or large device manufacturers, such as Hewlett Packard, are not comfortable with the applicability of this specification. Many major manufacturers are aiming at 260°C as the target temperature for both process and component qualification.

3.5.2.4. Cooling Rate. The impact of cooling rate is primarily on reliability, and not so much on the process defect. A rapid cooling rate reduces heat input, therefore reduces the extent of intermetallics formation, leaching, dewetting, and charring [1]. The impact on the former three issues is more significant at temperature above the melting temperature of solders.

In general, a rapid cooling rate tends to yield a small grain size. Joo and Yu [54] reported that, compared with a very slow cooling, a very fast cooling in lead-free solders creep sample preparation resulted in a one order of magnitude decrease in grain diameter, and two order of magnitude decrease in creep rate and two order of magnitude increase in time to rupture. Although a fast cooling is desirable at reflow soldering, the extent of improvement in reliability is limited by the maximum cooling rate allowed at SMT reflow. For most of the SMT reflow process, the maximum cooling rate allowed is often governed by the tolerance of

TABLE 3.16. Relation Among Flux Chemistry, Cooling Rate, and Number of Grains in 95.5Sn–3.5Ag–1.0Cu Solder Bump [55]

Cooling rate (227–206°C)	4.98°C/s	1.44°C/s
Flux A	2	1
Flux B	5	4
Flux C	3	3

components due to component cracking consideration, and is often set at 4–6°C/s. Table 3.16 shows the effect of cooling rate on number of grains in flip chip 95.5Sn–3.5Ag–1.0Cu solder bumps. The higher cooling rate did tend to result in an increase in number of grains per bump, therefore a decrease in grain size, although the effect appears to be fairly mild.

At times, a slow cooling rate may be desirable. For instance, upon cooling, BGA solder joints may crack or delaminate at interface at package side due to mismatch in thermal expansion. The phenomenon mostly observed at wave soldering, but may also happen at reflow cooling. A slow cooling will allow a reduced temperature gradient, hence lessen the chance of joint crack.

3.5.2.5. Profile Shape. Lee has discussed the optimization of reflow profile based on defect mechanism analysis in details [1, 56]. Although the examples given were mainly for eutectic Sn–Pb solders, the analysis and conclusion are equally applicable to lead-free reflow soldering, and are supported by many other works [26, 32, 49, 57]. Therefore, a tent profile with linear ramp up rate of 0.5–1°C/s, preferably 0.7°C/s, until reaching the peak temperature, then quickly cool down to ambient temperature is recommended for lead-free reflow process. Figure 3.14 demonstrates such a profile, and has been used by Motorola to successfully manufacture more than a million cellular phones already [26].

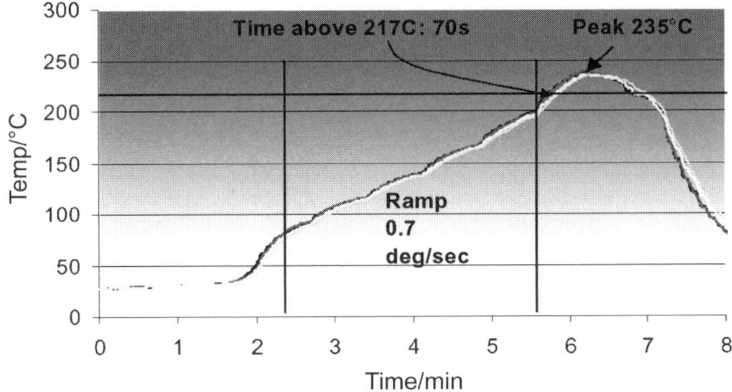

Figure 3.14. A linear ramp profile used for 95.5Sn–3.8Ag–0.7Cu no-clean solder paste reflow, where peak temperature = 235°C ± 5°C; time above liquidus = 70 s + /10 s [26].

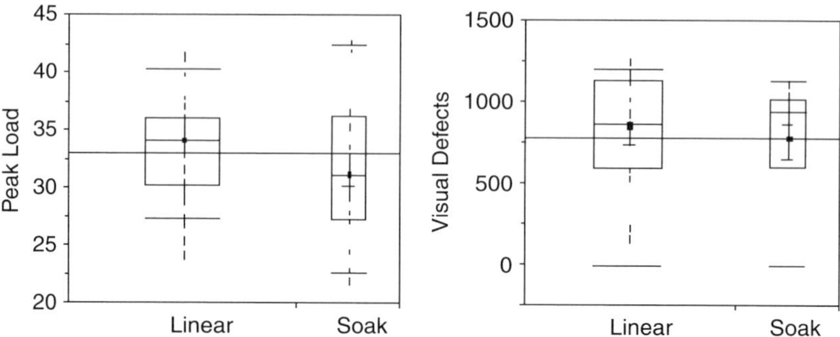

Figure 3.15. Lead-free ultimate tensile strength (left) and visual defects (right) results as a function of profile type [49].

Shina et al. compared the linear profile with conventional profile with a soaking zone by maintaining the same heat input. They reported that the linear profile having a peak temperature of 235°C is the recommended profile for their 4-in. × 5.5-in. test board [49]. It will minimize burn off of activators in the flux system and decrease throughput only by 5% relative to a tin/lead conventional profile, whereas utilizing a lead-free conventional profile will decrease throughput by 13%. Linear profile with a longer time above liquidus worked better than the traditional profile with a higher peak temperature and shorter time above liquidus. A linear profile showed a slight contribution to the maximum ultimate tensile strength obtained during the pull test but these profile types had no significance in the visual tests, as shown in Figure 3.15. The time above liquidus had only a small effect on visual defects, and no effect on pull strength.

Although a linear ramp profile is recommended for lead-free reflow, soak profile could also be used [2, 32, 57]. The profile with a long soak zone, as demonstrated in Figure 3.16, is of particular value when voiding is an issue [57], such as voiding in BGA. Voiding is attributed to outgassing within the solder joint when the solder is at molten state [58–66]. The effect of reflow profile on voiding at microvia for lead-free soldering is strongly dependent on the flux chemistry and solderability of

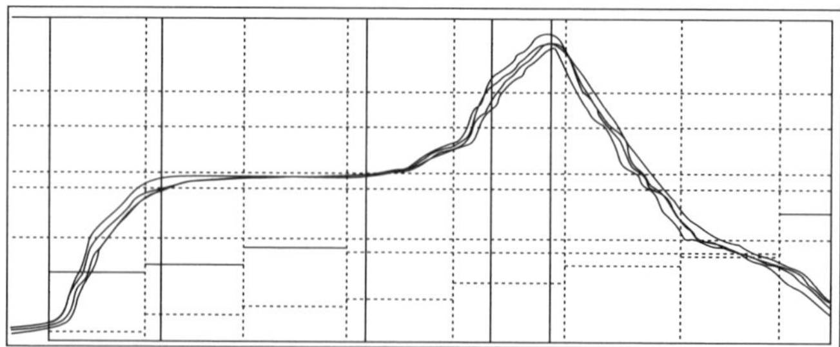

Figure 3.16. Profile with a long soaking zone is effective in reducing voiding in solder joints [57].

parts or boards. Employing a profile which will either suppress the outgassing when the solder is at molt state or enhance the wetting so that no flux will be entrapped within the solder joint promises a reduced voiding.

Although use of long soaking zone may have the disadvantage of compromising wetting and solder balling, it does allow drying out the volatile content of solder paste before reaching the melting temperature, thus minimizes outgassing when the solder becomes molten and consequently reduce the voiding extent.

The effective control of voiding with soaking profile is consistent with that of eutectic Sn–Pb experience [67]. Evans and Dahle [67] studied the effect of profile on voiding of BGA with 63Sn–37Pb solder bumps. The profiles tried differ in ramp rate and profile shape, and are shown in Figure 3.17. The time taken to reach melting temperature 183°C is approximately 250, 75, and 50 seconds for profile (a), (b), and (c), respectively. The extent of predrying can be approximated with the time taken prior to solder melting. Accordingly, the extent of predry can be ranked as profile (a) ≫profile (b) >profile (c). Figure 3.18 shows the relation between the profile type and the mean voiding of worst-case

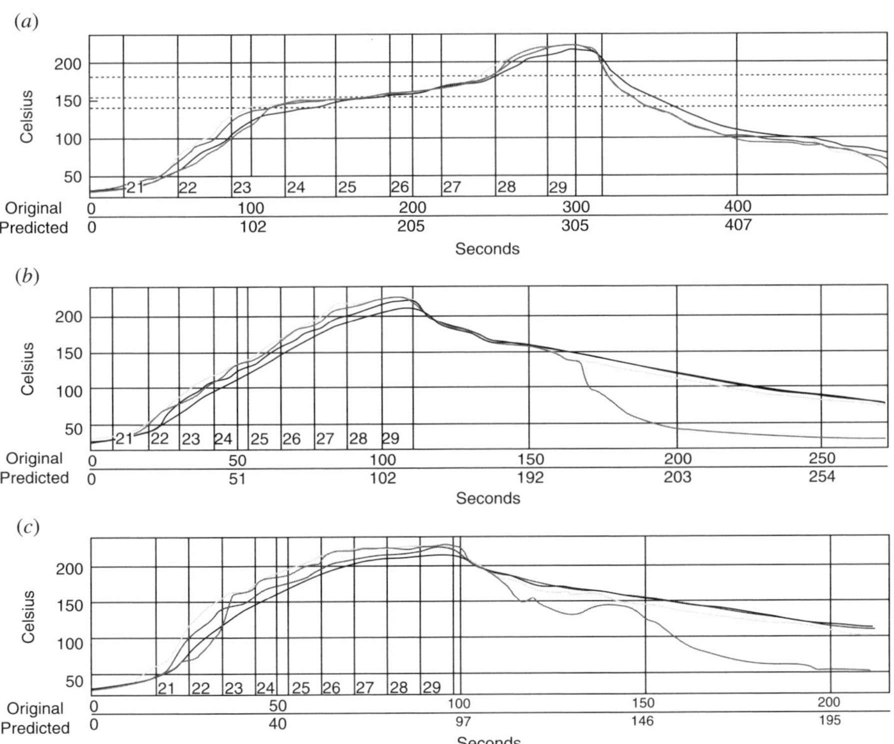

Figure 3.17. Reflow profiles with various predrying before reaching melting temperature 183°C [67]. The extent of predrying follows the order (*a*) 250 seconds > (*b*) 75 seconds > (*c*) 50 seconds.

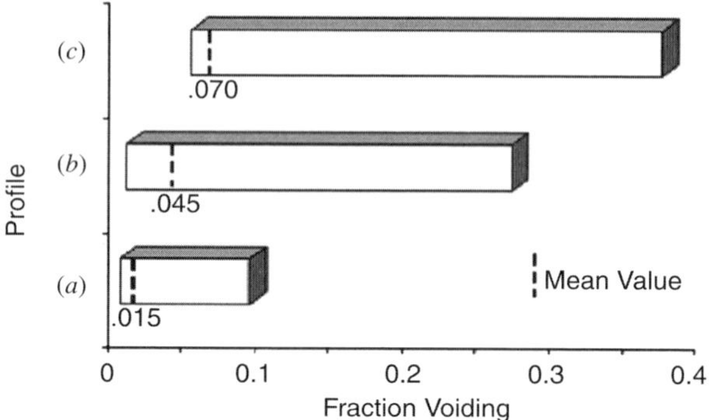

Figure 3.18. Effect of predrying on voiding in BGA, with predrying time (*a*) 250 seconds, (*b*) 75 seconds, and (*c*) 50 seconds [67]. The voiding is expressed as mean voiding of worst case BGA solder joints.

BGA joints. It is interesting to note that there is a strong relation between predrying extent and voiding, with more predrying correlated with less voiding.

The relation between profile and voiding is further elucidated by Liu et al. [64] and Lee [65, 66]. In their study, the impact of profile on voiding is quantitatively analyzed by defining soaking energy and melting energy, as exemplified in Figure 3.19 [64]. Here a soaking profile is developed, in addition to the reflow profile, in order to determine the volatile generated at each stage. The area under the soaking curve but above 150°C line is denoted as soaking energy and is exemplified by the shaded area. The melting energy is approximated by area below the reflow curve but above the soaking profile and is represented by the red area. In this study, the soaking energy is regulated by controlling the soaking time, while the melting energy is regulated by controlling the peak temperature.

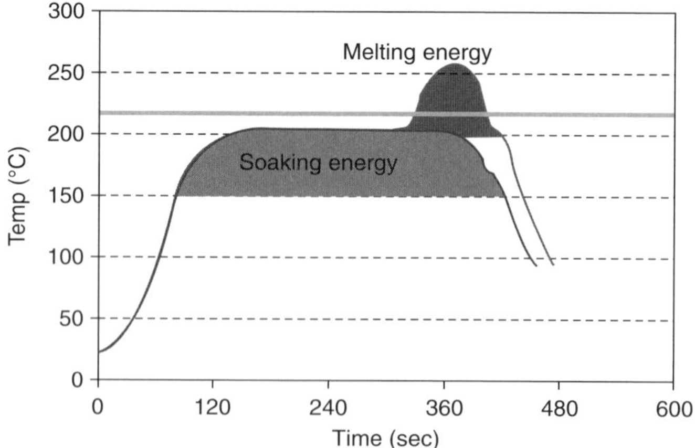

Figure 3.19. Melting energy and soaking energy exemplified for a reflow process [64].

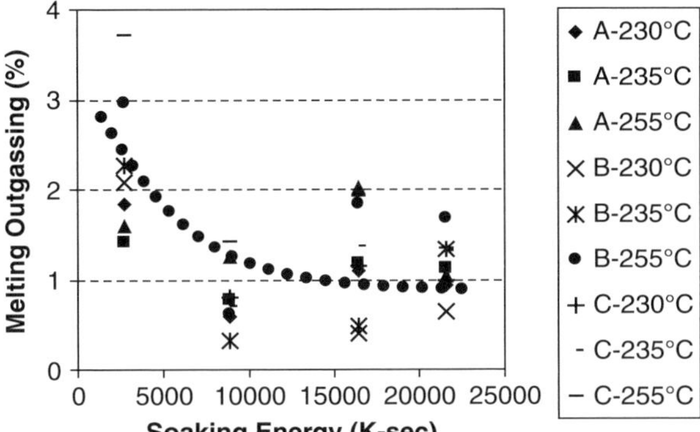

Figure 3.20. Relation between soaking energy and melting outgassing for solder pastes A, B, and C. The temperature denoted in legend represents the peak temperature, while the soaking energy is regulated by controlling the soaking time [64].

Increase in soaking energy does reduce melting outgassing, or the amount of outgassing when the solder is at molten state, as shown in Figure 3.20. The temperature denoted in legend represents the peak temperature, while the soaking energy is regulated by controlling the soaking time. Theoretically, this should result in a lower voiding.

However, the relation between soaking energy and voiding reveals a more complicate picture, as shown in Figure 3.21. Although considerable data scattering is present, a trend of decreasing voiding with increasing soaking energy can be

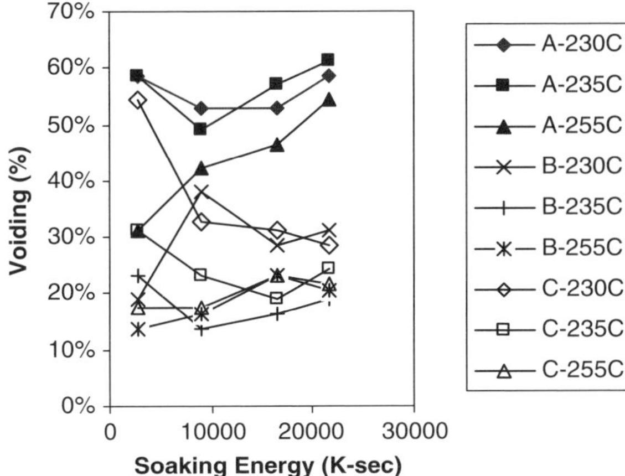

Figure 3.21. Relation between soaking energy and voiding for three solder pastes A, B, and C. The temperature denoted in legend represents the peak temperature, while the soaking energy is regulated by controlling the soaking time.

discerned for solder paste B and C. However, for solder paste A, an initial decrease followed by an increase in voiding with increasing soaking energy is observed for 230°C and 235°C peak temperature systems. For 255°C peak temperature system, the voiding of solder paste A merely increases with increasing soaking energy.

This complicate relationship is attributed to a combined effect of outgassing, oxidation, and flux chemistry. The melting outgassing does decrease with increasing soaking energy, therefore favors a reduced voiding. However, the oxidation of solder powder and metallization of parts also increases with increasing soaking. Increase in oxidation means a decrease in wetting, and consequently an increase in voiding. The two factors, outgassing and oxidation, conflict with each other in terms of impact on voiding, and are affect by flux chemistry. Fluxes with a greater oxidation resistance will benefit more from a longer soaking conditioning in voiding control.

Another important trend observed is a lower voiding associated with a higher peak temperature for all pastes studied. A higher peak temperature results in a better wetting, hence a reduced voiding.

The peak temperature is a double-sided sword. Depending on the relative wettability of parts, a higher peak temperature can have opposite effect on voiding. This is particularly true for BGA with microvia in pad, where entrapped flux in dead corner of microvia is almost inevitable, and the amount of entrapment is governed by the extent of wetting. For difficult to wet situation, such as combination of lead-free alloy and oxidized OSP surface finish, wetting dictates the voiding behavior, and a higher peak temperature helps suppressing the voiding.

On the other hand, for easy to wet situation, such as eutectic tin–lead on Ni–Au surface finish or design not involving microvia, melting outgassing dictates the voiding behavior. A low peak temperature reduces the amount of melting outgassing, hence promises a decrease in voiding. Similarly, a decrease in time above liquidus decreases the melting energy, and consequently can cause less voiding. This is supported by the work of Saiyed et al. [68].

It should be pointed out that the choice of profile is flux, design, and equipment dependent. No profile is perfect in minimizing every type of defects, and the objective should be set at minimizing the overall defect rate. A linear ramp up profile is a better choice for most situations. If the design poses major issues on certain defect

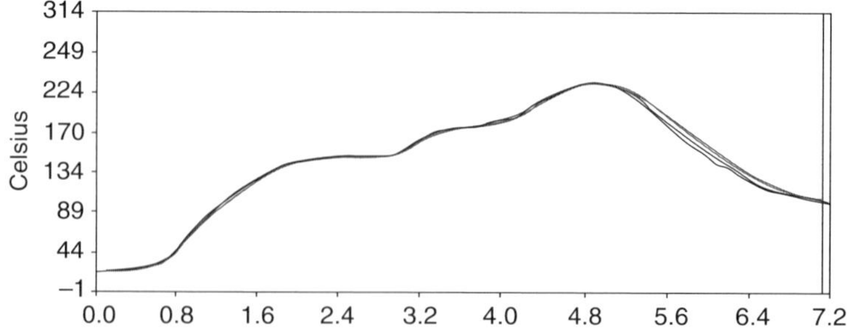

Figure 3.22. Double-step profile for Sn–Ag–Cu reflow [2].

types, then the profile should be adjusted accordingly [1]. In the event of a serious voiding potential, a long soaking profile often is a better choice, at the expense of a possible increase in some other defects.

Besides the two main stream profile types discussed above, some other profile types are also in use for Sn–Ag–Cu systems, such as double-step profile, as shown in Figure 3.22 [2]. Relatively, those profiles are considered as a tweaking of the two main profile types for specific applications.

3.5.3. Inerting

As discussed earlier, lead-free solders generally wet poorer than eutectic Sn–Pb solders [25, 26]. In order to compensate for this inferior wetting ability of alloys, fluxes with higher fluxing activity are desired. Furthermore, use of inert reflow atmosphere can further narrow down the gap of soldering performance between lead-free solders and eutectic Sn–Pb solders [26, 46–48]. The need for inerting is mainly due to the poor oxidation tolerance of many fluxes. Therefore, oven oxidation feature or inerting ability becomes an important consideration when selecting solder pastes [26].

Jensen et al. tested five lead-free solder pastes. The visual scoring for the solder joints was evaluated against solder paste type, print speed, peak temperature, temperature above liquidus, conveyor speed, and nitrogen inerting, with results shown in Figure 3.23 [69]. The relative importance of those variables are ranked below: nitrogen > paste brand > peak temperature > conveyor speed > print speed, time above liquidus.

In Shina et al. [49] work, the visual defects and ultimate tensile strength of lead-free solder joints were evaluated against materials and process parameters. In the

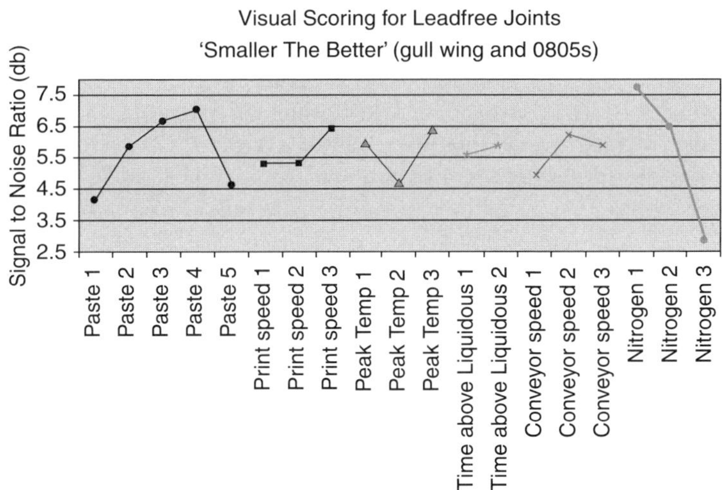

Figure 3.23. Response graph of visual scoring of wetting of lead-free joints [69].

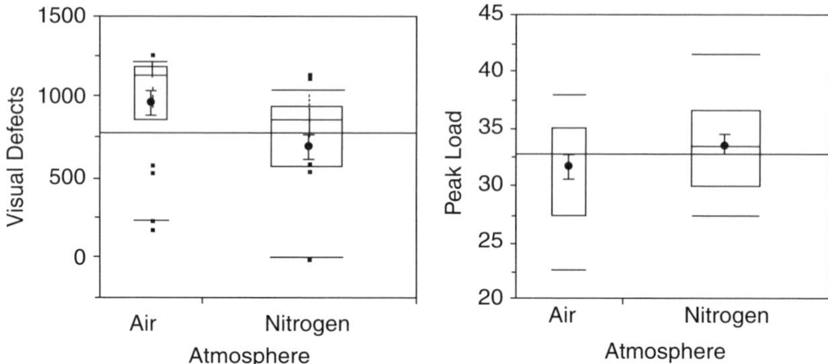

Figure 3.24. Effect of reflow atmosphere on the visual defects and ultimate tensile strength of Pb-free solder joints [49].

visual defects analysis the factors of significance included: surface finish (52.7%), the interaction between solder paste and surface finish (22.3%), nitrogen (10.7%), solder paste (7.2%), the interaction between surface finish and time above liquidus (1.5%), the interaction between solder paste and time above liquidus (0.7%), and time above liquidus (0.6%). Their results indicate that nitrogen is favorable for visual inspection criteria, as shown in Figure 3.24. Use of nitrogen allows the flux to clean the existing oxides from the metal surfaces and inhibits the effects of further oxidation at elevated temperatures. Hence, it expands the process window and facilitates solder wetting. This is especially critical for boards with large components where the temperature gradient across the board can be significant.

In the pull test analysis the important factors include: the solder paste (42.8%), surface finishes (7.3%) and conventional soak versus linear profile (3.1%). It is interesting to note that nitrogen did not hold any significance, and nitrogen rendered an average ultimate tensile strength comparable with that of air, as shown in Figure 3.24.

Hunt et al. [70] evaluated the solderability of several lead-free alloys in nitrogen. For Sn–Zn solder alloys it is found to be necessary to use nitrogen inerting with oxygen levels better than 100 ppm. When using Sn–Ag–Cu, Sn–Cu and Sn–Ag–Bi solder alloys good wetting times can be achieved with nitrogen inerting at 5000 ppm oxygen levels for superheats above 40°C. When soldering particularly difficult assemblies (e.g., multilayer boards), inerting to 50 ppm oxygen when using Sn–Ag–Cu, Sn–Ag–Bi and Sn–Cu solders is considered beneficial.

The higher surface tension and the lower superheats of lead-free alloys, compared to those of eutectic Sn–Pb, place a great demand on the flux capability. Although flux development may offset the need to inert, the use of nitrogen may become inevitable during the transition stage, especially if cleaning is to be avoided thus only a benign flux is allowed. This is mainly due to the insufficient oxidation tolerance of many current fluxes [26].

The relation between the flux oxidation tolerance and the demand for oven inerting ability has been studied by Jagger and Lee [71]. In his work, a semiempirical model was developed for predicting the soldering performance of solder pastes

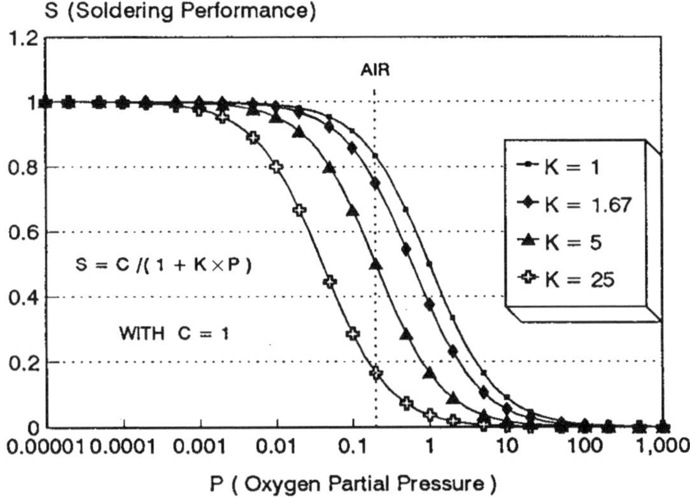

Figure 3.25. Theoretical relation between soldering performance S and oxygen partial pressure P [71].

with various levels of oxidation tolerance reflowed under various levels of inert atmosphere [71], as shown below.

$$S = C/\{1 + K \times P\}$$

where S is soldering performance of a given flux; C is a constant and can be set as 1; P is oxygen concentration, expressed as oxygen partial pressure; K is R_0/R, reflecting the relative tendency of solder oxidation, with a higher value indicating a higher oxidation tendency; R_0 is oxidation barrier content of a current high-performance full-residue flux; R is oxidation barrier content of a given flux.

The relation of S versus P and K is shown in Figure 3.25. In this graph, the relation is exemplified for oxidation tendency K value being 1, 1.67, 5, and 25. It can be seen that for low oxidation barrier content pastes ($K > 1$), the soldering performance increases rapidly first, then gradually levels off with decreasing oxygen concentration. The lower the oxidation barrier content, the lower the oxygen concentration at which it levels off. The soldering performance versus oxygen concentration curves are superimposable by moving the curve horizontally.

The soldering performance can be reflected by properties such as solder joint bond strength or solder spread area. The model is considered applicable to any solder systems. In Lee's work, the model was compared with the soldering performance, including both bond strength and spread area data for eutectic Sn–Pb. In general, the experimental data match this model fairly well, as exemplified by the correlation between bond strength data and model shown in Figure 3.26.

The model points out that the need of inerting at reflow for most of the fluxes mainly results from the insufficient oxidation barrier content of those fluxes.

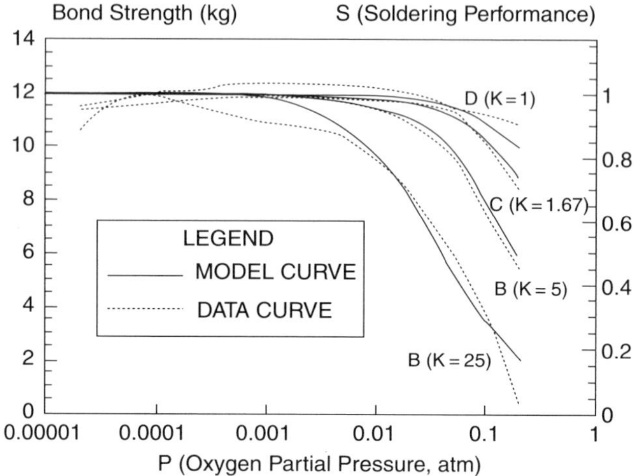

Figure 3.26. Correlation between bond strength data and model [71].

Although the current high-performance full residue flux ($K = 1$) performs relatively satisfactorily, it can be improved further by increasing the oxidation barrier content (so that $K < 1$) in the flux or by selecting a resin material with a lower oxygen permeation constant.

3.6. MICROSTRUCTURES OF REFLOWED JOINTS

The microstructures of cross-section of solder joints prepared by reflowing a series of no-clean lead-free solder pastes are exemplified in Figure 3.27 [25]. The pastes were made of solder powder ($-325/+500$ mesh) and reflowed in a forced convection oven under air with a peak temperature of 30°C above liquidus temperature of given alloy. 63Sn–37Pb was also processed as a control.

96.5Sn–3.5Ag. This alloy consists of very small amount of gray Cu_6Sn_5 particles and strings of bright Ag_3Sn intermetallic particles in Sn dendrite matrix. The Cu_6Sn_5 intermetallic is due to the reaction between the solder and the copper base metal.

99.3Sn–0.7Cu. This alloy consists of gray Cu_6Sn_5 intermetallic in a Sn dendrite matrix. The large, rod-shaped Cu_6Sn_5 intermetallic particles may be due to reaction between the solder and the base metal Cu, and the small Cu_6Sn_5 intermetallic particles may be precipitates of the alloy.

95.5Sn–3.8Ag–0.7Cu. Few gray Cu_6Sn_5 particles and strings of bright Ag_3Sn intermetallics scattered in Sn matrix are observed. The Cu_6Sn_5 may be formed either from the reaction between the solder and the base metal Cu, or from the 0.7% Cu in the alloy. It is speculated the large Sn–Cu particles

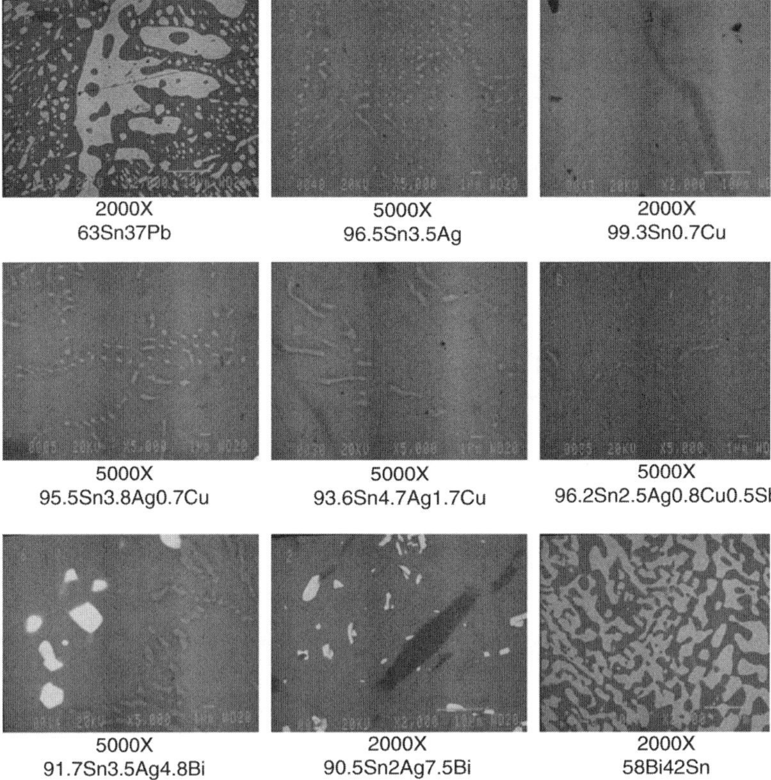

Figure 3.27. SEM images of cross sections of solder joints of no-clean solder pastes with −325/+500 mesh powder. The pastes were reflowed in forced convection oven under air with a peak temperature of 30°C above liquidus temperature [25].

may be due to reaction between solder and the Cu metallization. Handwerker also reported a similar microstructure for 95.5Sn–3.9Ag–0.6Cu for CSP joint, with tin dendrites separated by Cu–Sn and Ag–Sn intermetallic compounds [72].

93.6Sn–4.7Ag–1.7Cu. Similarly to 95.5Sn–3.8Ag–0.7Cu alloy, there were Cu_6Sn_5 and Ag_3Sn intermetallics in Sn matrix. However, the size of both Ag_3Sn (bright) and Cu_6Sn_5 (gray) intermetallics particles are larger here, perhaps due to the higher concentration of both Ag and Cu in this alloy.

96.2Sn–2.5Ag–0.8Cu–0.5Sb. The microstructure of this alloy is similar to Sn–Ag–Cu alloys discussed above.

91.7Sn–3.5Ag–4.8Bi. Some large elemental Bi precipitates (very bright) and strings of Ag_3Sn particles (bright) are present. Also observed were small amount of Cu_6Sn_5 particles (gray), due to the reaction between solder and copper pad.

90.5Sn–2Ag–7.5Bi. There were some elemental Bi precipitates (very bright phase) with various sizes, and a fairly small amount of Ag_3Sn phase (bright) present in the Sn matrix. The morphology of the Ag_3Sn phase was similar to that of Sn–Ag–Cu alloys. There were also some large Cu_6Sn_5 intermetallic phase (gray), which is due to the reaction between the solder and the Cu metallization on the board.

58Bi–42Sn. Elemental Bi precipitates (bright) in Sn phase observed. The volume ratio of Sn phase/Bi phase was reported to be 49:51 [23, 24].

The microstructures of solders are not only determined by solder composition, but also affected by heat history. Thus, for 95.8Sn–3.5Ag–0.7Cu, the size of the beta-Sn dendrite globule was 5–10 μm in the fast cooled specimens, while the size of the thermally stabilized specimens was 50–100 μm [73]. The fast-cooled specimens showed about two orders of magnitude lower minimal strain rate and longer rupture time than the thermally stabilized specimens in creep test.

Formation of Sn dendrites can have a significant effect on long-range order, including surface morphology. Figure 3.28 shows the cross-section of BGA joint formed with Sn–Ag–Cu ball and Sn–Ag–Cu paste, with obvious Sn dendrites alignment and surface roughness [74]. The surface roughness can be fairly significant, as demonstrated by Figure 3.29 [75]. Celestica characterized surface roughness and determined it to be a result of eutectic shrinking back during solidification exposing large primary beta-Sn dendrites. Importantly, no evidence was found that the roughness influences crack formation during thermal cycling [74].

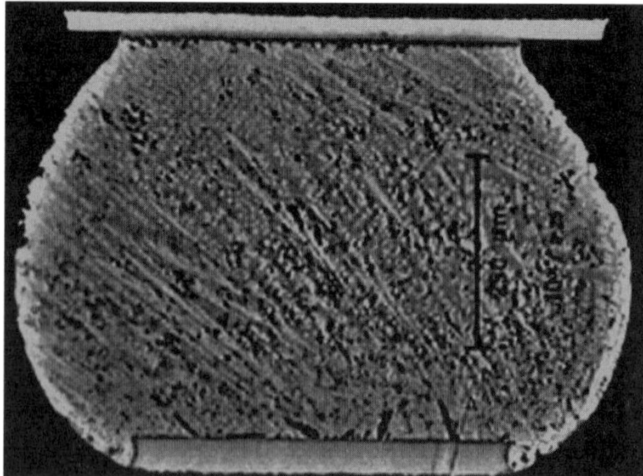

Figure 3.28. BGA joint with Sn–Ag–Cu ball and Sn–Ag–Cu paste showing obvious Sn dendrites alignment and surface cracks resulted [74].

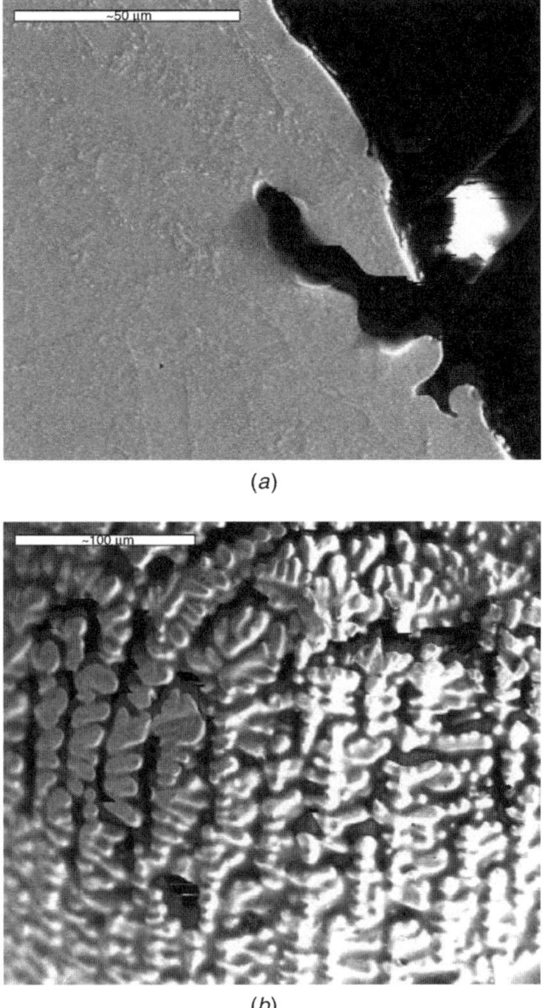

Figure 3.29. Sn–Ag–Cu paste Sn–Ag–Cu ball joint roughness. (*a*) Cross section, (*b*) top view, SEM [75].

3.7. CHALLENGES OF LEAD-FREE REFLOW SOLDERING

Lead-free reflow soldering has been successfully introduced to the electronic industry. Sn–Ag–Cu turns out to be the most favorite choice globally for reflow applications. Adequate flux chemistries have been developed, process windows have been defined, and varieties of lead-free materials have evolved to support the implementation of lead-free reflow soldering. As a result, millions and millions of lead-free electronic devices have been manufactured and are being used everywhere. However, the success of implementing lead-free soldering should at best be regarded as a good start, instead of achieving the ultimate goal. There are still a lot of challenges facing lead-free reflow soldering, and have to be ironed out one by one before lead-free reflow can truly become a mature process and technology. The major challenges of lead-free reflow soldering will be discussed one by one as listed below.

3.7.1. Voiding

Voiding tends to increase considerably with lead-free solder pastes, particularly with Sn–Pb components [1, 33, 76–78]. Figure 3.30 shows voiding performance of solder joints formed with Sn–Pb paste with Sn–Pb or lead-free components, Sn–Ag–Cu paste with Sn–Pb components, and Sn–Ag–Cu paste with Sn–Ag–Cu components [77]. Results indicate that Sn–Pb paste with Sn–Pb components or lead-free components exhibit the least amount of voiding, Sn–Ag–Cu paste with Sn–Ag–Cu components show more voiding, and Sn–Ag–Cu paste with Sn–Pb components display the highest voiding. In another study, the voiding behavior for BGA solder bumping using sphere attached with solder paste was investigated [76]. Sn–Pb paste/Sn–Pb balls showed no voiding at all. Sn–Pb paste/lead-free balls showed 0.1% voiding, while lead-free paste/lead-free balls and lead-free paste/Sn–Pb balls showed around 1.7% voiding.

The voiding is caused by flux outgassing, and aggravated by poor wetting and large joint coverage area [1]. Since lead-free solders generally wet poorer than Sn–Pb solders, higher voiding associated with lead-free pastes or lead-free components than Sn–Pb counterparts is expected. This expectation is consistent with results for system without large coverage area on the solders, such as BGA solder bumping process [76].

However, for system with large coverage area on solder joints, such as BGA assembly process, it is not very obvious why Sn–Ag–Cu paste with Sn–Pb components exhibit higher voiding than Sn–Ag–Cu paste with Sn–Ag–Cu components,

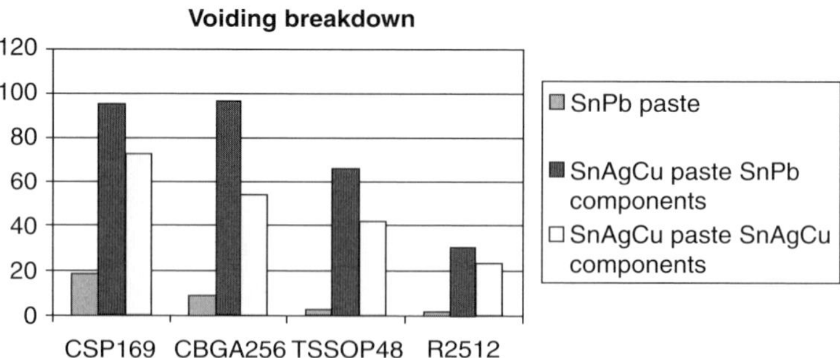

Figure 3.30. Voiding breakdown of solder joints formed with Sn–Pb paste/Sn–Pb or Sn–Ag–Cu components, Sn–Ag–Cu paste/Sn–Pb components, and Sn–Ag–Cu paste/Sn–Ag–Cu components [77].

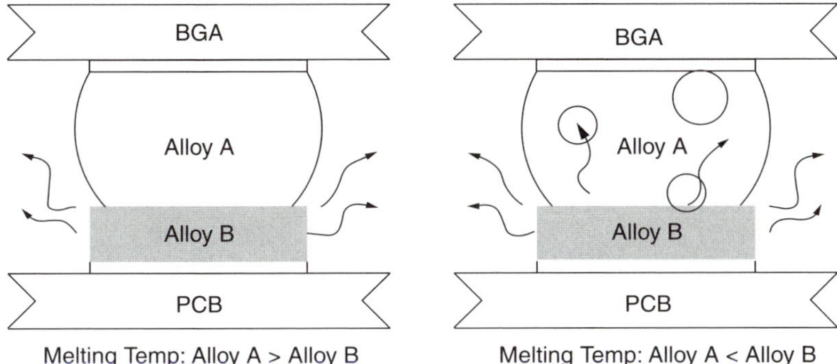

Figure 3.31. A model for voiding behavior at BGA assembly process involving two solder alloys [78].

which in turn is higher than Sn–Pb paste with lead-free components. If voiding is dictated by solder wetting, Sn–Ag–Cu paste with Sn–Ag–Cu components is expected to be the worst in voiding. This perplexing phenomenon can be explained by Lee's model, as shown in Figure 3.31 [78]. For systems with paste and components employing different alloys, if solder paste exhibits a higher melting temperature than the solder bump, the bump can turn into liquid solder sitting on top of solder paste which is still actively outgassing. Inevitably, some flux vapor will be "injected" into the interior of molten solder bump and form voids. Although eventually the solder paste also melts and excludes the flux from interior of solder, voids are already formed. On the other hand, if the solder paste exhibits a melting temperature lower than the solder bump, solder paste will melt, coalesce, and exclude flux from the joint interior before the bump melts. Consequently, no flux volatile will get into the solder bump, and voiding will be minimized.

Since voiding is sensitive to relative melting temperature of solder paste and solder on components, the best practice is not to use paste with higher melting temperature than solder of components. In the event that such combination is inevitable,

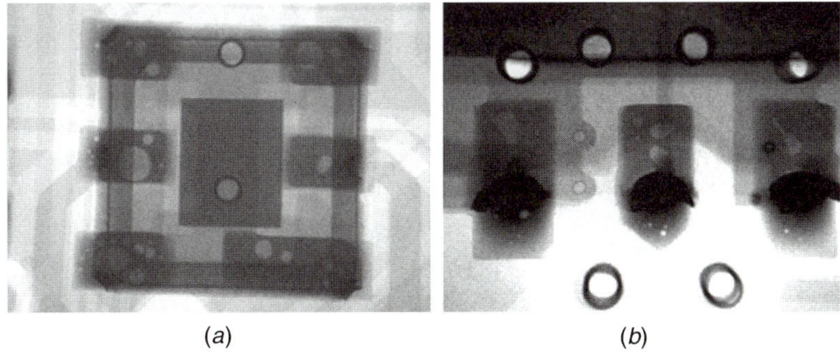

(*a*) (*b*)

Figure 3.32. Voids in (*a*) flash gold finished solder joints and (*b*) thick gold finished solder joints of 96.2Sn–2.6Ag–0.8Cu–0.5Sb solder after 500 thermal cycles [80].

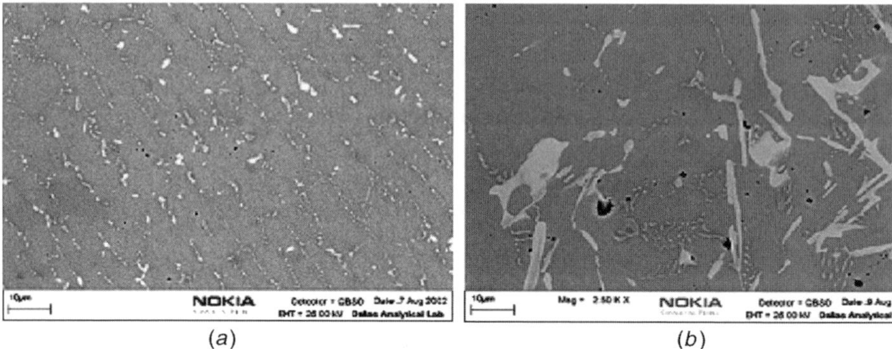

Figure 3.33. Microstructure of 96.2Sn–2.6Ag–0.8Cu–0.5Sb solder joints after reflow, (*a*) flash gold, and (*b*) thick gold [80].

the voiding can still be minimized with the use of long soaking profile and minimal dwell time above liquidus temperature, as discussed in the profile section [57, 68].

The voiding mechanism for lead-free solders is the same as that of Sn–Pb solder system [79]. Similar to Sn–Pb solders [1], voiding can also be affected by the type of surface finishes for lead-free solder pastes. For instance, the influence of gold plating on leadless organic and ceramic chip carrier component terminals on the performance of the 96.2Sn–2.5Ag–0.8Cu–0.5Sb solder joints was investigated with metallurgical analysis. A flash gold finish (0.02–0.05 μm) results in spherical voids, while a thick gold finish (0.5–1.2 μm) results in irregular voids, as shown in Figure 3.32 [80]. The spherical voids are simply shaped by surface tension of molten solder [1]. However, the irregular voids are not only affected by surface tension, but also interfered by the sluggish flow of solders with large content of $AuSn_4$ intermetallics, as shown in Figure 3.33 [80]. Similar irregular voids caused by the presence of large

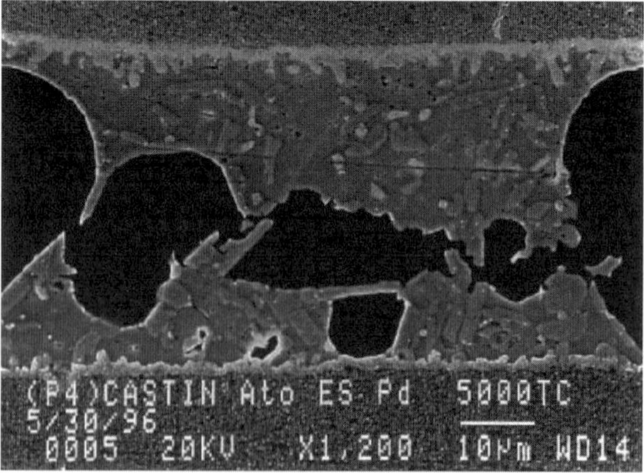

Figure 3.34. Secondary electron micrograph of the cross section of solder joint between SOIC lead and PWB on a Pd finished PWB after 2500 thermal cycles [81].

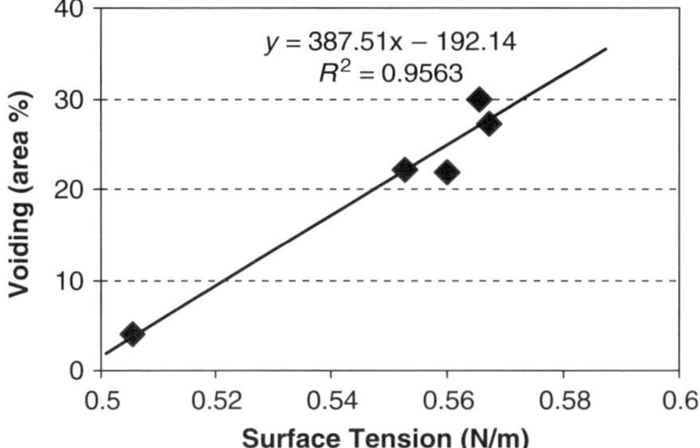

Figure 3.35. Relation between surface tension and voiding rate at microvia for 63Sn–37Pb and Sn–Ag–Cu alloys [63].

quantity of intermetallics are also observed in 96.2Sn–2.5Ag–0.8Cu–0.5Sb solder joints on Pd finish, as shown in Figure 3.34, where the void sizes range from several to one hundred microns in diameter [81].

Since voiding can be affected by solder wetting [1], solder alloys with a lower surface tension is expected to spread easier, thus promises a lower voiding. Indeed, this is what observed in Figure 3.35 [63, 65, 66]. The surface tension of some Sn–Ag–Cu alloys is determined, as shown in Figure 3.36 [63], with the ternary eutectic alloy 95.6Sn–3.5Ag–0.9Cu being the lowest among the SAC alloys studied. This ternary eutectic alloy is also the lowest in voiding among those alloys tested, as shown in Figure 3.37 [63].

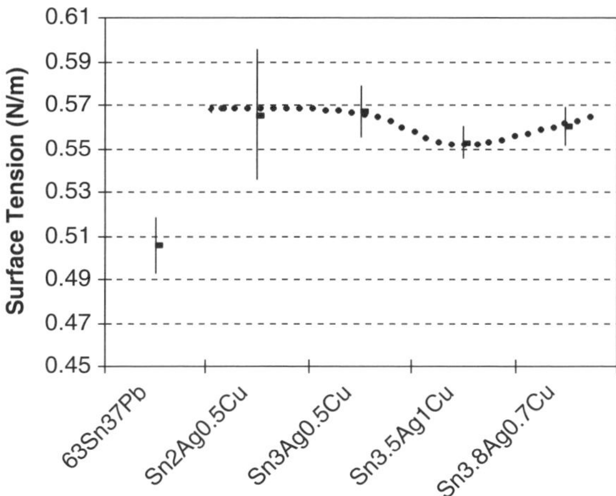

Figure 3.36. Surface tension of 63Sn–37Pb and Sn–Ag–Cu alloys determined at 245°C and 260°C, respectively [63].

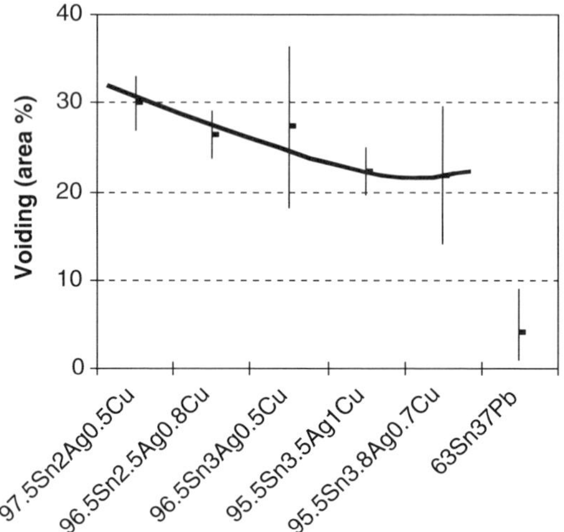

Figure 3.37. Effect of alloy composition on voiding at microvia [63].

Difference in surface tension probably is the most dominant factor causing lead-free alloys to wet poorer and accordingly have a poorer voiding than 63Sn–37Pb. The surface tension of 95.6Sn–3.5Ag–0.9Cu, 0.553 N/m, is about 10% higher than that of 63Sn–37Pb, 0.505 N/m. This difference appears to be sufficient to cause the voiding of 95.6Sn–3.5Ag–0.9Cu, 25% (of area), to be four times higher than that of 63Sn–37Pb, 5% (of area).

The voiding criteria for BGAs is defined in J-STD-001, and is required to be equal or less than 25%. Since this standard is based on Sn–Pb solders, and since lead-free solders may exhibit more voiding without compromising reliability [26], a new criteria should be established for lead-free solders accordingly. IPC also published a guideline IPC-7095 [82]. In this guideline, the process control criteria for void size limitation at BGA solder joint interface are set at 25%, 12%, and 4% of area for class 1, 2, and 3, respectively. Again, this guideline is based on Sn–Pb solders.

3.7.2. Tombstoning

Lead-free soldering has been reported to be more prone to have tombstoning than eutectic Sn–Pb, and was speculated to be due to the difference in surface tension and wettability [83]. However, since the number of tombstoning defects recorded is fairly low, with lead-free and Sn–Pb being 5 and 2, respectively, the statistical significance may be in question. Conflicting results have been observed, indicating a comparable tombstoning rate for 63Sn–37Pb and lead-free solder pastes at vapor phase reflow [84].

Regardless of the relative sensitivity toward tombstoning, lead-free solder pastes were found to be affected by surface tension of alloy, with lower surface tension being more prone to tombstoning, as shown in Figure 3.38 [84]. Solder with a

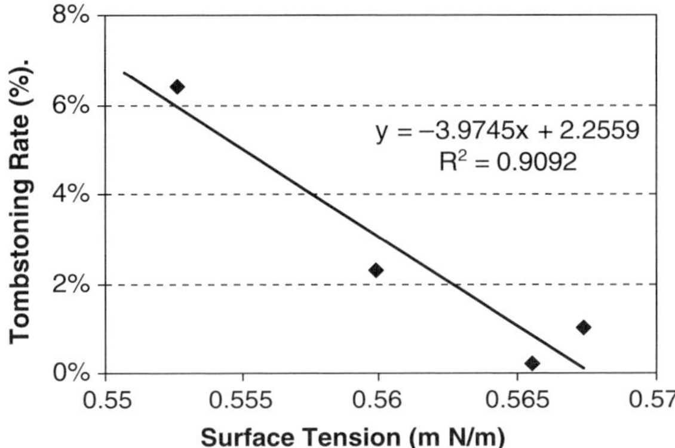

Figure 3.38. Relation between surface tension of Sn–Ag–Cu alloys and tombstoning rate [84].

lower surface tension spreads easier and faster, thus is more sensitive toward unbalanced temperature distribution, and consequently more prone to tombstoning.

Furthermore, the tombstoning is found to be governed by the melting behavior of solder [85]. Hypothetically, a pasty solder with large fraction of solid at onset of melting will exhibit a slow wetting, and consequently will not be able to develop a significantly unbalanced wetting force instantaneously. Accordingly, the greater the fraction of the solid at onset of melting, the lower the tombstoning rate will be expected. This fraction of solid can be estimated from differential scanning calorimetry data based on symmetry approximation, as exemplified by Figure 3.39.

The tombstoning rate indeed decreases with increasing fraction of solid at onset of melting, as shown in Figure 3.40 [63, 79]. This relationship allows the control of tombstoning with use of alloys with properly balanced melting behavior [85] and mechanical property.

The tombstoning can also be curtailed with pad design [1]. For instance, Motorola modified the pad shape to back to back U pads, with 11 mils spacing, 15.5-mil pad depth, and 22-mil width for 0402. This design reduces the fillet size, thus effectively controls the tombstoning of 95.5Sn–3.8Ag–0.7Cu solder paste caused by pads with unbalanced blind via underneath [86].

3.7.3. Fillet Lifting

Fillet lifting is lifting of solder fillet from pads at the end of soldering process. It mostly occurs at wave soldering. The basic cause of fillet lifting is thermal expansion mismatch. Alloys with larger pasty range are more susceptible to fillet lift, hence 96.5Sn–3.5Ag shows lowest tendency and 91.9Sn–3.4Ag–4.7Bi shows very severe tendency among the promising lead-free solder candidates investigated by National Center for Manufacturing Sciences (NCMS) [87, 88].

Fillet lifting is not only confined to wave soldering. At reflow soldering, fillet lifting, also termed as lift-off, has also been reported for Sn–Ag–Bi–Cu system

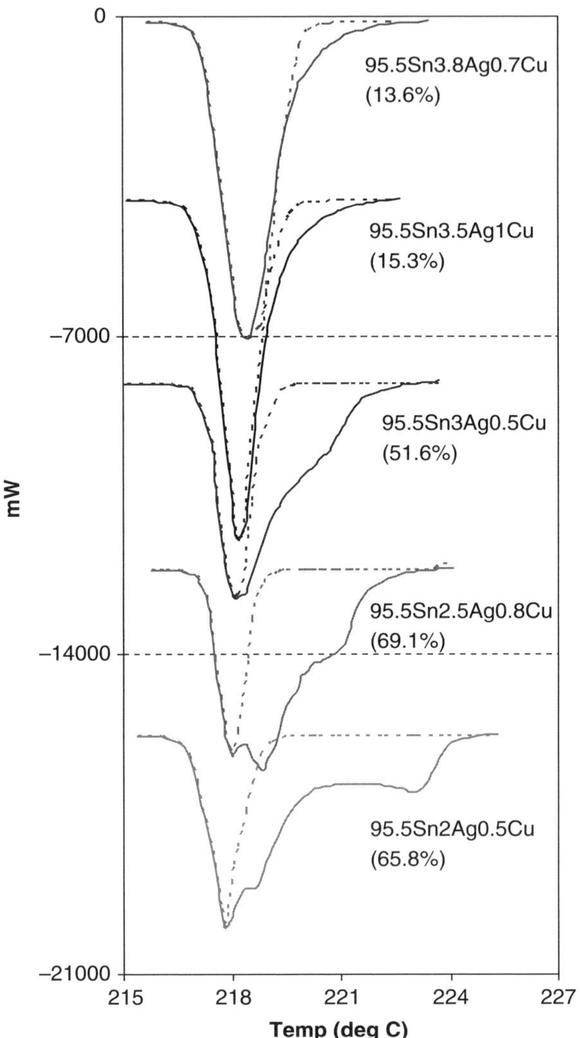

Figure 3.39. Determination of fraction of solid (expressed in percentage) of Sn–Ag–Cu alloys at onset of melting based on symmetry approximation [84].

[89, 90]. Figure 3.41 shows the solder joints of LSI packages, with whole row of fillets lifted from the pads on PCB. Upon close examination of contact area between solder fillets and pads, rich Bi redistribution in the vicinity of interface of solder/Cu pad was found for lifted joints or weak joints (<3N versus >6N) [89].

The mechanism of fillet-lifting for Bi-bearing alloy was investigated by Suganuma [90]. Although his work is mainly on wave soldering, the conclusion is considered equally applicable to reflow soldering. Upon solidification, Bi is enriched at the liquid interface region between solder fillet and Cu land by the dendrite

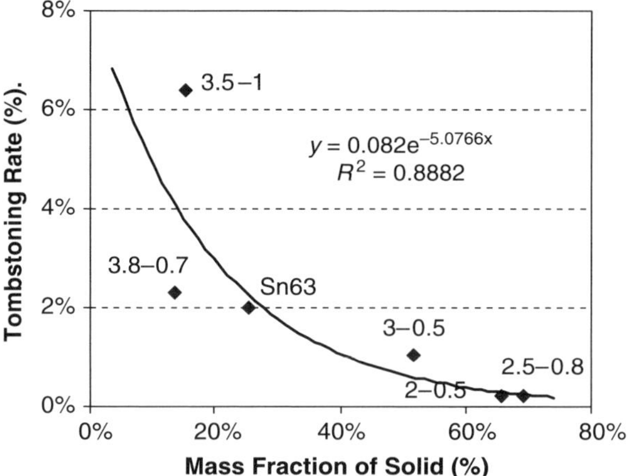

Figure 3.40. Relation between mass fraction of solid at onset of melting and tombstoning rate of Sn–Ag–Cu system. The x and y in data label refer to Ag and Cu content respectively of given SAC alloys [63, 79].

formation, as shown in Figure 3.42. The diffusion distance is only a few microns. Substantial amount of Bi along the interface results in a reduced melting temperature, as suggested by the Bi–Sn binary phase diagram (see Figure 3.43), thus delays the solidification of this liquid phase. Furthermore, formation of dendrite skeleton may suck residual liquid and aggravate this lifting process.

The liquid wicking back caused by dendrite skeleton formation is also reported by Harrison and Vincent [91] on Sn–Ag–Cu–Sb system. Since most of the main stream options of lead-free alternatives, such as eutectic Sn–Ag, eutectic Sn–Cu, and ternary eutectic Sn–Ag–Cu, are high in tin content and therefore have a strong tendency to form tin dendrites, it is reasonable to speculate that those alloys are more prone to exhibit fillet lifting when compared with eutectic Sn–Pb.

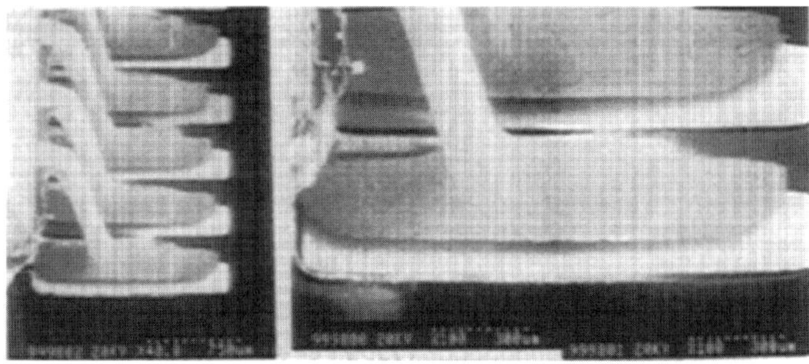

Figure 3.41. Ruptures in the vicinity of Sn–Ag–Bi–Cu solder/Cu pad interface [89].

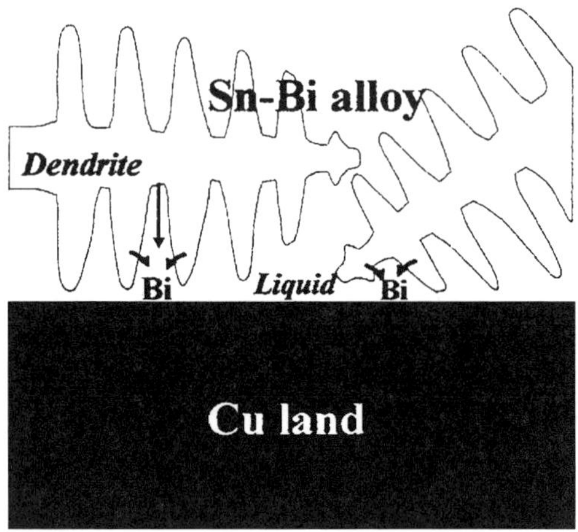

Figure 3.42. Enrichment of Bi at solder/copper pad interface due to dendrite formation [90].

3.7.4. Residue Cleaning

Cleaning of post-reflow flux residue of lead-free solder pastes is fairly challenging [92–95]. Mao et al. [92] initiated work using two no-clean flux formulations with a conventional eutectic Sn–Pb solder paste as a baseline. Reflowed boards were cleaned using a variety of different chemistries and processes, and cleanliness after defluxing was assessed by ionic contamination testing, and was expressed as cleaning efficiency.

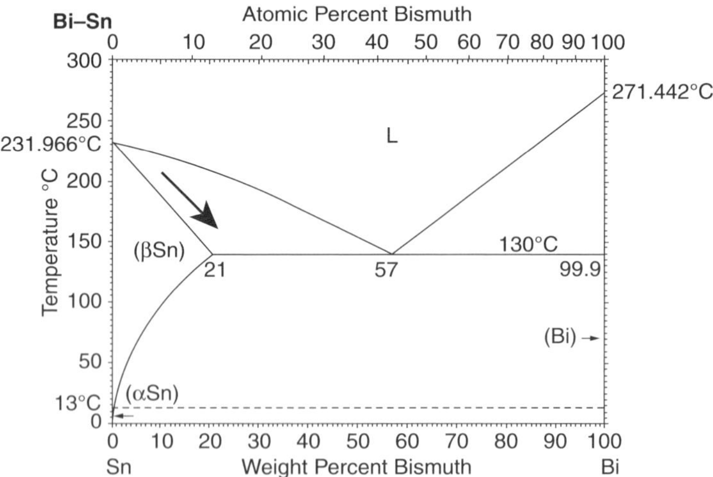

Figure 3.43. Bi–Sn binary phase diagram. Bi-enrichment will result in decrease in solidus temperature.

TABLE 3.17. Effect of Flux Chemistry/Alloy Type and Reflow Temperature on Cleaning Efficiency [92]

Paste	Reflow Temperature (°C)	CE_{IC}
Lead-containing	215	21.24
	235	14.35
Lead-Free	235	5.09
	255	1.12

Cleaning efficiency is a calculation relating the results from uncleaned boards with those of cleaned boards. Lower ionic contamination levels indicate improved cleaning. Mathematically, cleaning efficiency (CE) is expressed as shown below:

$$CE_{IC} = \frac{\text{Ionic level for uncleaned board}}{\text{Ionic level for cleaned board}}$$

If $CE = 1$, no change in cleanliness was found between the uncleaned and cleaned boards. If $CE > 1$, improvement in cleanliness was found, and if $CE < 1$, the cleaned boards were, in actuality, less clean than the unclean boards. The test data are shown in Tables 3.16 to 3.18.

Results indicate that higher reflow temperature results in a poorer cleaning efficiency (see Table 3.17), obviously due to the greater extent of residue charring. The deterioration of cleaning efficiency accelerates with increasing temperature. The cleaning efficiency of the lead-free system is poorer than the Sn–Pb system (see Table 3.17), due to higher reflow temperature, higher activity of tin, and more complicated flux chemistry of lead-free flux system. Overall, the semiaqueous cleaning system is the most efficient cleaning chemistry, followed in order by the neutral pH

TABLE 3.18. Effect of Cleaning Chemistry on Cleaning Efficiency [92]

Cleaner System	Process	CE_{IC} (Overall)	CE_{IC} (Overall Average)	CE_{IC} (Pb-Free)	CE_{IC} (Pb-Free Average)
Saponified aqueous	C1	13.19		0.88	
	C4	9.41		1.64	
	C7	4.00	8.86	0.80	1.11
Neutral pH aqueous	C2	13.33		1.15	
	C5	13.49		1.48	
	C6	8.40	11.74	1.73	1.46
Semiaqueous	C3	21.12	21.12	16.73	16.73
Co-solvent	C8	0.67	0.67	0.41	0.41

TABLE 3.19. Effect of Process Parameter on Cleaning Efficiency [92]

Process Parameter	CE_{IC} (Overall)	CE_{IC} (Pb-Free)
Centrifugal	15.88 (13.26[a])	6.26 (1.02)
Spray-in-air (batch)	11.45	1.56
Spray-in-air (in-line)	6.20	1.26
Vapor degrease	0.67	0.41

[a]Data include saponified aqueous and neutral pH aqueous systems only, same as spray-in-air processes. It does not include semiaqueous cleaning chemistry data.

aqueous system, the saponified aqueous system, and a co-solvent vapor degreasing system, as shown in Table 3.18. For lead-free cleaning, the semiaqueous chemistry is the only effective cleaning chemistry of those tested. As to the cleaning process, the centrifugal process was the most efficient, followed in order by spray-in-air (batch), spray-in-air (in-line), and vapor degreasing, as shown in Table 3.19. However, the ranking of process efficiency may be affected by the cleaner concentration and cleaning time factors. In the former case, a higher cleaner concentration is observed to be associated with higher cleaning efficiency as well, thus contributing to the ranking of process efficiency. For lead-free cleaning, the ranking of processes

TABLE 3.20. Effect of Reflow Temperature on Number of CAF [97]

Flux	#CAF at 201°C Reflow	#CAF at 241°C Reflow
Polyethylene glycol-600(PEG)	90	55
PEG/HCl	None	None
PEG/HBr	None	None
Polypropylene glycol 1200 (PPG)	None	455
PPG/HCl	None	379
PPG/HBr	1	423
Polyethyiene propyiene glycol 1800 (PEPG 18)	1	406
PEPG 18/HCl	10	135
PEPG 18/HBr	9	279
Polyethyene propyene glycol 2600 (PEPG 26)	None	91
PEPG 26/HCl	6	218
PEPG 26/HBr	None	51
Glycerine (GLY)	None	56
GLY/HCl	None	583
GLY/HBr	3	104
Ocyl phenol ethoxylate (OPE)	None	83
OPE/HCl	14	62
OPE/HBr	2	599
Linear Aliphatic Polyether (LAP)	None	Not tested
LAP/HCl	15	203
LAP/HBr	None	272

becomes batch spray > in-line spray > centrifugal > vapor degreasing. The cleaning process is not as critical as cleaning chemistry, and the semiaqueous cleaner is superior in cleaning efficiency when dealing with lead-free cleaning.

3.7.5. Thermal Damage

Perhaps the most challenging task of implementing lead-free soldering is assuring no thermal damage caused by the high soldering temperature. NEMI reported that a general increase in internal delamination damage was observed by C-SAM with elevated lead-free reflow temperature (247°C versus 218°C). It was stipulated that increase in baking time may be a solution to reduce delamination tendency for TSOP at higher reflow temperature [33]. NEMI also reported that some lead-free coupons on some laminate materials started to fail after three times reflow, and the low T_g materials were all open after six times reflow. In general, the moisture sensitivity level MSL was found to degrade by one level for every 5–10°C increase in reflow peak temperature, and degradation of MSL may increase with increasing profile dwell time above 200°C [96].

The thermal damage is also reflected by the conductive anodic filament (CAF) phenomenon. CAF formation is a failure mode for printed wiring boards in which a conductive filament forms along the epoxy/glass interface growing from anode to cathode. Presence of CAF often causes lowering in surface insulation resistance SIR value. Turbini et al. observed that a higher board processing temperature resulted in increased numbers of CAF for most of the fluxes tested, as shown in Table 3.20 [97].

With 260°C being the probable reflow peak temperature [96], it is obvious that newer molding compounds and laminate materials are desired in order to prevent thermal damage.

3.8. SUMMARY

Lead-free solder paste technology is a critical element in implementing lead-free soldering. At this stage, lead-free solder pastes have been successfully developed with satisfactory performance. Compared with eutectic tin-lead solder pastes, lead-free pastes in general exhibit a comparable deposition performance but a narrower soldering processing window. Challenges remain in order to widen the processing window.

REFERENCES

1. N.-C. Lee, *Reflow Soldering Processes and Troubleshooting: SMT, BGA, CSP, and Flip Chip Technologies*, published by Newnes, an imprint of Butterworth-Heinemann, pp. 270, 2002.

2. A. Rae, J. Belmonte, and L. Hozer, *Real-Life Tin–Silver–Copper Alloy Processing*, Apex, March 29–April 2, 2003, Anaheim, CA.
3. Y. Toyoda, The latest trends in lead-free soldering, ISEPT, pp. 434–438, August 8–11, 2001, Beijing, China.
4. K. Suganuma, Japan Leadfree 2001. ISIR, Osaka University.
5. N.-C. Lee, Lead-free soldering—Where the world is going, *Adv. Microelectroni.* September/October 1999.
6. N.-C. Lee, Getting Ready for Lead Free Solders, European Surface Mount Conference, Brighton, UK, 1996.
7. J. H. Lau, C. P. Wong, N.-C. Lee, and S. W. Ricky Lee, *Electronics Manufacturing with Lead-Free, Halogen-Free & Conductive Adhesive Materials*, McGraw-Hill, New York, 2003.
8. Japan Electronic Industry Development Association, Challenges and Efforts Toward Commercialization of Lead-Free Solder—Roadmap 2000 for Commercialization of Lead-Free Solder, version 1.3.
9. J-STD-006, General Requirements and Test Methods for Electronic Grade Solder Alloys and Fluxed and Non-Fluxed Solid Solders for Electronic Soldering Applications, 1994.
10. A. Lawley, Atomization—The Production of Metal Powders, Metal Powder Industries Federation, Princeton, NJ, 1992, p. 166.
11. E. E. de Kluizenaar, Surface oxidation of molten soft solder: An Auger study, *J. Vac. Sci. Technol.* **A1**(3), 1480–1485, 1983.
12. J. C. Ivankovlts, B. M. Adams, and Y. W. Loo, *Controlled Atmospheres for Soldering Processes*, Air Products and Chemicals, Inc., Pub. No. 325-9313, Allentown, PA, 1991.
13. *CRC Handbook of Chemistry and Physics*, 75th edition, D. R. Lide, CRC Press, Inc. 1994, pp. 5–21.
14. T. Yamaguchi and T. Enomoto, *Tin-Zinc Solder Paste*, Apex, San Diego, CA, January 14–18, 2001.
15. R. Kurz and E. Kleiner, Uber das Anlaufverhalten von flussigen Zinn-Blei-Loten, *Z. Werkstoffkunde/J. Mat. Technol.* **2**(8), 418–422, 1971.
16. R. J. Klein Wassink, *Soldering in Electronics*, Electrochemical Publications, 1984, p. 110.
17. D. D. Hillman and L. S. Chumbley, Characterization of tin oxidation products using sequential electrochemical reduction analysis (SERA), *Soldering & Surface Mount Technol.* **18**(3), 31–41, 2006.
18. D. Tench, *The Mechanics of Solder Alloy Wetting & Spreading*, Van Nostrand Reinhold, New York.
19. T. Baird, J. Fryer, and E. Riddell, Oxidation of thin tin films, *Surf. Sci.* **28**(2), 525–540, 1971.
20. A. Lin, N. Armstrong, and T. Kuwana, X-ray photoelectron/Auger electron spectroscopic studies of tin and indium metal foils and oxides, *Anal. Chem.* **49**(8), 1230–1238, 1977.
21. A. Z. Miric and A. Grusd, Lead-free alloys, *Soldering Surf. Mount Technol.* **10/1**, 19–25, 1998.
22. Data from Indium Corporation of America.
23. J. Glazer, Metallurgy of low temperature Pb-free solders for electronic assembly, *International Mater. Rev.* **40**(2), 65–93, 1995.

24. F. A. Mohamed and T. G. Langdon, *Philos. Mag.* **32**, 697–709, 1975.
25. B. Huang and N.-C. Lee, Prospect of Lead Free Alternatives for Reflow Soldering, IMAPS'99, Chicago, IL.
26. A. Butterfield, V. Visintainer, and V. Goudarzi, Lead Free Solder Flux Vehicle Selection Process, SMTA International, Chicago, IL, September 20–24, 2000.
27. G. R. Biard, Effect of alloy and composition on shelf life of water soluble solder pastes, Apex, March 29–April 2, 2003, Anaheim, CA.
28. X. Bao and N. C. Lee, Engineering solder paste performance via controlled stress rheology analysis, in *Proceedings of Surface Mount International*, San Jose, CA, September 1996.
29. W. Rubin and M. Warwick, Some developments in solder cream technology, *J. SMT*, 17–24, 1990.
30. M. Xiao, K. J. Lawless, and N. C. Lee, Prospects of solder paste applications in ultra-fine pitch era, in *Proceedings of Surface Mount International*, San Jose, CA, August 1993.
31. A. Bhave and D. Santo, Solder paste printing—A competitive study, A SPIR Project Report for Indium Corporation of America, July 2003.
32. Q. Sheng, C. Bradshaw, and S. Kwiatek, Properties of lead free alloy and performance properties of lead free no clean solder paste, Apex, January 19–24, 2002, San Diego, CA.
33. J. Bath, Lead-free reflow process experience (lead-free process team), IPC/NEMI Symposium on Lead-Free Electronics, September 18–19, 2002, Montreal, Canada.
34. C. Ashmore and R. Goldsmith, Investigating Mass Imaging Lead Free Materials Using Enclosed Print Head Technology, Nepcon West/Fiberoptic Automation Expo, S22P03, Jan Jose, CA, December 3–6, 2002.
35. R. Zhang and W.-B. Chen, R. Doraiswami, S. Sankararaman, et al., Preliminary study on lead-free sn42–Bi57–Ag1 35.66, in *Advances in Fine Pitch Lead Free Assembly Process*, 53rd Electronic Components & Technology Conference, S20P5C, New Orleans, LA, May 27–30, 2003.
36. Product data sheet on NC-SMQ81 solder paste with 58Bi42Sn of Indium Corporation of America.
37. R. S. Venkatesh, K. Srihari, A. Mazloom, S. Kamath, and J. Craik, Solder paste evaluation for lead-free assembly, Apex, San Diego, CA, January 14–18, 2001.
38. J.-F. Gong, G.-W. Xiao, P. C. H. Chan, R. S. W. Lee, and M. M. F. Yuen, A reliability comparison of electroplated and stencil printed flip-chip solder bumps based on UBM related intermetallic compound growth properties, 53rd Electronic Components & Technology Conference, S16P6C, New Orleans, LA, May 27–30, 2003.
39. R. Doraiswami, S. Sankararaman, W. Kim, et al., Advances in fine pitch lead free assembly process, 53rd Electronic Components & Technology Conference, S20P5C, New Orleans, LA, May 27–30, 2003.
40. N.-C. Lee, Lead-free soldering of chip scale packages, *Chip Scale Rev.*, March/April, 2000.
41. N.-C. Lee, Lead-free soldering and low alpha solders for wafer level interconnects, SMTA International, 2000, Chicago, IL.
42. B. Huang and N.-C. Lee, Low cost solder bumping via paste reflow for area array packages, SMTA International, Chicago, IL, September 2001.
43. Galden PFPE Vapor Phase Fluids, data sheet of Solvay Solexis Inc., modified 12/13/02.

44. B. Willis, Pin-Hole Intrusive Reflow Design and Assembly, short course in SMTA International, Chicago, IL, September 21–25, 2003.
45. T. Herzog, S. Rudolph, and K.-J. Wolter, Process Capability, Wetting Behavior and Temperature Dependent Shear Strength of Alternative Lead Free Solder Joints, SMTA International, September 22–26, 2002, Chicago, IL.
46. M. Miyazaki, K. Oki, S. Nomura, and T. Takei, Lead-free soldering system in reflow, wave and wire soldering technology under conventional soldering temperature, Apex, San Diego, CA, January 14–18, 2001.
47. P. N. Houston, B. J. Lewis, D. F. Baldwin, and P. Kazmierowicz, Taking the pain out of Pb-free reflow, Apex, March 29–April 2, 2003, Anaheim, CA.
48. H.-J. Albrecht, K. Wilke, and D. S. Brodie, Board Level Reliability of Lead-Free Soldered Interconnections, SMTA International, September 22–26, 2002, Chicago, IL.
49. S. Shina, H. Belbase, K. Walters, T. Bresnan, P. Biocca, T. Skidmore, D. Pinsky, P. Provencal, and D. Abbott, Selecting material and process parameters for lead-free SMT soldering using design of experiments techniques, Apex, San Diego, CA, January 14–18, 2001.
50. W. Johnson, R. Lugo, S. V. Sattiraju, and G. Jones, Improved thermal process control for lead-free assembly, Apex, San Diego, CA, January 14–18, 2001.
51. M. Kelly, D. Colnago, V. Sirtori, J. Bath, S. K. Tan, L. Hook Teo, C. Grosskopf, K. Lyjak, C. Ravenelle, and E. Kobeda, Component temperature study on tin-lead and lead-free assemblies, SMTA International, September 22–26, 2002, Chicago, IL.
52. N. Lycoudes and M. Freedman, JEDEC lead free position, International Conference on Lead-Free Electronic Components and Assemblies Technical Conference, San Jose, CA, May 1–2, 2002.
53. IPC/JEDEC J-STD-020B, July 2002.
54. D. K. Joo and J. Yu, Effects of microstructure on the creep properties of the lead-free Sn–3.5Ag–Cu Solders, 52nd ECTC, S29-P5, San Diego, CA, May 28–31, 2002.
55. W. Yin, N.-C. Lee, F. Dimock, and K. Mattson, Effect of flux and cooling rate on microstructure of flip chip SAC bump, SMTA International, Chicago, IL, September 2005.
56. N.-C. Lee, Optimizing reflow profile via defect mechanisms analysis, IPC Printed Circuits Expo'98.
57. K. Seelig and D. Suraski, Finally! Practical guidelines for achieving successful lead-free assembly, Apex, March 29–April 2, 2003, Anaheim, CA.
58. C. Chiu and N.-C. Lee, Options and concerns of BGA solder bumping, ISHM, pp. 408–416, Philadelphia, PA, October 14–16, 1997.
59. C. Chiu, N.-C. Lee, K. Randle, and C. Parrish, Voiding in BGA at solder bumping stage, ISHM, pp. 462–471, Philadelphia, PA, October 14–16, 1997.
60. W. O'Hara and N.-C. Lee, Voiding mechanism in BGA assembly, Best paper of Session for SMT-BGA, ISHM, pp. 24–30, Los Angeles, CA, October 24–26, 1995.
61. W. Hance and N.-C. Lee, Voiding mechanisms in SMT, China Lake's 17th Annual Electronics Manufacturing Seminar, 1993.
62. H. Jo, B. Nieman, and N.-C. Lee, Voiding of lead-free soldering at microvia, IMAPS, Denver, CO, September 4–6, 2002.
63. A. Dasgupta, B. Huang, and N.-C. Lee, Effect of lead-free alloys on voiding at microvia, Apex, Anaheim, CA, February 23–27, 2004.

64. Y. Liu, W. Manning, B. Huang, and N.-C. Lee, A model study of profiling for voiding control at lead-free reflow soldering, Nepcon Shanghai, China, April 11, 2005.
65. N.-C. Lee, Critical parameters in voiding control at reflow soldering, *Chip Scale Rev.*, August–September 2005.
66. N.-C. Lee, Love Triangle: How will lead-free affect the relationship among components, alloys and fluxes?, *Circuits Assembly*, October & December 2005.
67. J. L. Evans and B. Dahle, Evaluating the use of next generation reflow profile control for high volume electronics manufacturing, Apex, March 29–April 2, 2003, Anaheim, CA.
68. S. Saiyed, D. Santos, and J. A. McLenaghan, SMT assembly process comparison of Pb-free alloy systems, Apex March 29–April 2, 2003, Anaheim CA.
69. C. Jensen, F. Kuhlman, and M. Pepples, Robust optimization of a lead free SMT process, Apex, March 29–April 2, 2003, Anaheim, CA.
70. C. Hunt, D. Lea, and S. Adams, Evaluation of the comparative solderability of lead-free solders in nitrogen, Part II, SMTA International, September 22–26, 2002, Chicago, IL.
71. P. Jaeger and N.-C. Lee, A model study of low residue no-clean solder paste, Nepcon West, Anaheim, CA, February 1992.
72. C. Handwerker, Lead free alloy selection criteria, data, and modeling, IPC/NEMI Symposium on Lead-Free Electronics, September 18–19, 2002, Montreal, Canada.
73. D. K. Joo and J. Yu, Effects of microstructure on the creep properties of the lead-free Sn–3.5Ag-Cu solders, 52nd ECTC, S29-P5, San Diego, CA, May 28–31, 2002.
74. J. Sohn, Lead free solder joint reliability overview, IPC/NEMI Symposium on Lead-Free Electronics, September 18–19, 2002, Montreal, Canada.
75. P. Snugovsky, Z. Bagheri, M. Kelly, and M. Romansky, Solder joint formation with Sn–Ag–Cu and Sn–Pb Solder Balls and Pastes, SMTA International, September 22–26, 2002, Chicago, IL.
76. P. Snugovsky, Z. Bagheri, M. Kelly, and M. Romansky, Solder joint formation with Sn–Ag–Cu and Sn–Pb solder balls and pastes, SMTA International, September 22–26, 2002, Chicago, IL.
77. J. Jessen, X-ray imaging of lead free solder, Etronix, Anaheim, CA, March 1, 2001.
78. J. Lau, C. P. Wong, N.- C. Lee, and R. Lee, *Electronics Manufacturing with Lead-Free, Halogen-Free, and Conductive-Adhesive Materials*, McGraw-Hill, New York, 2002.
79. H. Jo, B. Nieman, and N.-C. Lee, Voiding of lead-free soldering at microvia, IMAPS, Denver, CO, September 4–6, 2002.
80. W. Peng, S. Dunford, P. Viswanadham, and S. Quander, Microstructural and performance implications of gold in Sn–Ag–Cu–Sb interconnections, 53rd Electronic Components & Technology Conference, S20P1C, New Orleans, LA, May 27–30, 2003.
81. U. Ray, I. Artaki, D. W. Finley, G. M. Wenger, T. Pan, H. D. Blair, J. M. Nicholson, and P. T. Vianco, Assessment of circuit board surface finishes for electronic assembly with lead-free solders, SMI 96, September 10–12, 1996, San Jose, CA.
82. IPC-7095 Design and assembly process implementation for BGAs. August, 2000.
83. D. Geiger, F. Mattsson, D. Shangguan, M. T. Ong, P. Wong, M. Wang, T. Castello, and S. Yi, Process Characterization of PCB Assembly Using 0201 Packages with Lead-Free Solder, Nepcon West/Fiberoptic Automation Expo, S18P03, San Jose, CA, December 3–6, 2002.

84. B. Huang and N.-C. Lee, Conquer tombstoning in lead-free soldering, Apex, Anaheim, CA, February 23–27, 2004.
85. B. Huang and N.-C. Lee, patent pending.
86. Lead-Free QuickStart Program provided by Indium Corporation of America and Motorola.
87. C. Handwerker, NCMS lead free solder project: A national program, NEMI Lead Free Solder Meeting, Chicago, May 25, 1999.
88. Lead-Free Solder Project Final Report, NCMS Report 0401RE96, August 1997.
89. T. Nakatsuka, K. Serizawa, T. Soga, H. Shimokawa, and A. Nishimura, Reliability of Pb-free solder joints of surface-mounted LSI packages after flow-soldering, IMAPS, pp. 330–335, Boston, MA, September 20–22, 2000.
90. K. Suganuma, Mechanism and prevention of lift-off in lead-free soldering, IMAPS, pp. 325–329, Boston, MA, September 20–22, 2000.
91. M. R. Harrison and J. H. Vincent, IDEALS: Improved design life and environmentally aware manufacturing of electronics assemblies by lead-free soldering, SSTC, 1999.
92. R. Mao, N.-C. Lee, B. A. Bivins, A. A. Juan, and J. Wadford, Latest developments in post-solder cleaning of lead-free solder paste residues, Apex 2002, San Diego, CA, January 20–24, 2002.
93. N.-C. Lee and M. Bixenman, Lead-free: How flux technology will differ for lead-free alloys & its impact on cleaning, *Etronics*, March, 2001.
94. B. A. Bivins, A. A. J., B. Starkweather, N.-C. Lee, and S. Negi, Post-solder cleaning of lead-free solder paste residues, SMT International 2000, Chicago, IL.
95. K. Seelig and D. Suraski, A study of lead-contamination in lead-free electronics assembly and its impact on reliability, SMTA International, September 22–26, 2002, Chicago, IL.
96. Lead-Free Component Team, Component implications of lead-free reflow assembly, IPC/NEMI Symposium on Lead-Free Electronics, September 18–19, 2002, Montreal, Canada.
97. J. Turbini, W. R. Bent, and W. J. Ready, Impact of higher melting lead-free solders on the reliability of printed wiring assemblies, SMTA International, Chicago, IL, September 20–24, 2000.

CHAPTER 4

Impact of Elevated Reflow Temperatures on Component Performance

RICHARD D. PARKER, JACK McCULLEN, NICK LYCOUDES, and
R. J. ARVIKAR

4.1. INTRODUCTION TO COMPONENT "LEAD-FREE" ISSUES

The transition to lead (Pb)-free assembly of electronic systems will affect components in two key ways: (1) The higher melting temperature of Pb-free solders will result in increased thermal exposure to components and (2) Pb-containing materials will be replaced with acceptable Pb-free alternatives.

The European Union (EU) directive on Pb-free electronics has been interpreted to mean that all Pb-containing alloys must be replaced with Pb-free materials in the component construction. For the last 30 years, component terminals have been plated with a Sn–Pb-solderable surface finish. This must now be replaced with alternatives, such as pure tin (Sn), gold (Au), nickel–palladium–gold (NiPdAu), or high-Sn alloys. Each of these finishes has its own set of potential issues for the end user, such as tin whiskering (see Chapter 7), added cost, and availability. If a package contains internal soldered connections, the solder alloy is also expected to be Pb-free. However, an anomaly to this situation is that in certain situations, very high Pb content (>85% Pb) solders are exempt under the European RoHS legislation where there is not an adequate high-temperature solder replacement currently available. In general, internal soldered connections—which are common in many types of package die, coils, and crystals—should be Pb-free.

The industry has not chosen one surface finish as universal, but, as of this writing, it appears that pure Sn will be the terminal surface finish of choice in some 85% (by volume) of electronic components produced, with NiPdAu or NiPd supplying about 15%.

The second major effect of the Pb-free transition on components is the increased thermal exposure the component will undergo during assembly onto a substrate.

Lead-Free Electronics. Edited by Bradley, Handwerker, Bath, Parker, and Gedney

While there are a large number of Pb-free solder alloy choices, alloys in the Sn–Ag–Cu family have been the focus of iNEMI and the rest of the microelectronics world. This family of solders has a liquidus temperature of 217–221°C, which leads to a minimum reflow temperature of ∼230°C and has been the focal point for much work with components. The actual reflow temperatures used in any particular assembly reflow process will vary, depending on many factors unique to each board/component/process combination. Joining temperatures between 230°C and 260°C will be common processing requirements for components. This is 30–40°C hotter than the typical Sn–Pb eutectic solder reflow profiles. This temperature excursion occurs during both reflow soldering and wave soldering and, unless designed to withstand these temperatures, can have a significant effect on the component's long-term reliability.

4.2. MOISTURE/REFLOW IMPACT ON PACKAGED INTEGRATED CIRCUITS

The iNEMI Component Team focused on the thermal issues that affect packaged integrated circuits (ICs). These components were believed to be the most sensitive to the increase in processing temperatures that would be needed for Pb-free solder alloys. The team initially set out to quantify the temperature effects on molded packaged ICs. The predominate impact identified was manifested in the moisture sensitivity rating, as defined in the IPC/JEDEC J-STD-020 [1] for these packages. As a result, a concerted effort was made to understand the impact of various thermal

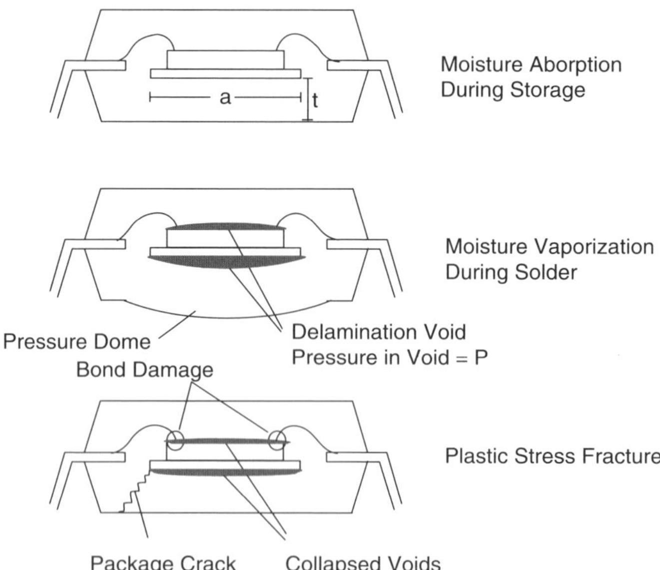

Figure 4.1. Schematic of moisture absorption and expansion in plastic package.

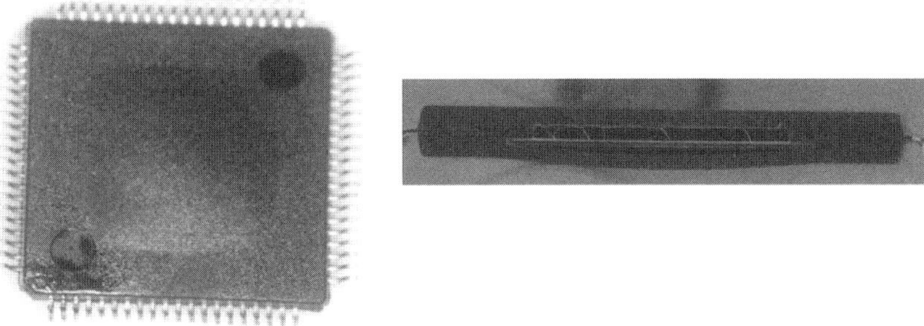

Figure 4.2. Plastic package bulge due to vapor pressure during solder reflow.

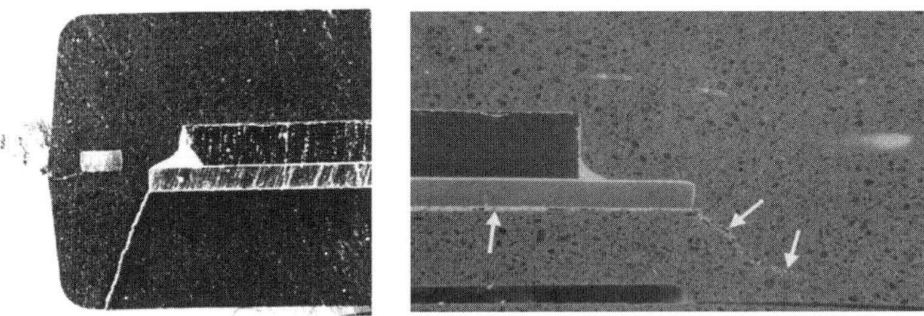

Figure 4.3. Plastic package cracks due to "popcorn" effect during solder reflow.

profile shapes, thermal mass, and the distributed thermal mass of circuit board on the component body temperature. The component body temperature was recognized as the critical parameter that controlled the reliability impact of absorbed moisture. As a result, it was necessary to minimize the body temperature while ensuring that the all component leads reached a high enough temperature to achieve good solder joints during the reflow soldering process. Several studies were conducted with a variety of component types and PWBs.

Solder reflow results in component temperatures going from room temperature to 230°C or more within minutes. This causes the vapor pressure from the moisture that is present in a molded package to increase dramatically. As shown in Figures 4.1–4.3, the increased vapor pressure can result in bulging, delamination, cracking, or other mechanical damage to the package, resulting in initial failure or compromised component life. The most severe case is when the pressure exceeds the strength of the package and a rupture occurs to vent the vapor pressure. This condition is called the "popcorn" effect.

4.3. IMPACT OF INCREASED SOLDER PEAK REFLOW TEMPERATURES ON MOISTURE SENSITIVITY LEVEL RATINGS

The Sn–Ag–Cu solder alloy family chosen to replace Sn–Pb solder has an approximate melting point of 217°C. This necessitates an increase in the soldering temperatures over that needed for the Sn–Pb reflow process where the solder has a 183°C melting point. To determine a reference temperature for testing the durability of molded IC packages as well as other electronic components, an extrapolation was done from existing knowledge of Sn–Pb eutectic alloy reflow processing. The typical Sn–Pb reflow process uses 210–225°C as a peak processing temperature, or a superheat of 27–42°C over the eutectic liquidus temperature. Applying this superheat to the new Sn–Ag–Cu alloy family liquidus temperatures yields an expected Sn–Ag–Cu

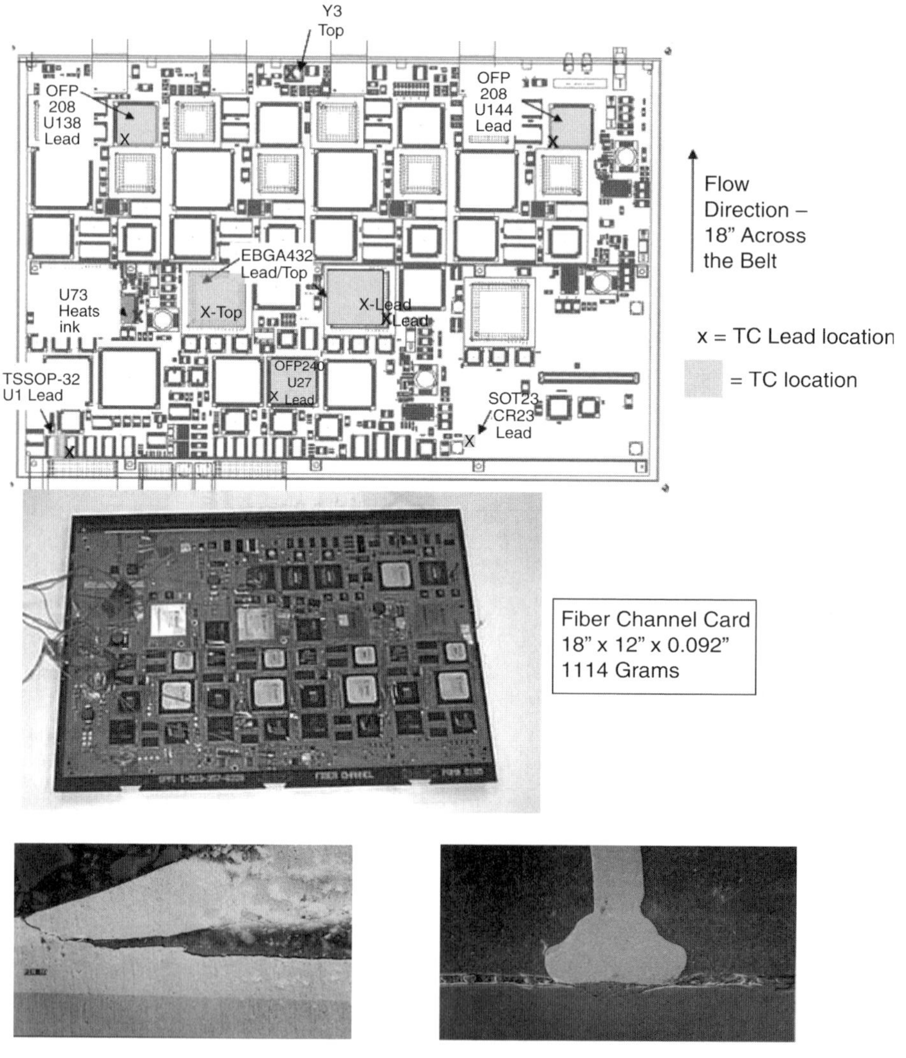

Figure 4.4. Diagram and photo of a test board showing thermocouple arrangements and component types evaluated.

process peak range of 244–259°C. From this, the iNEMI Component Team chose the upper limit, rounded to 260°C (+0/−5°C) as a starting point for evaluating the impact of the higher reflow temperatures on components. It was believed that this temperature would represent the worst-case condition for component assembled by the reflow process. Also of importance, but to a lesser degree, was the knowledge that a hotter reflow profile allows for a larger process window for the board assembler, by improving solder wetting and allowing a greater range of temperatures across the PCB while still meeting the minimum reflow temperature requirements [2]. Here the term "process window" refers to the temperature span that the circuit board and components will achieve during a reflow process. This is measured as the lowest solder joint temperature that is desired to the hottest component temperature that is allowed. Included in this span is the tolerance of the measurement system, the oven temperature accuracy, and so on. Therefore the higher the components temperature limit, the more "window" in temperature range an assembly can have.

The main challenge for an assembler is thermal management during reflow and, in particular, the minimization of temperature differences across the assembly during the soldering process. A balance must be achieved between board assembly costs (minimized by high throughput, high yields, and ability to build multiple products with minimal profile variations) and utilizing reflow profiles that are minimally destructive to the devices being soldered. By minimizing the temperature spread across the assembly, lower peak reflow temperatures can be used during assembly. The use of several thermocouples on the assembly will facilitate finding the coldest and hottest parts during reflow (Figure 4.4).

Using the coolest peak temperatures in the profile (Figure 4.12) will maximize a component's chance for surviving the reflow process. Research and experience with eutectic Sn–Pb assembly has shown that the higher the reflow temperature the worse the negative impact will be on plastic encapsulated surface mount packages and other components made with moisture-permeable materials [3, 4]. In some cases, older components simply cannot withstand the higher reflow temperatures even if

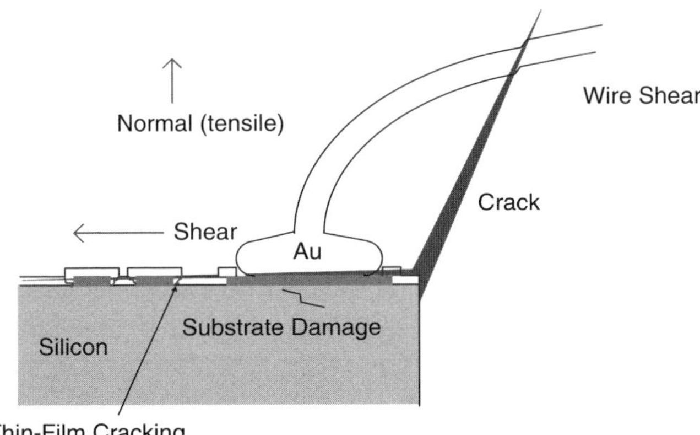

Figure 4.5. Effect of package cracking and delamination on wires, bonds and passivation.

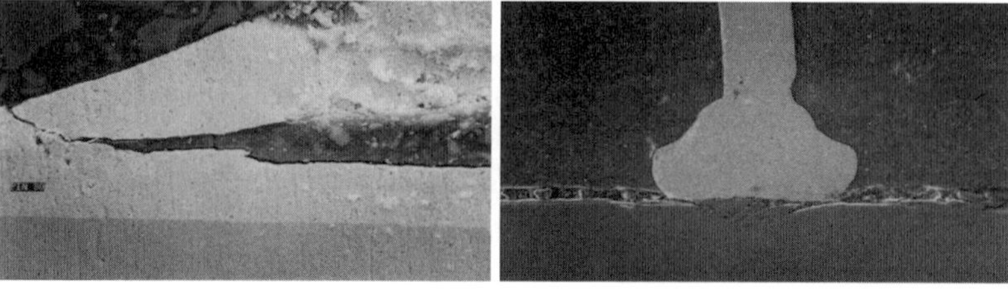

Die paddle delamination; broken down bond.

Die surface delamination; bond cratering.

Figure 4.6. Bond failures resulting from moisture/reflow induced delamination.

baked dry before reflow. The categories of parts susceptible to moisture-induced failure include all molded IC packages, both surface mount and leaded through-hole devices, and area array packages, such as organic substrate ball grid arrays (BGAs).

The water vapor pressure due to moisture inside a plastic IC packages increases rapidly when the package is exposed to the higher solder reflow temperatures. This pressure can cause internal delamination of the plastic body from the die and/or leadframe, internal cracks that do not extend to the outside of the package, bond damage, wire necking, bond lifting, die lifting, thin film cracking, and cratering beneath the bonds (Figures 4.5 and 4.6). Below are examples of bond failures resulting from delamination caused by excessive package moisture during reflow. Figure 4.7 shows the moisture/reflow effect on a BGA. Figure 4.8 shows bond wire damage resulting from an internal crack intersecting the wires.

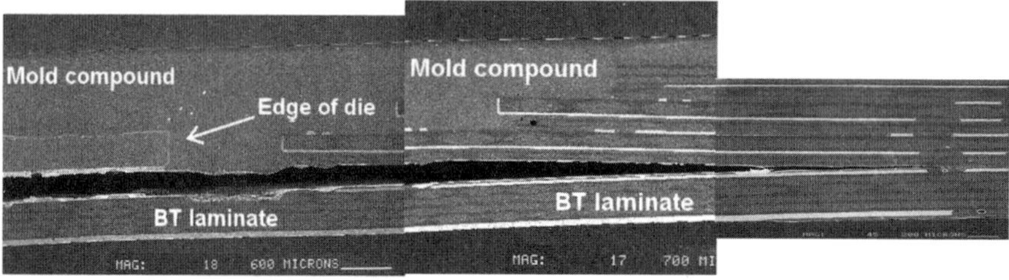

Figure 4.7. BGA substrate damage after reflow.

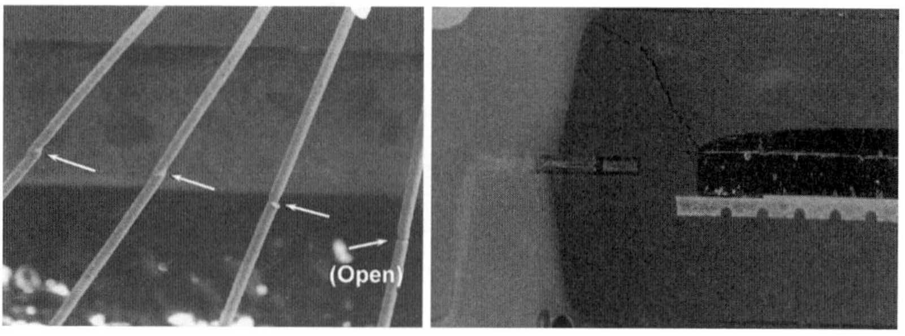

Figure 4.8. Bond wire damage from package crack.

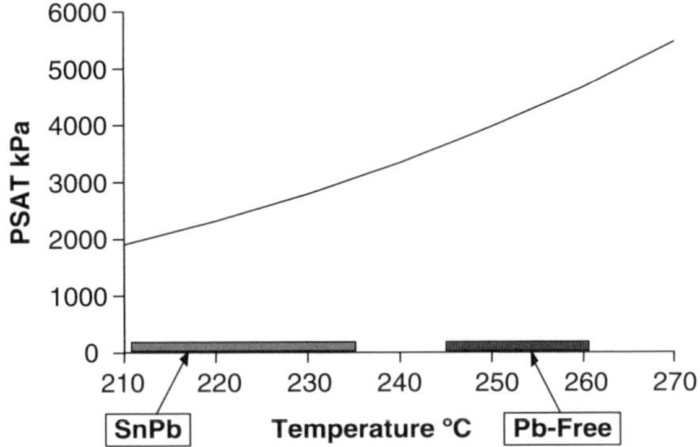

Figure 4.9. Vapor pressure versus temperature.

In the most severe case, the internal stress will exceed the mold compound/encapsulant strength and result in package cracks that extend to the outer surface of the package. This is commonly referred to as the "popcorn" effect because the internal stress causes the package to bulge and then crack with an audible "pop." Surface mount packaged devices soldered with a reflow cycle (which encompasses the majority of packages) are more susceptible to this problem than components exposed only to a wave soldering process. Component body temperatures can reach 200–260°C during reflow soldering, whereas topside component body temperatures during wave solder only reach temperatures of 130–150°C. The increased temperature needed to reflow Pb-free alloys significantly increases the vapor pressure inside the package (Figure 4.9) while reducing the strength of the mold compound (Figure 4.10). Using the following equation, we can determine the cavity pressure at an internal interface when we have certain relative humidity (RH) at the interface [4]:

$$P_{\text{cav}} \xrightarrow[\ell \to 0; w \to \infty \ (\text{or } t \to 0)]{} H_0 \times P_{\text{sat}}(T_0) \times \frac{S(T_0)}{S(T_1)}$$

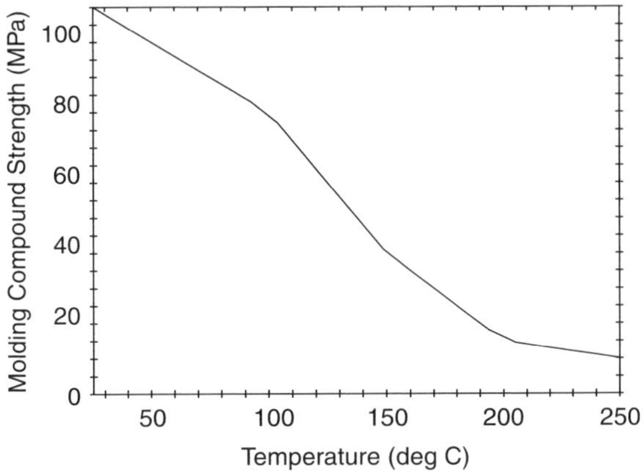

Figure 4.10. Mold compound strength versus temperature.

where H_0 is the relative humidity at T_0, T_0 is the temperature at which the RH is measured, T_1 is the reflow temperature, and S is the saturation coefficient. For example, if a package has reached equilibrium at 85°C and 85% RH and is exposed to reflow at 215°C, the pressure at the die paddle–mold compound interface is 15.3 atmospheres (1.6 MPa). If the same package is exposed to 260°C reflow, the cavity pressure is 34.2 atmospheres (3.5 MPa), a 123% increase.

Not all failures will be apparent after reflow. The "walking wounded" (components with internal damage) might not appear damaged in service until after many temperature cycles. Cracks and delaminations of the mold compound will redistribute stresses in the package. Shear stresses in the package increase, especially at die corners. Stresses normal to the die that are usually compressive become locally tensile at die edges, resulting in cracks and failures. Figure 4.11 shows the redistribution of stress in a package after cracking [3].

Degradation of the moisture sensitivity level (MSL) of plastic encapsulated semiconductor devices is the initial indicator for an increased probability of failures as the maximum reflow temperature increases. The degree of degradation of the MSL depends on many factors such as those listed in J-STD-020 [1]. This standard provides MSLs and the procedure used to rate packaged ICs for moisture exposure and has been adopted universally across the industry. The ratings then give the user a safe processing window relative to moisture exposure from the environment. The MSL rating determines the "floor life" (the allowable time period after removal from a moisture barrier bag, dry storage, or dry bake and before the solder reflow) of components in a typical factory environment of 30°C and 60% RH. The best MSL is level 1, which means that a component is not moisture-sensitive, MSL levels 2–6 indicate a stepped reduction in "floor life," with level 2 the best

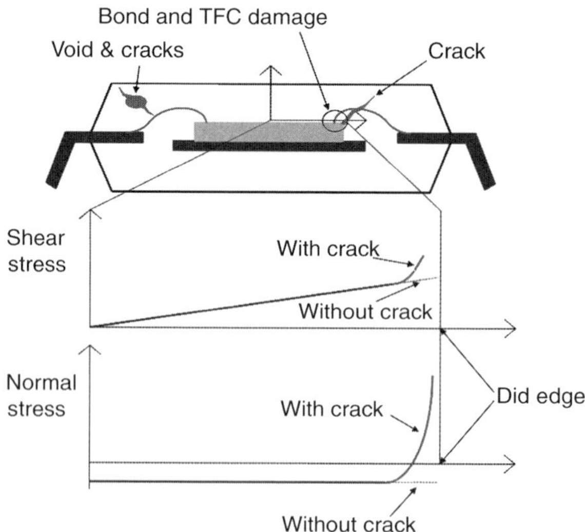

Figure 4.11. Package stress with and without crack.

TABLE 4.1. MSL Level Floor Exposure Limits from J-STD-020C

MSL Level	Floor Life	
	Time	Conditions
1	Unlimited	≤30°C/85% RH
2	1 yr	≤30°C/60% RH
2a	4 weeks	≤30°C/60% RH
3	168 h	≤30°C/60% RH
4	72 h	≤30°C/60% RH
5	48 h	≤30°C/60% RH
5a	24 h	≤30°C/60% RH
6	Time on label	≤30°C/60% RH

and level 6 the worst (see Table 4.1). Components that exceed their rated "floor life" must be baked prior to assembly using reflow.

4.4. IMPACT OF INCREASED SOLDER PEAK REFLOW TEMPERATURES

In discussions between IC suppliers and their customers, the degree of MSL degradation always comes up when discussing the highest temperature that can be used for successful soldering. As a result, the iNEMI Components Team set out to gather available information regarding the effect of higher process reflow temperatures (PRTs) on the MSL rating of semiconductor surface mount packages. In order to obtain data quickly, the team decided to obtain recent, existing data from members of iNEMI and JEDEC organizations rather than set up a new testing program. Four different suppliers provided test data from (a) their current production parts, (b) parts that were redesigned or had process modifications to enhance MSL ratings that could be implemented in the near term, or (c) significant design or material changes that were thought to significantly enhance MSL ratings, but would require implementation in the next generation of designs.

To understand the impact that an increase in reflow temperatures would have on molded packaged ICs, it was necessary to divide the total temperature increase into brackets—that is, temperature steps or ranges. These brackets (ranges) could then be used to test for pass/fail conditions following the procedures and criteria specified in J-STD-020. Passing ratings could then be associated with a peak reflow temperature limit.

The information obtained was in the form of MSL ratings for four peak reflow temperature (PRT) ranges. These ranges were (a) 220–225°C, (b) 230–235°C, (c) 250–255°C, and (d) 255–260°C. A variety of packages were tested using the MSL preconditioning and rating method per J-STD-020A, which was in force at the time. Although not all packages were evaluated at all PRTs, as it would have been if a full factorial design-of-experiments had been performed, the data indicate that MSL typically degrades by one level for every 5–10°C increase in PRT. The typical profile pattern referenced in J-STD-020A was used, but was modified to meet the PRTs described above. The failure criteria were from J-STD-020 and included the presence of cracking and/or internal delamination. Acoustic microscopy and, in some cases, electrical testing were employed to detect internal component damage.

The test data were compiled in a manner that protected the identity of the packaging house, which was irrelevant information for this benchmark study. This study served as a starting point to focus development in the appropriate areas. The data are summarized in Tables 4.3 and 4.4. The data contain various package type(s) (such as PLCCs or SOICs), with further delineation based on lead frame configurations and package structures. A package type with a given lead configuration in the tables is referred to as a "package test group."

The PRT shown per package type and structure are those that met that specific MSL. Note that as defined in J-STD-020, the PRT reflects the *component body temperature* and not the leadframe/board/solder joint temperature. Four temperature brackets were used in these tables: PRT of 220–225°C (standard for Sn–Pb soldering), PRT 235–240°C (minimum temperature for Sn–Ag–Cu soldering), PRT of 250–255°C, and PRT of 255–260°C (felt to be worst case for Sn–Ag–Cu soldering). A typical profile shape is shown in Figure 4.12, where Tp is the peak PRT.

The structure and design/materials of the packages were identified by using the symbol S (S stands for structure) with a number next to it indicating the version (vintage) of the structure. For example, S1 indicates the original (baseline structure existing at the time of the experiment) for a given package group. So S1 would be representative of parts that were in mass production at the time. S2 indicates an improved design—intended to enhance MSL ratings—but with reasonable

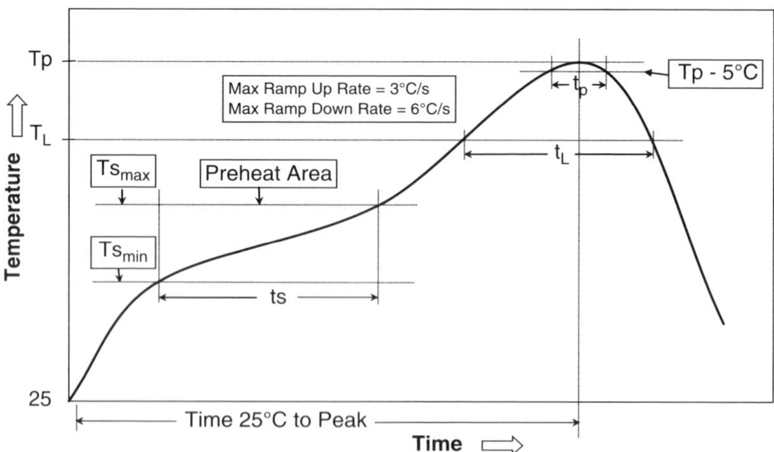

Figure 4.12. Typical reflow profile for MSL testing.

changes that might be implemented quickly on production parts. S3 is indicative of major design/material changes to substantially improve MSL levels. However, these more advanced improvements might only be practical to implement on new component releases. It should be noted that S1, S2, and S3 are relative and do not imply identical structures across all package test groups, because four different manufacturers supplied their current and improved designs. Thus S1 for supplier A is necessarily different from S1 for supplier B. It is also often true that S1 for two different part types from the same supplier are different. S1, S2, and S3 indicate the version (vintage) of a structure (design) within a particular package test group and are used to indicate the level of improvement of a structure. This method of grouping normalized the results between suppliers and part types and more fairly allowed comparisons of the performance between the different manufacturers represented. The four suppliers provided test data on a broad spectrum of components. Suppliers take some risk in supporting these kinds of tests because they are, by design, destructive and, taken out of context, can give a supplier a "black eye." Therefore, the data are necessarily presented without identification of which manufacturers supplied the data.

The data from the evaluation groups are shown in Tables 4.2 and 4.3. The data are presented by package type and lead count. For each structure tested the MSL at the tested temperature range is shown. For example, the 28-lead PLCC structure 1 (S1) MSL rating is level 1 at 220–225°C and degrades to a level 3 at 235–240°C. Components that can attain an MSL rating of 1 require little or no special handling during assembly. An MSL rating of 3 is widely considered to be the minimum acceptable for most manufacturing facilities. MSL ratings of 4 and lower require increasing levels of special handling, when used in a production environment. The reader should note that for most package types, the MSL level is decreased (higher number) as the solder PRT temperature is increased. There were many package types in this grouping that users would find undesirable if used at the higher PRT temperatures.

4.4.1 Discussion

Increasing the PRT degraded the MSL rating of nearly all molded IC packaged parts designated as S1 components—that is, parts that were then currently in manufacturing. The MSL of the various package types were evaluated, degraded by at least one level for every five to ten degrees centigrade increase of PRT.

The component manufacturers formed a small committee to analyze the results of the tests. This committee found that there was no single or simple solution to the problem. They agreed that any of the following attributes (factors) could affect the moisture sensitivity of a device and may require reclassification if they are changed:

- Encapsulation (mold compound or Glob Top) material/process
- Die attach material/process
- Die pad area and shape

TABLE 4.2. MSL Data Versus Component Type and Construction

			PRT Temperature Ranges						
			220–225	235–240	250–255	255–260			
			MSL Level						
Package Type	Number of Leads	Structure	1	2	2a	3	4	5	6
PLCC	28	S1	220–225			235–240			
		S2	235–240			255–260			
		S3	255–260						
PLCC	44	S1	220–225			235–240			
		S2	235–240			255–260			
		S3	255–260						
PLCC	44	S1				220–225	235–240	250–255	
PLCC	52	S1				220–225	235–240		
		S2				235–240			
		S3				255–260			
PLCC	68	S1				220–225	235–240		
		S2				235–240			
		S3				255–260			
PLCC	68	S1				220–225			235–240
SOIC-N	8	S1	220–225			235–240			
		S2	235–240			255–260			
		S3	255–260						
SOIC-N	14	S1	220–225			235–240			
		S2	235–240			255–260			
SOIC-N	16	S1	220–225			235–240			
		S2	235–240			255–260			
		S3	255–260						

SOIC-N	16	S1	220–225			
		S2				255–260
SOIC-W	16	S1	220–225		235–240	
		S2	235–240			
		S3	255–260			
SOIC-W	20	S1	220–225		235–240	
		S2	235–240			
		S3	255–260			
SOIC-W	20	S1	220–225			
		S2				255–260
SOIC-W	24	S1	220–225		235–240	
		S2	235–240		255–260	
		S3	255–260			
SOIC-W	28	S1	220–225		235–240	
		S2	235–240		255–260	
		S3	255–260			
SOP	24	S1	220–225			
		S2				255–260
SSOP	38	S1	235–240			
		S2				255–260
SSOP	56	S1	235–240			
		S2				255–260
TSSOP	20	S1	235–240			
		S2		255–260		
TSSOP	56	S1	235–240			
		S2		255–260		
TVSOP	56	S1	235–240			
		S2		255–260		

TABLE 4.3. MSL Data Versus Component Type and Construction

			PRT Temperature						
			220–225	235–240	250–255	255–260			
			MSL Level						
Package Type	Number of Leads	Structure	1	2	2a	3	4	5	6
Exposed die pad	28	S1	220–225						
		S2		255–260					
Exposed die pad	100	S1				220–225			
		S2					255–260		
QFP	44	S1	220–225						
		S2	235–240			255–260			
		S3	255–260						
QFP	44	S1				220–225		235–240	250–255
QFP	48	S1	220–225						
		S2	235–240			255–260			
		S3	255–260						
QFP	52	S1	235–240						
		S2				255–260			
		S3	255–260						
QFP	64	S1	220–225						
		S2	235–240			255–260			
		S3	255–260						
QFP	100	S1				220–225			
		S2							255–260
QFP	128	S1		220–225					
		S2		255–260					
LQFP	64	S1				220–225			
		S2							255–260

LQFP	100	S1				220–225			
		S2					255–260		
LQFP	144	S1		220–225					
		S2		255–260					
LQFP	208	S1	220–225						
		S2	255–260						
MQFP	208	S1			220–225				
		S2			255–260				
PQFP	132	S1				220–225			
		S2							255–260
TQFP	64	S1		235–240					
		S2				255–260			
TQFP	176	S1				220–225			
		S2					250–255		
PBGA	196	S1			220–225				
		S3				255–260			
PBGA	388	S1			220–225				
		S3			255–260				
PBGA	388	S2					235–240	250–255	
TBGA	352	S1					235–240		
		S2							250–255
uBGA	48	S1	220–225						
		S2	255–260						
uBGA	240	S1				220-225			
		S2					255–260		
M2CSP—Stacked	72	S1			220–225				
		S2			255–260				

- Body size/thickness
- Passivation/die coating
- Lead frame, substrate, and/or heat spreader design/material/finish
- Die size/thickness
- Wafer fabrication technology/process
- Interconnect
- Lead lock taping size/location as well as material

The overall package construction plays a large role in the results of the moisture sensitivity testing. No single solution was found that can be applied to solve all the problems. New mold compounds are being developed and qualified that improve the temperature capability and are also "halogen free," die coating is being implemented, package designs are changing, and other additional changes will increase IC package robustness. The user of these packages needs to be aware that nearly all the changes will increase cost. But, as new components are released with these changes and production volumes increase, the costs to the end user will not be visible.

4.5. OBSERVATIONS ON PROFILING FOR THE LEAD-FREE REFLOW PROCESSES

In the course of developing the recommendations on thermal limits for Pb-free components, the iNEMI Pb-Free Component Team initiated a research project focusing on the Pb-free reflow process. Tests were performed at the research facilities of two iNEMI member companies that manufacture reflow ovens. A temperature profiling equipment supplier also participated with one of the reflow oven manufacturers. Both evaluations focused on finding the lowest peak temperature, of the hottest component on the board, which was required to successfully solder thermally challenging assemblies. Oven Supplier A used a large optical fiber channel board provided by an iNEMI member OEM. The board was 18 in. × 12 in. × 0.092 in. thick and contained a range of components that represented ~80% of the component sizes (by mass) currently used in the Electronics Assembly Industry (see Figure 4.4). The other reflow oven manufacturer used a large (13 in. × 16 in. × 0.060 in., 1.25 kg) computer motherboard assembly provided by another iNEMI member OEM (see Figure 4.13). In both cases, the purpose of the experiment was to determine to what extent current reflow technology could successfully process large, complicated assemblies in a drastically reduced process window.

The experiments were run to determine what reflow conditions would be required to achieve body temperatures and solder joint temperatures between 232°C and 240°C. The lower peak limit (or minimum reflow temperature) was specified by Pb-free Sn–Ag–Cu solder paste suppliers, and the high peak limit represented the maximum temperature tolerance of many standard components. Both studies managed to achieve final ΔT values across the assemblies of 11–12°C, but this

Figure 4.13. Computer mother board.

still exceeded the process goals. To achieve these minimum ΔT values, variations to the standard tin–lead profiles were extensive and included:

- Drastically reduced belt speeds and special attention to air flow in the reflow oven
- Adding up to four heated zones to the reflow (furnace) section of the profiles (10 zone furnace utilized for evaluations)
- Ramp–soak–spike (traditional) profile versus "tent" profile (ramp to peak)
- Reorienting the board to the belt—turning it 90° so the long side of the board enters the oven rather the short side

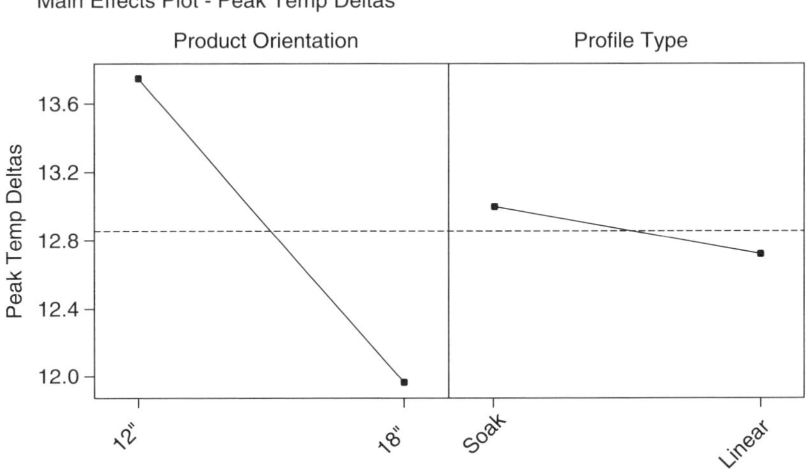

Figure 4.14. Main effects plot from BTU tests.

The factor that had the largest effect on ΔT was the orientation of the board as it went through the furnace. Orienting the board lengthwise across the furnace provided the greatest reduction in ΔT (see Figure 4.14).

One of the oven suppliers also utilized a special computer program developed to optimize oven profiles, which helped to more quickly achieve an optimum reflow profile. This extensive testing was done under optimal conditions and determined that the physical/thermodynamic limitations of the current process made reflowing large assemblies, containing high mixes of component densities, was not possible without exceeding 240°C maximum component temperature target.

General observations on the oven profiling were:

- Pb-free reflow processes will require greater levels of process control than traditional tin–lead soldering processes to avoid exceeding excessive component temperatures.
- Additional furnace zones may be required for low delta temperatures and process control.
- Doing the detailed temperature profiling for each product run will allow the lowest temperature profile to be used, thus reducing the thermal stress on the components.
- The use of software tools (oven profiling software) greatly simplifies the task of finding the optimal profile for a given process.
- The use of automated process monitoring systems may be required to keep Pb-free reflow processes in control.

It was the feeling of the iNEMI Components Team that electronics assemblers converting to Pb-free assembly need to be aware that there are serious process setup and control issues involved in this transition. In the near term, component manufacturers may have difficulty providing components with high enough thermal tolerances to alleviate the issues raised by even a reduced Pb-free reflow process window. It will be essential that component users focus on process improvement and control in order to achieve adequate quality and yield. This is discussed in more detail in the Lead-Free Reflow and Rework chapter (Chapter 8).

4.6. IC PACKAGE IMPROVEMENT OPTIONS FOR BETTER PACKAGE MSL AT HIGHER LEAD-FREE SOLDER REFLOW TEMPERATURES

The Pb-free initiative led to two tasks that needed to be undertaken by component suppliers:

1. Determine the applicability of the current BOM (Bill of Materials) set at higher temperatures (230°C, 240°C, 250°C, and 260°C) in terms of MSL ratings.
2. Determine BOM sets that are capable of meeting the high temperature requirements in the event of the current BOM set being unable to meet it. The new BOM set must be capable of meeting the current MSL levels but at higher temperature reflow.

TABLE 4.4. J-STD-020C Pb-Free Process—Package Classification Reflow Temperatures

Package Thickness	Volume mm^3 <350	Volume mm^3 350–2000	Volume mm^3 >2000
<1.6 mm	260°C + 0°C	260°C + 0°C	260°C + 0°C
1.6–2.5 mm	260°C + 0°C	250°C + 0°C	245°C + 0°C
≥2.5 mm	250°C + 0°C	245°C + 0°C	245°C + 0°C

Tolerance: The device manufactures/supplier shall assure process compatibility up to and including the stated classification temperature (this means peak reflow temperature +0°C—for example, 260°C + 0°C) at the rated MSL level.

J-STD-020 was updated to revision C in July 2004 to address the increased temperature stress imposed by Pb-free soldering profiles. Table 4.4 shows the temperature conditions that an overmolded package must meet based on its volume and thickness. Many legacy components could not meet these temperature processing requirements. Testing components to these metrics drove the need to develop improvements in thermal capability for most suppliers.

This section addresses the steps taken to improve the reliability of the plastic molded packages. The Pb-free initiative provided the opportunity to evaluate the previous material sets and find ways of improving the reliability of the packages, both at the stand-alone level and at the board assembly level. From the component standpoint, the primary concern was the high-temperature reflow, which will affect the substrate material (core, solder resist), die attach epoxy, mold compound, and substrate warpage/coplanarity.

The area array packages (PBGAs, CSPs, etc.), are manufactured in a variety of formats. Broadly, there are two: laminate-based and tape-based. Generally, the laminate-based components are overmolded-type packages and the tape-based components are encapsulant-based. The challenges for each of these families are varied.

The schematic of a laminate-based thermally enhanced PBGA package (Figure 4.15) illustrates the building elements of a package.

The constituents of the package that needed investigation were:

(a) *Pb-Free Solder Ball.* Experimental data such as that shown in Figure 4.16 revealed that Pb-free BGA solder balls using SAC [Sn–(3–4)Ag–(0.5–0.9)Cu] show improved solder ball shear strength on the order of 1.5- to 2-fold that of the eutectic 63Sn–37Pb solder ball after 1000 cycles of temp cycling between −65°C and 150°C.

(b) *The Substrate Core.* The investigation of substrate core material revealed that the currently used core materials were not all suitable for processing at the higher temperatures. Other materials with higher T_g values are available that show better performance in terms of warpage and board level performance. New materials are also becoming available in the halogen-free format, potentially addressing future recycling or restriction issues.

(c) *Substrate Solder Resist Material and its Method of Application.* The solder resist application process and the resist material itself needed to be changed

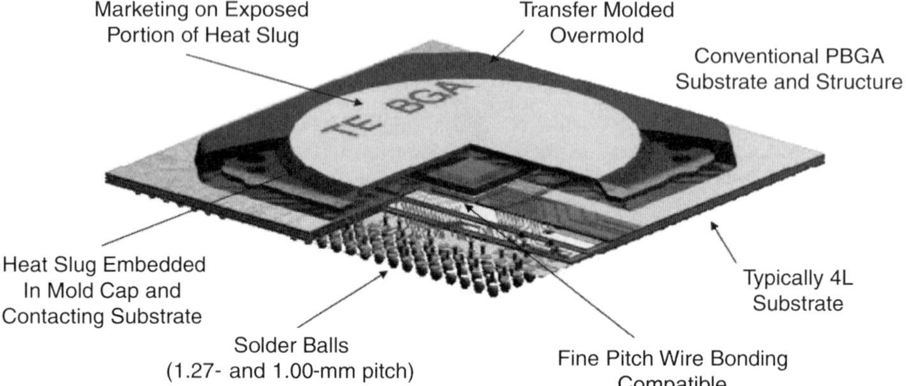

Figure 4.15. Schematic of a laminate-based thermally enhanced PBGA package.

to meet the new requirements consistently. The substrate manufacturing process in general needs to be enhanced with tighter process controls in place.

(d) *Vias.* The plated-through-hole via integrity needs to be maintained at higher temperatures. It is hoped that via hole plugging will provide sufficient margin to maintain the reliability of the vias after Pb-free reflow.

(e) *Die Attach Epoxy.* The previous die attach epoxies, in general, did not meet the new Pb-free reflow temperature requirements, mostly because of moisture absorption. High levels of moisture resulted in die attach cracking at the higher Pb-free reflow temperatures. Reducing the MSL rating could fix the problem, but these reduced levels were considered unacceptable by the industry. Modified materials developed by the epoxy vendors have better moisture absorption characteristics and provide improved MSL ratings at Pb-free reflow temperatures.

(f) *Molding Compound/Encapsulant.* The capability of the previous mold compounds to meet the higher reflow requirements was found, in general, to be

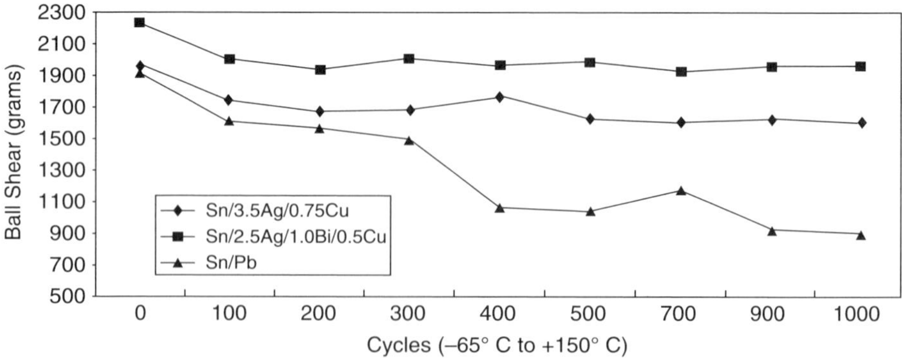

Figure 4.16. Ball shear strength versus temperature cycle for different solder ball compositions.

insufficient. Newer mold compounds have been developed that show improved adhesion, absorption characteristics, and warpage performance which results in improved MSL ratings at the higher reflow temperatures.

Every supplier of overmolded electronic packages will have to assess what changes are required to meet the additional stresses imposed on the components by the Pb-free soldering processes.

4.7. FREQUENCY CONTROL PRODUCTS

4.7.1. Introduction

Many types of frequency controlling components are also impacted by the higher soldering temperatures of Pb-free assembly. Crystals and crystal-based products are widely used in many applications for frequency control, such as for a clock or timing function. As such, they find applications covering a broad range of products in telecommunications, wireless, networking, data communication, and so on. Frequency control products (FCP) typically utilize a piezoelectric material, such as quartz, used in both low-frequency bulk acoustic wave and high-frequency surface acoustic wave (SAW) technology. Examples include (a) simple discrete crystals (strip or round) or SAW crystals packaged in metal or ceramic housings and (b) integrated high-functionality products such as oscillators. The latter device types cover a wide spectrum from simple fixed-frequency oscillators to voltage-controlled (VCXOs), high precision, and/or high-stability temperature-compensated (TCXOs) and ovenized oscillators (OCXOs). Typical examples of these products, from a simple discrete crystal to a complex oscillator illustrating the technology, are shown in Figures 4.17 through 4.19.

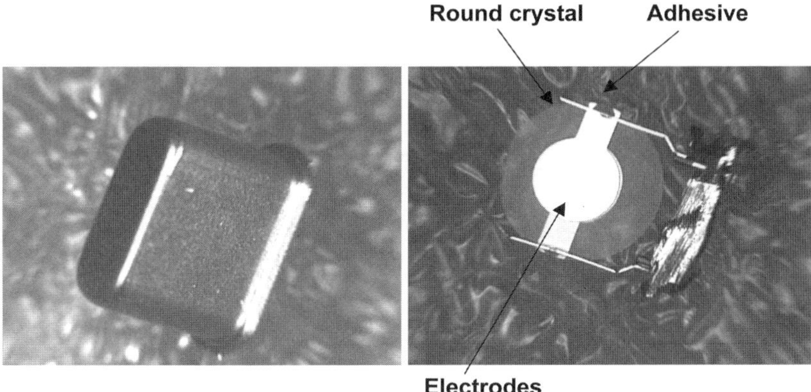

Figure 4.17. External and internal views of a discrete crystal unit in a UM-style sealed metal package.

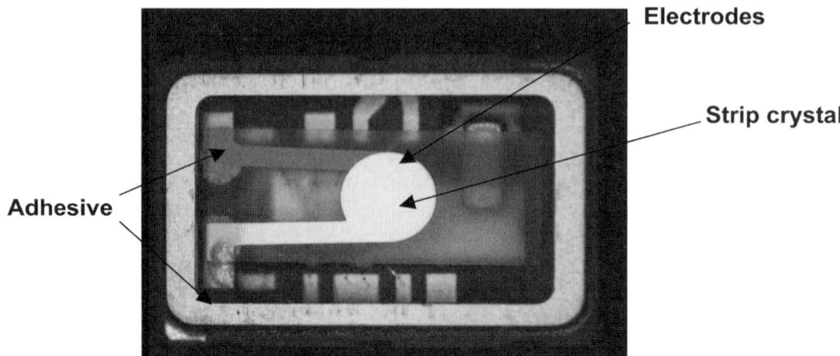

Figure 4.18. Open view of a ceramic-packaged crystal.

4.7.2. High-Temperature Assembly Issues

High-temperature exposure to FCPs during assembly with Sn–Ag–Cu solders can significantly degrade them in several ways. Such thermal exposure can affect the stress on the crystal mount or cause changes internally to the package from the outgassing of the adhesives used for crystal assembly. These sensitive components have great difficultly surviving peak reflow temperatures in the 230°C–260°C range, such as those expected to be encountered during the use of the new Pb-free solders.

High-temperature exposure can also affect the electrode characteristics and degrade the electrode resistance from structural or metallurgical changes to the evaporated electrodes. Examples of such changes to oscillator performance are illustrated in Figure 4.19. The chart represents changes in the nominal or design

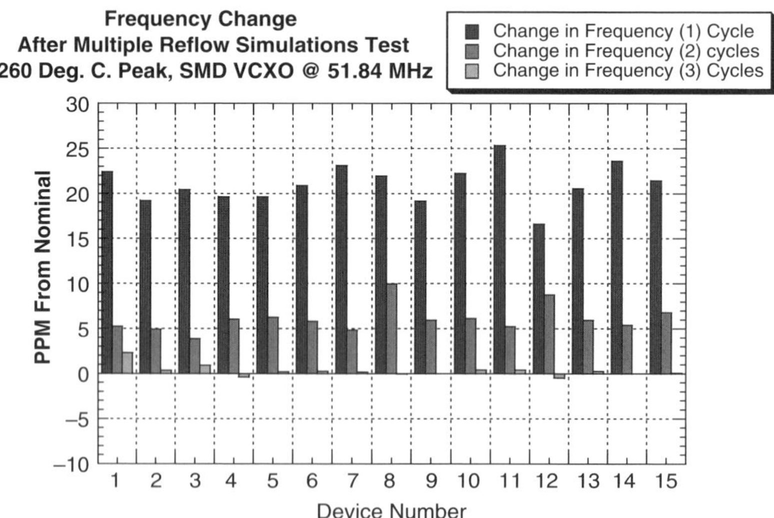

Figure 4.19. Change in the frequencey of a SMD VCXO to multiple simulated thermal reflow exposures, peak 260°C.

frequency of a miniature SMD VCXO when subjected to multiple simulated thermal reflow profiles with a peak of around 260°C.

If the product specification calls for an initial frequency stability of 25 ppm, for example, such large changes would cause the devices to fail the required specification when assembled on a customer board using the high-temperature reflow process. From a physical standpoint, this represents a change in frequency due to the effect on the crystal mounting stress. Interestingly, this parametric change is greatest after the first exposure, but virtually negligible after the subsequent reflows. This may actually offer a way to "precondition" the device prior to shipment so that from the stress relief during actual board assembly, the frequency change should be somewhat smaller and product will still meet specifications.

As mentioned earlier, the higher the fundamental crystal frequency, the greater the shift in frequency or other parametric changes. Such changes, if sufficiently large, can actually cause the product to be out of specification. Note that the magnitude of such change will depend on the package size and assembly technology.

While products that utilize adhesives for assembly may only show parametric changes, exposure to high temperatures of other FCP products that use conventional solder materials and processes may be catastrophic. Such products will require an exception to the current assembly processes.

4.7.3. Future Directions to Address Pb-Free Assembly

Many types of product design changes can be envisioned that will ensure that the frequency control products can meet the stringent requirements dictated by the use of new Pb-free high-temperature solders. Product that currently utilizes Pb-based solder for internal assembly will need to have Pb-free solders substitutes. Furthermore, the overall assembly must be sufficiently robust to survive the reflow process using the Pb-free solders. Often, the latter requirement means that the component must survive not simply one pass, as during the initial assembly, but also any subsequent or additional reflow exposures due to any board repair work that may be needed or due to the use of multiple passes for a dual-sided board assembly.

Use of adhesives that can withstand high-temperature exposure, both during assembly by the manufacturers and during customer board assembly, must be employed. Adhesive suppliers are beginning to offer adhesives suitable for high-temperature assembly with good outgassing and low-stress characteristics that should offer good long-term product reliability.

Manufacturers will need to evaluate such adhesives for compatibility with their current processes and equipment, as well as the ability to meet the required aging characteristics. Manufacturing process changes to develop a more robust product for Pb-free assembly are needed. These may include in-process stabilization bakes or curing, as well as any post-assembly (before or after final sealing) "simulated" exposure(s) to expected board assembly thermal profile by the customer. Some manufacturers currently use a post-sealing reflow exposure for certain components, particularly small SMD oscillators and crystals, as a way to "condition" the part and to ensure that it will not change significantly during actual board assembly.

4.8. "LEAD-FREE" COST IMPACT ON COMPONENTS

The conversion to Pb-free is not free, although any increase in the direct material cost of Pb-free materials *per se* is not the primary issue.

During the last few years, IPC/JEDEC standard J-STD-020 has been updated to address some of the issues associated with Pb-free reflow. For example, the reflow temperature specification has been increased because of the new requirements for Pb-free assembly. The standard prescribes a temperature-testing regime that is keyed to the IC package thickness and volume. There are three Pb-free temperature ranges identified in the standard: 245°C, 250°C, and 260°C. Extensive evaluations conducted by many IC suppliers (see previous MSL section), show that most components would drop 1–3 moisture sensitivity levels (MSL) when tested at 260°C peak reflow temperatures with the previously used designs and materials sets.

As described above, many factors could affect the MSL. Many of these have significant cost and performance impacts for the packaging house and the end user. The impact of some factors could be far greater that others. When trying to improve the MSL level, we have to look at the IC package as an entire system. There are complex interactions among all the materials used in a package construction. Just because material "A" did make a major improvement on one supplier's design, it does not mean that the same material will have a similar impact on the same part from another supplier.

Component suppliers, in cooperation with their material providers, are actively trying to find the best design and material combinations that will give them the MSL capability to meet the new requirements. Finding the best materials is not the only problem. For the vast majority of IC packages, the problem also consists of finding a design and material-set combination that will improve performance at an acceptable cost.

The impact on cost depends on the current MSL level of the package. Current MSL levels for many packages depend on the material technology used, the manufacturing method, and the reflow temperature used to classify them for MSL. The majority of these packages were classified as MSL-1, using either 220°C or 240°C (for Sn–Pb assembly). MSL-1 does not require any special process or packing after the final test operation and are shipped without any special packing protection. No special handling is required by the end customer before solder reflow operation. Some older packages like SOPs, PLCC, and QFPs use less advanced materials and processes and may drop 1–2 MSL levels if tested at 260°C. Anything Level 2 or below requires Dry-Bake and Dry-Pack.

The dry-bake and dry-pack processing processes, at the IC packaging house, need additional floor space for ovens and for dry packing equipment. Additional space is also required for the supporting materials and labor, thus increasing the packaging and assembly costs.

Since several hundred billion semiconductor packages are produced annually, the cost to add these operations could be in billions of dollars. In order to avoid the investment in capital, space, and labor required to dry-bake/pack, component suppliers are looking for design changes and better materials that will get them back to MSL-1 at 260°C.

For those packages that are classified MSL less than 2, dry bake/pack is already standard handling procedure. The packages may drop to MSL-4 or less at 260°C reflow, but since they are already doing a dry bake/pack, the problem will be transferred to the board assembler. The assembler will have to deal with the additional restrictions on these packages, such as strict control on the time available after the package container is opened up until units are used, or the need to dry-bake the units just prior to use. OEMs and assembler subcontractors may have to install dry-bake ovens and special storage areas that will keep the components dry prior to use on a reflow line. In this case, the cost is transferred from the component supplier to the board assembler. However, not all suppliers are the same. Many users can switch suppliers of the same part and improve MSL level, which makes MSL capability a competitive consideration for the supplier. Preliminary figures from suppliers show an increase in mold compound price in the range of $2/kg to $8/kg (US). This increase will represent billions of dollars in additional cost to suppliers at current volumes produced in the industry. However, a very small amount is used by each component, so it will be a relatively small impact at the component level.

It is difficult to quantify all the other cost increases that will be required to be capable of Pb-free processing. More energy will be needed because dry-bake and reflow ovens will use more energy due to the increased reflow temperature. The need to increase the reflow temperature up to 260°C may require more furnace zones and longer processing. But, as future material sets and packaging process improvements find their way into high volumes, costs will simply be buried into the new package designs, not visible to the end user.

Eventually, Pb-free component volumes will be high enough that the cost penalty will swing back to the incumbent Pb-based component finishes. Suppliers will not want to maintain two surface finishes and two assembly lines, so users (such as defense contractors, medical systems and aero-space) who are not required to utilize Pb-free finishes will see increased prices for Sn–Pb finishes, if they are still available at all.

4.9. PACKAGING IDENTIFICATION OF LEAD-FREE PACKAGED ICS

As the PCB assembly industry transitions from a tin–lead to a Pb-free soldering world, there is a need to differentiate the IC components containing lead from those that are Pb-free. This is the message that component suppliers are getting from their customers who are preparing to start producing Pb-free products. (See Chapter 10.)

The transition from tin–lead to Pb-free has been shown to typically consist of three phases:

Phase I. Pb-free solders (Sn–Ag–Cu) will be used in assembly with current component finishes (either Pb-free or Sn–Pb finish).

Phase II. Some applications will start to convert to 100% Pb-free (all the elements in and/or on the PCB board).

Phase III. Tin-lead solder is entirely eliminated from all electronics.

During the Phase I transition period, it may be difficult for OEMs and PCB subcontractors to procure all their components Pb-free. Phase I will require component suppliers to meet the higher reflow temperatures (~260°C) required for Pb-free processes. The challenge for component suppliers during this phase is to develop better materials that can withstand the higher reflow temperature and/or eliminate and protect their components from gaining moisture prior to use (reflow). This will mainly be a cost issue. There will be no need during this phase to differentiate Pb-free from Sn–Pb components as long as they are "forward compatible" and can survive the higher reflow temperatures.

Phase II is the period where both Sn–Pb and Pb-free components will be available and used on the same PCB board or at least at the same factory. During this phase there will be a mixture of products; some will be produced 100% in a Pb-free processes (100% of all the elements of the PCB will be Pb-free) and some will be produced using either 100% Sn–Pb or a mixture of Pb-free and Sn–Pb components.

This period may require some form of identification of the components and a separation of the inventories at the PCB manufacturing floor.

The choices for component suppliers to support their customers' needs are as follows:

Option 1. Provide two finishes for their components: Sn–Pb and Pb-free. There is a need to be able to differentiate clearly and easily between Sn–Pb and Pb-free components.

Option 2. Selectively produce Pb-free components upon request and expand their Pb-free portfolios as demand starts to ramp up. This option will also require different identification of Pb-free and Sn–Pb components.

Option 3. Convert 100% to Pb-free and make sure all their components are 100% backward compatible. This option may not require differentiation, because all the components are Pb-free and backward compatible.

TABLE 4.5. Process Compatibility Matrix

	Reflow Soldering Process	
Component Finish	Pb-Bearing Solder (205–220°C peak)	Pb-Free Solder (235–260°C peak)
Pb-bearing component	Current/historical	Forward compatibility[a]
Pb-free component	Backward compatibility[a]	Future

[a]The term "forward compatibility" refers to use of Pb-bearing component finishes in a Pb-free soldering process. Likewise, "backward compatibility" refers to use of Pb-free components in a Pb-bearing soldering process.

Option 1 has major implications for component suppliers. With thousands of parts in their portfolios, it will require twice the number of parts. In addition to the enormous amount of manufacturing floor space required to maintain two part numbers, the logistics to handle two component finish combinations, starting from the time that a customer inputs an order, through fabrication until it is delivered, will have a significant impact on the cost of the components.

Option 2 is a more acceptable solution, especially for component suppliers who have not yet transitioned to a Pb-free finish. Assigning "special" part numbers to a few parts is doable. However, the problem can get out of control when the request for "specials" starts to ramp-up; this could eventually end as in Option 1.

Option 3 is the best solution. Since the components are Pb-free and most are backward compatible (apart from Pb-free Sn–Ag–Cu BGA/CSP), there is not a major need to identify the parts. The key to this solution is to select a Pb-free finish that is backward compatible and make the conversion now, before the start of Phase II.

There are IC components that are Pb-free already and have been backward compatible for many years. Ni–Pd and Ni–Pd–Au terminal finishes have been in use for many years. Both of these finishes have some challenges in the reflow soldering process that require addressing by the user. This may require slightly higher reflow peak temperatures or different solder pastes fluxes. In addition, these finishes may not be suitable for alloy 42 lead frames because of potential corrosion field issues. Pure tin as a terminal finish has also been in production for several years, mainly in discrete chip capacitors and through-hole components. Both Ni–Pd and pure tin component coatings have been proven to be backward compatible with Sn–Pb soldering processes. There are other alloys being developed at this time; some of them are showing promising results for backward compatibility.

As stated, Options 1 and 2 will require some level of identification. Inputs from some OEMs and PCB subcontractors suggest several alternatives:

(a) Clear identification via a symbol on the component itself. This option requires a different symbol for Sn–Pb and Pb-free components. Parts will need a different ordering part number and fabrication instructions; this presents the same logistics issues as Option 1 described above.
(b) Parts may not need to be labeled with symbols, but need to have different ordering part numbers and different shipping boxes with different shipping labels. The issue is not as severe as in (a), but still presents the same logistics problems to the supplier as in Option 1.

4.10. CONCLUSIONS

It should be obvious that the best alternative is Option 3: Convert the components (apart from BGA/CSP) to Pb-free with a finish that is backward compatible as quickly as possible. This will provide the least painful transition for both the component suppliers and the end customers.

ACKNOWLEDGMENTS

The authors would like to acknowledge the following iNEMI Component Team members for their contributions to the success of the team: Edgar Zuniga, TI; Swaminath Prasad, ChipPAK; Dennis Barbini, Vitronics Soltec; Greg Jones, Kic Thermal; Karen Walters, formerly with BTU International.

REFERENCES

1. IPC/JEDEC J-STD-020C, Moisture/Reflow Sensitivity Classification for Nonhermetic Solid State Surface Mount Devices.
2. Improved Thermal Process Control for Pb-Free Assembly, IPC SMEMA Council APEX 2001, Wayne Johnson, Roger Lugo, Seshu V. Sattiraju, Auburn University, Greg Jones, KIC Thermal Profiling.
3. Jack McCullen and Thomas Moore, Reliability of Electronic Packaging, tutorial at the 47th Electronic Components & Technology Conference, May 18–21, 1997.
4. Jack McCullen and Richard Shook, Understanding Moisture/Reflow Sensitivity for IC Packages: Achieving Pb-Free Assembly Classification and Handling, short course at JEDEX San Jose, April 14–16, 2004.
5. APEX 2000 iNEMI, Pb-free component group presentation, available at www.iNEMI.org
6. APEX 2001 iNEMI, Pb-free component group presentation, available at www.iNEMI.org

CHAPTER 5

Lead-Free Assembly Reliability—General

EDWIN BRADLEY

5.1. INTRODUCTION

The widespread introduction of lead-free solder assembly is a challenge to the electronic equipment design and manufacturing communities. The comfort level of decades of experience using tin–lead solder and plating is being disrupted by the rapid implementation of lesser known materials. The field-use reliability of products using these new solder materials is expected to be different than tin–lead, and generating the basic level of understanding of the performance of these materials is a daunting task that is underway at companies and universities around the world. A further complicating factor is the range of reliability requirements for various products and applications, from those that may require minimal reliability (e.g., "disposable" electronics such as one-time-use cameras) and those that are in critical-use areas where failure can result in deaths (e.g., pacemakers or aviation electronics). In addition, there are the different end-use conditions that have widely varying levels of stress and environments. The goal of this chapter is to compare the reliability of tin–lead and lead-free solders under various test conditions and to illuminate the potential impact (positive or negative) to specific applications.

There are general solder property databases available online (including those managed by NIST [1] and Matweb [2]) that have Pb-free solder properties, but much of these data references basic physical properties that affect reliability but not actual reliability data of soldered assemblies. This chapter aims to bring together representative data in a number of different areas to compare and contrast the performance of Pb-free alloys to that of eutectic Sn–Pb solder.

Lead-Free Electronics. Edited by Bradley, Handwerker, Bath, Parker, and Gedney

5.1.1. Solder Joints

Solder joints are complex, composite structures that can vary greatly in microstructure and mechanical properties, depending on the base metal to which the solder is bonded, the cooling rate at solidification, and subsequent thermal aging that can recrystallize or coarsen the solder microstructure. When the solder wets and bonds to the base metal (most commonly copper or nickel), there is a reaction between the solder and the base metal that forms a layer of one or more intermetallic compounds (IMC) on the base metal. With tin-based solder alloys, the IMCs typically contain tin. Moving away from the IMC layer and it may into the solder proper, the solder composition may be slightly depleted of tin due to the formation of the IMC layer, and it may also be enriched with the elements of the base metal(s) that may have diffused into the solder. There can also be intermetallic compounds in the main of the solder joint, and these can take the shape of rods, plates, and more equiaxed particles (Figure 5.1). In cross sections, these three-dimensional features have their two-dimensional sections visible, so that plates will appear as rectangular shapes, while rods will appear as circular/oval or rectangular shapes, depending on their aspect ratio and orientation to the plane of the section.

5.1.2. Alloys of Focus

This chapter will focus on the tin-rich lead-free alloys (primarily Sn–Ag–Cu) and compare them to the eutectic Sn–37Pb. The datasets in the open literature have evaluated numerous different solder alloys, and oftentimes these are slight variations in composition for certain alloy families such as Sn–Ag–Cu. The goal here is to show the general reliability trends of the lead-free solders compared to tin–lead. None of these data can replace specific product-level testing, because variations in assembly,

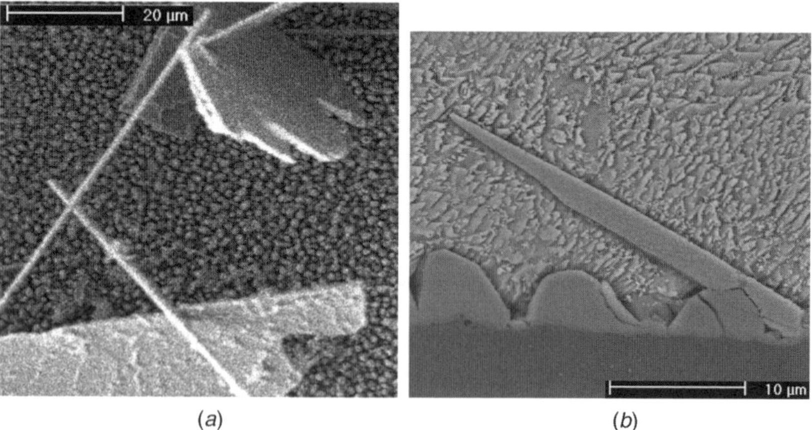

Figure 5.1. (*a*) BGA solder joint microstructures with solder matrix etched away and (*b*) a cross-section of a similar joint. Three-dimensional intermetallic structure after surrounding solder was etched away. Note the rods and plates along with the bumpy surface of the IMC layer on the base metal.

materials, and design can greatly affect the absolute reliability of solder joints, whether tin–lead or lead-free.

5.1.3. Sample Size and Type

A major issue to be aware of when referencing solder properties is the physical size of the test specimens. Most solder joints found in electronic assemblies are very small compared to the size of standard mechanical test samples, and the size decreases monotonically as passive devices get smaller and BGAs have narrower pitches (ball-to-ball spacing). Thus, the width of a small solder joint may only be a few grains wide, as well as containing intermetallic plates that are microstructural discontinuities and that can affect failure modes and overall deformation behavior.

5.2. BASIC PHYSICAL PROPERTIES OF SOLDER

The attractive property of both tin–lead and lead-free solders that drive their use in electronic assemblies is their relatively low melting points, which allows for metallurgical bonding at temperatures that are safe for components. Referring back to Chapter 1, the melting point of eutectic Sn–Pb solder is 183°C, and for Pb-free the most popular alloys from the tin–silver–copper (Sn–Ag–Cu or SAC) family have melting points at or near 217°C. Ashby developed the concept of homologous temperature (T_h), which is the ratio of the application or use temperature over the alloy melting point in kelvin:

$$T_h = T/T_m$$

Thermally activated deformation mechanisms begin to be noticeable at $T_h = 0.3$ and are often dominant at $T_h = 0.5$. Because solders have relatively low melting points, even room temperature (here taken as 25°C) is an appreciable fraction of the melting point (T/T_m) of most solders and of all the ones used the vast majority of electronics assemblies (Table 5.1). Even at −55°C, the T_h is 0.48 for tin–lead and 0.44 for Sn–Ag–Cu alloys. Thus, the low melting point that allows for assembly is also is the Achilles' heel of solder, in that the solders experience thermally activated deformation mechanisms at the use conditions of the electronic assemblies.

TABLE 5.1. Homologous Temperatures of Various Solder Alloys

Alloy (by Weight)	Melting Temperature (°C)	T_h at 25°C
Sn–37Pb	183	0.65
Sn–3.8Ag–0.7Cu	217	0.61
Sn–48In	118	0.76
Sn–58Bi	138	0.73
Au–20Sn	280	0.54
Sn–0.7Cu	227	0.59

TABLE 5.2. Room Temperature Dynamic Elastic Modulus Values

Alloy	Modulus (GPa)	Reference
Sn	49.9	3
Sn–3.5Ag	56	4
Sn–2.5Ag–0.8 Cu–0.5Sb	51.2	5
Sn–37Pb	33.6–35	4, 5

5.2.1. Elastic Modulus

The elastic modulus values of tin-rich Pb-free solders are generally higher than those of eutectic tin–lead. Most reported values of solder modulus are made with acoustic methods and are often referred to as dynamic. Table 5.2 shows the reported elastic modulus values at room temperature of various alloys over several different studies. Overall the Sn–Ag alloys have higher moduli than eutectic Sn–Pb. The value of the modulus for solder materials has been found to depend on the strain rate [6, 7] and is not a constant at a given temperature as shown in Figure 5.2. The static modulus values were taken at a strain rate of 8×10^{-4}/s.

5.2.2. Uniaxial Tensile Properties

One of the most basic representations of mechanical properties is the uniaxial tensile curve. From this curve, one can calculate the elastic modulus, yield strength, ultimate tensile strength, strain-to-failure, and toughness. It is also used in mechanical simulation. The mechanical properties measured from any test are dependent on the microstructure

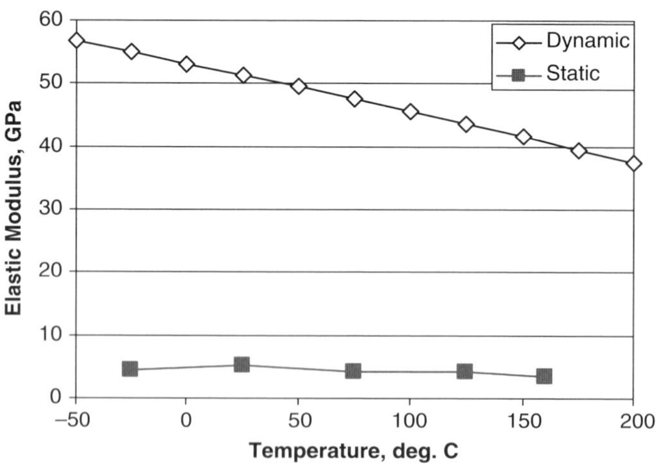

Figure 5.2. Elastic modulus values for Sn–3.9Ag–0.6Cu using dynamic and static methods. (Adapted from Vianco et al. [6]).

of the sample, and large samples may or may not be relevant to the surface mount solder joint. In any case, the data shown here are a combination of large and small samples that show both trends and relative performance of the solders tested. The fundamental equation describing the plastic tensile deformation [8] is

$$\sigma = K\varepsilon^{n}\dot{\varepsilon}^{m}$$

where σ is true stress, ε is true strain, $\dot{\varepsilon}$ is true strain rate, n is strain hardening exponent, m is strain-rate sensitivity, and K is a material constant. Solders are very sensitive to strain rate and temperature at the field usage conditions, so any tensile test data must be taken at the strain rate and temperatures of interest. The variables m, n, and K can also be sensitive to the microstructure, so these must be known when evaluating the relevance of tensile data for any particular application.

Bradley and Sharda [9] carried out tensile tests on Sn–3.8Ag–0.7Cu and Sn–37Pb using samples with 1.5-mm gauge length and at a strain rate of 0.001/s. The reductions of area for the two alloys were close, with the Sn–Pb having a slightly more ductile response. Fracture surfaces of the alloys are shown in Figure 5.3 and the ductile features are clearly evident.

Although all material properties are temperature dependent to varying degrees, solder mechanical properties are very sensitive to changes in temperature, given that the melting point is relatively close to room temperature. Figure 5.4 is a plot of the ultimate tensile strength as a function of temperature for several lead-free alloys from solder tape [10] as well as bulk cast samples [11, 12], and they all drop rapidly as the melting temperature is approached. The drop in strength with temperature is due to the increasing amount of thermally activated deformation mechanisms.

Solders are strongly sensitive to the applied strain rate, and tensile tests run at appreciably different strain rates will typically have increasing yield and ultimate strengths as the strain rate increases. Figure 5.5 shows data from Pang and Xiong

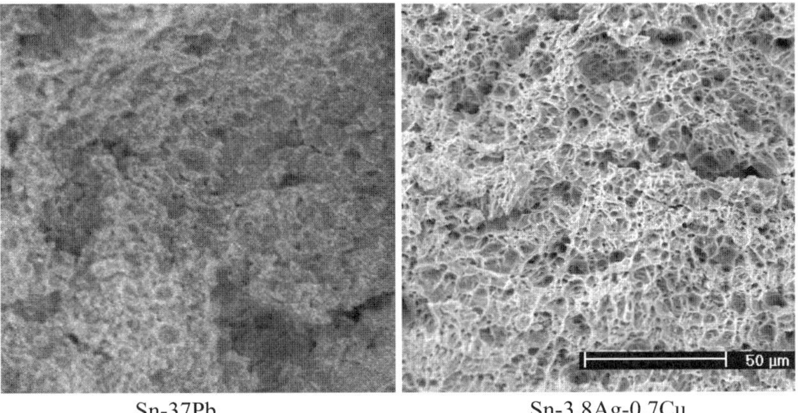

Figure 5.3. Fracture surfaces of solder samples pulled at 0.001/s. Both are at same magnification.

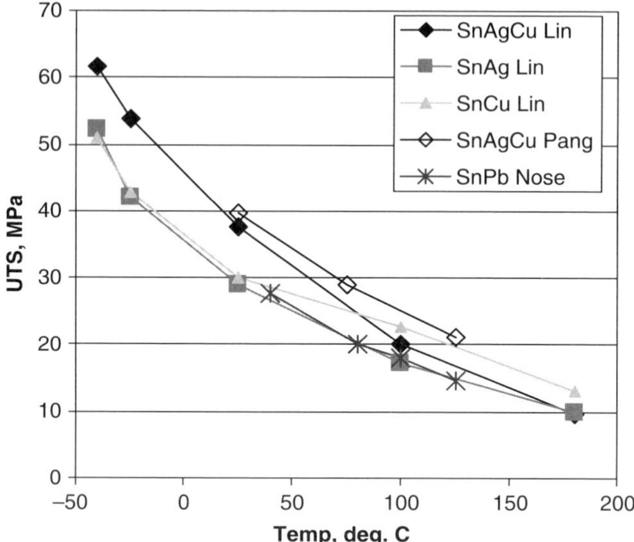

Figure 5.4. Ultimate tensile strength of several lead-free solders as a function of temperature from −40°C to 180°C. Strain rates are as follows: Lin et al., 0.001/s; Pang and Xiong, 0.0005/s; Nose et al., 0.001/s (adjusted from 0.005/s).

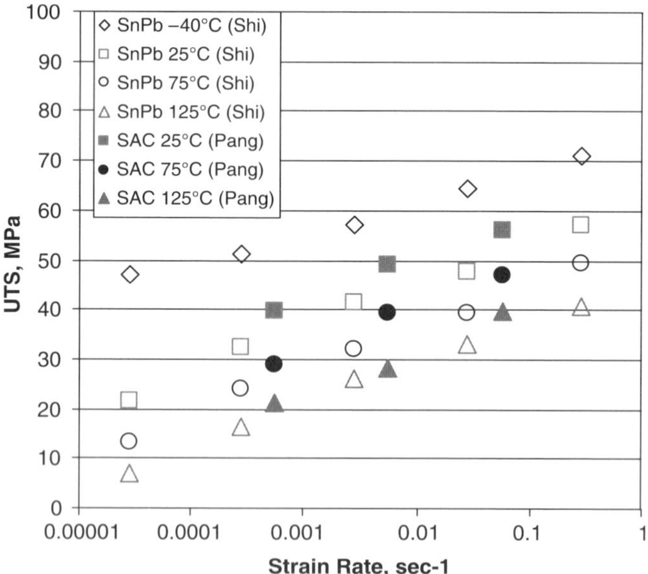

Figure 5.5. Ultimate tensile strength versus strain rate at various temperatures from [11] and [13]. This shows the strong effect of strain rate on the solder strength for both Sn–Ag–Cu and Sn–Pb solders.

[11] and Shi et al. [13], and both the Sn–Pb and Sn–Ag–Cu alloys show strong trends with strain-rate at all test temperatures.

5.3. CREEP DEFORMATION

Because solders have relatively low melting points, even room temperature is an appreciable fraction of the melting point (T/T_m) of most solders and of all the ones used the vast majority of electronics assemblies. Thermally activated deformation mechanisms begin to dominate at homologous temperatures above 0.5; at 25°C, every solder in Table 5.1 meets that threshold.

Figure 5.6 shows a set of creep data for small, thin solder joints (3-mm diameter and 0.18-mm thick) from Zhang et al. [14], and the data show the smaller strain-rate sensitivity of the Sn–Ag–Cu compared to the Sn–Pb. Wiese et al. [15] performed creep tests on flip-chip samples and found that the strain-rate sensitivity (m) for the Sn–Ag and Sn–Ag–Cu alloys were much higher than that for Sn–Pb. The result showed that at lower stresses, the effective strain rate of the lead-free solders is much smaller and as a result the stresses are not fully relaxed except in the limit of extremely long test/dwell times.

Amagai et al. [16] tested large bulk samples in creep (Figure 5.7) and found trends similar to those of Zhang et al. [14]. The room temperatures showed less separation between the Sn–3.5Ag–0.7Cu and Sn–37Pb, while the 125°C data were more spread out at the lower strain rates. Generally the data show that the Sn–Pb alloy is much less creep-resistant than Sn–Ag–Cu at the lower stress–strain rates but is comparable at the higher strain rates, especially at the higher temperatures.

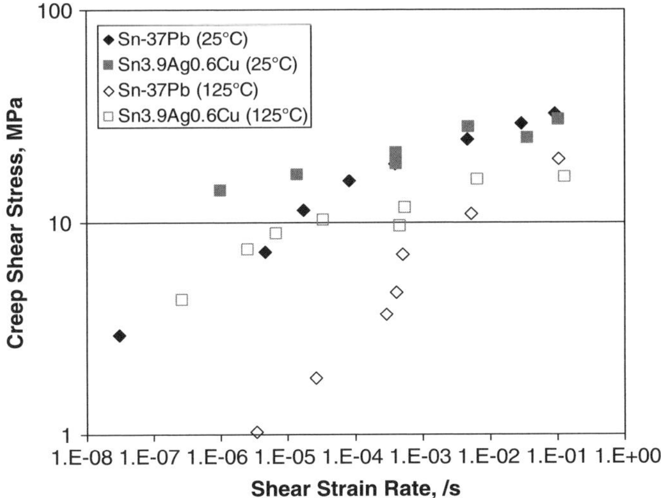

Figure 5.6. Shear creep data from Zhang et al. [14] comparing Sn–Ag–Cu to Sn–Pb using thin solder samples soldered to copper.

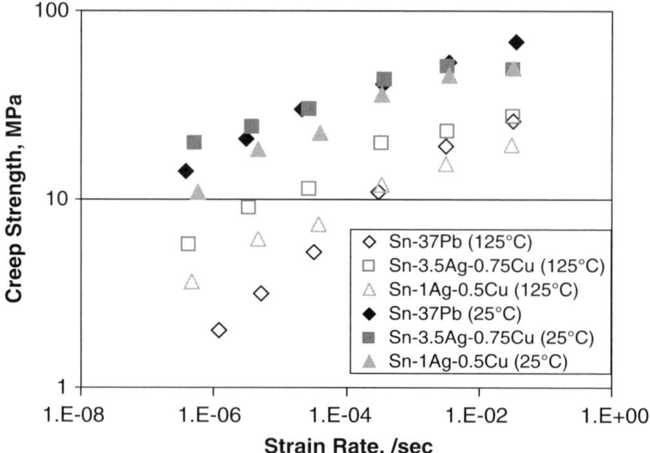

Figure 5.7. Creep data developed from bulk samples by Amagai et al. [16].

For most thermal cycle tests, the effective strain rate of the joint will be determined by the temperature ramp rate during the heating and cooling periods. Generally, the time required to change the temperature is at least 10 minutes (600 seconds). Taking an extreme case that the strain applied to the joint is 5%, the strain rate in the joint can be estimated to be no greater than 0.0001/s.

5.4. THERMAL FATIGUE

Thermal fatigue characterization has been the area of the most testing for tin–lead as well as lead-free solders. There have been numerous studies on this mechanism for lead-free solders, as seen in many journal and conference proceedings. The driving force for damage in thermal fatigue is the coefficient of thermal expansion (CTE) mismatch between the different materials that make up an electronics assembly, along with the fact that the solder plastically deforms to accommodate a portion of that strain mismatch. Damage accumulates within the solder joint with each temperature cycle, and eventually a crack is nucleated that will propagate through the joint and create an electrical open. The parameter that most effectively relates to thermal fatigue is ΔW, which is the strain energy density change per cycle. Darveaux [17] showed that the crack nucleation (N_0) in thermal fatigue follows the function

$$N_0 = C_1(\Delta W)^p \tag{5.1}$$

where C_1 and p are material constants. Once the crack is nucleated, the lifetime of the solder joint is determined by the propagation rate da/dN and joint length a, and that typically follows the equation

$$da/dN = C_3(\Delta W)^q \tag{5.2}$$

TABLE 5.3. Room Temperature CTE Values for Common Surface Mount Materials

Materials	CTE (ppm/°C)	Reference
Halogen-free FR-4 PCB (x, y)	16–22	18
Silicon	5	19
LTCC	3–8	20
Sn	22, 23.8	2, 21, 22
Sn–Ag–Cu	23	6
Pb	29.3	20
Sn–37Pb	24.7	3
Cu	17	22
Alloy 42	5	23
PBGA component	9–11	

where C_3 and q are material constants (some microstructurally dependent) that determine the relative life of an assembly.

In Table 5.2 there is a list of common materials used in electronic packaging; when the solder is used to attach two different materials that have a large differential in CTE (Δ_{CTE}), the strain will be greater. For example, an LTCC module soldered to a laminate PCB will have a Δ_{CTE} of 8–16 ppm/°C. Flip-chip assemblies have even greater Δ_{CTE}, and with very small solder joints it is very susceptible to thermal fatigue if no other strain mitigation strategy is used. The following section will cover the performance of various package types, as well as the performance of solders both with and without strain mitigation.

Looking at Table 5.3, it is clear that the maximum CTE mismatch occurs when combining conventional FR-4 PWBs with either silicon or ceramic chips/modules. That differential in strain will be transferred to the solder joint; also, the greater that strain, the fewer thermal cycles the assembly will be able to withstand. For fatigue testing, decreasing thermal fatigue life (in terms of cycles) correlates with increased ΔT, greater peak temperature, longer cycling periods, and lower cyclic frequency.

Greater overall stiffness in the solder assembly will also reduce the fatigue life for a given component/solder combination, and this has been shown with the reduction in life of double-sided BGA assemblies compared to single-sided [24]. As the temperature deviates from ambient for a single-sided assembly, the board and component have some freedom of bending (similar to a bimetallic strip) which accommodates a portion of the strain that would otherwise be taken up by the solder joint. Another example of the effect of increasing stiffness on fatigue life is from Farooq [25], where CBGA devices assembled to PCBs showed monotonic decrease in thermal fatigue characteristic life with increasing ceramic substrate thickness.

Analysis and Plotting of Reliability Data. Many solder reliability tests are performed to component failure, and they are plotted on probability plots that are fitted with Weibull or Lognormal distributions. These distributions are described by equations that are more flexible in fitting different types of datasets. The Weibull fit often used is

$$F(t) = 1 - e^{-\left(\frac{t}{\alpha}\right)^{\beta}} \tag{5.3}$$

where F is the fraction failed at time t, α is the scale parameter,* and β is the shape parameter. The α parameter is also referred to as the characteristic life and is the time to reach 63.2% failure rate. This is close to the median or N50 value. As β increases, the failure distribution becomes tighter.

Ultimately, the value of accelerated tests lies in its relationship to reliability in the field. The expectation is that differences in the accelerated tests are consistent with that in the field. Thus, understanding and predicting the early failures is often of great interest because those failures can incur warranty costs and are more likely to reduce customer satisfaction. Given that the early failures are at the tail of the failure distribution, the confidence in those data is less than that near the mean, and that is reflected in the wider confidence limits around the tails of the distribution. In experiments with small sample size, the confidence range of β can be quite large. Thus, comparisons of experimental variables using estimates of the low percentile failures can potentially be misleading. In this chapter, the analysis of Weibull data will be primarily by comparing α values and then supplementing that with β values.

Effect of Thermal Fatigue Test Conditions. There are many variables that can affect thermal fatigue test results such as the PCB finish, solder joint composition, reflow profile (including peak temperature, time-above liquidus, and cooling rate), substrate CTE, and the thermal cycling profile (temperature ramp rate, length of dwell time at temperature). Testing method can also be with shock or cycling. In shock testing, the boards are transferred between the cooler chamber and the hotter chamber, so that the ramp rates are faster and more cycles can be completed per day. In some cases the medium is an inert liquid, which results in extremely rapid temperature ramp rates. For thermal cycling, the ramp rate is held to 8–14°C/min, and often the boards remain in the same chamber and the temperature of the chamber is cycled between the T_{max} and T_{min} settings. The liquid media method has the fastest ramp rate and thus the shortest time per cycle for a given dwell time. There are JEDEC specifications that document the methods of thermal cycling and shock [26, 27].

General test results at various conditions have been reported by numerous studies. Syed [28] studied PBGA packages and found better results for Sn–Ag–Cu versus SnPb across the range of conditions tested, with better results at the more benign regimes. The NEMI results (see Chapter 6) also tend to confirm that result, with equivalent or better results in a range of components for Sn–4Ag–0.5Cu versus Sn–Pb.

*Sometimes the Weibull scale parameter is referred to as eta (η), but η is equivalent to α.

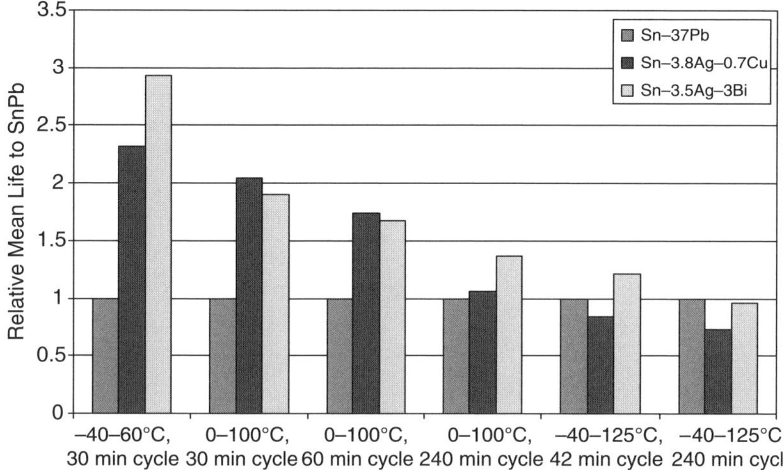

Figure 5.8. Bartelo data for CBGA assemblies showing decreasing relative performance for Pb-free alloys versus Sn–Pb as the ΔT and dwell time increase.

Bartelo et al. [29] tested ceramic BGA assemblies with three different solder ball alloys and showed that increasing either dwell time or ΔT reduced α for Sn–Ag–Cu relative to Sn–Pb (Figure 5.8). The best performance for the Pb-free alloys relative to Sn–Pb was for the 0–100°C thermal cycle. All of the Pb-free data was normalized relative to Sn–Pb for each test condition. This prevents comparison of the effect of test condition within a given alloy, but it can be inferred that the Sn–Ag–Cu reliability is degrading more severely as $T_{\max}$ and ΔT increase.

Yoon et al. [30] studied the effect of changing the mean cycling temperature and dwell time [with the ΔT held constant (in this case at 100°C)] for assembled LCCC packages. These ceramic packages have a high stiffness, low CTE, and thin solder joints, so they are a worst case in terms of the applied strain per thermal cycle when assembled to standard FR-4 printed circuit boards. As shown in Figure 5.9, the Pb-free alloys outperformed Sn–Pb at the lower mean temperature, but were at parity or worse at the higher mean temperature. Given that the ΔT was the same (100°C) for all tests, the differences in the relative performance of the Sn–Ag–Cu alloy versus Sn–Pb must be attributed to the damage accumulation at the higher temperature. It is also interesting that contrary to Figure 5.8, at both ranges where $T_{\max} = 125$°C, the Pb-free assembly fatigue life was not degraded with additional dwell time, while the Sn–Pb assembly life was degraded.

Overall, Sn–Ag–Cu alloys tend to perform better at more benign thermal cycling environments. The reason could be that those alloys are more creep resistant than Sn–Pb. As shown in Section 5.3, the strain-rate sensitivity of Sn–Ag–Cu alloys is much lower than for Sn–Pb eutectic, and the creep strain rate is lower at most applied stresses. This means that for Sn–Ag–Cu, thermal cycles will accumulate

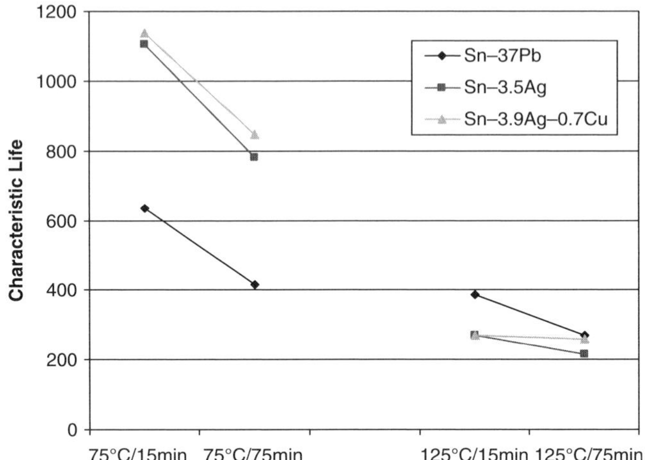

Figure 5.9. Thermal cycle results for 68 I/O LCCC. Labels show $T_{\max}$/dwell time and $\Delta T = 100°C$ for all tests. (Adapted from Ref. 30.)

more damage as the dwell time is increased, because the full creep strain created by the CTE mismatch is able to be realized within the solder over longer time periods.

Effect of Printed Circuit Board Finish on Thermal Fatigue. There are many different PCB finishes in use in the electronics industry including organic solderability preservative (OSP), electroless nickel immersion gold (ENIG), electrolytic gold over nickel, immersion silver, immersion tin, and hot air solder level (HASL). All but the ENIG and electrolytic gold are plated onto copper base metal. The PCB finish can affect the microstructure of the solder joint for a given solder alloy due to the dissolution of the finish into the solder joint and from the intermetallic layer that forms on the metal layer. These effects will be combined and are difficult to separate.

Che et al. [31] compared BGA, PQFP, and TSSOP components assembled with Sn–3.8Ag–0.7Cu solder on ENIG and OSP boards, and found the characteristic life of the OSP assemblies was worse than the ENIG by at least 30%. Amagai et al. [16] reported that Sn–Ag–Cu BGAs performed better than Sn–Pb BGAs on Ni–Au-finish PCBs, but the reverse was observed on Cu-OSP finish boards. Thus, results on one type of PCB finish cannot be predicted to hold for other finish types.

Effect of Silver Content on Sn–Ag–Cu Alloy Thermal Cycle Reliability. Many evaluations of Sn–Ag–Cu assemblies have focused on alloys between 3–4% Ag and 0.5–1.0% Cu. There has been some work conducted to measure the sensitivity of the Sn–Ag–Cu composition on reliability. Kang et al. [32] tested CBGA components* with solder balls ranging from 2.1% Ag to 3.8%

*These components appear to be very similar to the ones tested by Bartelo and Farooq.

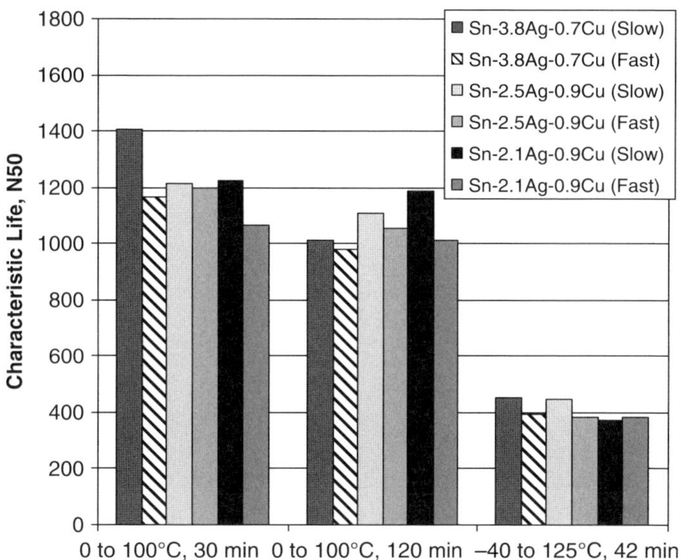

Figure 5.10. Thermal fatigue data from Kang et al. [32] on CBGAs with various solder ball alloys and reflow cooling rates. The data are grouped by test ΔT and total cycle time.

Ag and found that the α values were fairly consistent (no β values were reported) as shown in Figure 5.10. The effect of reflow cooling rate also appeared to be relatively insignificant as well. However, the effect of dwell time for the 0–100°C tests was greatest for the 3.8% Ag alloy, while the 2.1% Ag alloy showed almost no change. It should be noted that the reduction in life for 3.8% Ag at the longer dwell only served to bring its performance to parity with the other Sn–Ag–Cu alloys.

The IPC Solder Product Value Council (SPVC) also found statistically insignificant differences between Sn–3Ag–0.5Cu, Sn–3.8Ag–0.7Cu, and Sn–4Ag–0.5Cu alloys in thermal fatigue on CSP 84 I/O and TSOP 44 I/O parts tested at 0–100°C with 10-minute dwells. Other components were tested, but with very small samples sizes. The original NEMI study did compare Sn–3Ag–0.5Cu and Sn–4Ag–0.5Cu solder paste assembly for a 208PBGA with Sn–4Ag–0.5Cu solder balls. The Sn–3Ag–0.5Cu paste cell performed statistically worse than the Sn–4Ag–0.5Cu paste cell, with the α reduced approximately 8%.

Syed et al. [33] tested TSOP packages and found that the Sn–3.9Ag–0.6Cu alloy outperformed the lower silver Sn–3Ag–0.5Cu alloy for both alloy 42 and copper lead frames, with either Sn–Pb- or Sn-plated lead frames. The parts were cycled for 30-min cycles (with 12-min dwells) from −55°C to 125°C, and Table 5.4 shows the results as a function of component type and leadframe material. As expected, the copper lead frames outperform the alloy 42 given their higher CTE (17 versus 4). The ratio α to the first failure was calculated, because the ratio should trend with β in that lower values indicate less variability in the data.

TABLE 5.4. Thermal Fatigue Results from Syed et al. [33] on Lead-Frame IC Packages

Component/ Lead Frame	Paste/Plating	α (Cycles)	First fail (Cycles)	Ratio of α to First Fail
TSOP/A42	SAC1/Sn–Pb	1291	1031	0.80
	SAC2/Sn–Pb	1007	794	0.79
	SAC1/Sn	2263	1979	0.87
	SAC2/Sn	1374	1339	0.97
TSOP/Cu	SAC1/Sn–Pb	2389	2011	0.84
	SAC2/Sn–Pb	2053	1568	0.76
	SAC1/Sn	3916	3095	0.79
	SAC2/Sn	3474	2526	0.73

SAC1 stands for Sn–3.9Ag–0.6Cu and SAC2 stands for Sn–3Ag–0.5Cu.

It should be noted that the Sn-plated parts performed consistently better than the Sn–Pb-plated parts within an alloy type, and this included Sn–Pb paste assemblies (not shown). For the Sn–Pb paste parts, it would be surprising if there was a difference due to plating chemistry alone since the dilution of the Sn–Pb paste is very small for the Sn finish.

Strain Mitigation Strategies for Thermal Fatigue. The most common method of minimizing the strain on a solder joint experiencing thermal fatigue is through the use of underfill. Underfills are polymer-based materials that fill the gap between an IC package and the substrate that it is soldered to and support some of the strain that would otherwise be sustained by the solder joints. The underfill can be a composite that contains other materials that change the CTE, modulus, and T_g due to the rule of mixtures of the constituents. For example, silica is often added to underfill materials to reduce the CTE, and this has the effect of more closely matching the CTEs of the other materials and lowers the overall stress on the solder joint. It is also critical for the underfill to wet the surfaces of the solder balls, the component, and the PCB, so that there is adequate transfer of strain to the underfill that would otherwise be sustained by the solder.

5.4.1. Acceleration Versus Field Correlation

Reliability testing should be designed to re-create actual field failure modes and to exercise the product or assembly such that problems that could surface in the field are uncovered. Ideally, the reliability test will be run long enough to generate failures so that relative performance to other subsequent tests can be calculated. This may also allow the performance of the reliability test to be related to life in the field. If such tests are very expensive or time-consuming, oftentimes tests are designed to run for a fixed amount of time, and a certain number of failures (or none) are allowed. These types of tests are more often used (a) to ensure that

TABLE 5.5. Values for the Norris–Landzberg Equation

Solder	a	b	E_a (eV)
Sn–Ag–Cu	2.65	0.136	0.188
Sn–Pb	2.00	0.333	0.122

some minimum level of reliability is achieved and (b) give a means to identify the presence of infant failures that would show up earlier in the field. The drawback is that there is often no way of knowing if the reliability is changing if there are no failures to allow a reliability analysis to be run.

Tests that are used at a product level are often changed and optimized as a result of product experience in the field so that a better level of correlation to the test is achieved.

In thermal fatigue, there have been relationships developed that allow the solder joint life to be estimated for various thermal profiles. The Norris–Landzberg equation [34] predicts the acceleration factor (AF) of tin–lead solder joints for differing thermal cycling conditions.

$$\mathrm{AF} = \left(\frac{\Delta T_1}{\Delta T_2}\right)^a \left(\frac{f_2}{f_1}\right)^b \exp\left(\frac{E_a}{k}\left(\frac{1}{T_2} - \frac{1}{T_1}\right)\right) \tag{5.4}$$

where the test variables are ΔT_i (temperature delta of the cycle), T_i (maximum temperature of the cycle), and f [frequency of thermal cycles per day (minimum of six)], and the material constants are the exponents a and b, E_a (activation energy), and k (Boltzmann constant $= 8.617 \times 10^{-5}$ eV/K). Pan et al. [35] published values for these constants for Sn–Ag–Cu solder and these are shown along with those for Sn–Pb in Table 5.5. The Sn–Ag–Cu constants predict that Sn–Ag–Cu alloys will perform relatively better than Sn–Pb as dwell time shortens and the ΔT is reduced as reflected in the ratio values in the last column of Table 5.6. It should be noted that these values in Table 5.4 are the best fit over a range of components, so any comparison of the AF values is more qualitative and reflective of general trends.

Liang et al. [36] showed a relationship of creep work damage versus cycles to failure for several solder alloys. A portion of the figure is shown in Figure 5.11 and is used to explain the general trend of Pb-free solders that outperform Sn–Pb

TABLE 5.6. Calculated Acceleration Factors for Sn–Ag–Cu and Sn–Pb for the Same Assumed Test and Field Conditions Using Constants from Ref. 35

Cycle Comparison	Frequency Ratio	AF for Sn–Ag–Cu	AF for Sn–Pb	$\mathrm{AF}_{\mathrm{Sn-Ag-Cu}}/\mathrm{AF}_{\mathrm{Sn-Pb}}$
−40 to 125°C versus −10 to 60°C	4	23.5	7.0	3.4
−40 to 125°C versus 25 to 60°C	4	75.7	16.9	4.5
0 to 100°C versus −10 to 60°C	4	4.3	2.0	2.2
0 to 100°C versus 25 to 60°C	4	13.9	4.9	2.8
0 to 100°C versus 25 to 60°C	8	12.6	3.9	3.2

in many thermal fatigue tests, but sometimes underperform at extreme test conditions (high inelastic strain energy density). Apparently, the damage accumulation is greater in typical Pb-free solders as the strain density increases. A flat line would indicate a solder that is very sensitive to strain energy density (and test/field conditions), while a vertical line would be insensitive to changes in those conditions (i.e., the solder has identical life for all strain conditions). This obviously has implications for the relationship of test results to field survival. Given that most tests are run at accelerated conditions with respect to temperature, this means that the relative thermal–mechanical performance of Pb-free assemblies in the field is likely to be better than indicated in the accelerated test compared to Sn–Pb.

Clech [37] plotted characteristic life/crack area versus cyclic shear strain range for Sn–Pb and Sn–Ag–Cu joints and the plot is very similar to that in Figure 5.11, with the Sn–Ag–Cu having a steeper slope than Sn–Pb. Based on the crossover point, Sn–Ag–Cu should outperform Sn–Pb for cyclic strain values lower than 6.2% strain. Based on the variability in the test results, the crossover is more of a range than a precise value. But the expected result of both the Clech and Liang analysis is that Sn–Ag–Cu will tend to perform better relative to Sn–Pb in thermal fatigue for milder use conditions relative to the accelerated tests. Thus, the Pb-free accelerated test results will likely be conservative and add a factor of safety when compared to Sn–Pb.

Effect of Solder Voids on Thermal Fatigue. Surface-mount solder joints always have some degree of voiding due to the entrapment of gases resulting from the use of solder paste. Sn–Ag–Cu alloys have shown to have greater amounts of trapped voids compared to Sn–Pb (refer to Chapter 3). Qi et al. [38] measured the solder joint voids of thermally cycled 1206 resistors assembled with Sn–3.8Ag–0.7Cu and Sn–37Pb solder paste, and they compared when those

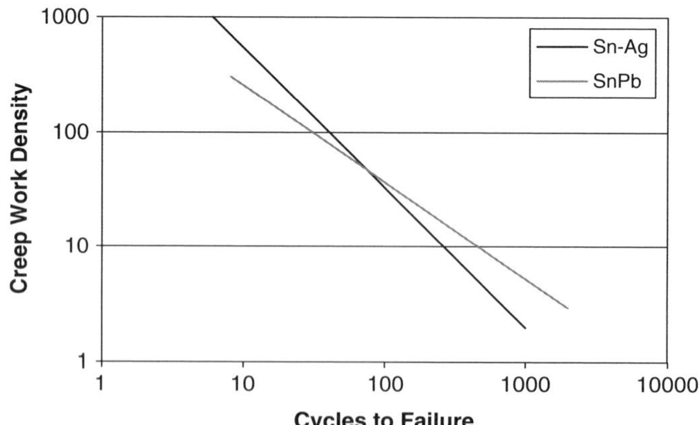

Figure 5.11. Relative fatigue life as a function of creep work density. (Adapted from Liang et al. [36].)

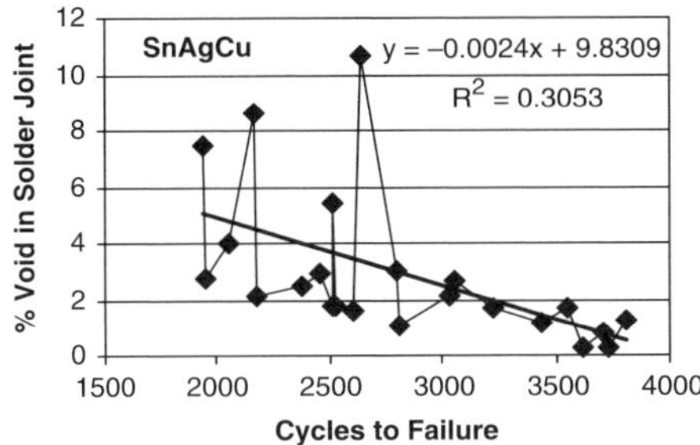

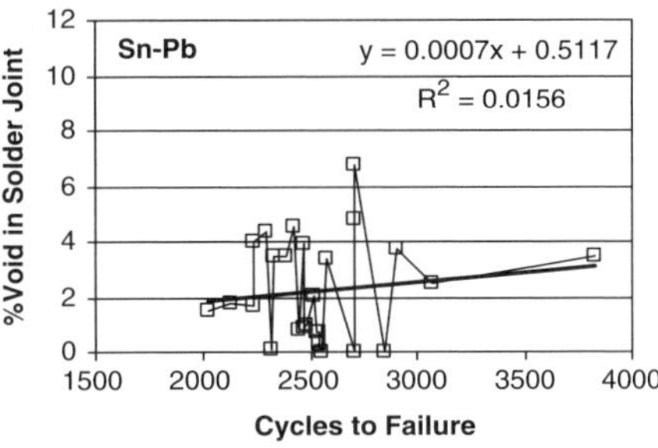

Figure 5.12. Data by Qi et al. [38] on the effect of solder voids on fatigue life for 1206 resistors. There is a weak trend for the Sn–Ag–Cu alloy and essentially no trend for Sn–Pb.

components failed as a function of the measured void level (Figure 5.12). The Sn–Ag–Cu assemblies demonstrated a weak trend of improving thermal–mechanical performance with reduced voiding, while the Sn–Pb assemblies showed no trend, indicating a random effect (almost zero slope with low R^2). The author's experience is that thermal fatigue is not negatively affected by voids except in two cases: (a) when the voids are very large, such as one seen in Figure 5.13, or (b) numerous, very small voids that collect at the solder interface (Figure 5.14) which effectively reduces the solder area where the crack propagates.

5.4.2. Effect of Pb Contamination on Thermal Fatigue of Lead-Free Solders

In the conversion to lead-free, there will be a substantial period where the existing tin–lead infrastructure will be coexisting with the emerging lead-free infrastructure

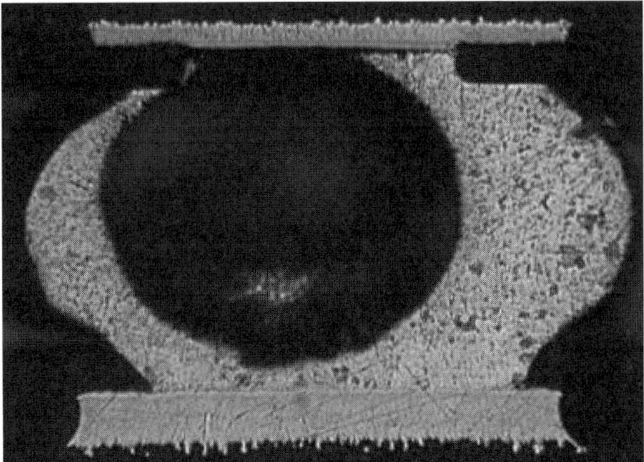

Figure 5.13. Cross section of a mixed BGA assembly (Sn–Pb ball with Sn–Ag–Cu paste) demonstrating the large voids that can form due to the differential melting points of the paste and ball.

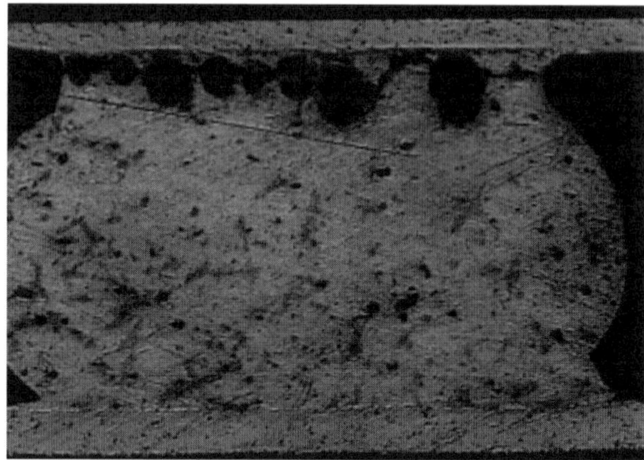

Figure 5.14. Cross section of a BGA solder joint showing small voids due to outgassing from the PCB finish. Identical parts assembled to boards with a different PCB finish did not exhibit this voiding.

as it relates to component and PCB finishes, as well as to assembly processes and rework. In these cases, whether by design or accident, there will be cases of soldering lead-containing parts with lead-free pastes and the converse. These are referred to as forward compatibility (Sn–Pb plating/balls with lead-free solder paste) and backward compatibility (lead-free plating/balls with Sn–Pb solder paste). The concern addressed here is about the solder joint integrity/reliability of such "mixed" alloy compositions rather than the potential internal component damage associated with higher reflow temperatures. For more details regarding the thermal effects on components, please refer to Chapter 4.

The change in solder composition created by the mix varies with component type. BGA solder balls have much more solder volume than the solder paste deposited to bond them to the PCB, so the balls composition will dominate the overall composition. The converse is true for lead-frame packages because the plating thickness is typically less than 10 μm while the nominal solder deposited is between 50 and 80 μm, which is roughly 50% of the stencil thickness owing to the flux content of the solder paste. Table 5.7 shows the relative Pb content for various mixed assemblies from the first NEMI project where the solder paste composition was Sn–3.9Ag–0.6Cu.

Although backward compatibility assemblies were not manufactured in the NEMI project, the comparable situation can be calculated using the same ratios of solder from the above case and assuming the paste is Sn–37Pb, while the BGAs are Sn–3.9Ag–0.6Cu and the components are plated with pure Sn. As expected, the Pb-free BGAs would have much lower levels of Pb incorporated from the paste, from 3.7 to 5.8 wt% Pb, while those for the lead-frame devices are hardly different from the paste, with the final composition from 35 to 36 wt% Pb.

TABLE 5.7. NEMI Solder Composition of Forward Compatible Mixed Technology

Component	Average Solder from Part (%)	Average Solder from Paste (%)	Nominal wt% Pb in Ball or Plating	Nominal wt% Pb in Mixed Joint
208 CSP	88.3	11.7	37	33.3
169 CSP	86.5	13.5	37	32.7
256 PBGA	91.6	8.4	37	34.4
256 CBGA	87.3	12.7	36	32.1
2512 Resistor	5.3	94.7	10	0.56
48 TSOP	3.3	96.7	10	0.37

Forward Compatibility (SnPb Plating/Balls with Lead-Free Solder Paste). Forward compatibility refers to the assembly of Sn–Pb components in a lead-free soldering process. NEMI has studied this effect in thermal fatigue for several component types, and there are substantial differences in the composition depending on whether the components are BGA-style or with leads.

In the case of mixing the Sn–Ag–Cu paste alloy with Sn–Pb components, the melting of the plating or solder balls is not an issue because they will be molten well before the paste in a reflow assembly. The NEMI data show that there were no noticeable compositional gradients within the BGA joints, because the lead-free paste fully mixed with the Sn–Pb solder balls. There was some Pb segregation due to the pasty range that results from the mixture, because of the ternary Sn–36Pb–2Ag eutectic which melts at 179°C forms [39, 40].

Chalco [41] compared the four combinations of lead-free balls and paste for a 0.5-mm BGA tested at 0–100°C with 40-min cycles. The results are shown in Table 5.8 and the Pb-free performed better than all other combinations, including the standard Sn–Pb combination. All three sets of assemblies containing Pb were very similar in performance, including α (characteristic life) and β.

Clech [37] summarized data from several papers to compare the life of mixed assemblies versus standard Sn–Pb assemblies. The data used were early failure data (typically to 1% or 5% failure rate) taken from Weibull curves, and there was a mix of PCB finishes and component types. Using that criterion, there was not a strong trend comparing BGA assemblies of Sn–Pb ball + Sn–Ag–Cu paste

TABLE 5.8. Chalco Data for Mixed and Unmixed BGAs

Ball/Paste[a]	α (Cycles)	1% Fail (Cycles)	β
Pb-free/Pb-free	2078	1345	8.4
Pb-free/SnPb	1365	873	10.3
Sn–Pb/Pb-free	1114	773	12.6
Sn–Pb/Sn–Pb	1235	861	12.8

[a]Pb-free ball is Sn–3.3Ag–0.7Cu and paste is Sn–3.9Ag–0.6Cu; Sn–Pb ball and pastes are Sn–37Pb.

versus Sn–Pb ball + Sn–Pb paste. The mixed technology generally underperformed standard Sn–Pb for the 0–100°C test conditions.

Hot Cracking with Mixed Technology. One potentially critical issue with introducing Pb into Pb-free solders such as Sn–Ag–Cu is solder crack formation due to hot tearing, which occurs during solidification as a result of the expanded melting range. This effect was documented for pin-in-hole wave-soldered joints in the 1997 NCMS study [42] and was referred to as *fillet lifting*. The degree of fillet lifting/hot cracking correlated with the melting range of the solder, and Pb additions to Sn–Ag–Cu alloys create a 30–40°C delta between liquidus and solidus temperatures. An example of this phenomenon is shown for a lead frame with Sn–20Pb plating on a surface mount component that was reflowed to an OSP board with Sn–Ag–Cu solder paste. The solder joints had varying degrees of cracking after reflow assembly, and these cracks promoted early failures in stress testing. An optical micrograph (Figure 5.15) of a part after reflow shows the crack forms close to the IMC layer, where the Pb level is higher due to the reaction of Sn with the copper substrate. Figure 5.16 is an SEM micrograph showing the

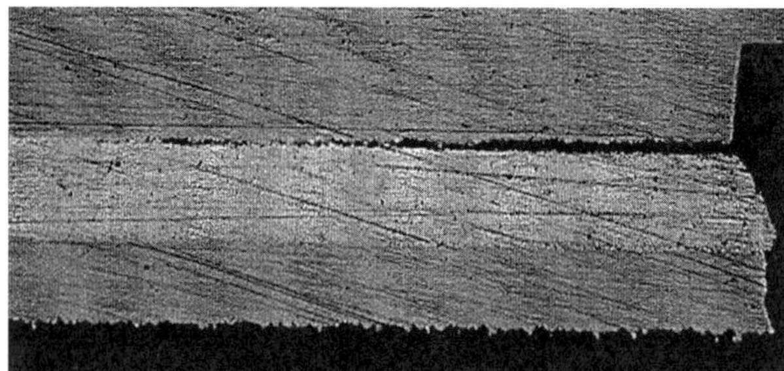

Figure 5.15. Example of a mixed solder joint (Sn–Pb plating with Sn–Ag–Cu solder) demonstrating hot tearing of after reflow assembly.

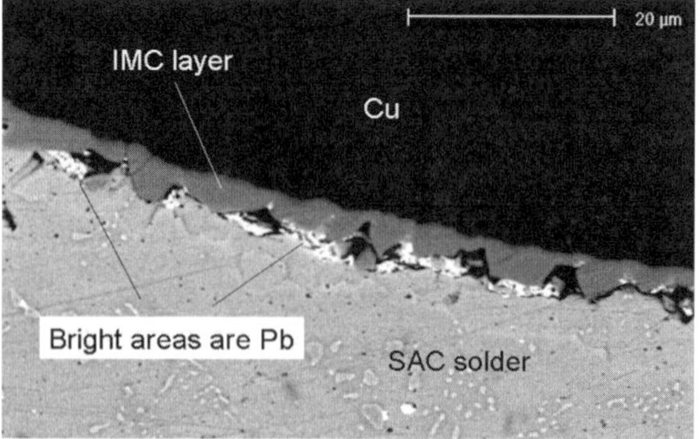

Figure 5.16. SEM micrograph showing solder separation from the IMC layer due to Pb enrichment after reflow assembly.

TABLE 5.9. Table Comparing Reflow Cracking of Sn–Pb Versus Pure Sn Plating

Lead-Frame Plating	Average Post Reflow Crack Area (%)
Sn–20Pb	12.6
Sn	0.0

cracks associated with the Pb-rich areas just ahead of the main crack. Components with a pure tin finish did not form cracks when assembled in the same reflow environment (Table 5.9).

Backward Compatibility (Lead-Free Plating/Balls with Sn–Pb Solder Paste). Backward compatibility refers to the ability for lead-free components to be successfully implemented into a Sn–Pb assembly process. The use of lead-free finishes has been widespread for years, because many PWB finishes are lead-free such as organic solderability preservatives (OSP), immersion gold, immersion silver and immersion tin. Also, components such as connectors and contacts have been plated with gold and/or palladium and used in Sn–Pb surface mount applications, and many passives and shields have pure tin terminations. This is not to say that some issues could not come up, but there is an extensive experience base for this combination that bodes well for most applications.

Intel published results comparing Sn–Ag–Cu and Sn–Pb BGAs assembled with Sn–Pb paste tested from −40°C to 125°C [43]. The data for 15 mm × 15 mm BGA are shown in Table 5.10 and the Sn–Ag–Cu packages were assembled using two reflow conditions, 208°C peak and 220°C peak, while the Sn–Pb packages were processed using an undisclosed "standard" reflow profile. The 208°C peak generally did not fully melt the Sn–Ag–Cu ball while the 222°C peak did so. The comparative reliability between Sn–Ag–Cu and Sn–Pb packages was slightly variable, with the higher peak reflow condition performing worse when comparing α values.

TABLE 5.10. Data for Backward Compatibility from Hua et al. [43]

Solder Ball	Finish[a]	Profile (°C)	α(cycles)	β
Sn–Ag–Cu	HASL	208s	3007	5.6
Sn–Ag–Cu	HASL	220s	2621	6.2
Sn–Ag–Cu	HASL	208r	2978	6.4
Sn–Pb	HASL	Control	3055	10.4
Sn–Ag–Cu	OSP	208s	3313	15.0
Sn–Ag–Cu	OSP	220s	2509	7.5
Sn–Ag–Cu	OSP	208r	3355	5.9
Sn–Pb	OSP	Control	2549	12.4

[a]HASL was of Sn–Pb composition.

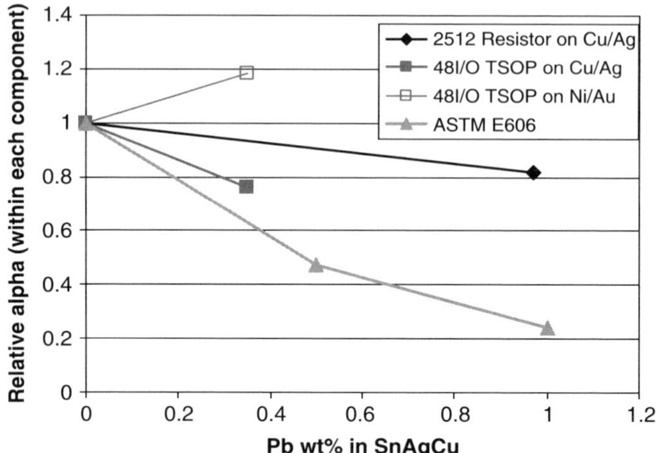

Figure 5.17. Relative fatigue life for surface mount components and E606 bulk solder fatigue samples as a function of Pb content. The comparison used for the components is the Weibull α.

However, the Weibull slopes (β) are generally shallower for the Sn–Ag–Cu packages, indicating greater variability in the part-to-part reliability. An interesting thing to note is that the mixed alloy was comparable to the fully Sn–Pb control.

The impact of Pb in the joint is not consistent and appears to depend on the local stress in the joint and the joint geometry. Seelig performed an ASTM E606 low-cycle fatigue test on bulk solder samples and found that increasing Pb content in Sn–Ag–Cu decreases the number of cycles at which the failure criterion of a 20% load drop is reached [44]. The NEMI reliability results on leaded devices also show that generally the introduction of Pb to the Sn–Ag–Cu solder joint is deleterious to the fatigue life, but it is not always the case. Figure 5.17 has the relative fatigue life of several types of components tested by NEMI project and the Seelig data as a function of Pb content. Note that the life is degraded, though not as severely as found by the low-cycle fatigue test, but for the 48 I/O TSOP component reflowed to Ni–Au, the mixed alloy performed better than the pure Sn–Ag–Cu.

Clech [37] summarized data from a number of papers to compare the life of assemblies with Sn–Ag–Cu ball + Sn–Pb paste versus Sn–Pb ball + Sn–Pb paste. The data used were cycles to low failures (typically to 1% or first failure) taken from Weibull curves, and there was a wide mix of PCB finishes and component types. Using that criterion, the data were scattered with the mixed assemblies both underperforming and outperforming standard Sn–Pb for a range of test conditions.

One other issue with backward compatibility is excessive voiding with Sn–Pb solder balls. Although many evaluations have successfully soldered such BGAs, the process is susceptible to excessive voiding that can create poor solder joints and potential shorting. An example of this is shown in Figure 5.13 and is discussed more fully in Chapter 3.

5.5. CREEP RUPTURE

Creep rupture is a failure mode that is active when solder joints are under continuous monotonic stress, and those joints are able to deform sufficiently to completely open and create electrical failure. Although solder is capable of measurable creep in many use conditions due to its low T_h, this is more of an issue at elevated temperatures/stresses which are able to create sufficient strain to cause failure. Given that Sn–Ag–Cu alloys have lower creep rates than Sn–Pb (Figures 5.6 and 5.7), this is expected to increase the rupture time of these alloys relative to Sn–Pb. The slope of time to failure versus applied stress for creep rupture shows that the Pb-free alloys have a shallower slope compared to Sn–Pb; that is, the lower stresses result in much better rupture properties for Pb-free solders. This is analogous to the stress exponents of the creep equations.

The data in the literature [45–48] show that Sn–Ag–Cu tends to perform better in terms of time to rupture versus Sn–Pb for most at elevated temperature conditions, and this is illustrated in Figure 5.18. Wu [48] tested both bulk solder samples and through-hole solder joint samples in creep rupture and found that the comparative performance of the Pb-free and Sn–Pb alloys was different for the two types (Figure 5.19). For the through-hole joints, the Pb-free alloys (Sn–0.5Cu and Sn–3.5Ag) were much better than Sn–37Pb at 130°C, but there was a crossover point where Sn–Pb was better at 30°C at the highest load tested ($\approx$ 30 MPa). For the bulk samples, Sn–Pb has a higher slope and there is a crossover at 20 MPa with Sn–3.5Ag. Shin and Yu [49] found that bismuth affects the creep rate of Sn–Ag solders and also greatly affects the rupture times. A moderate amount of Bi (around 2 wt%) gives the best results, while increasing the Bi degrades the performance, especially at lower applied stresses.

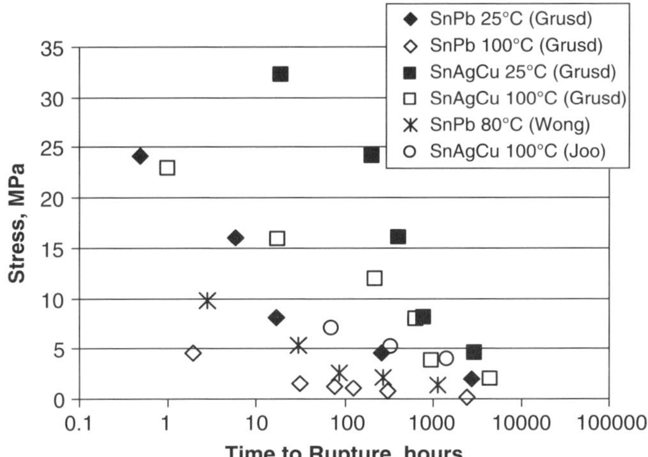

Figure 5.18. Creep rupture data for bulk samples comparing Sn–Ag–Cu and Sn–37Pb.

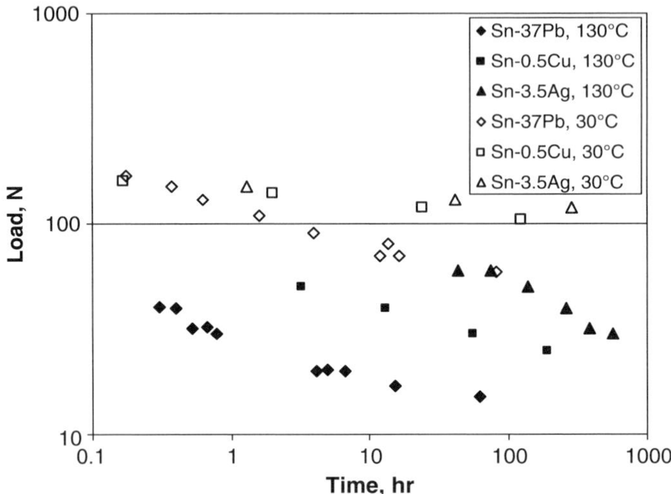

Figure 5.19. Creep rupture data on through-hole solder joints. The nominal area of the joints was 5.2 mm^2, not including the fillet. (Adapted from Wu et al. [48].)

Bradley and Mulligan ran dead weight load tests test on leadframe devices assembled to printed circuit boards [50]. The tests were run at various temperatures and stresses with in three material combinations: (1) Sn–Pb solder paste with Sn–10Pb lead-frame plating, (2) Sn–Ag–Cu paste with Sn–10Pb plating, and (3) Sn–Ag–Cu paste with copper lead frames (the Sn–Pb plating was etched away). The stress was nominally calculated by the applied load/solder joint area, and the time to failure was measured via automatic data acquisition through a daisy chain circuit. As shown in Figure 5.20, the fully lead-free Sn–Ag–Cu assembly had

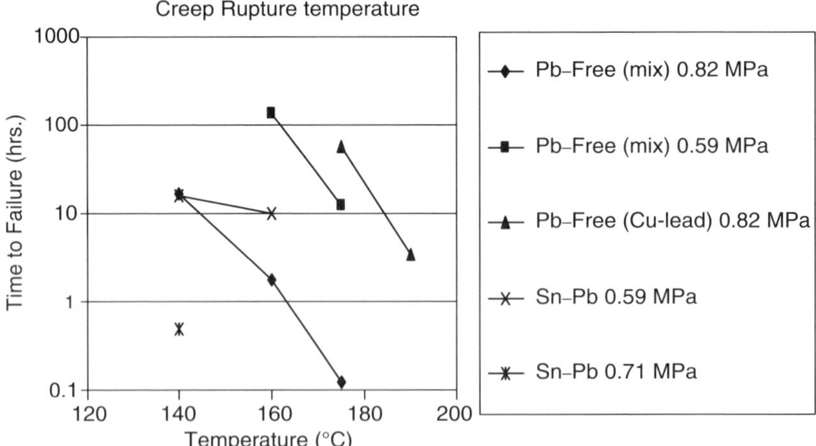

Figure 5.20. Creep rupture data for SMT joints showing the better performance of the Pb-free (Sn–Ag–Cu) joints compared to mixed technology and Sn–Pb.

substantially greater times to failure at a given temperature and stress level. The lifetimes of Pb-doped Sn–Ag–Cu fell between the pure Sn–Ag–Cu and pure Sn–Pb. Thus, the creep rupture data trends for SMT joints follow those from the bulk samples. The failure mode for the joints was typically separation at the solder/intermetallic interface for all three material combinations.

5.6. ISOTHERMAL (MECHANICAL) FATIGUE

Mechanical fatigue is defined here as a repeated applied stress or strain in a constant temperature environment. This is a less published topic in the literature compared to thermal fatigue, but can be important in certain applications such as vibration in automobiles as well as keypad/button actuations in cell phones and radios. In the case of keypads and buttons, components are often on the opposite side of the PWB from the keypad, and the pressing of the keypad flexes the PWB and mechanically cycles the solder joints, potentially causing failure.

Kariya and Otsuka [51] and Kanchanomai et al. [52, 53] have evaluated several Sn–Ag alloys in isothermal fatigue (Δ_{strain} varying from 0.5% to 2%) using bulk samples and found that Sn–Ag–Cu outperformed Sn–Ag–Bi, with increasing Bi content degrading the performance. Similar testing was performed comparing Sn–Pb to Sn–Ag, and the eutectic Sn–Pb was outperformed by Sn–Ag [54]. Figure 5.21 summarizes the data from Kanchanomai et al. [52, 53]. The interesting factor is that the data for all of these alloys collapse on a straight line when normalized for true ductility. The true ductility was measured in uniaxial tension at a strain rate of 0.04/s on samples of similar size as the mechanical fatigue samples. Thus, the more ductile solder alloys perform better in isothermal mechanical fatigue.

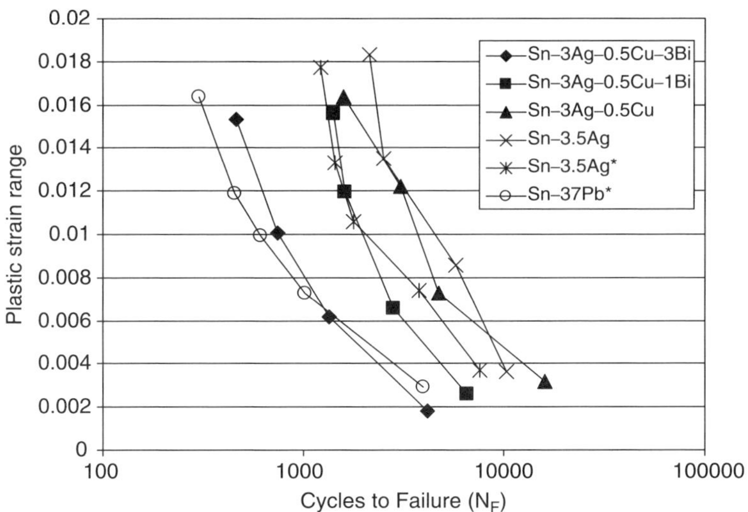

Figure 5.21. Mechanical cycling data from Kanchanomai et al. [52, 53].

TABLE 5.11. True Ductility of Solders as Measured by Kanchanomai et al. [52, 53]

Alloy	Tensile Strength, MPa	True Ductility
Sn–3.5Ag	37.5	1.60
Sn–3Ag–0.5Cu	50.6	1.23
Sn–3Ag–0.5Cu–1Bi	72.6	0.80
Sn–3Ag–0.5Cu–3Bi	91.3	0.36

Kariya and Otsuka [51] tested a number of Sn–3.5Ag alloys with various ternary additions in order to measure their effect on mechanical fatigue performance. Copper and zinc had little effect, while indium moderately reduced the fatigue performance and bismuth was by far the most deleterious. All of these Sn–Ag ternary alloys were better in mechanical fatigue than Sn–Pb, with the exception of the higher Bi-level additions.

Data on actual soldered assemblies shows similar trends, but has the added influence of the intermetallic layer between the solder and the base metal. Lin et al. [55] published data on flip-chip assemblies and found that Sn–0.7Cu outperformed Sn–Ag–Cu and Sn–37Pb in isothermal fatigue. The Sn–Cu was consistent on both Ni and Cu substrates, while the Sn–Ag–Cu was much better on Cu than Ni.

Effect of Pb-Contamination on Mechanical Fatigue. Oliver et al. [56] tested pin-in-ring solder joints doped with Pb (Figure 5.22). Overall displacement in the solder joint was between 2 and 3 microns, or approximately 5% strain. Small amounts of Pb (up to 2%) did sharply degrade performance of Sn–3.5Ag–0.7Cu solder when mechanically cycled, but only to the level of eutectic Sn–Pb.

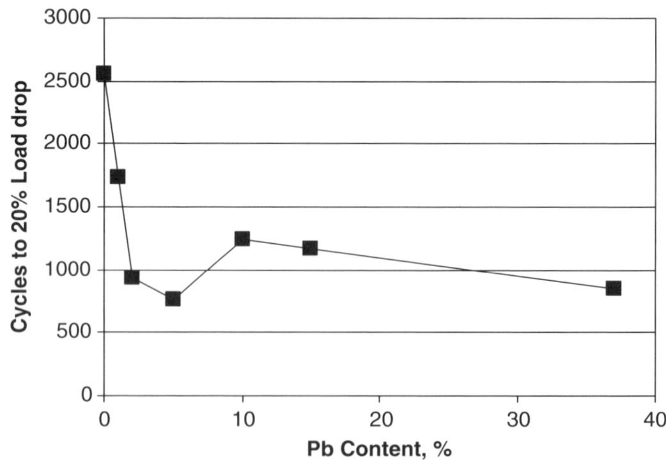

Figure 5.22. Mechanical fatigue results for pin-in-ring samples. The alloys tested were Sn–3.5Ag–0.7Cu doped with varying amounts of Pb. (Adapted from Oliver et al. [56].)

Isothermal Mechanical Cycling on BGA Assemblies. Although the trends identified in larger bulk solder samples can be valuable, data generated with actual solder assemblies is almost always more relevant as a means of understanding how actual SMT assemblies will behave. Park and Lee [48] performed mechanical fatigue tests on BGA-style samples in various orientations, ranging from 0° (pure tension) to 90° (pure shear). The balls were 0.76 mm in diameter, soldered to a PCB substrate with 0.58-mm SMD Ni–Au finish pads. The results from 0° and 90° tests are shown in Figure 5.23 and the Sn–Ag–Cu alloy outperformed the Sn–Pb eutectic by a significant amount at both orientations, agreeing with the trends shown for the bulk sample data shown above. The mixed-mode tests showed similar trends.

Vibration Tests. Vibration applies small repeated deformations to solder joints and is associated with high-cycle fatigue. Often in the field, the bulk of cycles will

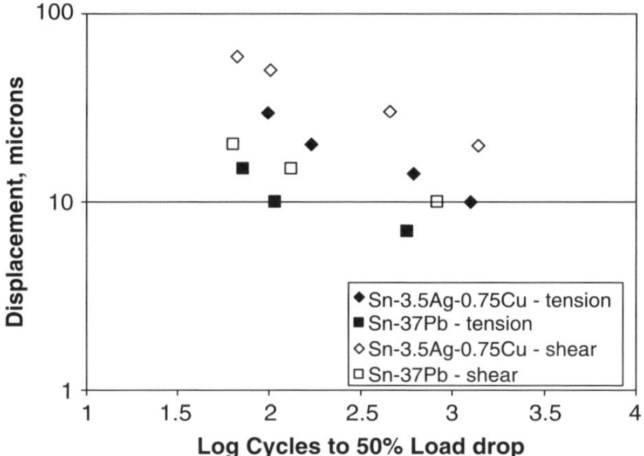

Figure 5.23. Isothermal fatigue results in tension and shear for BGA-style samples. The failure criterion is a 50% load drop during the test. (Adapted from Park and Lee [57].)

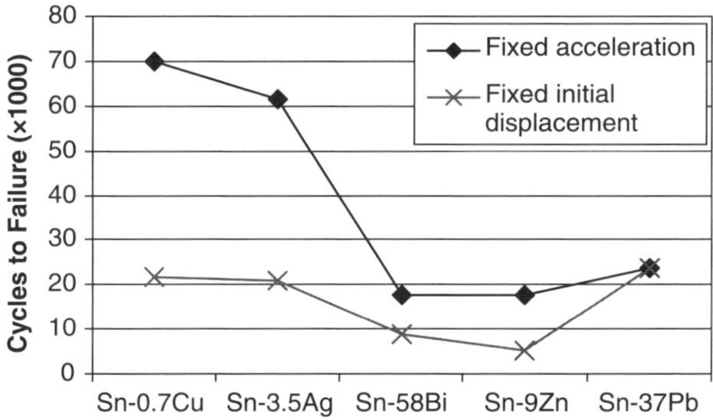

Figure 5.24. Data from Song et al. [58] on vibration testing of bulk solder samples.

be at a specific temperature, so the stress is nominally isothermal. Song et al. [58] tested bulk samples of numerous Pb-free alloys and Sn–Pb and saw that Sn–0.7Cu and Sn–3.5Ag greatly outperformed Sn–37Pb in fixed acceleration conditions, but were comparable at fixed displacement.

5.7. OUT-OF-PLANE BENDING

Out-of-plane bending (hereafter referred to as bending) is an area that has gathered more interest in the literature in recent years. Bending can overload the solder joints and break them outright if a board is deformed sufficiently or the components are in an area that experiences a high amount of strain due to other nearby components and the location of mounting screws or snaps. A monotonic load or strain can cause a creep rupture failure, while repeated bending such as induced by repeated pressing of a button mounted on a PCB can fatigue the solder joint to failure. The PCBs of portable electronics are often subjected to bending in product test and assembly as well as in customer usage. Stiffer components such as ceramic filters will have stress concentrations at the edges of solder joints, and BGAs with the barrel shaped joints are susceptible to interfacial fracture during bending [59, 60]. BGAs typically have a barrel shape that results in the highest strain and stress very close to the component or PCB interface.

There are several mechanisms that can cause a decrease in impact performance. These include nucleation and growth of secondary intermetallic layers at the solder/metal interface [61], black pad issue associated with electroless nickel/immersion gold PWB finish [62], and micro-voids that collect at the interface. An example of micro-voiding is shown in Figure 5.14 and is due to organics incorporated into plating deposits that outgas into the solder during reflow [63].

5.7.1. Gold Embrittlement

Gold embrittlement is a well-known issue with Sn–Pb alloys. In the classic case of gold embrittlement in Sn–Pb solder, $AuSn_4$ intermetallic plates form upon

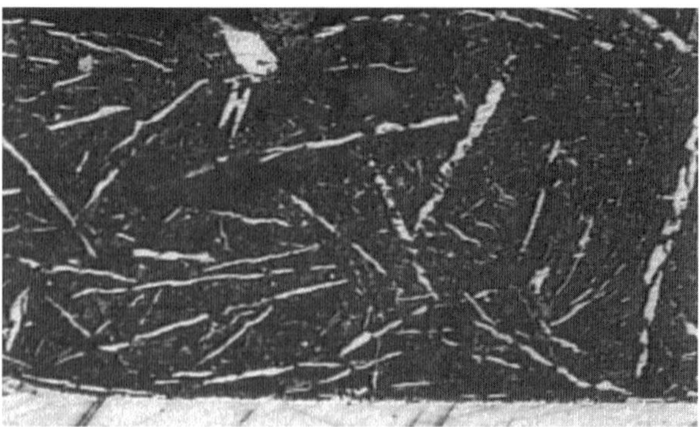

Figure 5.25. Micrograph of $AuSn_4$ plates in a Sn–Pb solder joint.

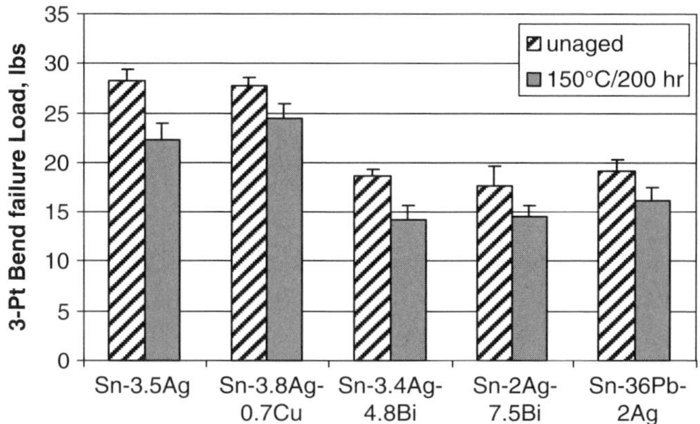

Figure 5.26. Three-point bend results for a 28 I/O LCCC device used to evaluate gold embrittlement susceptibility. The Sn–Ag and Sn–Ag–Cu alloys performed better than the Sn–Pb alloy, even after thermal aging at 150°C.

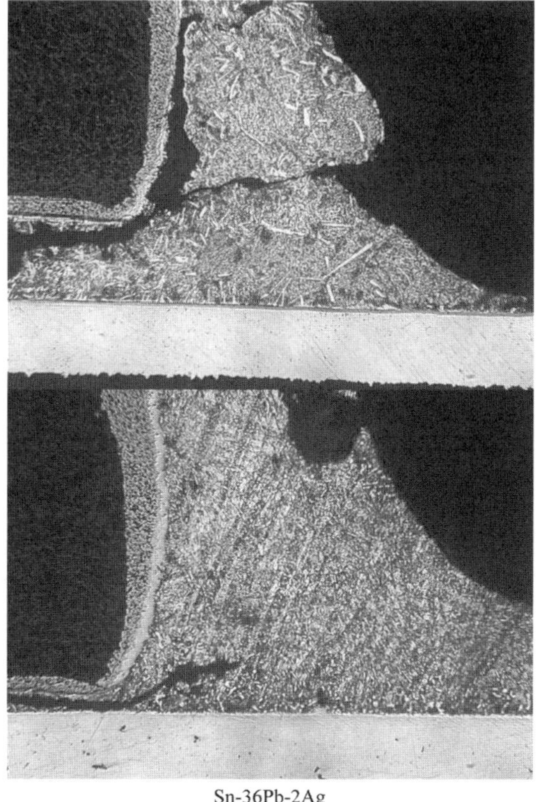

Sn-36Pb-2Ag

Figure 5.27. Comparison of LCCC solder joints showing the presence of $AuSn_4$ plates in the Sn–Pb–Ag and not in Sn–Ag–Cu for bend test samples.

solidification (Figure 5.25) and the interface between the Au–Sn_4 plates and the solder is weak mechanically [64].

In order to evaluate the effect of gold on the comparative bending performance of Pb-free alloys to Sn–Pb, an LCCC module assembly was characterized in three-point quasi-static bending. The LCCC module has a nominal 50 μin. (1.2 μm) of gold over nickel, and this resulted in a nominal gold content of about 3 wt% in the solder joint. Figure 5.26 is a plot showing the relative bending performance of the alloys for un-aged and aged assemblies. The Sn–Ag(Cu) alloys are much stronger even with aging compared to the Sn–Pb and Sn–Ag–Bi alloys. This result agrees with the microstructures, because the Sn–Ag–Cu does not have the $AuSn_4$ plates in the joint [65] at this gold composition and reflow cooling rate (Figure 5.27). Also note that in Figure 5.28 there are numerous examples of fracture

Sn-36Pb-2Ag

Figure 5.28. Perspective view of LCCC fracture surfaces after three-point bend test, and note the shiny $AuSn_4$ intermetallic surfaces on the Sn–Pb sample. The Sn–Ag–Cu sample has a much more ductile fracture surface.

along the $AuSn_4$ plates in the Sn–Pb joints (shiny features on the surface), while the Sn–Ag–Cu does not show that phenomenon and has a much more matte fracture surface appearance. Jacobson and Humpston [66] showed that phase diagrams predict that compared to eutectic Sn–Pb, eutectic Sn–Ag solder can accept up to twice the gold content before $AuSn_4$ plates will form (10 wt% versus 5 wt% Au). The level of gold in the solder joint needed to create the IMC plates will be affected by the actual solder composition and solidification kinetics.

5.8. IMPACT/SHOCK LOADING

Portable or handheld products are prone to being dropped during use, which can cause large amounts of printed circuit board bending at substantial deformation rates that can result in fractured solder joints. This is often referred to as impact loading and is more likely to affect stiff components and BGAs. Although such handling by the consumer is not encouraged, there may be a benefit to designing products that are reasonably robust with regard to this loading condition. The susceptibility of a solder joint to sustain damage during drop is related to many variables including the PCB thickness, arrangement of PCB supports, component types and their layout on the board, and solder joint quality. The amount of localized deformation resulting from drop can be larger than expected given the variable stiffness on the printed circuit board created by the components. Figure 5.29 is a plot showing the printed circuit board strain versus time on a cellular phone type product in the vicinity of a BGA component. In addition to the amount of board flexure created by the drop, the strain rate of the board deformation is on the order of $3\ s^{-1}$.

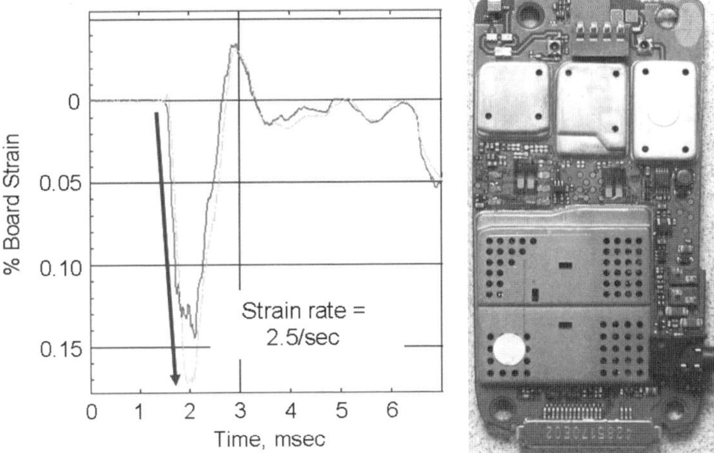

Figure 5.29. The strain versus time plot for a drop test from a board assembled into the phone housing. Also shown is the tested board with the attached strain gages. The components of interest are on the opposite side of the board.

Together, the strain and strain-rate effects can confound conclusions drawn from tests conducted at slower strain rates.

Product-level tests are the ultimate criteria to evaluate component assembly reliability in drop. One drawback to this method of testing is the variability of the impact orientation induced in releasing the product, whether manually or by machine due to rotation occurring in the interval between release and impact. It is very difficult to not induce some degree of rotation during drop, and that rotation will vary from drop to drop. Although this variability is a real-life phenomenon, it makes achieving test repeatability difficult, especially if there is a specific impact orientation that is much more prone to creating solder joint damage. One means to minimize this variation is by use of a drop tower that physically holds the sample in the same orientation and results in a more repeatable test.

There has been effort to develop test methods for component assemblies, such as JESD22-B111, to address this type of loading phenomenon [67]. However, this test requires a daisy-chained test board and a shock tower and can have anywhere from 1 to 15 components on the test board. The B111 test does not apply a large amount of board strain per impact, and the dependent test variable is number of impacts to electrical failure. The board-bending strain near the component will vary depending on its position on the board for setups with more than one component.

In order to more consistently characterize BGA assemblies for their performance in drop-like conditions and allow measurement of precise failure criteria, a novel test was developed to measure the board strain-to-failure for assembled packages during impact [68]. The test board is mounted in a four-point bend fixture, and a weight is dropped onto the upper span to create the rapid board bending and stress the solder joints (Figure 5.30). The strain versus time signature is shown in Figure 5.31, and the strain rate is close to that seen in product level

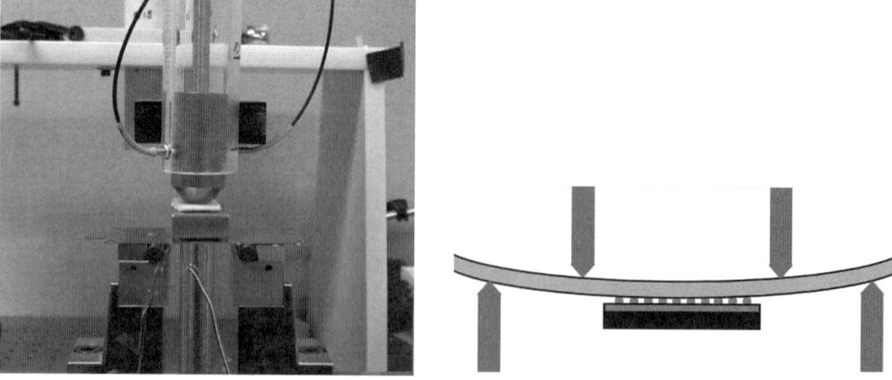

Figure 5.30. Four-point bend impact test setup, with the weight resting on the upper span and the board assembly sandwiched between the upper and lower spans. A schematic showing the strain on the board is shown.

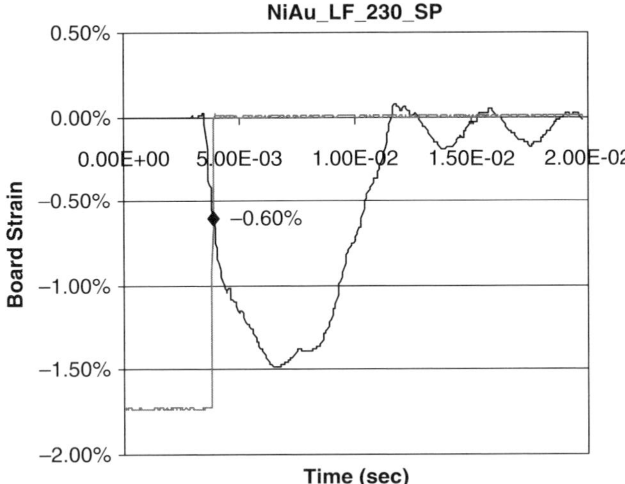

Figure 5.31. Strain versus time plot generated by the test fixture and board shown in Figure 5.30. The step function is the continuity plot of the BGA package and indicates the first fracture of a solder joint during the impact event. The intersection of the two curves is the amount of board strain required to break the first BGA solder joint.

drop tests. In the example shown, daisy-chain samples are used to measure the precise PCB bending strain when the first solder joint fails. However, non-daisy-chained samples can be used because the impact strain can be varied with the drop height.

An interesting result from this test is that Pb-free BGAs often fail at a lower PCB strain than same parts with Sn–Pb solder joints when subjected to impact-type loading [67, 69]. Figure 5.32 shows a probability plot of Sn–Pb and Sn–Ag–Cu BGAs and the Sn–Pb outperforms by a factor of about 2x. A comparison of the

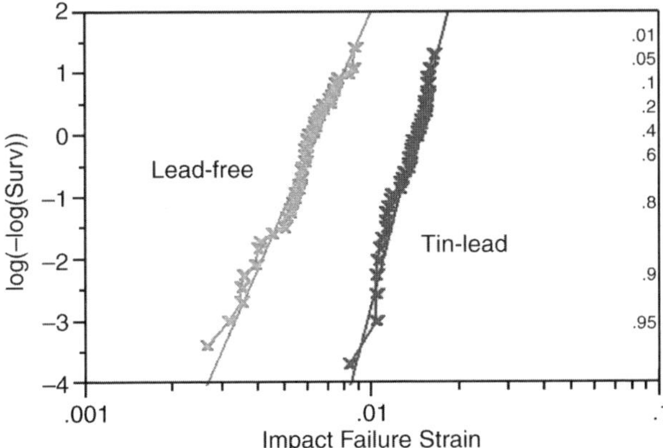

Figure 5.32. Board strain at failure for tin–lead and lead-free BGAs tested at two different deformation rates.

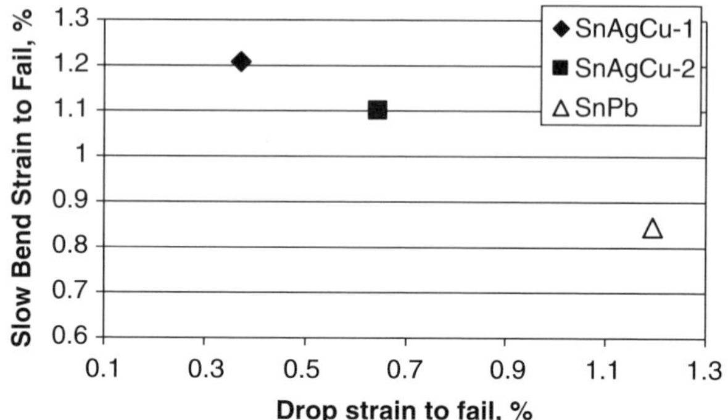

Figure 5.33. Comparison of BGA assemblies in quasi-static bending versus impact loading showing that the bend results do not correlate with drop/impact performance.

PCB board failure strain at a quasi-static strain rate versus the four-point bend impact test shows that there is a substantial difference in the performance of the Pb-free joints relative to the lead-free solder joints.

The reason a high strain rate impact test is needed to characterize BGA impact performance is shown in Figure 5.33 where BGA assemblies were tested at two different bend speeds. The Pb-free BGAs outperform Sn–Pb at quasi-static bending (0.001/s) yet are substantially worse in impact-type conditions than SnPb. Thus the quasi-static bend results cannot be used to correlate with impact performance.

The solder joint failure mode for impact type events is often within the IMC layer at the solder/BGA-side metal pad, with most of the intermetallic layer breaking cleanly away with the solder. Figure 5.34*a* is a cross-section view showing the crack in the IMC layer next to the nickel base metal of the BGA pad. Figure 5.34*b* is a perspective view taken in an SEM of a fractured BGA solder

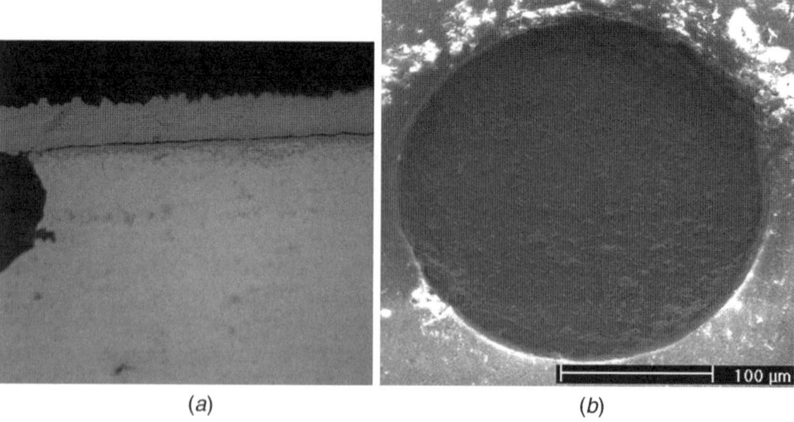

Figure 5.34. (*a*) Cross section of a fractured BGA solder joint showing the crack at the IMC layer and (*b*) perspective of the pad showing the relatively clean fracture surface with little residual IMC or solder remaining.

pad exposed after impact. The separation is very clean, with only a limited amount of intermetallic material remaining on the pad.

Effects of Substrate Metallization on BGA Impact Performance. Given that the fracture location for BGAs in impact loading is often within the IMC layer at the base metal/solder interface, the substrate metallization can be a critical variable related to impact performance. BGA substrates have typically been finished with electroplated nickel–gold, with the gold needed for wire-bonding on the reverse side of the substrate. This thickness of gold is minimized for solder reliability purposes, but must be thick enough for reliable wire bonding. The standard thickness of gold is around 0.5 μm, and this creates a gold content in the solder ball of about 0.5 wt% for a 0.5-mm pitch BGA. Given that the PCB finish is often Cu-based, it makes sense that the BGA solder strength could be improved by using a Cu-based finish, such as OSP. Lal et al. [70] compared the impact performance of BGA packages finished with OSP to Ni–Au and found that the results were comparable at lower peak reflow conditions, but the Ni–Au improved relative to the OSP at higher peak reflow temperatures (Figure 5.35). Additional work in this area to study the sensitivity of impact performance to plating/finish variables would be valuable to the industry.

Effects of Solder Alloy Composition on Impact Performance. Documenting the differences between BGAs with Sn–Pb and Sn–Ag–Cu solder balls, there is interest in how the Pb-free solders behave in order to optimize the strength of the assemblies. Amagai et al. [71, 73] have evaluated BGAs under impact conditions.

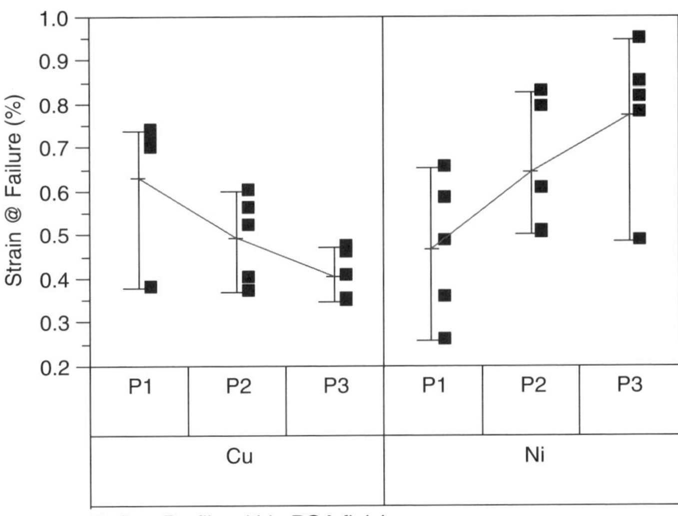

Figure 5.35. Impact performance of Sn–Ag–Cu BGAs as a function of peak reflow temperature and BGA substrate metallization. P1 = 230°C, P2 = 246°C, P3 = 262°C. (Adapted from Ref. 70.)

In one study, BGAs bumped with Sn–Ag–Cu alloys having varying silver content of 0–3 wt% along with 1% Cu were evaluated for repeated impact performance. The lower silver content alloys performed better as shown in Figure 5.36, with a greater number of impacts required to create open solder joints. This type of repeated impact test is more of a fatigue test, given that at least 30 impact events were needed to create a solder joint failure. However, the failure mode is still apparently brittle, because it is located at the solder–BGA base metal interface.

Interfacial Voiding and Effect on Impact Loading. One concern with lead-free solder joints relates the development of interfacial voids in the IMC layer(s) next to the metal pad substrate. Kirkendall voids form due to differential diffusion rates of atomic species as they migrate across a boundary under thermodynamic driving forces. In a solder joint, the solder typically forms an intermetallic compound with the base metal that was soldered, forming a covalent crystal structure between components of the solder and the base metal. If that solder joint is exposed to sufficiently high temperatures over a sufficiently long time, the migration of material may form voids at that interface. This is of interest if enough voids form at the interface that the interface is weakened and fails at lower stresses relative to a nonvoided solder assembly.

A number of papers that studied intermetallic growth have commented on the presence (or lack thereof) of voids in the interfacial intermetallic layer(s). The one trend that appears is that when voids do form, they are present at an interface where the solder bonded to a copper substrate [72–74]. Chiu et al. [73] found severe voiding in a copper-substrate BGA soldered to a copper finish PCB that was aged at 125°C for 1000 hours, and the number of drops to failure was reduced by 80% after only 240 hours of aging. Mei et al. [74] performed aging

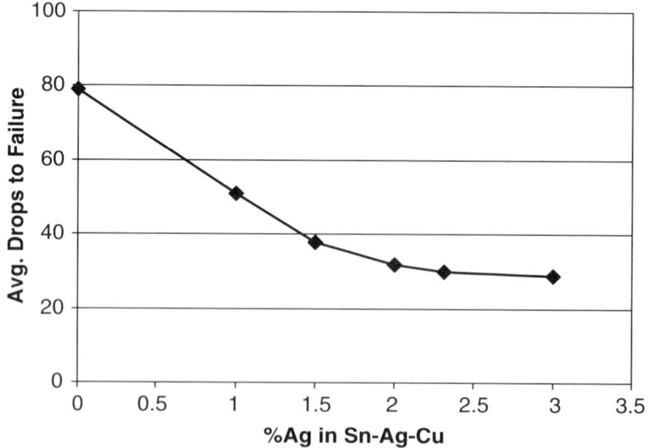

Figure 5.36. Plot of silver content versus number of impact events to cause solder joint failure in as-reflowed BGA assemblies using Sn–Ag–Cu solder balls. Note that the curve flattens out above 2% Ag. (Adapted from Ref. 71.)

experiments on various components with both Sn–Pb and Pb-free solder balls and plating, and in general they found lower levels of voiding compared to Chiu et al. and did not find impact performance degradation in his tests. Mei et al. developed an equation using Chiu et al.'s data to make a worst-case estimate of interfacial voiding as a function of temperature and aging time.

$$A = Ct^{0.5} \exp(-Q/RT) \tag{5.5}$$

where A is the fraction of void length to the total length of the interface in a cross section, C is a material constant, t is time in days, Q is the activation energy, T is the absolute temperature in kelvin, and R is the gas constant. For the Chiu et al. dataset, Mei et al. calculated that $C = 145{,}590\ \text{day}^{-0.5}$ while $Q = 45{,}413$ J/mol.

What Eq. (5.5) shows is that the concern with interfacial voids is with high temperature applications. Table 5.12 shows the predicted length of time to reach 50% void area at the interface using Eq. (5.5) for various temperature exposures. Mei et al.'s results showed much less interfacial voiding than predicted by Chiu et al.'s data, so Chiu et al.'s data were assumed to be a worst-case upper limit.

This author also has studied interfacial voiding in BGA assemblies, and the data shows that there is variation depending on the part supplier. As shown in Figure 5.37, there is some IMC voiding for Cu substrate BGAs aged at 175°C at 70 hours. There were no voids seen in parts with Ni substrates. Based on Mei's equation, the estimated amount of voiding at 175°C/70 hours would be extremely high, so this is additional evidence that the solder interfacial voiding is very specific to the type of plating and the specific solder joint configuration and may not be specifically due to the Kirkendall mechanism. Any application at high temperatures should be examined to characterize void formation and ensure, if present, that it is not an issue.

In any interfacial voiding study conducted on actual solder joints (especially for BGAs), it must be noted if the solder joint is single-substrate/unassembled (e.g., a loose BGA component) or double-substrate/assembled (e.g., BGA assembled to a PCB). This is important in that the interaction between the two interfaces can lead to vastly different results than that found with single interface solder samples due to the different chemical gradients developed by diffusion.

TABLE 5.12. Mei et al.'s Calculation Using Chiu et al.'s Data

Temperature (°C)	Time to 50% Interface Voiding (days)
50	5747
75	506
100	62
125	9.8
150	1.9

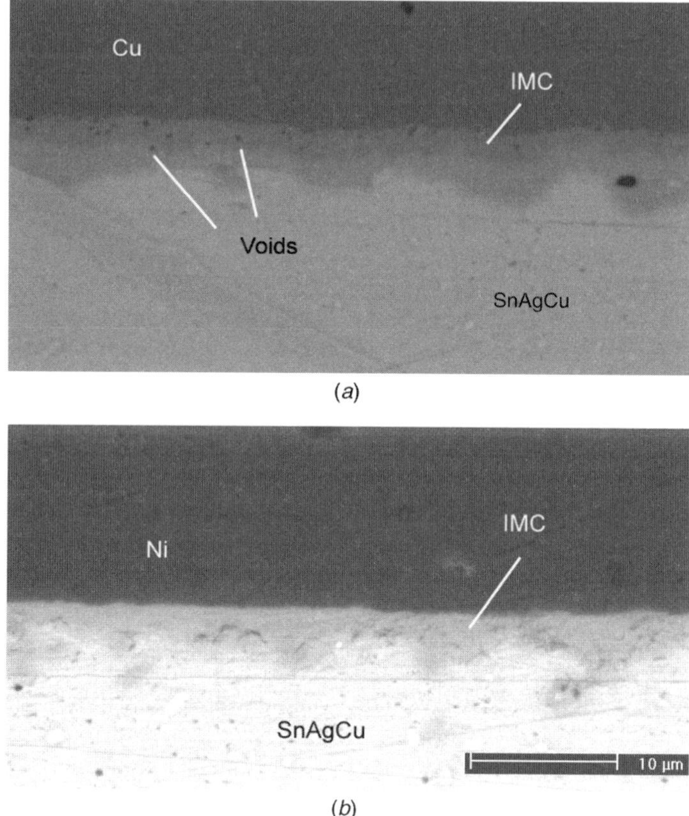

Figure 5.37. Cross sections of Sn–Ag–Cu-bumped BGAs assembled to OSP-finish printed circuit boards and aged for 70 hours at 175°C. Photos were taken at the BGA side of the solder joint. Image in (*a*) is a BGA with Cu substrate showing numerous voids in the IMC layer, while (*b*) is a BGA with Ni–Au metallization and no voids are visible.

Interfacial voids were not seen for single-substrate samples on either Ni or Cu by Yoon et al. [75], while Anderson and Harringa [76] found voids in aging of dual-substrate Cu bulk samples at 150°C. Anderson and Harringa reported reducing the voiding with the addition of Co and Fe. Both Pang et al. [77] and Vianco et al. [78] tested single-sided Cu samples up to 205°C and only saw minor voiding for Sn–3.9Ag–0.6Cu, and none at the lower temperatures.

5.9. EFFECT OF REWORK ON RELIABILITY

Given that any component/assembly process is not perfectly robust, a certain fraction (however small) of those components are subject to repair processes that create solder joints that are different to some degree from those produced in the reflow process. These differences are to be expected when comparing a soldering iron or a heat gun to a reflow oven since the heating and cooling rates are often quite different. When solder is added to the joint by means of solder wire, the differences can be even greater. To date, the best attempt at characterizing the variation introduced by rework on solder joint reliability is from the NEMI Pb-free rework project. A summary of those results can be found in Chapter 9.

5.10. HIGH-TEMPERATURE OPERATING LIFE (HTOL)

This type of testing applies a high current density (J) to assembled chips at elevated temperatures to drive electromigration in solder joints. The electric current enhances diffusion and can open voids, accentuate intermetallic compound formation, and thus create solder joint failures. Several reports show that Pb-free is better than Sn–Pb in these types of environments.

Balkan et al. [79] reported on HTOL tests run on flip chips soldered to Ni–Au finished PCBs, and the results showed much better performance with the Sn–Ag–Cu alloys than with Sn–Pb or Sn–Ag. Gee et al. [80] also found that Sn–4Ag–0.5Cu was equal to or better than Sn–Pb at similar test conditions.

5.11. ELECTROCHEMICAL MIGRATION

Electrochemical migration (EM) is the process of metal redistribution in the presence of an electrolyte and an applied electric field [81]. All metals are susceptible to this phenomenon, although the degree varies greatly and depends on the applied voltage, the current, and the type of electrolyte present. Elemental silver is known for being very sensitive to migration [82], and thus there has been some concern that the Sn–Ag family of alloys may be more at risk than eutectic Sn–Pb. As shown in Chapter 6, samples of the Sn–3.9Ag–0.6Cu solder paste reflowed onto coupons and were tested in SIR (surface insulation resistance), and EM did not show any signs of migration under those conditions. The author's experience with numerous no-clean Sn–Ag–Cu solder pastes under SIR test conditions is that they are not any more susceptible to metal migration than similar Sn–Pb solder pastes.

The reason for this is that the silver in the Sn–Ag–Cu alloy is almost totally bound as Ag_3Sn intermetallic, and very little silver remains in solid solution in the tin matrix. Given that the reflow requirements for most lead-free alloys requires

TABLE 5.13. HTOL Results at $T = 163°C$ and $J = 4700\ A/cm^2$

Alloy	Weibull α (Hours)	β (Slope)
Sn–Ag	166	1.52
Sn–Ag–Cu–Sb	550	3.3
Sn–Ag–Cu	774	3.3
Sn–37Pb	208	4.0

Source: Adapted from Ref. 79.

TABLE 5.14. Electrochemical Migration of Solder Alloys

Alloy	Time to Short Circuit (Seconds)	Current at Short Circuit (Milliamperes)
Sn–3.5Ag	120	0.022
Sn–9Zn	330	0.25
Sn–37Pb	10	5

Source: Adapted from Ref. 83.

the flux chemistry to retain activity at higher temperatures and that for longer times, one must always test any new paste to ensure that the specific combination of process, materials do not cause a problem. But this is holds true for conventional Sn–Pb solder processes as well.

Tanaka [83] has tested several lead-free alloys for ionic migration susceptibility. The samples were solder-plated copper lines on printed circuit board material. The gap was 0.32 mm and 5VDC was applied, with a de-ionized water droplet placed onto the lines. The Sn–3.5Ag and Sn–9Zn outperformed Sn–37Pb, with the data shown in Table 5.14. The time to shorting was much longer with the Pb-free alloys compared to the Sn–37Pb, and the amount of dissolution was much lower as measured by the current. This caused the Sn–Pb to form numerous growing dendrites, while the Sn–Ag and Sn–Zn alloys had single dendrites that created the ultimate short.

The better performance of the Pb-free alloys was attributed to (a) the greater degree of passivation of the tin and (b) the fact that the alloying elements were bound to the tin as intermetallic compounds. Environmental conditions that reduce the passivation of the tin would change the relative performance of these alloys.

5.12. TIN WHISKERING

At the time of publication, many component suppliers are announcing that their lead-free finishes for leaded components (e.g., QFP, TSOP, DIMM) will be tin-rich plating. Japanese companies are leaning toward tin–bismuth plating with low amounts of bismuth ($<4\%$), because some data suggest that the elemental additive lowers the stress in the tin-plating layer and mitigates whisker growth. U.S. and European companies have migrated toward matte tin without other elemental additives. This topic is covered in much greater detail in Chapter 7.

5.13. TIN PEST

Tin pest refers to the solid-state transformation of elemental tin from a body-centered cubic (β) structure to diamond cubic (α). Thermodynamically, the transformation is energetically favorable at 13°C, but a sufficient driving force is necessary for the kinetics to occur in a reasonable amount of time. The literature contains no reports of tin pest occurring within Sn–Ag–Cu or Sn–Ag solder joints that have been thermally cycled down to −55°C. Roubaud et al. [84] reported storing an DIMM assembly for 6 months at −50°C and no evidence of pest was observed. Kariya et al. [85] showed some pest formation in a tensile bar of Sn–0.5Cu that was stored at −18°C. The pest began nucleating after about one year of storage, and the machined areas of the bar showed greater fraction of α-tin. Lasky [86] pointed out that elements that are soluble in tin such as Bi and Pb are the most effective in suppressing pest, with silver and copper less so. The author is not aware of any reports of pest in wave or surface mount solder joints to date using Sn–Ag–Cu alloys.

5.14. SUMMARY

The substantial amount of published work on the reliability of Pb-free solders over the last several years shows that the Pb-free solders are as reliable as Sn–Pb eutectic for a large range of use conditions. Mixed technology is not as clear cut, and there is risk of hot tearing at low Pb concentrations, and voiding when Sn–Pb BGAs are assembled with Pb-free paste. Additionally, there are a few areas where more care must be taken, such as with high strain-rate impact and high temperature exposure for copper only substrates. This summary is to be used as a guideline, and any conversion to Pb-free must be associated with product-level reliability evaluations so that the specific performance is known and can be compared directly with the original Sn–Pb solder assembly recipe.

Although much excellent work on solder reliability has been conducted to date, additional research is needed to more fully understand the acceleration factors of Pb-free solders in thermal fatigue as they relate to actual field-type usage conditions. There is also a need to understand the plating systems used in component manufacture and how they influence interfacial microstructure at the intermetallic layer and subsequent solder joint performance in high-strain-rate conditions.

ACKNOWLEDGMENTS

The author would like to thank Pete Gilmore, Carol Handwerker, and Ron Gedney for reviewing the chapter.

REFERENCES

1. NIST Boulder Solder database: www.boulder.**nist**.gov/div853/lead%20free/**solders**.html
2. Matweb database: www.matweb.com
3. E. A. Brandes and G. B. Brook, Eds., *Smithells Metals Reference Book*, 7th edition, Butterworth-Heinemann, Boston, 1992, pp. 22-96–22-97.
4. R. J. McCabe and M. E. Fine, Athermal and thermally activated plastic flow in low melting temperature solders at small stresses, *Scripta Mater.* **39**(2), 189, 1998.

5. Karl Seelig and David Suraski, The status of lead-free solder alloys, in *Proceedings, 50th IEEE 2000 Electronic Components and Technology Conference*, May 21–24, 2000, Las Vegas, NV.
6. P. T. Vianco, J. A. Rejent, and A. C. Kilgo, Time-independent mechanical and physical properties of the ternary 95.5Sn–3.9Ag–0.6Cu solder, *JEM* **32**(3), 142–151, 2003.
7. A. F. Fossum, P. T. Vianco, M. K. Neilsen, and D. M. Pierce, A practical viscoelastic model for lead-free solder, *J. Electron. Packaging* **128**, 71–81, 2006.
8. E. W. Hart, The theory of the tensile test, *Acta Metall.* **15**, 351–355, 1967.
9. E. Bradley and J. Sharda, Internal Motorola Report.
10. J.-K. Lin, A. De Silva, D. Frear, Y. Guo, J.-W. Jang, L. Li, D. Mitchell, B. Yeung, and C. Zhang, Characterization of lead-free solders and under bump metallurgies for flip-chip package, *ECTC*, 455–462, 2001.
11. J. H. L. Pang and B. S. Xiong, Mechanical Properties for Sn–3.8Ag–0.7Cu Solder Alloy, *IEEE CPMT*, 1–11, 2005.
12. H. Nose, M. Sakane, Y. Tsukada, and H. Nishimura, Temperature and strain rate effects on tensile strength and inelastic constitutive relationships of SnPb solders, *J. Electron. Packaging* **125**, 59–66, 2003.
13. X. Q. Shi, W. Zhou, H. L. J. Pang, and Z. P. Wang, Effect of temperature and strain rate on mechanical properties of 63Sn/37Pb solder alloy, *ASME J. Electron. Packaging* **121**(3), 179–185, 1999.
14. Q. Zhang, A. Dasgupta, and P. Harwell, Viscoplastic constitutive properties and energy-partitioning model of lead-free Sn–3.6Ag–0.6Cu solder alloy, *Proc. ECTC*, 1862–1868, 2003.
15. S. Wiese et al. Constitutive behaviour of lead-free solders vs. lead-containing solders —experiments on bulk specimens and flip-chip joints, *Proc. ECTC*, 2001.
16. M. Amagai, M. Wantanabe, M. Omiya, K. Kishimoyo, and T. Shibuya, Mechanical characterization of Sn–Ag-based lead-free solders, *Microelectron. Reliability* **42**, 951–966, 2002.
17. R. Darveaux, Solder joint fatigue life model, in *Proceedings, TMS Annual Meeting*, Orlando, FL, 1997, pp. 213–218.
18. T. Fisher, Y. Takeda, N. Desai, and J. Zollo, Characterization of environmentally-friendly halogen-free materials for radio frequency electronics use, *Proc. SMTA*, 485–490, 2002.
19. C. R. Barrett, W. D. Nix, and A. S. Tetelman, *The Principles of Engineering Materials*, Prentice-Hall, Englewood Cliffs, NJ, 1973.
20. N. Chandler et al., in *MCM C/Mixed Technologies and Thick Film Sensors*, W. Kinzy Jones et al., Eds., Klumer Academic, Dordrecht, 1995, pp. 195–208.
21. Y. S. Touloukian, R. K. Kirby, R. E. Taylor, and P. D. Desai, *Thermal Expansion: Metallic Elements and Alloys*, Plenum, New York, 1975.
22. H. E. Boyer and T. L. Gall, *Metals Handbook*, American Society for Metals, Metals Park, OH, 1985.
23. D. R. Edwards and S. K. Groothuis, Mechanical Stress in IC packages, in *Characterization of Integrated Circuit Packaging Materials*, T. M. Moore and R. G. McKenna, Eds., Butterworth-Heinemann, Boston, 1993, pp. 57–77.

24. R. Darveaux, Optimizing the reliability of thin small outline package (TSOP) solder joints, in *Advances in Electronic Packaging 1995, Proceedings, ASME Interpack '95*, pp. 675–685.
25. M. Farooq, L. Goldman, G. Martin, C. Goldsmith, and C. Bergeron, Thermo-mechancial fatigue reliability of Pb-free ceramic grid arrays: experimental data and lifetime prediction, 2003 ECTC, pp. 827–833.
26. JESD22-A106-B, Thermal Cycling, July 2000.
27. JESD22-A106-A, Thermal Shock, April 1995.
28. A. Syed, Reliability of Lead-Free Solder Connections for Area-Array Packages, APEX 2001.
29. J. Bartelo, S. R. Cain, D. Caletka, K. Darbha, T. Gosselin, D. W. Henderson, D. King, K. Knadle, A. Sarkhel, G. Thiel, and C. Woychik, Thermomechanical Fatigue Behavior of Selected Lead-Free Solders, 2001 APEX, LF2-2.
30. S. Yoon, Z. Chen, M. Osterman, B. Han, and A. Dasgupta, Effect of stress relaxation on board level reliability of Sn based Pb-free solders, *ECTC*, 1210–1214, 2005.
31. F. X. Che, J. H. L. Pang, B. S. Xiong, L. Xu, and T. H. Low, Lead free solder joint reliability characterization for PBGA, PQFP and TSSOP assemblies, *Proc. ECTC*, 916–921, 2005.
32. S. K. Kang, P. Lauro, D.-Y. Shih, D. W. Henderson, T. Gosselin, Jay Bartelo, S. R. Cain, C. Goldsmith, K. J. Puttlitz, and T.-K. Hwang, Evaluation of thermal fatigue life and failure mechanisms of Sn–Ag–Cu solder joints with reduced Ag contents, *Proc. ECTC*, 661–667, 2004.
33. A. Syed, W.-J. Kang, T.-S. Kim, and J. Cannis, Board level reliability of QFP and QFN type packages with matte Sn lead finish, *SMTA*, 325–329, 2004.
34. K. Norris and A. Landzberg, *IBM J. Res. Dev.* **13**, 266, 1969.
35. N. Pan, G. A. Henshall, F. Billaut, S. Dai, M. J. Strum, R. Lewis, E. Benedetto, and J. Rayner, An acceleration model for Sn–Ag–Cu solder joint reliability under various thermal cycle conditions, *Proc. SMTAI*, 876–883, 2005.
36. J. Liang, S. Downes, N. Dariavach, D. Shangguan, and S. M. Heinrich, Effects of load and thermal conditions on Pb-free solder joint reliability, *JEM* **December**, 1507–1515, 2004.
37. J. P. Clech, Lead-free and mixed assembly solder joint reliability trends, APEX, S28-3-1, 2004.
38. Y. Qi, A. R. Zbrzezny, M. Agia, R. Lam, H. R. Ghorbani, P. Snugovsky, D. D. Perovic, and J. K. Spelt, Accelerated thermal fatigue of lead-free solder joints as a function of reflow cooling rate, *JEM* **December**, 1497–1506, 2004.
39. E. Bradley III and J. Hranisavljevic, Characterization of the melting and wetting of Sn-Ag-X solders, *IEEE Trans. Electron. Packaging Manuf.* **24**, 255–260, 2001.
40. U. R. Kattner and C. A. Handwerker, Calculation of phase equilibria in candidate solder alloys, *Z. Metallkd.* **92**, 740–746, 2001.
41. P. Chalco, Solder fatigue reliability issues in lead-free BGA packages, *Pan Pac SMTA*, 2002.
42. National Center for Manufacturing Science, *Lead-Free Solder Project: Final Report*, 1997.

43. F. Hua, R. Aspandiar, C. Anderson, G. Clemons, C. Chung, and M. Faizul, Solder joint reliability assessment of Sn–Ag–Cu BGA components attached with eutectic Pb–Sn solder, *SMTA*, 246–252, 2003.

44. K. Seelig and D. Suraski, Lead-free electronics assembly for the real world: joint failures, lead contamination and other important considerations, APEX, 2002.

45. A. Grusd, Lead-free solders in electronics, *Proc. SMI*, 648–661, 1999.

46. B. Wong, D. E. Helling, and R. W. Clark, A creep rupture model for two-phase eutectic solder, *IEEE Trans. CHMT* **11**(3), 284–290, 1988.

47. D. K. Joo and J. Yu, Effects of microstructure on the creep properties of the lead-free Sn–3.5Ag–Cu solders, *Proc. ECTC*, 1221–1225, 2002.

48. K. Wu, M. Aoyama, N. Wade, J. Cui, S. Yamada, and K. Miyahara, Creep and rupture behavior of Cu wire/lead-free solder-alloy joint specimen, *JEM* **32**(12), 139–197, 2003.

49. S. W. Shin and J. Yu, Creep deformation of microstructurally stable of Sn–3.5Ag–*x*Bi solders, in *2001 International Symposium on Electrical Material Packaging*, pp. 229–234.

50. E. Bradley and R. Mulligan, Motorola Internal Report, 2003.

51. Y. Kariya and M. Otsuka, Effect of bismuth on the isothermal fatigue properties of Sn–3.5mass%Ag solder alloy, *J. Electr. Mater.* **27**(7), 866–870, 1998.

52. C. Kanchanomai, Y. Miyashita, and Y. Mutoh, Low-cycle fatigue behavior of Sn–Ag, Sn–Ag–Cu, and Sn–Ag–Cu–Bi lead-free solders, *JEM* **31**(5), 456–465, 2002.

53. C. Kanchanomai, Y. Miyashita, and Y. Mutoh, Low-cycle fatigue behavior and mechanisms of a lead-free solder 96.5Sn/3.5Ag, *JEM* **31**(2), 142–151, 2002.

54. C. Kanchanomai, S. Yamamoto, Y. Miyashita, Y. Mutoh, A. J. McEvily, Low cycle fatigue test for solders using non-contact digital image measurement system, *Int. J. Fatigue*, **24**, 57–67, 2002.

55. J.-K. Lin A. De Silva, D. Frear, Y. Guo, J.-W. Jang, L. Li, D. Mitchell, B. Yeung, and C. Zhang, Characterization of lead-free solders and under bump metallurgies for flip-chip package, *Proc. ECTC*, 455–462, 2001.

56. J. Oliver, M. Nylén, O. Rod, and C. Markou, Fatigue properties of Sn–3.5Ag–0.7Cu solder joints and effects of Pb-contamination, SMTA, 2002.

57. T. S. Park and S. B. Lee, Isothermal low cycle fatigue tests of Sn/3.5Ag/0.75Cu and 63Sn/37Pb solder joints under mixed-mode loading cases, *Proc. ECTC*, pp. 979–984, 2002.

58. J. M. Song, T. S. Lui, L. H. Chen, and D. Y. Tsai, Resonant vibration behavior of lead-free solders, *JEM* **December**, 1501–1508, 2003.

59. E. Bradley and K. Banerji, Effect the relationship of PCB finish to the reliability and wettability of ball grid array packages, *IEEE Trans. CPMT, Part B: Adv. Packaging* **19**, 320–330, 1996.

60. E. Bradley, P. Lall, and K. Banerji, Effect of thermal aging on the microstructure and reliability of ball grid array solder joints, in *1996 Proceedings, Surface Mount International*, pp. 95–106.

61. R. Darveaux, K. Banerji, A. Mawer, and G. Dody, Reliability of plastic ball grid array packages, in *Ball Grid Array Technology*, McGraw-Hill, New York, 1995, Chapter 13.

62. Z. Mei, M. Kaufmann, A. Eslambolchi, and P. Johnson, Brittle interfacial fracture of PBGA packages soldered on electroless nickel/immersion gold, in *Proceedings, ECTC*, 1998, pp. 952–961.

63. P. Roubaud, G. Ng, G. Henshall, S. Prasad, F. Carson, R. Bulwith, R. Herber, S. Kamath, and A. Garcia, Impact of intermetallic growth on the mechanical strength of lead-free BGA assemblies, in *Proceedings, APEX* 2001, IPC.

64. J. Glazer, P. A. Kramer, and J. W. Morris Jr., Effect of Au on the reliability of fine pitch surface mount solder joints, *Circuit World* **18**(4), 41–46, 1992.

65. E. Bradley, Lead-free solder assembly: Impact and opportunity, in *2003 Proceedings, ECTC*, pp. 41–46.

66. D. M. Jacobson and G. Humpston, Gold coatings for fluxless soldering, *Gold Bull.* **22**(1), 9–18, 1989.

67. JESD22-B111, Board Level Drop Test Method of Components for Handheld Electronic Products, available at www.jedec.org

68. D. Reiff and E. Bradley, A novel mechanical shock test method to evaluate Lead-free BGA solder joint reliability, in *Proceedings, 55rd Electronic Components and Technology Conference*, 2005, pp. 1519–1525.

69. T.-Y. Tee, H.-S. Ng, C.-T. Lim, E. Pek, and Z. Zhong, Board level drop test and simulation of TFBGA packages for telecommunication applications, in *Proceedings, 53rd Electronic Components and Technology Conference*, 2003, pp. 121–129.

70. A. Lal, E. Bradley, and J. Sharda, Effect of reflow profiles on the board level drop reliability of Pb-free (SnAgCu) BGA assemblies, in *Proceedings, 55th Electronic Components and Technology Conference*, 2005, pp. 945–953.

71. M. Amagai, Y. Toyoda, and T. Tajima, High solder joint reliability with lead free solders, *Proceedings, 53rd Electronic Components and Technology Confercence*, 2003, pp. 317–322.

72. M. Amagai, Y. Toyoda, T. Ohnishi, and S. Akita, High drop test reliability: lead-free solders, in *Proceedings, 54th Electronic Components and Technology Conference*, 2004, pp. 1304–1309.

73. T.-C. Chiu, K. Zeng, R. Stierman and D. Edwards, and K. Ano, Effect of thermal aging on board level drop reliability for Pb-free BGA packages, in *Proceedings, ECTC*, 2004, pp. 1256–1262.

74. Z. Mei, M. Ahmad, M. Hu, and G. Ramakrishna, Kirkendall voids at Cu/solder interface and their effects on solder joint reliability, in *Proceedings, ECTC*, 2005, pp. 415–420.

75. J.-W. Yoon, S.-W. Kim, J.-M. Koo, D.-G. Kim, and S.-B. Jung, Reliability investigation and interfacial reaction of ball-grid-array packages using the lead-free Sn–Cu solder, *JEM* **33**(10), 1190–1199, 2004.

76. I. E. Anderson and J. L. Harringa, Elevated temperature aging of solder joints based on Sn–Ag–Cu: Effects on joint microstructure and shear strength, *J. Electr. Mater.* **33**(12), 1485–1496, 2004.

77. J. H. L. Pang, L. Xu, X. Q. Shi, W. Zhou, and S. L. Ngoh, Intermetallic growth studies on Sn–Ag–Cu lead-free solder joints, *J. Electr. Mater.* **33**(12), 1219–1226, 2004.

78. P. Vianco, J. A. Rejent, and P. F. Hlava, Solid-state intermetallic compound layer growth between copper and 95.5Sn–3.9Ag–0.6Cu solder, *JEM*, 991–1004, 2004.

79. H. Balkan, D. Patterson, et al., Flip chip reliability: Comparative characterization of lead-free (Sn/Ag/Cu) and 63Sn/Pb eutectic solder, in *Proc. ECTC*, 1263–1269, 2002.
80. S. Gee, N. Kelkar, J. Huang, and K.-N. Tu, Lead-Free and PbSn Bump Electromigration Testing, IPACK, 2005.
81. G. DiGiacomo, *Reliability of Electronic Packages and Semiconductor Devices*, McGraw-Hill, New York, 1997.
82. S. Chada and E. Bradley, Investigation of immersion silver PCB finishes for portable product applications, in *SMTAI Proceedings*, 2001.
83. H. Tanaka, Factors leading to ionic migration in lead-free solder, ESPEC Technology Report, 2002, pp. 1–9.
84. P. Roubaud, G. Henshall, E. Hernandez, and A. Classen, Reliability of Pb-free DIMMs assemblies on OSP and Ni–Au PWB finishes, in *Proceedings, IPC APEX*, 2002.
85. Y. Kariya, C. Gagg, and W. Plumbridge, Tin pest in lead-free solders, *Solder Surf. Mount Tech.* **13**, 39–40, 2001.
86. R. Lasky, *Tin Pest: A Forgotten Issue in Lead-Free Soldering*, SMTA International, 2004, pp. 838–840.

CHAPTER 6

Lead-Free Assembly Reliability: iNEMI Evaluation and Results

ELIZABETH BENEDETTO and JOHN SOHN

6.1. RELIABILITY TEAM GOALS

The mission of the iNEMI Reliability Team was to assess solder joint reliability by performing reliability testing for selected solders, components, and board finishes. The team achieved the following deliverables:

- Defined reliability test requirements
- Developed reliability test matrix
- Designed and fabricated test vehicles
- Performed reliability tests
  - Thermal cycling
  - Three-point bend testing
  - Electrochemical migration testing
- Carried out failure analysis and determined root cause
- Performed statistical analyses on failure data
- Documented results

In addition, the team promoted modeling for reliability by providing the necessary data and information to the modeling community. These data and information included the material properties of the components and boards used, thermal cycling conditions and tolerances, raw failure data, and root causes of the failures.

Lead-Free Electronics. Edited by Bradley, Handwerker, Bath, Parker, and Gedney

6.2. RELIABILITY TEST MATRIX

The Reliability Test Plan was formulated in the fall of 1999, based on the best information available at that time [1, 2]. Key to formulating the plan was the anticipated availability of components, boards, solder pastes, assembly facilities, and reliability testing facilities. As such, it was decided to have separate test boards for each of the components selected so that testing would not necessarily have to

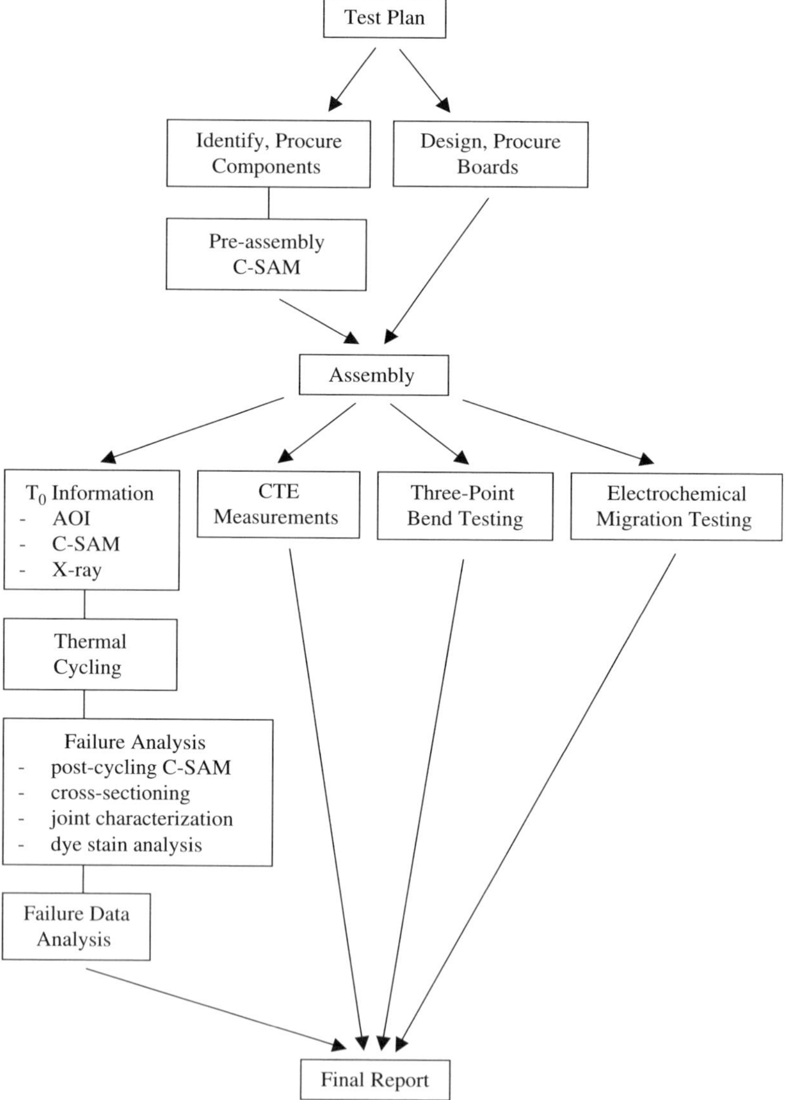

Figure 6.1. Flow chart depicting all of the activities performed by the Reliability Team.

wait until all of the components were available. Also, it was decided to distribute the thermal cycling among multiple facilities, so that resource constraints at any one facility would not hold up the entire testing program.

In addition to comparing the relative reliability performance of Pb-free solder joints to the performance of Pb-containing joints, a "mixed" cell involving assembly of Pb-containing components with Pb-free solder paste was included. The reasoning behind this choice of mixed cell was that there would be a transitional period when not all component termination finishes or ball compositions would be Pb-free, but that Pb-free solder paste was likely to be available.

Assuming a good solder joint at time zero (beginning of test), the team decided to focus primarily on thermal fatigue of the solder joints, with most of the activity centered on thermal cycling as the means to assess thermal fatigue. In addition to thermal cycling, three-point bend testing was used to simulate damage that occurs from out-of-plane deformation during drop test or assembly. Since the selected solder contains silver (Ag), electrochemical migration testing was performed by several iNEMI member solder paste companies on the iNEMI-recommended Pb-free alloy (Sn–3.9Ag–0.6Cu) to determine whether there were any concerns with the alloy regarding electrochemical migration.

All of the activities are depicted in Figure 6.1.

6.3. COMPONENT-PASTE-BOARD FINISH COMBINATIONS

In order to characterize a wide variety of component types that might be significantly affected by the proposed change in solder material, the team looked for components that would span across the various types of solder interconnects used in the industry.

The test matrix, showing components (both Pb-free and Pb-containing), solder pastes (Sn–3.9Ag–0.6Cu, and Sn–37Pb alloys), and thermal cycling conditions used with each cell, is shown in Table 6.1. Inclusion of the JEITA alloy (Sn–3Ag–0.5Cu) was added after the initial test matrix was finalized in order to have additional data on this material.

The following conventions are used in Table 6.1 and throughout this chapter:

Pb: Pb-containing component finish/ball composition, assembled with Sn–Pb paste

Pb-Free: Pb-free component finish/ball composition, assembled with Pb-free paste

Mixed: Pb-containing component finish/ball composition, assembled with Pb-free paste

Several 208CSP and 256PBGA test boards were supplied to the Assembly Process Team for preliminary rework studies. A small number of reworked parts were included in the reliability study, but rework processes were not optimized and there were not sufficient quantities evaluated to make statistically meaningful comparisons.

TABLE 6.1. Component Test Matrix

Component (Im–Ag Board Finish Unless Otherwise Indicated)	−40°C to 125°C			0°C to 100°C		
	Pb	Mixed	Pb-Free	Pb	Mixed	Pb-Free
48 I/O TSOP	X	X	X			
48 I/O TSOP, Ni–Au boards	X	X	X			
2512 Resistor	X	X	X			
2512 Resistor, Ni–Au boards	X					
169 CSP	X	X	X	X	X	X
208 CSP	X	X	X	X	X	X
208 CSP JEITA alloy			X			
256 PBGA	X	X	X	X	X	X
256 CBGA				X	X	X
Reworked 256 PBGA	X	X	X			
Reworked 208 CSP	X		X			

6.4. COMPONENTS

Six component types were used in the testing, with both Pb-free and Pb-containing ball composition or lead finish. The components and the companies that provided them are shown in Table 6.2. Also shown are thermal cycling conditions to which the components were subjected and the test facilities where thermal cycling took place.

- The Pb-free balls were provided by Hereaus and had a nominal composition of Sn–4.0Ag–0.5Cu.

TABLE 6.2. Component Descriptions and Conditions

Component	Source	Description	Reliability Testing: −40°C to 125°C	Reliability Testing: 0°C to 100°C
Type 1 TSOP	AMD	48 pin, leads on short side	Solectron	
2512 Resistor	Koaspeer	Zero-ohm chip resistor	Sanmina-SCI	
169 CSP	Lucent	0.8-mm pitch, 11 × 11-mm package, 7.7 × 7.7-mm die	Kodak	Lucent
208 CSP	Chippac	0.8-mm pitch, 15 × 15-mm package, 8.1 × 8.1-mm die	Kodak	Sanmina-SCI
256 PBGA	Amkor	1.27-mm pitch, 27 × 27-mm package, 10 × 10-mm die	Celestica	Sanmina-SCI
256 CBGA	Topline component, IBM ball attach	1.27-mm pitch, 27 × 27-mm package, no die		Motorola

- The Pb-containing balls had either Sn–37Pb or the commonly used Sn–Pb–Ag alloy Sn–36Pb–2Ag.
- The lead termination finishes were Sn–10Pb for both the TSOP and resistor.
- The Pb-free platings were Ni–Pd for the TSOP and pure Sn for the resistor.
- All solder and plating compositions are listed in weight percent.

6.5. TEST VEHICLES

Six different boards, one for each component type, were used for the reliability testing. The circuitry in each of the six board types allowed each component to be electrically monitored (one loop) during thermal cycling.

The boards used for area array packages were designed by StorageTek (see Figures 6.2 and 6.3). These boards were eight-layer, 0.062-in. FR-4 with a T_g of approximately 170°C, with four component sites per board.

The TSOP and R2512 boards (shown in Figures 6.4 and 6.5), provided by Motorola, were four layer, 0.062 in. standard FR-4 boards (T_g of approximately 125°C), with 16 component sites per board.

Depending on the fixturing at the different test facilities, the boards utilized either connector fingers or terminals for hard wiring. All of the area array test boards used MacDermid's immersion silver (ImAg) surface finish (16.8 μin. thick, with a standard deviation of 1.5 μ in.). Some of the TSOP and R2512 boards used the same immersion silver finish as the area array boards; the rest used electroless nickel–immersion gold (NiAu) finish.

The boards are shown in Figures 6.2–6.5, including cross-sectional information for the area array boards.

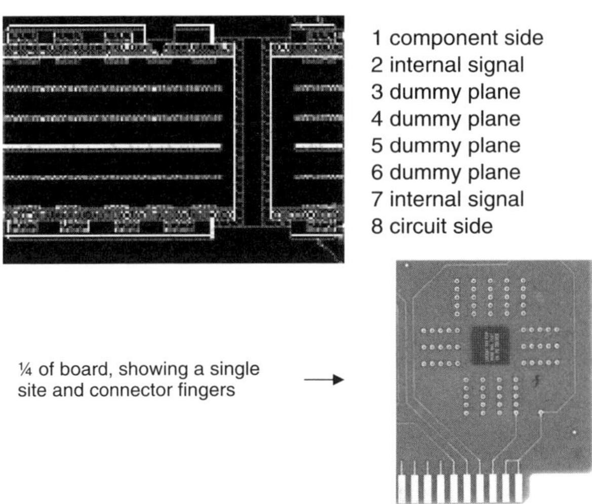

Figure 6.2. 169CSP Test Board, with layer cross section.

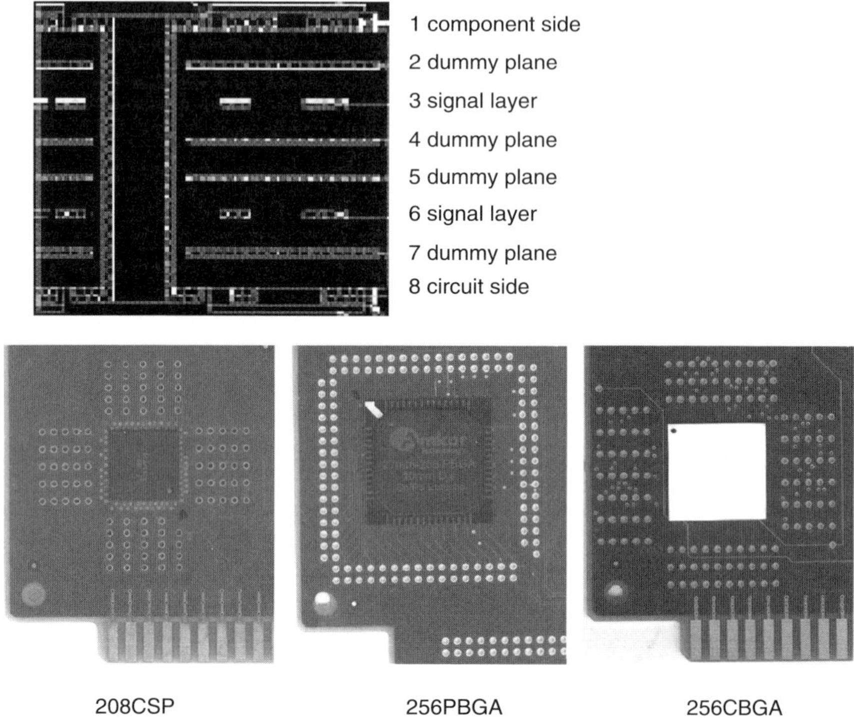

Figure 6.3. Representative structure of 208CSP, 256PBGA, and 256CBGA Test Boards, with layer cross section.

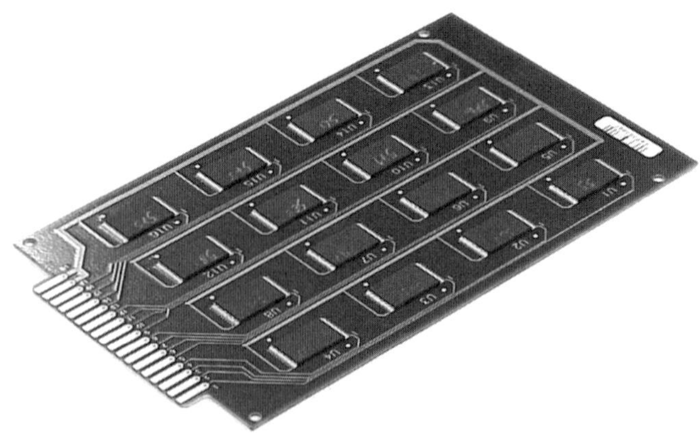

Figure 6.4. 48 TSOP Test Board.

Figure 6.5. R2512 Test Board.

6.6. PRE-TEST/POST-ASSEMBLY INFORMATION

Prior to assembly, the CSP, PBGA, and TSOP packages were analyzed using acoustical imaging (C-SAM). After board level assembly, these packages were again analyzed via C-SAM in order to document effects of assembly on package integrity, as well as to provide reference points for use in post-thermal cycling failure analysis. The solder joints of the area array assemblies were also characterized using transmission X-ray inspection.

To provide baseline information, cross sections were made of one component from each of the test cells. This detailed information is contained in the failure analysis reports and is briefly discussed below.

Examples of the information collected are shown in Figures 6.6 and 6.7. Figure 6.6 shows cross sections of representative 256PBGA Pb-free solder joints,

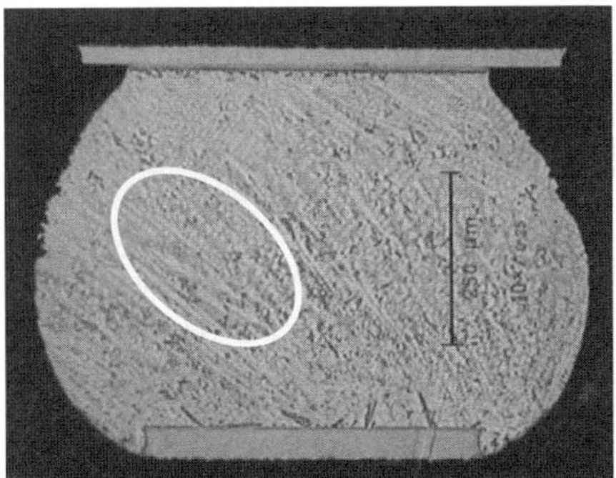

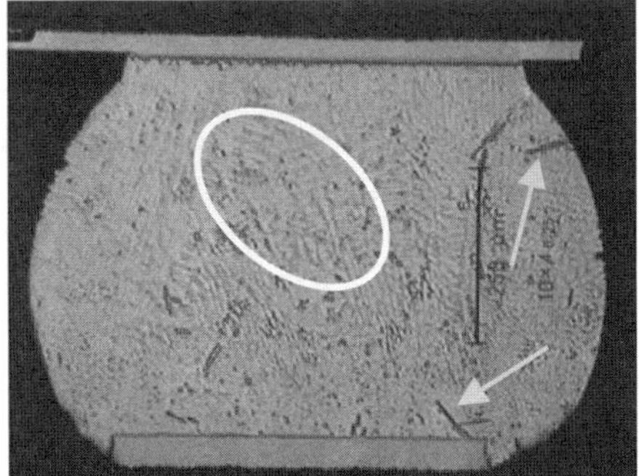

Figure 6.6. Cross sections of representative Pb-free 256PBGA joints after assembly, before thermal cycling.

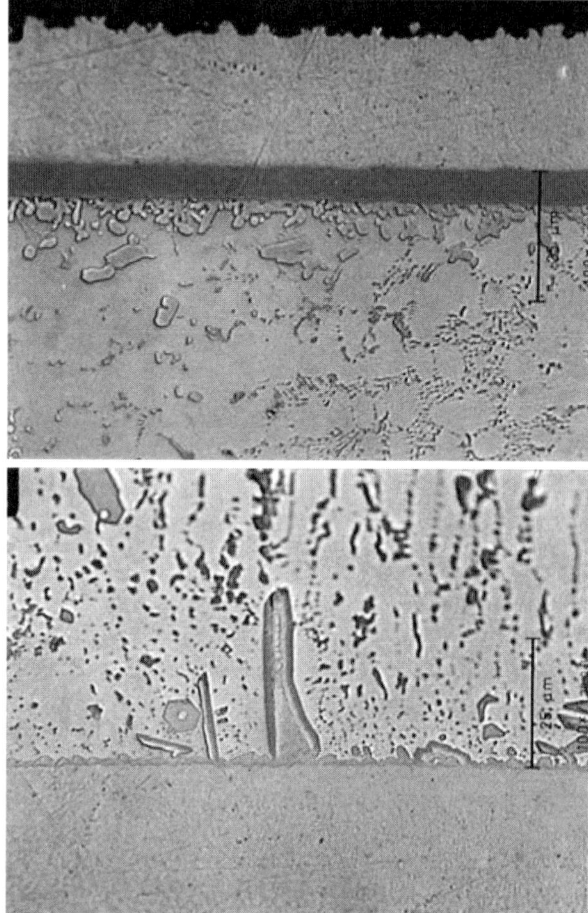

Figure 6.7. Solder–component and board–component interfaces from cross sections of representative Pb-free 256PBGA joints after assembly, before thermal cycling. Micrographs showing representative solder–component interface (top) and solder–board interface (bottom).

both joints being well-formed with no apparent defects, while Figure 6.7 presents views of the intermetallic compounds at the solder–component and the solder–board interfaces.

The solder joints appear well-formed with no apparent defects. The aligned Sn dendrite structures are apparent in both micrographs, and they are circled. Silver intermetallic particles (probably platelets) formed at the solder-board interface and at the solder–air interface as seen in the micrograph on the right, indicated by the arrows.

6.7. CTE DETERMINATION: COMPONENT AND BOARDS

The coefficient of thermal expansion (CTE) of component and board materials has a significant effect on the response of a solder system to thermal excursions. Excessive

TABLE 6.3. CTE Array Packages

Component	CTE (ppm/°C)
169 CSP	13–18
208 CSP	8–13
256 PBGA	10–17
CBGA (no die)	5.5

joint fatigue may be partially due to CTE mismatch rather than a specific property of the solder alloy. In order to assist in the assessment of thermal cycle results, CTE testing was performed on the materials used for this program.

Moiré interferometry was used to determine the effective CTE of each of the area array packages as a function of distance from the neutral point (DNP) [3]. The results of this test varied as a function of DNP for those packages with die (see Table 6.3).

Thermomechanical analysis was performed in order to determine the CTEs of the TSOP and CBGA boards [4]. The team believed that the R2512 should have a CTE comparable to that of the TSOP board and that the CBGA board should be representative of all of the area array boards. The values for the TSOP board ranged from 15 to 17 ppm/°C, and those for the CBGA board ranged from 17 to 19 ppm/°C.

6.8. THERMAL CYCLING CONDITIONS

The team decided to cycle the assemblies under either of two conditions (in several cases, a component set was evaluated under both conditions—see specific test plan in Table 6.1). The two conditions, −40°C to 125°C and 0°C to 100°C, were chosen to match the conditions used by member companies in their own testing, allowing for the failure results to be compared by each company directly to their own historical data. Testing was conducted to at least 50% failure in most cases, even when it exceeded 3000 cycles. In some cases, testing was carried out to >8000 cycles. The two thermal cycling conditions used were per JEDEC JESD22-A104B (July 2000) "Temperature Cycling," with the following specific requirements:

- −40 (+0, −5) to 125 (+5, −0)°C (Condition G of Table 1 of JESD22-A104B)
- 0 (+0, −5) to 100 (+5, −0)°C (Condition J of Table 1 of JESD22-A104B)
- Ramp: 10–14 C°/minute, calculated linearly between 10% and 90% of Delta-T range
- Dwell: Soak mode 2 of Table 2 of JESD22-A104B (5-minute minimum dwell) defined with "5-degree" dwell clock. Additional dwell criteria: At least 1/2 of the dwell time (2.5 minutes) must be at or beyond the target temperature.
- Measurement: All specified temperatures are of the test samples, not the chamber air, and were verified *in situ* on a sufficient number of samples prior to start of test.

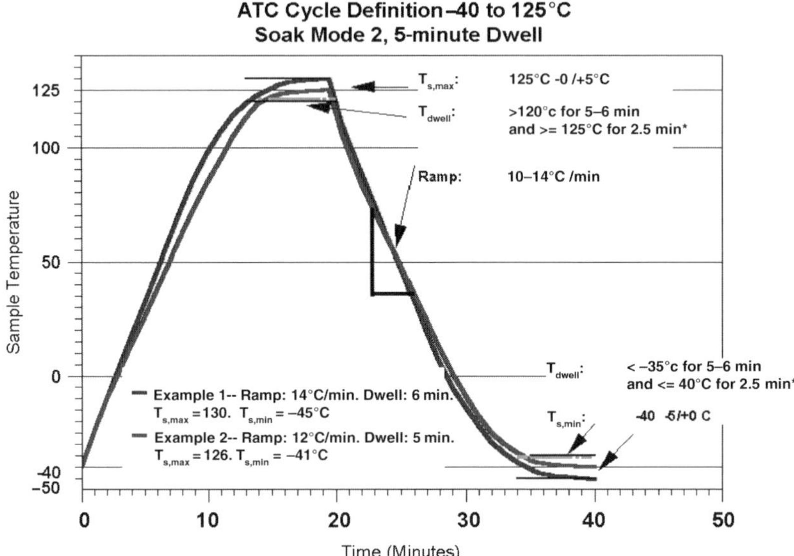

Figure 6.8. Example of a thermal cycle profile.

These additional requirements were imposed to minimize differences between test facilities such that results obtained at one facility might be extrapolated to another. An example of a thermal profile is shown in Figure 6.8. Each test facility ran chamber calibration tests, and it verified that it met the thermal profile requirements above [5–9].

6.9. FAILURE CRITERIA

In order to ensure consistency and reproducibility in the thermal cycling, a thermal cycle failure was defined using the requirements in IPC-SM-785 "Guidelines for Accelerated Reliability Testing of Surface Mount Solder Attachments," which are reproduced here:

Failure detection shall be by continuous monitoring of daisy-chain continuity test loops such that:

- At least one continuity interruption of one microsecond or less duration can be recorded for each test loop during any polling interval of two seconds or less.
- At least 10 such interruptions can be recorded per test loop to confirm the first failure indication.
- The monitoring current does not exceed 2 mA at no more than 10 V, and an electrical discontinuity is indicated by a loop resistance of 1000 ohms or more. False failure indications due to electrical noise can be a problem, particularly for loop resistance thresholds lower than 1000 ohms.

Each test facility's event detectors and data collection processes were calibrated to ensure that they were capable of meeting these requirements [10–15].
At the time the thermal cycling conditions were developed, the current industry standard, IPC-9701, "Performance Test Methods and Qualification Requirements for Surface Mount Solder Attachments" (January 2002), was still in development. The team supports IPC-9701 and recommends its use in future work.

6.10. THERMAL CYCLE RELATIVE PERFORMANCE

The relative performance of the different test cells in thermal cycling is shown in Table 6.4. In all cases, the Pb-free cells performed equivalent or superior to the Pb benchmark (Sn–Pb paste, Pb-containing component). This was determined by statistical analysis of the failure data using 95% confidence bounds in comparing the characteristic life, η [16]. In this table, the Pb column is shown to be zero, meaning statistically equivalent to the Pb benchmark, as it is being compared to itself.

6.11. FAILURE DATA, ANALYSIS PACKAGES

The failure data, provided by each test site in a standardized data format to aid in statistical analysis of the failure data, were analyzed at four facilities, each using a different analysis package. The failure data files contained the following information: stress test conditions, component type, component-paste combination (Pb, mixed, Pb-free), board and component ID, cycles and censoring (component failure cycle number, test complete cycle number where the component survived), cause of failure, and initial and final resistance values as determined by a four-point

TABLE 6.4. Thermal Cycling Relative Performance

Component (Im–Ag Board Finish Unless Otherwise Indicated)	−40°C to 125°C			0°C to 100°C		
	Pb	Mixed	Pb-Free	Pb	Mixed	Pb-Free
48 I/O TSOP	0	−	0			
48 I/O TSOP, Ni–Au boards	0	+	+			
2512 Resistor	0	0	0			
2512 Resistor, Ni–Au boards	0					
169 CSP	0	+	+	0	0	+
208 CSP	0	0	+	0	+	+
208 CSP JEITA alloy			0			
256 PBGA	0	0	0	0	0	0
256 CBGA				0	−	+

Legend: 0, statistically equivalent to the Pb benchmark; −, statistically worse than the Pb benchmark; +, statistically better than the Pb benchmark.

measurement under ambient conditions. The statistical software packages used are listed here:

- ReliaSoft Weibull++ 5.0
- MINITAB Release 13.1 (Intel version)
- WinSmith Weibull 1.0F (Fulton Findings)
- JMP Discovery, version 4.0.5 (SAS Institute)

In all cases, no significant differences in the survival analysis results arose from use of the different packages.

6.12. WEIBULL ANALYSES

Table 6.5 and Table 6.6 provide the Weibull statistics η(N63) and β using the results from the analyses performed for each of the components. Figure 6.9 shows the Weibull plot for 256 PBGA components as tested from $-40°C$ to $+125°C$.

In this example, we can see that the Weibull slope, β, is relatively high for all three datasets, supporting the conclusion that the parts were in the wear-out phase of their life. There do not appear to be multiple failure modes present (all data points fit the Weibull function well). Higher beta values are preferred and, as such, the Pb–Pb cell will have the latest first failure occurrence of the three datasets. Although the Pb-free joints show a higher characteristic life, η(63.2% failed), this was not the case for all component types, and statistically they varied in

TABLE 6.5. Characteristic Life, η(N63)

Component (Im–Ag Board Finish Unless Otherwise Indicated)	−40°C to 125°C			0°C to 100°C		
	Pb	Mixed	Pb-Free	Pb	Mixed	Pb-Free
48 I/O TSOP	7298	5627	7391			
48 I/O TSOP, Ni–Au boards	6358	(8237)	7592			
2512 Resistor	670	834	1018			
2512 Resistor, Ni–Au boards	551					
169 CSP	1944	3071	3254	3321	3687	8343
208 CSP	3138	3056	3555	(8355)	(8355)	(8355)
208 CSP JEITA alloy			3242			
256 PBGA	6960	7754	7268	(6000)	(6000)	(6000)
256 CBGA				X	X	X

Legend: Numbers in parentheses indicate the cycle number where the unfailing samples were removed from testing.
X—These parts did not perform typical of high-performance CBGAs; their intended use is assembly trials, not reliability testing. The CBGA components used in this project were found to be unrepresentative of high-performance CBGAs currently used in the industry [17].

TABLE 6.6. Weibull Slope, β

Component (Im–Ag Board Finish Unless Otherwise Indicated)	−40°C to 125°C			0°C to 100°C		
	Pb	Mixed	Pb-Free	Pb	Mixed	Pb-Free
48 I/O TSOP	6.6	9.7	8.3			
48 I/O TSOP, Ni–Au boards	14.4	[a]	16.7			
2512 Resistor	3.1	2.2	2.7			
2512 Resistor, Ni–Au boards	2.5					
169 CSP	6.6	10.6	7.5	7.5	2.9	4.1
208 CSP	8.4	5.0	9.5	[a]	[a]	[a]
208 CSP JEITA alloy			7.9			
256 PBGA	12.8	4.9	7.2	[a]	[a]	[a]
256 CBGA				X	X	X

[a]Insufficient number of failures to calculate β.

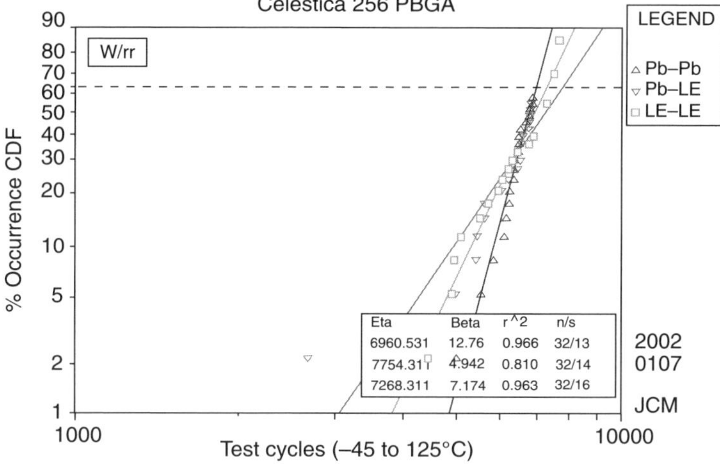

Figure 6.9. Weibull plots for 256 PBGA, cycled −40°C to +125°C at Celestica.

performance. However, in the thermal cycle test program performed by the NEMI team, the Pb-free material performed as good as or better than standard Sn–37Pb.

6.13. POST-CYCLING FAILURE ANALYSIS

After thermal cycling, both failed and surviving parts were characterized using the following analysis methods:

- Visual inspection (10–30×)
- C-SAM (scanning acoustic microscopy, C-mode) analysis

- Metallographic cross-sectioning to determine standoff height and joint geometry including:
  - wetting,
  - identification of first failure locations,
  - correlation of observed damage with electrical failure,
  - identification of crack initiation and type of damage,
  - characterization of microstructural coarsening of the solder,
  - characterization of intermetallic growth in the joint and in the solder,
  - characterization of the grain size and structure, and
  - compositional gradient mapping of the solder joints
- Surface microstructure characterization and relationship to crack formation
- Layout dye staining (dye and pry)

6.13.1. Visual Inspection and C-SAM

Visual inspection did not show any significant or unexpected results in any of the test cells. Scanning acoustic microscopy, C-mode (C-SAM), analysis showed some package-die delamination in some of the parts. However, none of these resulted in electrical failure during thermal cycling. More information on C-SAM analysis can be found in Chapter 8.

6.13.2. Metallographic Cross-Sectional Analyses

As noted in Section 6.6, cross sections were prepared from each component set prior to thermal cycling. After thermal cycling, cross sections were again prepared from each of the test cells (component–paste–thermal cycling condition). Detailed failure analyses were performed on each of the component sets. Highlights from each of the analyses follow.

6.13.2.1. 169 CSP. The 169CSP component set [18] was investigated in the greatest detail because it includes all three cells cycled under both thermal cycling condition and because all thermal cycling of the 169CSPs was completed first.

For all three test cells, failure during thermal cycling always occurred first in the solder at the component side. No pronounced differences in joint geometry were seen during the evaluation. Microstructural changes occurred as a result of thermal cycling.

Pb Cell. The time-zero joints showed considerable fraction of coarse Pb dendrites. As was expected for this material, wetting was found to be acceptable, void formation was not pronounced, and a good intermetallic formation was seen at both the component–solder and board–solder interfaces.

As expected, the Pb solder coarsened considerably as a result of thermal cycling. Failure occurred preferentially in the solder on the component side, with the failure

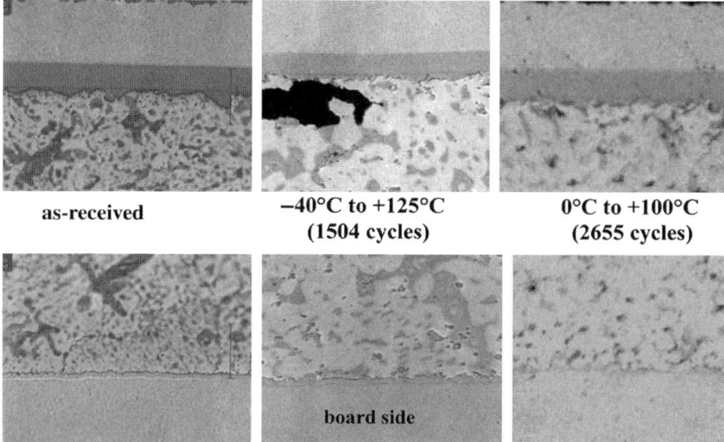

Figure 6.10. CSP169 Pb cell. Component-board pair on the left are as-received. Intermetallic layers are seen on both sides of the solder joint.

path appearing to travel between grains. Cross sections were made through a complete row of one component that failed during thermal cycling.

A mapping of full cracks in cross section against electrical opens was performed for one component from each of the two thermal cycling conditions. In both cases, matches were found between electrical opens and full cracks. An example of this methodology is shown in Figure 6.13 using a Pb-free component. Comparisons of microstructures, before thermal cycling and after cycling under each cycling condition, are shown in Figure 6.10. The same approach was used to characterize the components in the mixed cell (Figure 6.11) and the Pb-free cell (Figure 6.12).

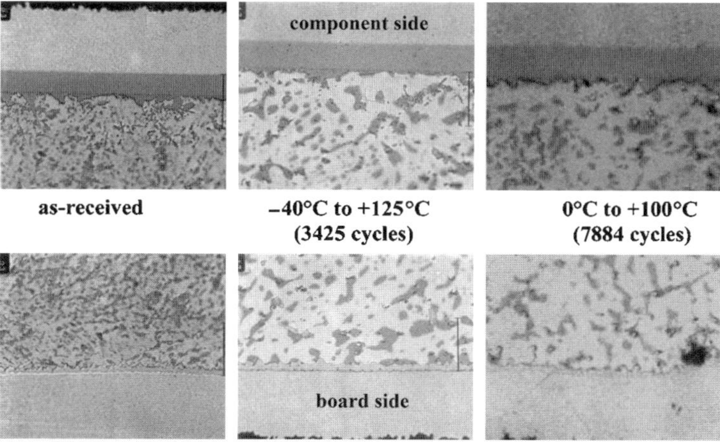

Figure 6.11. CSP169 mixed cell. Component–board pair on the left are as-received. Intermetallic layers are seen on both sides of the solder joint for all three conditions. Considerable coarsening has occurred in the solder after thermal cycling.

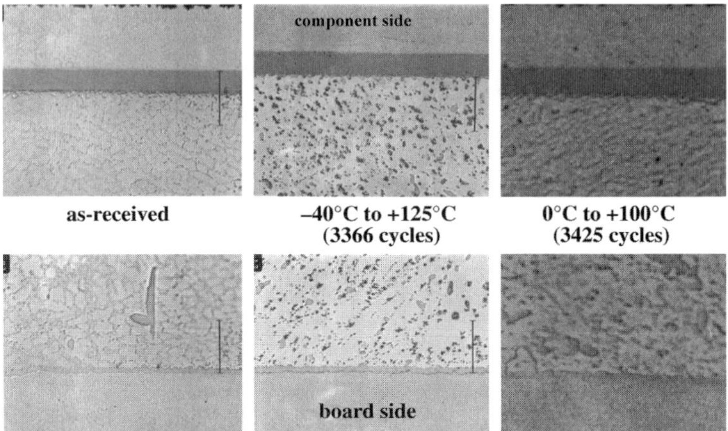

Figure 6.12. CSP169 Pb-free test cell. Component–board pair on the left are as-received. Intermetallic layers are seen on both sides of the solder joint for all three conditions. Considerable coarsening has occurred in the solder after thermal cycling.

Mixed Cell. Similar results were found as with the Pb cell for the pre-cycling components as well as for those exposed to thermal cycling.

Pb-Free Cell. The time-zero joints showed Sn dendrite structures with a significant fraction of intermetallic plates, primarily in the bulk of the solder but sometimes also growing/touching the land or the connection on the component. The contrast in the Sn phase indicates dendrite "single crystals," where all the dendrites are of the same orientation. Wetting was found to be acceptable and void formation was not pronounced.

Figure 6.13 shows Pb-free joints after cycling −40°C to +125°C. Cracks are seen near the component interface. Pronounced microstructural coarsening of the solder can be seen near the top of the component, in the "cone-shaped" region of the solder. Coarsening occurs both in the Sn phase and in the intermetallic phase. A "map" was constructed of each ball grid array component after failure analysis to compare electrical and mechanical opens. Incidences where full cracks observed in the cross sections correspond to electrical opens are shown in the schematic.

Most failures occurred on the component side, which may be because the component land pad was smaller than the corresponding land pad on the board. No attempt was made to optimize the joint design for these evaluations; both Pb-free and Pb joints had the same dimensions.

The fracture path appears to be affected by the presence of intermetallic particles at the interface on the component side. The fracture stays in the solder, but the path appears to be deflected by nearby intermetallic particles. The roughness associated with this top interface parallels the roughness of the intermetallic layer. The

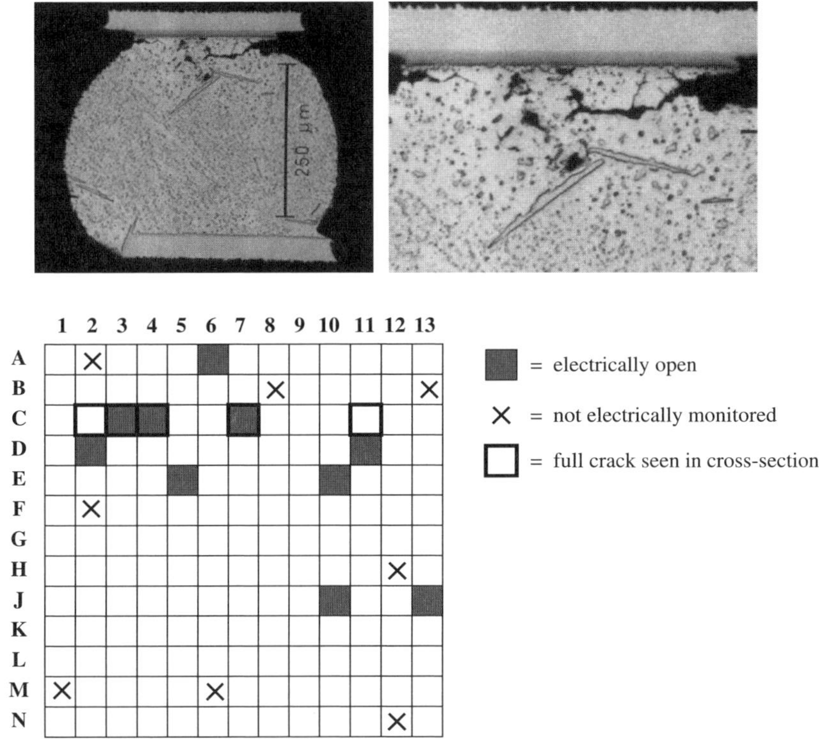

Figure 6.13. Failure Pb-free joints, 169CSP, cycled −40°C to +125°C, joint failure map.

intermetallic layers on both sides of the solder joint for all three conditions can be seen in Figure 6.12. Considerable coarsening has occurred in the solder after thermal cycling.

Polarized Light Study. Polarized light was used to further explore the degree of grain orientation in the Sn dendrite colonies in pre-cycling components from each of the test cells. In all three test cells, Sn colonies for all grains within a region have the same grain orientation. The number and size of the colonies varied from joint to joint in all test cells, with both pre-cycling and thermally cycled components. Figure 6.14 shows an example of the polarized light results from the Pb-free cell.

6.13.2.2. 208 CSP [19]. As with the 169CSPs, failure always occurred first in the solder at the component side in all three cells (Pb, mixed, Pb-free). Microstructural changes occurred as a result of thermal cycling. Coarsening occurred in the solder adjacent to the crack path for both thermal cycles in all three test cells. There was no apparent reason (obvious different failure modes) for the pronounced bimodal distribution in the failure data for the mixed cell.

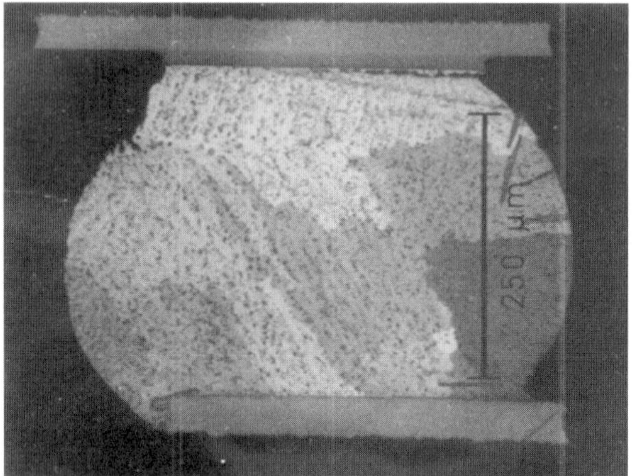

Figure 6.14. Polarized light micrograph, 169CSP Pb-free.

6.13.2.3. R2512 [20]. For all three material combinations, failure always occurred first in the solder. Gross microstructural damage occurred in all three cells, with little difference seen among the cells. Standoff heights were variable even within the same composition/surface finish group. Microstructural changes

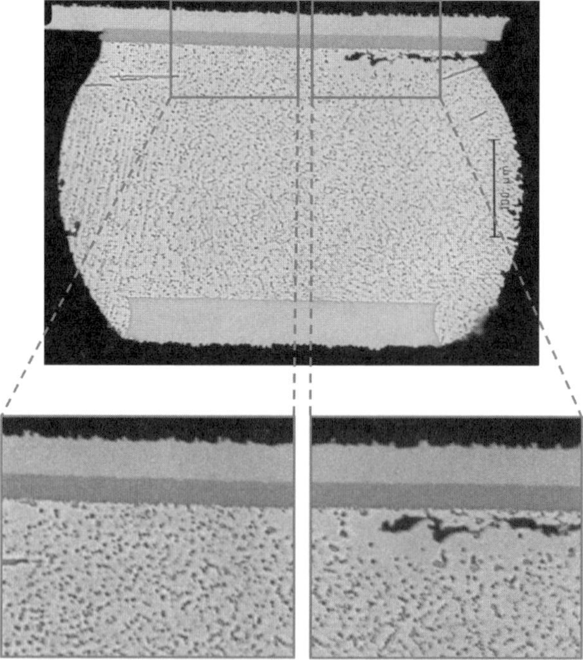

Figure 6.15. Significant coarsening of the solder has occurred near the crack path but does not extend far into the solder.

in the solder and in the thickness of the IMC on the board side occurred as a result of thermal cycling.

6.13.2.4. 48 TSOP [21]

Observations on As-Received Joints

- Solder rise on the Pb-free joints was not as great as for the other two cells. It was likely due to wetting issues as well as exposed copper at the tip of the lead frame.
- Reaction between the copper and the Ni–Au plating led to porosity and corrosion of the copper board. This was seen in Ni–Au–Cu areas not covered by solder.

Observations on Thermally Cycled Joints

- For all three material combinations, cracks ran from the heel of the lead along the component–solder interface and/or at the solder–board interface underneath the lead. Failure under the lead occurred primarily at the solder–intermetallic interface, but there was widespread fracture of the solder in the fillet area underneath the heel as the cracking proceeds.
- The reaction layer between the copper and the Ni–Au surface finish did not appear to grow or affect crack nucleation or propagation during thermal cycling.
- Microstructural changes, particularly coarsening of the solder, occurred as a result of thermal cycling.

In general, and with both board finishes, the amount of solder wicking up the gull wing joints in the Pb-free set of samples appeared to be limited as compared with joints in the other two cells. Standoff height between component leads and boards were variable, with similar variability for all compositions and for both board types.

A series of micrographs shows changes in microstructure from bulk fillet to under the component with the Ni–Au boards. Underneath the component, less eutectic and more Sn dendrites were seen, which means less Cu and Ag in the solder locally. In all three cells, along the board side there was a reaction within the copper layer forming voids in the copper near the interface; no such layer was seen with the immersion silver board finish.

6.13.2.5. 256 PBGA [22]. For all three material combinations, failure always occurred first in the solder at the component side. There were no pronounced differences in joint geometry among the solders either post-assembly or after thermal cycling. Microstructural changes in the solder occurred as a result of thermal cycling. All of the solders showed coarsening following thermal cycling. Coarsening of the Sn dendrites in the solder and growth of intermetallic plates at the solder–board interface were most pronounced for Pb-free joints thermally cycled from $-40^{\circ}C$ to $+125^{\circ}C$. Platelet growth was observed in Pb-free joints after thermal cycling from $0^{\circ}C$ to $+100^{\circ}C$, but little Sn dendrite coarsening was apparent. In the as-received joints, there appeared to be slightly more intermetallic at the component–solder interface for the Pb-free than for the other two cells. The amount of intermetallic in the "uniform" layer at all the interfaces did not appear to be affected by thermal cycling.

A comprehensive set of micrographs is included in the detailed report, including sections through an entire row of joints for each cell (Pb, mixed, Pb-free), comparison of joint failures, and interfaces and microstructures for both pre-cycling and post-cycling components, all for both thermal cycling conditions.

6.13.2.6. 256 CBGA [23]. Cross sections of the components before thermal cycling showed that the substrate metallization was very thin, and the solder mask led to pronounced undercutting of the solder. After reaction between the solder and the metallization, essentially nothing remained of the metallization. Fracture occurred at the metallization–substrate interface as well as in the solder near the interface. The fracture was straighter and less granular for the Pb-free solder than for the other two compositions. The combination of the low number of cycles to failure and the failure in the metallization means that the failure data are of reduced relevance to commercially used components.

An additional complicating factor for this component set is that the components were assembled rotated by 90° from the intended orientation. The circuit that was tested during thermal cycling was a zigzag pattern in outer row joints. This mistake was not catastrophic, because it was expected that the first joints to fail would be the external rows, due to the component not containing a die. Failure was observed to occur first on or near the expected rows.

It is important to note that the CBGA components used for this study were dummy components intended for use in solderability studies, not reliability studies. See "IBM Report on CBGA Reliability Performance" [17] for discussion of failure in high-performance CBGA components.

6.13.3. IMC Characterization

Detailed studies of the intermetallic compounds (IMCs) present in 169CSP solder joints, both before and after thermal cycling, were conducted and the composition and stoichiometry of the IMCs were determined [24–26].

As expected, the traditional Cu–Sn intermetallics, Cu_6–Sn_5 and Cu_3–Sn, formed at the interfaces in the various test vehicles. There were many similarities among the three solders and the various components, but also a number of differences in the growth rates and morphologies. These were found to depend on a variety of parameters, such as those discussed in Section 6.13.3.1.

6.13.3.1. IMC Observations Regarding Surface Finish

- In general, rapid formation and growth of Cu_6Sn_5 on the Cu pad were found, and much slower and more uneven growth of Cu_6Sn_5 was seen on the nickel finish. (Both the immersion Ag finish on the Cu and the Au on the Ni dissolved in the solder during reflow.) A macrograph of a typical 169CSP joint is shown

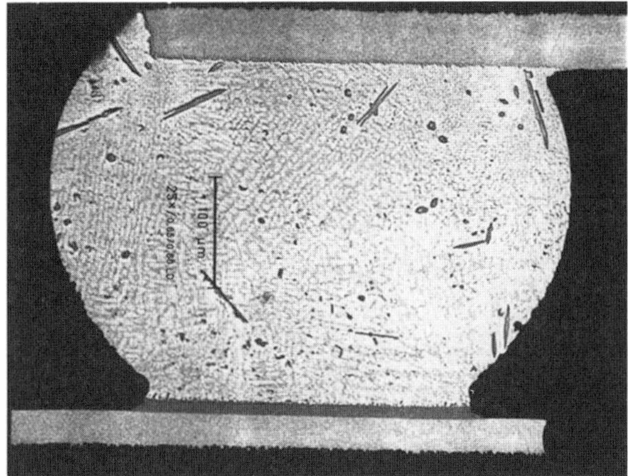

Figure 6.16. Micrograph of a 169CSP Pb-free joint prior to thermal cycling. Board side (copper pad with Ag finish) is to the top, and component side (Ni pad with Au finish) is at the bottom.

in Figure 6.16. The board interface (copper pad) is at the top, and the component interface (nickel pad) is at the bottom. Higher magnification images, to show typical microstructures at the interfaces just after reflow, are shown in Figures 6.17 and 6.18.

- The intermetallic reaction layers (Figure 6.17) between the board-side Cu pads and the solder for all three combinations of materials (Pb-free, mixed, Pb)

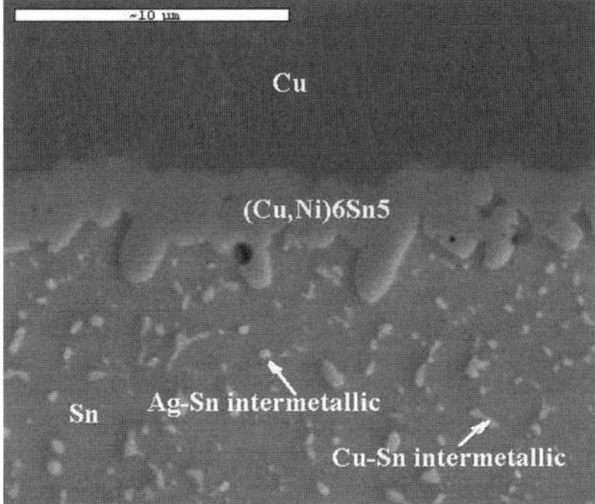

Figure 6.17. Typical structure at board interface, 169CSP Pb-free joint, before thermal cycling Micrograph was taken at the upper interface in Figure 6.16.

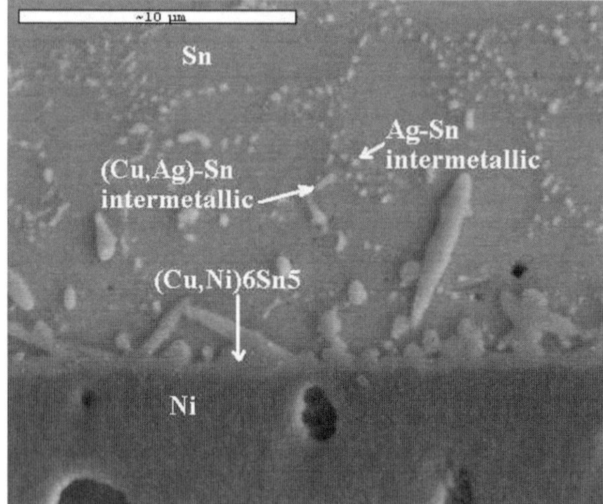

Figure 6.18. Typical microstructure at component interface, 169CSP Pb-free joint, before thermal cycling. Micrograph was taken at the lower interface in Figure 6.16.

are similar. The main intermetallic reaction layer between the Cu on the board side and the solder contains Cu, Sn, and up to 3 at% Ni (from dissolution at the other pad). The atomic ratio between the sum of the Cu and Ni and the Sn is very close to the expected 6:5 ratio (about 45 at% Sn). A range of about 37–48% Sn was found and was attributed to a combination of calibration uncertainty and the difficulty in excluding other phases, such as Cu_3Sn, during the microanalysis.

- The intermetallic on the nickel side (Figure 6.18) is most accurately described as Me_6Sn_5, where Me refers to the arbitrary (nonstoichiometric) mixture of nickel and copper that was present there. It was found that the nickel content of the intermetallic on this side was much higher, ranging from 13 to 55 at%. The average thickness for any intermetallic on the nickel side was much less than half of that for the copper side due to the lower availability of copper and the slower growth of nickel intermetallics. The intermetallic closest to the Ni layer seems to contain less Sn than the bulk intermetallic. An Me-to-Sn ratio near 3:2 was measured, which may indicate the formation of an Me_3Sn_2 intermetallic. The Ni–Sn phase diagram does indicate the presence of a gamma phase with 40 at% Sn, so a nickel interface may promote a phase not seen in Cu–Sn intermetallics. On the other hand, the accuracy of the analysis of such a small volume of intermetallic is questionable. The intermetallic formed on electrolytic Ni/electrolytic Au in Pb solder joints has less Cu and more Ni than in the Pb-free and mixed joints.
- Very little Cu_3Sn was observed at the Cu layer prior to thermal cycling.
- The pad and finish material (Cu, Ni, Ag, and Au) can partially (and for Au and Ag finishes, totally) dissolve in the solder while it is molten. This additional

material may stay in solution until the solder cools, or may form some phase (such as an intermetallic) while the solder temperature is still high. As the solder cools and solidifies, these dissolved elements may form phases other than those predicted for the pure solder, and so they change the microstructure from a simple two-phase eutectic. Sn, Ag_3Sn, and Cu_6Sn_5 primary crystals (dendrites) were found in many of the solder joints. Ag_3Sn particles had a rod-like shape, while the Cu_6Sn_5 crystals were hexagonal hollow rods. The eutectic between the dendrites contained Ag–Sn and (Cu, Ag)–Sn intermetallic. The microstructure of the mixed joints consisted of eutectic solder, Sn and Pb crystals, and a small amount of Ag–Sn and Cu–Sn intermetallic particles. Thus, the final microstructure is often complicated by the fraction of the pad material that is dissolved during soldering.

6.13.4. Thermal Cycling Effect on Solder Microstructure

- In general, thermal cycling caused a coarsening of the microstructure and introduced new phases. Figure 6.19 shows this increased complexity.
- The Cu_6Sn_5 intermetallic continued to grow at the Cu pad during cycling, with a thin Cu_3Sn layer appearing later in the cycling.
- Cycling of Sn–Pb solder/Ni joints leads to the formation and growth of Ni_3Sn_4 intermetallic. The presence of the Cu in solder dramatically alters the phase formation at the solder/Ni interface. Any available Cu combines with Ni and Sn to form $(Cu,Ni)_6Sn_5$, which grows instead of the Ni_3Sn_4. Ni_3Sn_2 may form either during longer soldering exposure time or during aging.

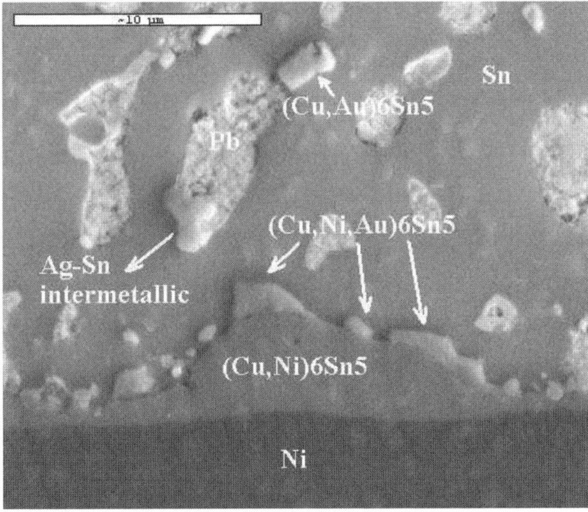

Figure 6.19. Typical structure at component interface, 169CSP Pb joint, cycled 0–100°C. Note that the structure is coarser and more complex than in Figure 6.18, and Au-containing phases have appeared.

- The Au layer that dissolved during reflow may reappear as an Au-containing intermetallic during aging. The outer intermetallic layer of both the board and the component side after thermal treatment contains significant amount of Au and is consistent with $(Cu,Au)_6Sn_5$.
- The Pb-free solder microstructure coarsens during aging, and large Cu_6Sn_5 and Ag_3Sn particles appear. The mixed cell solder structure coarsens and reveals the Ag_3Sn and Cu_6Sn_5 particles segregated in the boundaries between Sn and Pb crystals.

6.13.5. Gradient Mapping

NIST investigated whether there were any Ag or Pb compositional gradients across mixed-cell solder joints, using a 256PBGA joint [27]. Both area maps and line scan on a microprobe showed no systematic gradients in the joint compositions. Both the Ag and Pb concentrations varied, but these were randomly distributed throughout the joint.

Motorola calculated the relative nominal compositions of the mixed-cell solder joints to provide for comparison to the Pb and Pb-free solder joints [28]. The metal paste volume was calculated using paste volume data from the assembly of the test vehicles and the metal load of each paste. The solder ball volume was calculated from the nominal ball diameter. Dissolution of the base metal(s) from the solderable surfaces into the solder joint was ignored. It was found that there was approximately a 5% increase in the Sn levels in the joints as compared to the initial ball composition (Sn–37Pb), along with a 10% decrease in the Pb levels. With the R2512 and TSOP joints, the Sn level increased approximately 5%, but the Pb level decreased about 20-fold as compared to the plating composition (Sn–10Pb).

6.13.6. Layout Dye Staining (Dye-Pry)

The Nondestructive Testing community has used layout dye testing for many years. In fact, ASTM Standard E 165 “Liquid Penetrant Examination” was first issued in 1960. Since then, many varieties of these penetrating dyes have been developed to cover lipophilic, hydrophilic, and water-washable versions. Application of these products are described in ASTM Test Methods E 165, E 1208, E 1209, E 1210, E 1219, E 1220, and E 1418. They also have a Standard Practice Guide (E1417) and Reference Photographs (E433). These techniques were initially developed to find cracks in machined parts or on welds, but over the years they have been perfected for use in printed circuit-board technology. Figure 6.20 shows a typical dye-stained sample.

Layout dye staining was performed on 169CSP, 208CSP, and 256PBGA assemblies after thermal cycling [29–33]. Since each component was not pulled immediately at failure, it was not possible to determine which open ball was the initial failure point. However, these analyses were used to determine the relative extent and location of joint failure, as well as any unusual failure morphology.

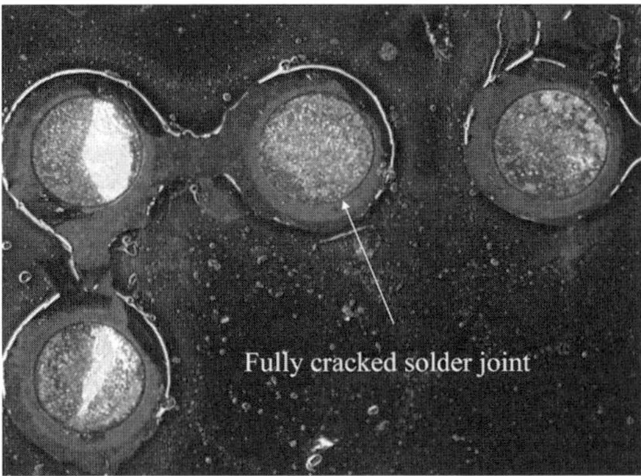

Figure 6.20. Dye-stained sample showing full and partial cracks in solder.

Dye Stain Conclusions

- The majority of failures were seen between the solder ball and the component.
- The Pb–Pb components exhibited more failures overall than did the Pb-free–Pb-free components.
- In general, SnPb–SnPb and SnPb–SAC joints showed far more cracking than the SAC–SAC joints.
- The SnPb–SAC combination led to the greatest amount of solder voiding, followed by SAC–SAC then SnPb–SnPb.

169CSP

- Pb-free wear-out (electrical) failures without corresponding dye-stain related solder joint failures appeared to be caused by via-in-pad failures.
  - One hypothesis for this is that the stiffer and stronger Pb-free solder joint applies a greater force on the pad than the Pb or mixed solder joints, making via-in-pad failures more likely.
  - Another hypothesis is related to the fact that the Pb-free solder joints solidify at a higher temperature than either the mixed or Pb joints. This may cause the Pb-free joints to apply a higher force on the solder pad as it cools through a larger temperature delta compared to the mixed and Pb joints.
- Several Pb-free parts had in-test electrical failures correlated to via-in-pad failures.
- The limited number of mixed and Pb parts examined showed no via-in-pad failures, but some via cracking was present.
- Conventional via-in-pad may not be compatible with a Sn–Ag–Cu Pb-free solder process.
- Solder crack area measurements correlate with survival reliability data.
- Some of the SAC–SAC (CSP169, 0–100°C) fails appear to be tied to a few large voids.

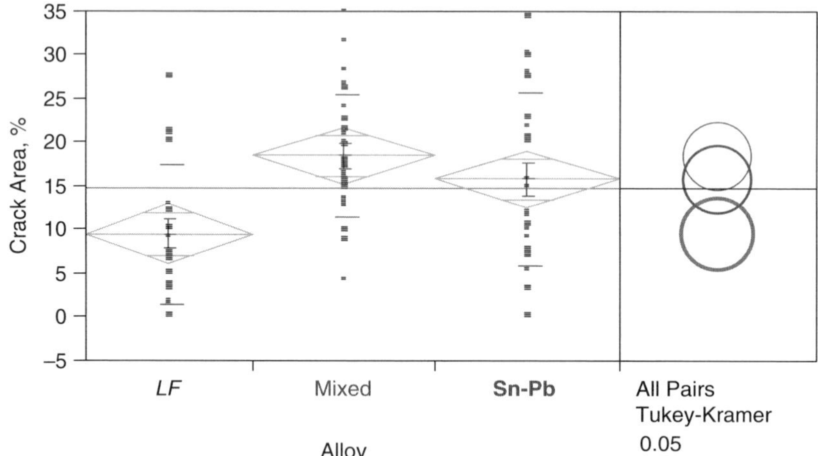

Figure 6.21. Statistical analysis of PBGA crack area.

208CSP

- The 208CSP components exhibited the majority of failures in the balls immediately surrounding the open center of the array, apparently nearest the edge of the die.

256PBGA

- Crack area data for 0–100°C samples (which did not fail during testing) shows less damage with Pb-free versus mixed and Pb cells. This agrees with −40–125°C data in terms of expected performance. Figure 6.21 shows a statistical analysis of the crack area seen on the inner rows of the PBGA.

6.14. BEND TESTING

Three-point bend testing was performed on 208CSP and 256PBGA assemblies [34]. This testing was undertaken in order to determine if there were any relative differences between Pb-free assemblies and Pb-containing ones concerning out-of-plane deformation that can result during board flexure associated with drops, testing, or assembly. A schematic of the test fixture is shown in Figure 6.22; this figure is not drawn to scale.

All failures occurred between the board (FR4) and the land pad on the board for both the 208CSP and the 256PBGA assemblies (an example of this type of failure is indicated in Figure 6.23). No differences in failure mode were observed between the various test cells, demonstrating that the Pb-free systems provide equivalent or better performance when compared to the Pb benchmark, as determined by statistical analysis.

Figure 6.24 shows a statistical comparison of normalized deflection by ball metallurgy and solder paste for the PBGA [35].

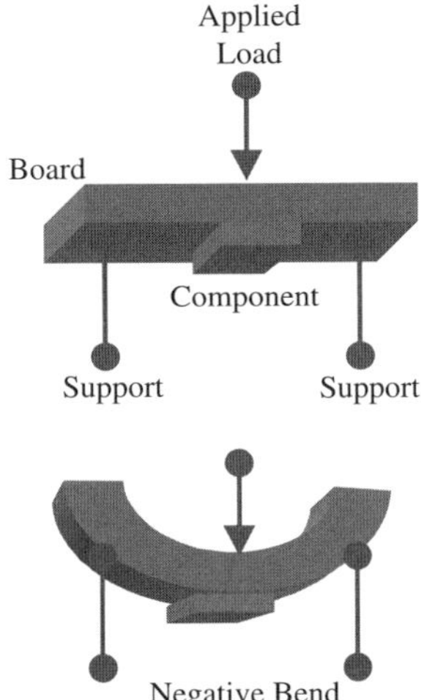

Figure 6.22. Schematic of three-point bend testing.

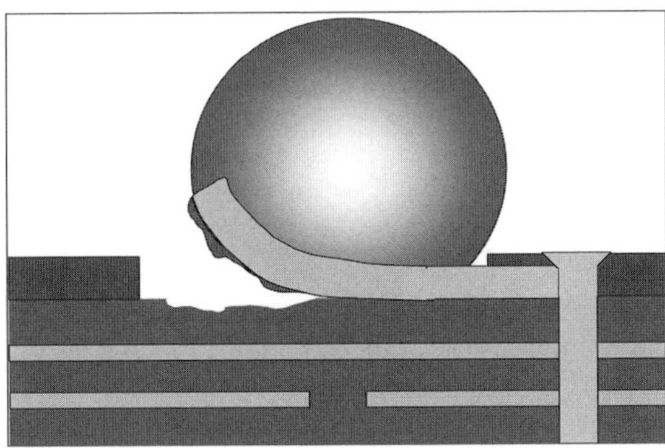

Figure 6.23. Schematic of pad-to-board failure interface. Not to scale.

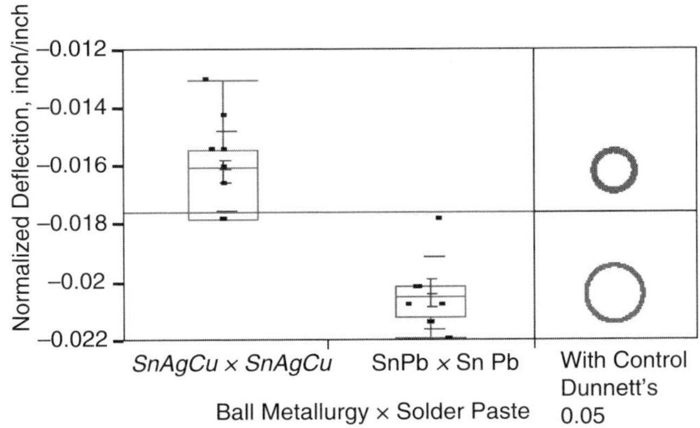

Figure 6.24. Statistical comparison of normalized deflection.

6.15. ELECTROCHEMICAL MIGRATION TESTING [36, 37]

The combinations of some solder and flux materials promote the growth of conductive crystalline filaments that may result in electrical shorts on a printed circuit assembly. Electrochemical migration (ECM) is an industry standard test that is performed to determine whether a material will grow these filaments across a nonmetallic substrate over time under the influence of an electrical bias.

Due to a concern regarding Ag causing Ag migration, solder pastes were assessed for electrochemical migration resistance using IPC-TM-650 Method 2.6.14.1, 65°C/85% RH, 10 V, 500 h. The test pattern used was IPC B25A, Pattern D (0.0125-in. lines and spaces). Four paste suppliers (Alpha, Heraeus, Indium, and Kester) performed the testing on their pastes containing the NEMI-recommended Sn–3.9Ag–0.6Cu alloy, with Sn–37Pb as the benchmark [38–41]. All Pb-free and Sn–Pb pastes passed, indicating that no electrochemical migration issues are inherent in the Sn–3.9Ag–0.6Cu when evaluated using the IPC test method described above. It is important to note that some industry experts believe that the IPC standard tests may not be sufficient to determine the reliability of solder and flux materials.

6.16. iNEMI TEAM CONCLUSIONS

The reliability of Pb-free solder joints (Sn–3.9Ag–0.6Cu) was assessed by three methods:

1. Thermal cycling, two temperature ranges
2. Three-point bend testing
3. Electrochemical migration testing

The performance of the Pb-free joints was compared to the performance of Pb-containing joints made with (1) current material sets and (2) current Pb-containing components assembled using Pb-free paste.

The reliability of the Pb-free solder joints was found to be equivalent or superior to the reliability of the Pb-containing joints made using current material sets as assessed by thermal cycling and three-point bend testing. No electrochemical migration issues were identified for the Sn–3.9Ag–0.6Cu alloy.

6.17. OVERALL SUMMARY, CONCLUSIONS

The Sn–3.9Ag–0.6Cu alloy selected provides solder joints for electronic component assembly that are equal to or better than the Sn–37Pb solder it replaces. More work is needed to define adequate rework processes and the effect on reliability of the higher processing temperatures on thicker boards (e.g., >0.100 in. thick). It is expected that fluxes for this higher-temperature alloy will be optimized as production increases. This SAC alloy is recommended to the industry for use in Pb-free surface mount electronics assembly.

6.18. ASTM TEST METHODS

ASTM Test Methods are listed below. These methods are available at www.astm.org. Click on the "Standards" tab, then "Individual Standards," and then the "E" tab of the alphanumeric listing.

E 165: E165-02 Standard Test Method for Liquid Penetrant Examination

E 1208: E1208-99 Standard Test Method for Fluorescent Liquid Penetrant Examination Using the Lipophilic Post-Emulsification Process

E 1209: E1209-99 Standard Test Method for Fluorescent Liquid Penetrant Examination Using the Water-Washable Process

E 1210: E1210-99 Standard Test Method for Fluorescent Liquid Penetrant Examination Using the Hydrophilic Post-Emulsification Process

E 1219: E1219-99 Standard Test Method for Fluorescent Liquid Penetrant Examination Using the Solvent-Removable Process

E 1220: E1220-99 Standard Test Method for Visible Penetrant Examination Using the Solvent-Removable Process

E 1418: E1418-98 Standard Test Method for Visible Penetrant Examination Using the Water-Washable Process

E1417: E1417-99 Standard Practice for Liquid Penetrant Examination

E 433: E433-71 (1997) Standard Reference Photographs for Liquid Penetrant Inspection

REFERENCES

1. *NCMS Lead-Free Solder Project Final Report*, NCMS, National Center for Manufacturing Sciences, 3025 Boardwalk, Ann Arbor, Michigan 48108-3266, Report 0401RE96, August 1997, and CD-ROM database of complete dataset, including micrographs and raw data, August 1999.
2. M. R. Harrison and J. Vincent, "IDEALS: Improved design life and environmentally aware manufacturing of electronics assemblies by lead-free soldering, *Proc. IMAPS Europe'99* (Harrogate, GB), June 1999.
3. B. Hunter, Storage Technology Corporation, CTE Measurements of Area Array Components Using Moire Interferometry, December 2001.
4. John Manock, Lucent Technologies, Measurement of Coefficient of Thermal Expansion of 48 TSOP and CBGA Boards, May 2003.
5. Adam Zbrzezny, Celestica, NEMI Pb-Free Reliability Test, March 2001.
6. Eastman Kodak, NEMI Pb-Free Reliability Test, March 2001.
7. J. Sohn, Lucent, NEMI Profile, February 2001.
8. Motorola, Motorola 0 to 100C Chamber Profile, April 2001.
9. C. Fieselman, Solectron, NEMI Lead-Free Reliability Project Thermal Cycle Profile, March 2001.
10. A. Zbrzezny, Celestica, personal correspondence, November 2000.
11. K. Fallon, NEMI Pb-Free Reliability Testing, Eastman Kodak, October 2000.

12. J. Sohn, Lucent Technologies, personal correspondence, November 2000.
13. E. Bradley, Motorola, personal correspondence, November 2000.
14. SCI Sytems Inc., personal correspondence, December 2000.
15. C. Fieselman, Solectron, personal correspondence, April 2001.
16. Overall Summary of NEMI Pb-Free Testing, August 2001–December 2001.
17. M. Farooq, C. Goldsmith, R. Jackson, and G. Martin, IBM Microelectronics, Lead (Pb)-Free Ceramic Ball Grid Array (CBGA):Thermo-Mechanical Fatigue Reliability, July 2002.
18. C. Handwerker and L. Smith, NIST, and P. Snugovsky, Celestica, Failure Analysis—CSP169, July 2002.
19. C. Handwerker and L. Smith, NIST, Failure Analysis—CSP208, August 2002.
20. C. Handwerker and L. Smith, NIST, Failure Analysis—R2512, August 2002.
21. C. Handwerker and L. Smith, NIST, Failure Analysis—TSSOP48, July 2002.
22. C. Handwerker and L. Smith, NIST, Failure Analysis—PBGA256, July 2002.
23. C. Handwerker and L. Smith, NIST, Failure Analysis—CBGA, August 2002.
24. P. Snugovsky, Celestica, NEMI Report on Intermetallic Composition and Stoichiometry, May 2003.
25. T. Siewert and C. McCowan, NIST, NEMI Report on the Stoichiometry of Intermetallic Layers, April 2002.
26. T. Siewert and C. McCowan, NIST, NEMI Report on the Stoichiometry of Intermetallic in TSOPs, October 2002.
27. T. Siewert and C. McCowan, NIST, NEMI Report on concentration Gradients in Mixed Cells, April 2002.
28. E. Bradley, Motorola, Calculated Mixed Cell Compositions for NEMI Reliability Test Board Assemblies, April 2002.
29. Elizabeth E. Benedetto, Hewlett-Packard Co., Dye Stain Analysis—NEMI Pb-Free Thermal Cycle Samples, November 2001.
30. E. E. Benedetto, Hewlett-Packard Co., Dye Stain Analysis—NEMI Pb-Free Thermal Cycle TSOPs, April 2002.
31. E. Bradley, Motorola, Red Dye Analysis of NEMI 169 CSP Parts from Thermal Cycling Exhibiting Both Early and Wearout Failure, June 2002.
32. E. Bradley, Motorola, Red Dye Analysis of NEMI 256 PBGA Thermal Cycled Parts, June 2002.
33. J. Bartelo, IBM, NEMI Lead-Free Dye and Pry IBM Conclusions, July 2002.
34. E. E. Benedetto, Hewlett-Packard Co., Reliability Test Methods—Three Point Bend, June 2000.
35. C. Fieselman, Solectron, Statistical Analysis of NEMI 3 point Bend Test, May 2001.
36. M. Romansky, Celestica, NEMI Pb-Free Reliability Subgroup Electromigration Test Plan Request (2nd edition), June 2000.
37. M. Romansky, Celestica, NEMI Pb-Free Reliability Subgroup Electromigration Test Report, September 2001.
38. Alpha Report, September 2001.
39. Hereaus Report, January 2001.
40. Indium Report, February 2001.
41. Kester Report, November 2000.

CHAPTER 7

Tin Whiskers: Mitigation Strategies and Testing

HEIDI L. REYNOLDS, C. J. LEE, and JOE SMETANA

7.1. INTRODUCTION

Whiskers have commonly been observed in electroplated metal layers of tin (Sn), zinc (Zn), and cadmium (Cd). Whiskers are metal, needle-like filaments that may emanate spontaneously from the finish surface. Whiskers have uniform diameters on the order of 1–5 μm and can be straight or kinked. To be considered a whisker, the length-to-width ratio is generally >2. Examples of tin whiskers [1] are shown in Figure 7.1.

The focus of this chapter will be on tin whiskers due to the significant impact that the removal of lead (Pb) has on the electronics industry. Tin-based plating—the current industry-preferred surface coating for the leads on electronic components—can be susceptible to whisker formation, especially when pure-tin or high-tin-content, Pb-free alloys are used. Tin (Sn) whiskers have been measured at maximum lengths of up to 10 mm [2]. As such, these whiskers could seriously impact product reliability. If they grow to critical lengths in service, whiskers could cause electrical shorts, disruption of moving parts, and/or degraded RF/ high-speed performance. For years, electronics manufacturers controlled tin whiskers by adding small amounts of lead (Pb), but now that the electronics industry is moving rapidly toward Pb-free products, the risks associated with the use of pure tin must again be addressed.

Tin whiskers have been studied for over 50 years [3] with a renewed interest in the last several years. Recently, an extensive review of theories and experiments related to tin whisker formation and growth was done [4]. A clear mechanism of whisker growth has not been identified. However, some observations have been made, including the fact that whiskers can grow in response to a regenerating stress [5–8] and often an incubation time precedes whisker formation and growth [5, 9, 10]. For ambient conditions, incubation times on the order of years have

Lead-Free Electronics. Edited by Bradley, Handwerker, Bath, Parker, and Gedney

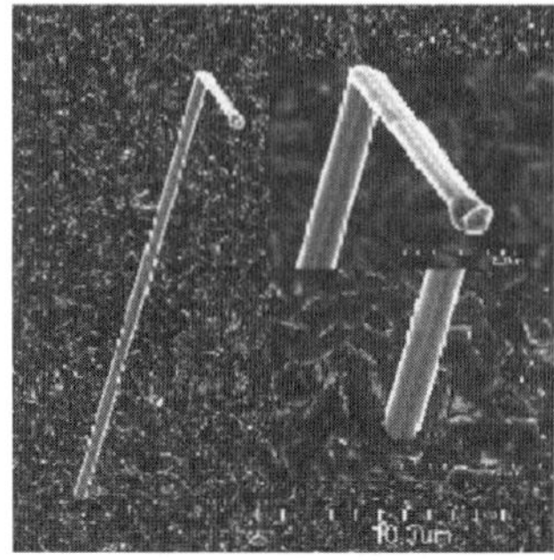

Tin whisker filaments
Photo courtesy of M. Williams, NIST

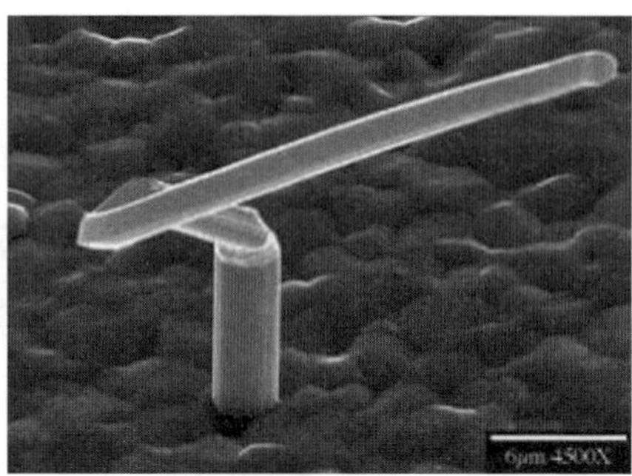

Kinked whisker
Photo courtesy of P. Bush. SUNY

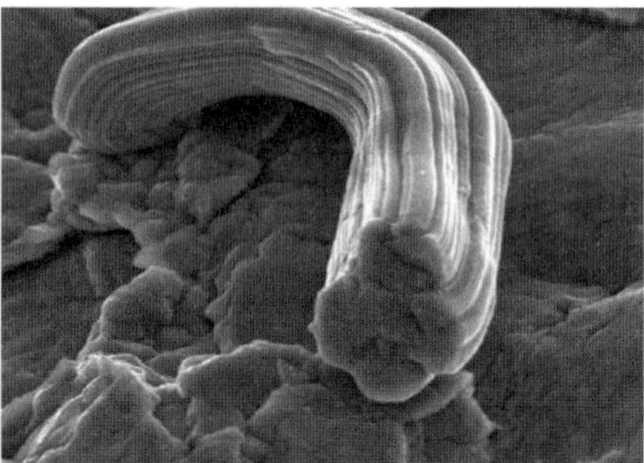

Tin whisker filament with striations
Photo courtesy of Rudy Wagner

Figure 7.1. Whiskers can be kinked, straight, or even curved. These tin whisker SEM images were taken from JEDEC standard JESD 22A121. The striations on the surface are a characteristic feature of whiskers. (From Ref. 1.)

been observed. For high-temperature/humidity conditions, incubation times of the order of 3 months have been observed.

Although there is no scientific consensus on whisker formation and growth fundamentals, there is some agreement that compressive stress in the tin film is the fundamental driving force behind whisker growth. Certain factors have also been identified that promote or accelerate their growth. High-humidity conditions result in increased oxygen penetration in tin finishes with associated increases in tin oxides and resulting stress. It has been theorized that high humidity may also increase the rate of tin grain boundary and/or surface diffusion, which affects tin whisker growth. Temperature may either promote or inhibit tin whisker growth. Too high a temperature results in stress relief by mechanisms other than whisker growth. At temperatures between 20°C and 50°C, the Cu_6Sn_5 intermetallic forms quite rapidly (within a few days), adding to stress in the plated film when tin is plated over copper-based substrates. Mismatch of the coefficients of thermal expansion (CTE) between tin and base materials results in stress during thermal cycle conditions, which drives whisker growth. The mechanism(s) for the necessary movement of tin atoms within the structure is still under considerable debate.

Since repeatable test methods for evaluating whisker performance did not exist that could be related directly to field conditions or performance, well-defined, standardized test methods were needed to facilitate the development and qualification of lead (Pb)-free, tin (Sn)-based finishes. Such standardized test methods would allow meaningful comparison of whisker propensity for different plating systems and processes in tests performed by different labs.

To fill this gap, iNEMI formed a Tin Whisker Accelerated Test Group to investigate and standardize tin whisker test methods in 2001. The past and continuing work of this Test Group and the development of tin whisker testing standards are described in Section 7.3.

It quickly became apparent that standard accelerated test conditions (e.g., higher temperature, humidity, and thermal cycling) were not going to provide a nice, clean set of tests that would predict whisker growth. However, driven by pending legislation in Europe, the electronics industry was racing to introduce lead-free electronics and needed some methodology for assuring reliability of Sn coatings. After much discussion, iNEMI members embarked on a serial methodology to minimize reliability exposure. The first step was to define a set of test conditions that would promote tin whisker growth and to recommend a protocol for inspecting whisker growth and data recording. The second step was to work on gaining fundamental understanding of whisker formation. The third step was to develop acceptance test criteria and mitigation practices that would provide an interim auxiliary tool to minimize reliability exposure of long-life, reliability-sensitive electronic systems.

The factors that have been observed to promote or accelerate tin whisker growth and the observed incubation periods under test conditions provided the foundation for JEDEC standard JESD22A121, "Measuring Whisker Growth on Tin and Tin Alloy Surface Finishes" [1]. The iNEMI project team recognized that, without knowing acceleration factors, it would not be possible to write a qualification specification; but using what is known and critical engineering judgment, it should be possible to write an acceptance specification. However, extending these test methods into an acceptance test standard proved to be a challenging task. Even though JESD22A121 specifies a standard set of tests for measuring tin whisker growth, there is no way at this time to relate these tests to field conditions with any reasonable degree of certainty.

The iNEMI Tin Whisker User Group published recommendations for acceptance test requirements in July 2004 [11], and it subsequently submitted these specifications to JEDEC and IPC for creation of a formal standard. JEDEC standard JESD201, "Environmental Acceptance Requirements for Tin Whisker Susceptibility of Tin and Tin Alloy Surface Finishes" [12] was eventually published in March 2006 after rigorous scrutiny by both the "User" and "Supplier" communities.

Since the current plating and whisker test methods neither guarantee the prevention of tin whisker growth nor accurately predict whisker growth or whisker lengths, iNEMI's recommendations, which are integrated into JESD201, take a threefold approach to reducing the risk of tin whiskers for high-reliability applications. The first requirement is that a viable mitigation practice—known techniques or uses of materials that prevent the formation of whiskers, or delay or retard their growth—be used with the component finish. The second requirement is that acceptance testing be conducted. This testing of the surface finish material set and manufacturing processes includes a defined set of base metals, underplating metals, surface finish alloy, surface finish bath chemistry, and process flow steps. Finally, process control of the tin plating at the supplier is a key variable. Certain factors about the plating processes are known to affect whisker growth, but definitive information on all the needed controls is not available, thus making it difficult to define process controls in a specification. JEDEC standard JESD201 provides guidelines for an ongoing monitoring program with an eye toward minimizing exposure.

While the recommended approach cannot eliminate the chance of a whisker-related failure in service, it will significantly reduce the risk of whisker-related problems. The remainder of this chapter will discuss, in depth, both recommended mitigation practices (Section 7.2) and tin whisker test development (Section 7.3).

7.2. MITIGATION STRATEGIES

As mentioned, tin whisker formation theory and growth mechanism are still under considerable debate. Even for the prevailing compressive stress theory, the whisker growth rate, absolute length, incubation time, and whisker density are a convolution of various factors, such as substrate materials, plating finish materials, plating process parameters, as-plated stress, post-plating trim and form operations. Due to the complexity of the material combinations and the process history of each component, it is not possible to recommend a single mitigation strategy that will work under all circumstances. Instead, a list of tin whisker mitigation guidelines will be presented here with sufficient background information for the reader to evaluate the applicability and select the appropriate strategy to suit their unique needs in terms of reliability risk and cost benefits in the market place. It is also advisable that these techniques be treated as general guidelines. The effectiveness of these practices may require optimization and assessment using JEDEC standard JESD201. Many "End Users" require tin whisker testing in conjunction with the use of some of the mitigation strategies discussed below. It should be noted that the various mitigation practices and techniques discussed here should not be interpreted as whisker prevention or elimination methods, but rather as whisker risk reduction methods when implemented effectively.

7.2.1. Non-Tin Plating: Avoid Tin-Plated Parts

The best strategy to eliminate any failure caused by tin whiskers is to avoid using pure-tin or high-tin-content plating on any part of a component, such as lead finishes, RF shields, mounting hardware, and electronic enclosures, regardless of whether its function is to establish electrical connections.

7.2.1.1. Ni–Pd–Au/Ni–Pd. Nickel–palladium–gold (or nickel–palladium) should be strongly considered for lead-frame applications. This plating has more than a 10-year history of field application. Early solderability issues have been resolved to a great extent. In addition, Ni–Pd–Au is not prone to whisker growth in most environments (gold has been observed to grow whiskers in certain environments). Users should be aware that molding compounds do not adhere as well to noble metals such as Pd and Au as they do to copper. As such, it may be more difficult for Ni–Pd–Au packages to achieve MSL 1 and 2 performance at the higher temperatures associated with Sn–Ag–Cu Pb-free assembly. Ni–Pd–Au has shown corrosion in accelerated tests using high-hydrocarbon-and-sulfur atmospheres. This corrosion has not been noted in actual field conditions.

7.2.1.2. Ag Finish. Silver (Ag) finishes are not prone to whisker growth in most environments. However, rapid growth of silver whiskers or dendrites may form in the presence of H_2S (found in some cases where the environmental air pollution contains SO_2). Additionally, users sometimes avoid silver finishes due to potential issues with electromigration and solderability shelf life.

7.2.1.3. Au or Ni–Au. Gold (Au) is not only a very good conductor but provides excellent corrosion resistance. Gold plating, though relatively expensive, is a well-established technology developed over the past several decades. Electrolytic gold or electrolytic gold over electrolytic nickel (not ENIG) are good alternatives to mitigate whisker formation. So far, there is only one documented gold whisker report, and the formation is most likely caused by rubidium contamination in the gold plating [13]. When plated directly on copper, a nickel barrier between copper and gold is required to prevent the gold–copper interdiffusion, especially during the assembly and solder reflow operations above 150°C. One common concern with gold plating is its incompatibility with tin-containing solder and the subsequent "gold embrittlement": The solder joint weakens during thermal cycling when the gold concentration in the solder joint is more than 3%. In reality, as long as the gold-plating thickness is small (20–30 μin., preferably less than 10 μin.), gold embrittlement is not considered a problem.

7.2.2. Alloying

The most effective strategy to reduce tin whisker propensity in the past 50 years has been addition of a minimum of 3% lead to tin plating. The addition of lead does not completely eliminate the whisker growth, but it delays the onset of whisker growth significantly as well as suppresses the growth rate dramatically. All observed whiskers from tin–lead (Sn–Pb) alloys as of today are sufficiently small (~20 μm) and do not pose a significant risk for today's microelectronics.

This strategy, however, is no longer viable for most products after July 1, 2006 due to current/pending European Union, California, and China regulations prohibiting the use of lead (Pb). In response to the ban on lead (Pb), the industry showed renewed interest in identifying other alloy elements to replace Pb and mitigate the whisker formation and growth. Recommended alloying approaches for tin whisker mitigation are discussed here. An alloying approach that is not recommended as a whisker mitigation is discussed in Section 7.2.6.1.

7.2.2.1. Sn–Bi. Tin–bismuth (Sn–Bi) alloy finishes are controversial when used in conjunction with eutectic Sn–Pb solder. When added to tin in amounts of 2–4% by weight, bismuth may aid in suppressing whisker growth. There is a low-melting-point ternary eutectic formed between Sn–Pb–Bi with a melting point at 96°C. However, it is not thermodynamically possible to form this ternary eutectic with small (1–5% by weight) additions of Bi to Sn finishes when soldered with Sn–Pb. There is a ternary Sn–Pb–Bi peritectic that is thermodynamically viable for small additions of Bi, and this peritectic has a melting point of 135°C. With lead-free solder, Sn–Bi is a viable candidate for component finishes. With eutectic Sn–Pb solder, it will be necessary to control the bismuth content of the finish between 2% and 4% so as to have enough bismuth to suppress whisker formation without getting into the compositional range of the ternary eutectic. In addition, keeping the Bi content low is required to retain solderability of formed leads.

7.2.2.2. Sn–Ag. Plated tin–silver (Sn/2–4% Ag) alloys in limited testing have shown promise in reducing tin whisker growth. Further investigation of this finish is needed.

7.2.3. Heat Treatment

Two types of heat treatments—fusing and post-plate baking—are usually considered for tin whisker mitigation. Pure tin has a melting point of 232°C. Fusing is performed at temperatures above the tin melting point to melt and re-solidify the tin plating under relatively slow cooling conditions, whereas post-plate baking (150°C for one hour within 24 hours of plating) subjects tin-plated parts to temperatures that are lower than the melting point primarily to alter the intermetallic structure and secondly to promote recrystallization and grain growth.

7.2.3.1. Fusing. Fusing is usually done (typically within one week after plating) by dipping the Sn-plated surfaces into a hot oil bath. Due to the melting and re-solidification process, stresses within the tin layer are reduced. This mitigation practice appears to have a good field history. It should be noted, however, that past performance of fused tin is based on assembly with tin–lead solder, which tends to have better wetting than Pb-free solders, and thus the exposed region of the tin finish was reduced.

7.2.3.2. Post-Plate Baking. When tin (Sn) is plated directly on copper (Cu)-based substrates, it is known that the Cu_6Sn_5 intermetallic grows along the grain boundaries rapidly (within a few days) when temperatures are below 60°C. The irregular intermetallic layer results in compressive stress in the Sn layer and

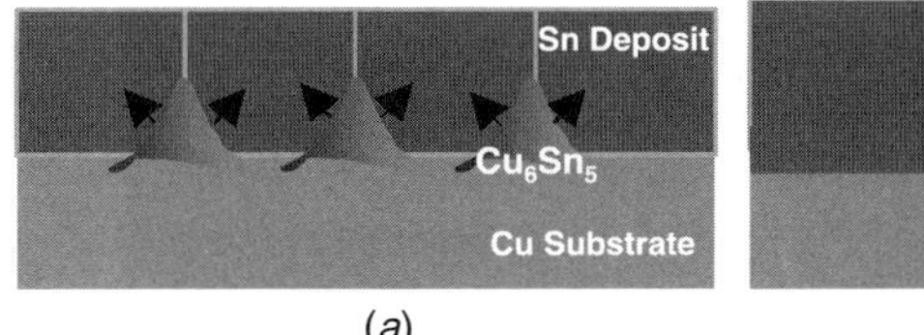

(a)

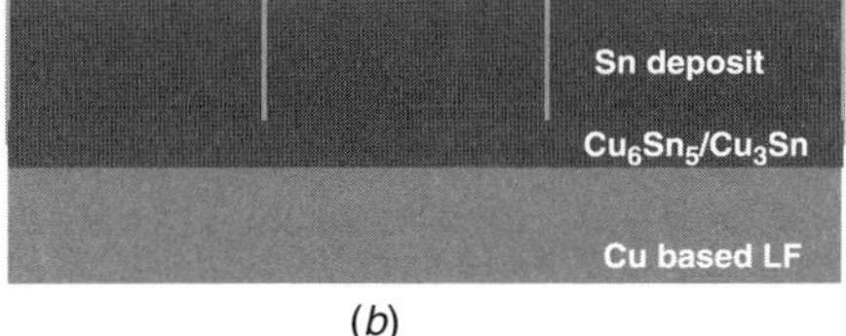

(b)

Figure 7.2. (*a*) Schematic drawing showing Cu_6Sn_5 growth in the grain boundaries and the resulting irregular intermetallic structure. (*b*) Schematic drawing showing a uniform, double layer of Cu_3Sn–Cu_6Sn_5 structure formed by bulk diffusion due to the post-plate bake (150°C for 1 hour within 24 hours of plating). (From Ref. 14.)

promotes tin whisker growth. One way to suppress the grain boundary diffusion-dominated irregular intermetallic growth is to perform a one-hour 150°C bake within 24 hours of the plating. At this higher temperature, the intermetallic formation proceeds primarily by bulk diffusion and the resulting intermetallic layer is more uniform and continuous, thus reducing the stress from irregular intermetallic formation. Furthermore, a second intermetallic compound, Cu_3Sn, grows between the Cu_6Sn_5 and Cu layers. This Cu_6Sn_5–Cu_3Sn intermetallic (IMC) double layer serves as a diffusion barrier reducing further Cu-diffusion into tin. The effect of this post-plate heat treatment on the intermetallic structure of tin over copper is illustrated in Figure 7.2.

In addition to modifying the intermetallic growth, this post-bake step can also provide the benefits of a typical annealing process: increasing grain size, reducing grain boundaries, releasing the stress in the plating, and annealing out the imperfections in the lattice of the Sn [15].

A post-plate bake, applied shortly after plating for tin over copper, has a long history of positive reports demonstrating its mitigation effectiveness. In many cases though, it is suggested that this heat treatment increases the incubation time significantly but does not completely eliminate the whisker formation and growth. Nonetheless, available historical data often indicate that the maximum length of the whiskers is reduced by the use of a post-plate bake. This post-plate heat treatment should only be used for components with tin plated directly on copper, not in conjunction with an underlayer as described in the next section.

7.2.4. Underlayer: Elimination of Cu_6Sn_5 Intermetallic Formation

Cu_6Sn_5 intermetallic formation has often been credited as the primary source of compressive stress in Sn plating over copper-based substrates. A diffusion barrier layer or underlayer between the plating and the substrate has therefore been proposed to retard the intermetallic growth and reduce tin whisker propensity. Commonly employed underlayer materials with Sn plating include both nickel and silver.

In general, the use of an appropriate underlayer is an effective measure in negating the Cu_6Sn_5 intermetallic-induced compressive stress at the tin–copper substrate interface. On the other hand, underlayer plating does little or nothing to address other sources of compressive stress. Coefficient of thermal expansion (CTE) mismatch (most commonly seen with pure Sn plating on Alloy 42 substrates), oxidation, or mechanical stress can still lead to whisker formation even when an underlayer is present.

The post-plate heat treatment described in the previous section is intended to alter the Cu–Sn intermetallic structure and, therefore, will not be effective in mitigating whisker formation in conjunction with a "non-copper" underlayer or substrate material. Furthermore, misapplication of a post-plate bake could lead to increased whisker growth and/or solderability problems [16].

7.2.4.1. Ni Underlayer. Adding an Ni underlayer between Sn plating and a Cu base metal is a popular recommendation [among users] to mitigate whisker formation. An Ni underlayer as thin as 0.2 μm has been shown to be effective in suppressing whisker formation for more than 350 days under ambient condition [10]. In practice, a minimum thickness ranging from 0.5 to 1.27 μm is generally recommended. In addition to its primary function as a copper diffusion barrier, a plate-like Ni_3Sn_4 intermetallic compound gradually forms when the Sn atoms diffuse into the Ni underlayer [32]. The vacancy-rich zone left behind in the Sn plating results in a tensile stress which contributes to the overall whisker mitigation effectiveness. The thickness, porosity, and ductility of the nickel plating are very important to ensure an effective barrier layer to hinder copper diffusion. It may be necessary to verify that the integrity of the underlayer is not affected after lead forming. Similarly, the control of tin bath impurities, particularly copper, is also critically important to make this mitigation method effective.

7.2.4.2. Ag Underlayer. Adding an Ag underlayer between Sn plating and Cu base metal has been proposed as a method to mitigate whisker formation, similar to Ni as noted above. Studies have shown that an Ag underlayer of thickness >2 μm on Cu-based substrates could completely suppress whisker growth under ambient condition for over 350 days [10]. Further examination revealed that only a very thin and almost regular layer of Ag_3Sn is formed at the interface and therefore causes no stress buildup from the intermetallic formation. In addition, it is evident that the Ag underlayer has effectively prevented Cu from diffusing into the Sn plating. Nonetheless, there is limited additional whisker test data demonstrating the effectiveness of an Ag underlayer for whisker mitigation. Further investigations, including stress effects due to molar volume changes and Kirkendall voiding effects, are still required for a thorough evaluation on the potential of this mitigation practice.

7.2.5. Varying Plating Methods and Processes

Two methods are predominantly used for commercial tin or tin alloy coatings: hot dipping and electroplating.

Hot dipping is the most economic way to coat a substrate with tin: The substrate is cleaned, coated with an activation flux, and then dipped into or fed through a bath of molten tin. The excess tin is removed by either mechanical wipers or air knives. If hot air is used as the air knives, the resulting tin layer is usually called HALT (hot-air-leveled tin).

Electroplating is another low-cost process suitable for large volume production as well as automation to increase the throughput. Nonetheless, electroplating involves a large number of process variables, such as current density, pH value, electrolyte concentration, impurity level, and bath temperature, which may significantly influence the plating quality and should be closely monitored. For electroplating of tin, two main groups of electrolyte baths are used and result in tin coatings with either a matte or bright appearance.

The tin whisker performance of hot-dipped tin and matte tin will be discussed in the following subsections. See Section 7.2.6.2 for more discussion on bright tin as a termination finish.

7.2.5.1. Hot-Dipped Tin. Hot-dipped tin is a commonly used process for connectors, some axial-leaded-through hole components, and devices such as relays, but less prevalent for lead-frame construction intended for electronic components. Hot-dipping is usually considered a favorable method to mitigate whisker growth due to the following reasons:

- The tin layer retains little internal stress because the hot-dipping process helps to relieve the internal stress.
- Hot-dipped tin has very large grain size (0.5–5 mm). As a result, the grain-boundary diffusion responsible for Cu_6Sn_5 intermetallic growth under normal storage of service conditions is suppressed, and little compressive stress is built in the plating as a function of time to promote whisker growth [17].
- When hot-dipping is used on copper-based substrate, an Sn–Cu intermetallic layer is formed immediately during the process via bulk diffusion. This uniform layer not only ensures that the tin coating adheres very well to the base metal, but also serves as a diffusion barrier against further grain-boundary growth of irregular Cu_6Sn_5 IMC at lower temperatures [18].

It is reported that no whiskers were observed under various test conditions [17, 19, 20]. Nonetheless, it is important to note that hot dipping does not always completely eliminate tin whiskers, as demonstrated by Pitt and Henning [21]. Pitt and Henning found that the use of hot dipping of Sn on Cu substrates considerably reduced the number of whiskers. These results suggest that hot-dipped tin is an effective mitigation practice to suppress tin whisker growth, although it does not guarantee complete elimination of whiskers.

7.2.5.2. Thick Electroplated Matte Tin. When electroplating is preferred over hot-dipping to meet the device configurations and manufacturing specification requirements, such as plating uniformity and thickness control for fine-pitch components, matte tin plating is strongly recommended.

Bright tin finishes are generally shiny in appearance, with a submicron grain size, high carbon content (0.2–1.0% by weight), and high internal stresses. In contrast, matte tin finishes generally have a dull appearance, with grain sizes ranging from 1 μm to 5 μm, carbon contents <0.05%, and relatively low internal stresses. Matte tin plating tends to be less prone to whisker formation and growth than bright tin plating. (See Section 7.2.6.2 for more discussion on bright tin as a termination finish.) Many current suppliers tout a proprietary version of this type of [matte] tin as "whisker-free." It must be restated here that the electroplating process involves many variables that may severely compromise the plating quality if they are not well-controlled. These claims of "whisker-free" chemistries are not supported. Adequate testing of a specific plating process, including plating parameters, is required in order to minimize the risk of whisker failures.

Industry data indicate that thicker tin finishes show a lower propensity for tin whiskers and/or a greater incubation time before tin whiskers occur. It is recommended that the tin layer thickness for components without a nickel or silver underlayer be at least 7-μm minimum (10 μm nominal) [10, 22–24]. In addition, it is strongly encouraged to incorporate the post-plating bake procedures mentioned in Section 7.2.3.2 to further reduce whisker growth propensity.

7.2.5.3. Apply a Conformal Coating. At the system level, the application of a conformal coating may be used to mitigate the risk caused by tin whiskers if pure or high-tin-content alloy finished parts cannot be avoided. In using a conformal coating, the goal is to obtain one or more of the following effects:

- Increasing the incubation period
- Suppressing the whisker growth rate
- Reducing the whisker growth density
- Containing the whiskers under the coating to prevent electrical shorts

NASA Goddard has designed a set of experiments to investigate the effectiveness of Uralane 5750 conformal coating in these areas. So far, it has been found that Uralane 5750 does not prevent the formation of tin whiskers. In fact, whiskers start to grow earlier on samples covered by the coating compared with uncoated samples [25]. Nonetheless, the conformal coat does reduce the whisker growth rate. As for the robustness of the conformal coating, it has been found that whiskers were able to penetrate ~0.25-mil-thick Uralane 5750 after 2.5 years under ambient storage conditions. For a coating with nominal thickness of 2 mil, there are extensive whiskers growing underneath the coating after 3 years. None of the whiskers have been able to grow through the coating yet, but they are pushing and stretching the conformal coating. These whiskers are expected to eventually penetrate the conformal coating.

It has been shown [26] that an exposed whisker cannot penetrate another coating layer as the whisker begins to buckle. If the dielectric strength and thickness of the coating material are judiciously chosen, conformal coating could still reduce the risk of a whisker creating a short between adjacent conductors even though the whiskers may not always be effectively contained.

The effectiveness of a conformal coat for tin whisker risk mitigation is a function of numerous variables that are dependent on the conformal coating material chosen, such as the adhesion strength, material toughness, shelf life and pot life, dielectric constant, coefficient of thermal expansion, modulus, and so on. In addition, other engineering considerations are necessary to properly select and apply a conformal coat to the specific locations. These factors include, but are not limited to, the configuration, geometry, and functionality of the components. Another important issue that users should be aware of is the ability to rework conformal coated assemblies. The reworkability is dependent on the coating material.

In summary, conformal coating is a potentially effective tin whisker risk mitigation practice at the assembly level. However, it should be treated as a secondary strategy to the other methods mentioned above. Conformal coating is a process that has been utilized to protect the printed circuit boards from dust, moisture, debris, and thermal shock. Various conformal coating products are commercially available for different process methods, including dip-coating, spray on, brush-on, or spin-on. The remaining considerations are related to selecting the appropriate coating material, application method, and optimal thickness. At present, more investigation is still needed to provide these general guidelines.

7.2.6. Lead (Pb)-Free Finishes that are not Recommended as Mitigation Practices for Tin Whiskers

7.2.6.1. Electroplated Sn–Cu. Tin and copper forms a eutectic alloy with a copper content of 0.7% and a melting point at 227°C. This composition is a good alternative to Sn–Ag–Cu for wave soldering, due to its relative low cost. However, as a electrical terminal plating finish, it is not preferred because it is prone to whisker growth. Various research reports show that whiskers grow faster on electroplated eutectic Sn–Cu finish than on pure tin finishes [27–30], and whisker lengths can reach a few hundred micrometers to $\sim$1600 μm after one year under ambient storage condition [10, 31].

NIST had investigated the correlation between Sn whisker growth and Sn deposit microstructure as a function of Cu content and found that Cu content has significant impact on the whisker density and length; therefore precise control of the Cu% in the deposit film is critical, and the consistency is not easy to achieve. Additionally, NIST found that Sn–Cu alloys had significantly greater stress as plated than did similar pure Sn alloys. Furthermore, when a Cu-based lead frame is used, measuring and monitoring the Cu content in the plating becomes problematic.

Due to the technical challenges associated with Sn–Cu plating process control as well as the concerns of excess Sn–Cu intermetallic formation in the plating layer, it is advisable to avoid electroplated Sn–Cu as an alternative Pb-free plating finish.

7.2.6.2. Bright Tin. As previously mentioned in Section 7.2.5.1, bright tin finishes are generally shiny in appearance, with a submicron grain size and high carbon content (0.2–1.0% by weight). The high carbon content comes from the co-deposited organic brightener used in the process to make the tin plating level/smooth and bright/shiny. As a result of the additives, bright tin plating often has inherent compressive internal stresses and is prone to whisker formation. Historically, bright tin is known to have high whisker propensity and can grow whiskers on the order of several millimeters to as long as 10 mm [2].

Bright tin has been used for many years in the electronic industry for its various favorable properties, such as the aesthetic appearance and hardness. Recently, it has been presented that bright tin can be made to be whisker-free under test conditions per JEDEC22A121 by using an Ni underlayer and careful plating process control [32]. Inspection of the Sn–Ni interface revealed orderly plate-like Sn–Ni intermetallic structure. This study demonstrated that tin whisker risk in bright tin could be effectively mitigated, nonetheless, the author also pointed out that the reasons for the improved whisker test performance were unclear. An Ni underlayer alone does not seem to guarantee the effectiveness: It has been documented that bright tin used with Ni underlayer may still grow tin whiskers in a relatively short period of time [33].

It may be possible that bright tin is an overlooked method to mitigate tin whisker growth, as is suggested in Ref. 32. However, more research is needed to develop viable, repeatable engineering solutions for suppressing whisker growth from bright tin finishes. At present, an overwhelming amount of data show that whiskers from bright tin are much longer in length and grow faster than that from matte tin. This fact coupled with the possibility that tin whiskers may break off and be transported to another part of a printed circuit board may significantly increase the failure rate of the system when bright tin, instead of matte tin, is used. Therefore, it is recommended to avoid bright tin if possible unless the effectiveness of the mitigation strategy on bright tin is thoroughly evaluated for long-term reliability.

7.2.7. Closing Remarks Regarding Mitigation Practices

Plating process parameters and process repeatability are critical. Impurities in the plating bath can have a significant effect on stresses in the finish which may result in different tin whisker growth behavior. It is possible to get conflicting tin whisker test data on platings using seemingly identical mitigation practices but different plating processes/parameters. It is clear that both a viable mitigation practice *and* testing of the finish for a specific plating process are required to reduce the risk of tin whisker failures for the end user.

7.3. TIN WHISKER TEST DEVELOPMENT

Based on the early work of the Tin Whisker Accelerated Test Group (Phase 1 and 2 evaluations) [34–36], iNEMI proposed a Tin Whisker Test Method document in 2003. This test method exposed the surface finishes to three environmental conditions, including ambient, 60°C and 93% relative humidity (RH), and thermal cycling between −55°C and +85°C. Phase 3 evaluation was intended to validate and verify these test methods and to link short- and long-term testing. The results of this work in combination with other industry studies have led to changes in the environmental conditions in the Tin Whisker Test Method as well as changes to sample size recommendation for inspection. These changes have been incorporated in the JEDEC standards JESD22A121 and JESD201 [1, 12]. In order to obtain a true accelerated test, further optimization of the JEDEC tin whisker test methods is needed. Phase 4 and Phase 5 evaluations investigate additional aspects of the tin whisker tests. Details regarding the evaluations (Phases 1–5) from the iNEMI Tin Whisker Accelerated Test Group are discussed in this section.

7.3.1. Phase 1 and 2 Evaluations

The initial evaluations (Phase 1) were performed on laboratory-plated bright Sn samples. These tests provided some insight into which environments would cause a plating finish to whisker. The results of Phase 1, however, were inconclusive because very few occurrences of whiskers were observed. A Phase 2 evaluation was performed with production plated matte Sn from two integrated circuit (IC) suppliers. The results of Phase 2 revealed that several environments were very effective in promoting whisker growth.

7.3.1.1. Phase 1 Evaluation. In Phase 1, samples [brass coupons and 8-lead small outline integrated circuit packages (SOICs)] were prepared with bright Sn plating along with standard Sn–Pb plating as a control group. Based on past literature, bright Sn was more prone to whisker growth and was, therefore, chosen to improve the likelihood of whisker growth. The samples were then subjected to assorted combinations of environments identified through industry bench-marking. Two preconditioning methods (thermal cycling from −40°C to 90°C for 500 cycles and one week ambient storage) in combination with five storage environments (55°C/ambient humidity, 55°C/85% RH, 85°C/ambient humidity, 85°C/85% RH, and ambient office condition) were evaluated. Ambient humidity was uncontrolled, ranging from 20% to 60% RH, and ambient storage/office temperature was 20–25°C. Sample size was 40 units per each test condition. Storage time was 4 weeks. Optical microscopy (50–100× magnification) and scanning electron microscopy (300×) were used for the inspection. Higher magnifications (~2000×) were used for measuring whisker lengths. Five leads per unit were inspected.

The results of the Phase 1 study were inconclusive. Whiskers formed only on the bright Sn-plated brass coupons and were few in number—much less than expected. Some odd-shaped eruptions formed on the Sn-plated 8 lead SOICs but no confirmed whiskers. These results were surprising. It is speculated that because the samples were plated in a laboratory, the level of impurities and/or contamination were maintained very low and, thus, helped to retard whisker growth. Additionally, later investigations suggested that the test time was insufficient to exceed the incubation periods for whisker growth.

7.3.1.2. Phase 2 Evaluation. The Phase 2 test method evaluation was also performed with 8-lead SOIC packages, but these packages were plated at assembly houses using production plating baths. One IC supplier (Supplier A) provided samples plated with a methane sulfonic acid (MSA) bath and samples plated with a sulfate-based electrolyte. A second supplier (Supplier B) provided samples plated with a second MSA bath. Thick (10–12 μm) and thin (2–3 μm) matte Sn samples, as well as Sn–Pb samples, were included in the evaluation. The list and description of samples are presented below:

A = 2–3 μm, matte Sn (sulfate) on OLIN194 Cu SOIC molded/singulated
B = 10–12 μm, matte Sn (sulfate) on OLIN194 Cu SOIC molded/singulated
C = 2–3 μm, bright Sn on brass coupon
D = 10–12 μm, 90 Sn/10 Pb on OLIN194 Cu SOIC molded/singulated (control)
E = 2–3 μm, matte Sn (MSA) on OLIN194 Cu SOIC molded/singulated
F = 10–12 μm, matte Sn (MSA) on OLIN194 Cu SOIC molded/singulated

The samples described above were subjected to the different environmental stress conditions presented in Table 7.1 for 4 weeks. A parallel study was performed with chip fuses by a supplier of passive components.

The Phase 2 results were more conclusive. The environmental stress conditions evaluated in this study were sufficient to create whiskers on 8-lead SOICs and on chip components. In general, more whiskers grew with the −55°C to 85°C or −40°C to 90°C thermal cycle methods. Ambient, 60°C/95% RH, and 30°C/90% RH storage methods also created a few whiskers but were not as effective as the temperature cycle methods. It was observed that the subsequent exposure to

TABLE 7.1. Test Conditions and Sample Description for Whisker Tests with 8 Lead SOICs for Phase 2 Evaluation

Legs	Temperature Cycle (°C)	Temperature (°C) and Relative Humidity (%)	Supplier Plating Site	Remarks
6	—	60, 95	A	Temperature and humidity
7	—	60, 95	B	Temperature and humidity
8	—	30, 90	A	Humidity
9	—	30, 90	B	Humidity
10	−55 to 85	30, 90	A	Temperature cycle + humidity
11	−55 to 85	30, 90	B	Temperature cycle + humidity
12	−55 to 85	Ambient	A	Temperature cycle
13	−55 to 85	Ambient	B	Temperature cycle
14	Ambient	Ambient	A	Ambient
15	Ambient	Ambient	B	Ambient

temperature and humidity did not add significantly to the whisker length or frequency when thermal cycling was performed first. In addition, the bath chemistry/plating process appeared to have the most significant influence on whisker growth as compared to the environments.

7.3.2. Phase 3 Evaluation

7.3.2.1. Phase 3: Experimental Objective. Phase 3 evaluation was performed to validate and also to verify some proposed iNEMI tin whisker tests to determine whether tests could differentiate between surface finishes and mitigation methods. The intent was also to investigate any connection between the performance in 1-month tests (used in Phase 1 and Phase 2 evaluations) and longer-term tests (1 year), in an attempt to correlate short- and long-term performance. Another objective was to observe incubation times and saturation of whisker growth to determine optimal inspection intervals and test durations.

7.3.2.2. Phase 3: Experiment. Three environmental test conditions were evaluated at three durations, as described in Table 7.2. Surface finishes were inspected at various intervals for the durations given in Table 7.2. For the thermal cycling and high-temperature/humidity conditions, inspection intervals were consistent at 500 cycles and 1000 hours, respectively. However, for the ambient conditions, surface finishes were inspected every 1000 hours until 5000 hours. Then, these finishes were inspected at 8000 hours and 10,000 hours. Inspection intervals were increased in this case because of the apparently long incubation times and slow growth in the ambient conditions, as discussed in Section 7.3.2.3.

In the Phase 3 study, whisker inspection was performed exclusively with scanning electron microscopes (SEMs). For the thermal cycling and the high-temperature/humidity conditions, separate parts were removed from the chambers at each

TABLE 7.2. Basic Description of Whisker Tests for Phase 3 Evaluation

Environmental Condition	Inspection Interval	Total Duration	Finishes Evaluated
Uncontrolled 20°C to 25°C 20 to 60% RH	1000–3000 h	10,000 h	A–J
	Irregular (@2 kh and 10 kh) 3 kh for N 7 kh for L,M	10,000 h	K–O
60°C (+5) 93% RH (+2, −3)	1000 h	9000 h	A–J
	2000 h	9000 h	K–O
−55°C (+0, −10) to +85°C (+10, −0) air-to-air temperature cycle (20 min per cycle)	500 cycles	3000 cycles	A–J

inspection interval. For the ambient condition, the same parts were inspected by SEM at each inspection interval.

Notes for Tables 7.3 and 7.4. Most of the part types and lead-frame alloys listed in Tables 7.3 and 7.4 and used in the experiment are common and commercially available in the electronics industry. Abbreviations for component package types are as follows: "QFP" stands for quad flat pack, and "SOIC" stands for small outline integrated circuit. Copper (Cu)-based, lead-frame alloys are abbreviated as "C194" (which contains ~2% iron) and "C7025" (which contains ~3% nickel). Iron–nickel (Fe–Ni) alloy lead frames are abbreviated as "Alloy42," which contains 42% Ni and a balance of Fe. "JEITA" represents the Japan Electronics and Information Technology Industries Association.

The number of leads per part varied for the different finishes as shown in Table 7.3. When the experiment started, iNEMI was advocating inspection of three random leads per part; however, it was clear at the start of the experiment that this would not be sufficient, so all leads were inspected. All leads on three parts were inspected in the "live bug" position. ("Live bug" means that the part is standing on the feet of the leads, as it would be when mounted on a board.) For each inspection interval and surface finish, the longest whisker lengths from three parts were recorded. The average maximum whisker length was also calculated from these three measurements for each inspection interval and surface finish as well.

To validate and verify the whisker test methods, a range of surface finishes and mitigation methods were included in the experiment, for the parts tested as described in Table 7.4. The plating characteristics that were varied were the surface finish alloy, lead-frame material, deposit thickness, heat treatment, and underplate, as

TABLE 7.3. Description of the Parts Used in the Phase 3 Experiment

Label	Part Type	Number of Leads	Plating Method
A	QFP	32	Production
B	QFP	64	Production
C	QFP	64	Production
D	QFP	64	Production
E	QFP	64	Production
F	QFP	44	Production
G	SOIC	16	Production
H	QFP	64	Production
I	SOIC	8	Experimental
J	QFP	64	Production
K	JEITA sample	N/A (coupon)	Experimental
L	JEITA sample	N/A (coupon)	Experimental
M	JEITA sample	N/A (coupon)	Experimental
N	SMT connector		Production
O	SOIC	8	Production

TABLE 7.4. Description of the Surface Finishes Used in the Phase 3 Evaluation

Label	Alloy	Type	Nominal Thickness (μm)	Lead-Frame Alloy	Mitigation Practice
A	Sn	Matte	10	C7025	None
B	Sn	Matte	10	C194	None
C	Sn	Matte	10	C194	1 h @150°C
D	Sn	Matte	10	C194	245°C peak reflow
E	Sn	Matte	3–5	C194	None
F	Sn–Bi	Matte	10	C194	None
G	Sn–Cu	Matte	10	C194	None
H	Sn–Ag	Matte	10	C194	None
I	Sn	Hot dip	5–15	C194	None
J	Sn–Pb	Matte	10	C194	None
K	Sn	Bright	10	C194	None
L	Sn	Bright	10	C1020	None
M	Sn	Bright	10	Alloy42	None
N	Sn	Matte	2.5	Be–Cu	Ni layer (1.27 μm)
O	Sn/Phase 2	Matte	10	C194	

well as matte and bright tin-based electrodeposit and solder dip. These finishes were acquired from a variety of sources. Specifically, a single production source was used for A–E, separate production lines were used for F–H and M, a bench-top process was used for hot dipping I, and the JEITA consortium created the experimental samples for J–L. All surface finishes were Pb-free, except for J, which was an Sn–Pb alloy, described in Table 7.5. For the tin-alloy samples, the concentration of the alloying element was measured using energy dispersive and/or X-ray fluorescence spectroscopy. The results are listed in Table 7.5. As shown, there is some discrepancy/disagreement in the target and measured values. This discrepancy/disagreement may be explained by (1) plating that is actually off-target composition, (2) the fact that energy dispersive spectroscopy is being used at or near its threshold

TABLE 7.5. Composition of the Sn–Bi, Sn–Cu, Sn–Ag, and Sn–Pb Alloy Surface Finishes in % Mass Fraction for Phase 3 Evaluation

Label	Alloy Element	Target Composition	EDS Measurement	XRF
F	Bi	2 ± 1%	1.2%	1%
G	Cu	2–3%	0.5%	
H	Ag	2–4%	0.3%	0–2.5%
J	Pb	5–15%	9.5%	

detection limit of approximately 1%, (3) uneven alloy distribution for the X-ray fluorescence result of the Sn–Ag electrodeposit, or (4) inability to measure Cu in a thin Sn–Cu electrodeposit on a Cu-based substrate.

All surface finishes listed in Table 7.4 were exposed to the isothermal conditions described in Table 7.2 of an uncontrolled office ambient condition and 60°C/93% RH. Surface finishes A–J were also exposed to the thermal cycling condition.

7.3.2.3. Phase 3: Results and Discussion. As the results of the Phase 3 evaluation are presented and discussed, it is important to bear in mind that only one example of each surface finish/substrate was used for this evaluation. Therefore,

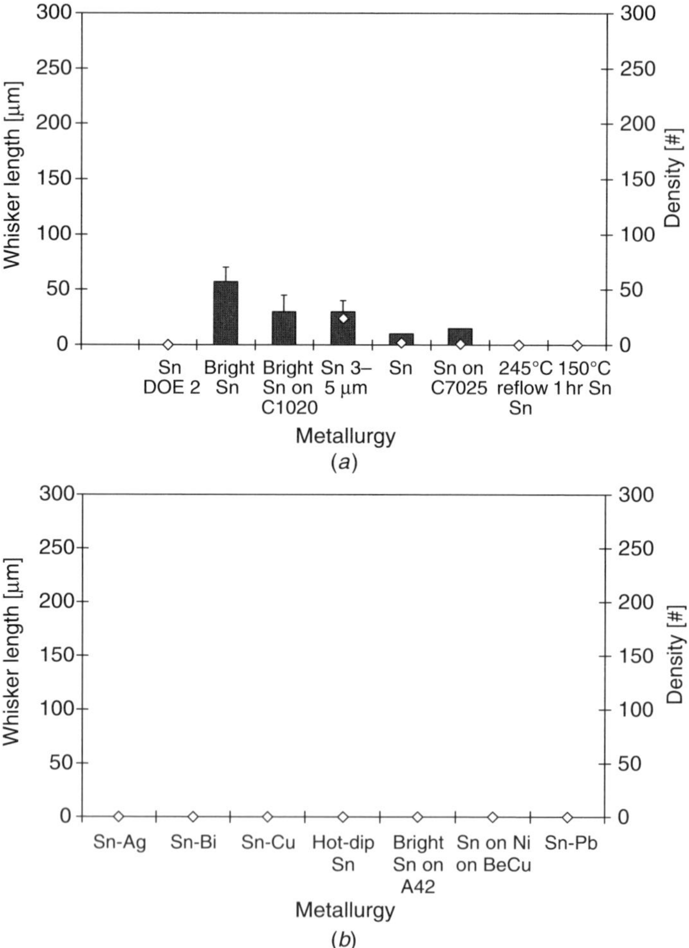

Figure 7.3. Whisker occurrence and length observations for various surface finishes exposed to the ambient environmental condition after 10,000 hours. *Note*: Whisker density was not measured for every plating type.

this study alone cannot be used to prove or disprove the effectiveness of a particular surface finish, substrate, underplating or heat treatment. For the objectives of the Phase 3 evaluation, one example of each type of metallurgy was sufficient. Industry data suggest that multiple examples of a specific metallurgy are required to understand any general trends.

7.3.2.3.1. Ambient Condition. In an air-conditioned office environment, whiskers grew on 5 out of 15 (~33%) platings by 10,000 hours with lengths up to 50 μm as shown in Figure 7.3. The longest whiskers were found on bright tin and thin tin finishes (3–5 μm thick). However, the majority of finishes (67%), including Sn–Pb, did not grow whiskers.

Notes for Figures 7.3, 7.4, and 7.7. The average maximum whisker lengths are represented by columns, and the absolute maximum whisker lengths are represented by a line above the column. Both of these measurements reference the left vertical axis. The number of whiskers observed is represented by diamonds and reference the right vertical axis. Moreover, the density measurement is defined as the total number of whiskers counted in nine fields of view, approximately 0.56 mm^2 of surface area.

For the surface finishes that exhibited whiskers after exposure to the ambient condition, whisker growth began between 5000 and 10,000 hours. It is unclear

TABLE 7.6. Incubation Times for Various Surface Finishes for the Ambient and High Temperature/Humidity Test Condition During Phase 3 Evaluation[a]

Test Duration (h)	First Appearance of Whiskers by Surface Finish Type (Ambient Test)	First Appearance of Whiskers by Surface Finish Type (60/93 Test)
0		
1,000		
2,000		3–5 μm Sn, Sn, Sn on C7025, Sn solder dip
3,000		Annealed Sn, Sn–Cu, Sn–Ag, Sn–Bi
4,000		Reflowed Sn at 245°C
5,000	3–5 μm Sn	
6,000		Sn–Pb
8,000		
10,000	Bright Sn on C194 & C1020, matte Sn on C7025	
>10,000	Sn solder dip, annealed Sn, Sn–Cu, Sn–Ag, Sn–Bi, reflowed Sn at 245°C, Sn–Pb	None

[a]Note that only 10 finishes (A–J) are included in the incubation study because inspection was made at regular intervals for these finishes, but not for finishes K–O. The ambient test was ended at 10,000 hours. The last row of the table includes all electrodeposits that did not exhibit whiskers during the test.

whether the surface finishes that did not exhibit whiskers in 10,000 hours will never form whiskers or will form whiskers after longer incubation periods. The incubation periods for surface finishes A-J are shown in Table 7.6. For surface finishes K-O, inspection was made at less frequent intervals, so incubation times are not listed.

During the 10,000 hours that were investigated, whiskers did not grow on the majority of Pb-free finishes and did not grow on the Sn–Pb finish. At the uncontrolled, air-conditioned, office-ambient condition, whisker occurrence and growth is too slow to be a viable test for surface finish acceptability. Other investigators have observed faster whisker growth at 23°C/93% RH, 30°C/60% RH [14], and 30°C/90% RH [37]. In addition, an uncontrolled warehouse-ambient condition may yield faster whisker growth than the uncontrolled office ambient [38].

7.3.2.3.2. High Temperature/Humidity Condition. All Pb-free, Sn-based surface finishes produced whiskers in the high-temperature/humidity condition, as shown in Figure 7.4. During exposure to the high-temperature/humidity condition, most surface finishes grew whiskers by 3000 hours and by 6000 hours all surface finishes, including Sn–Pb, grew whiskers, as shown in Table 7.6 and Figure 7.6. Characterization of the Sn–Pb parts revealed that the observed whiskers only grew in regions where the Pb was not present. This illustrates the fact that variations in Sn–Pb production platings can also result in some amount of whisker growth.

At high temperature and humidity, whiskers grew significantly longer and growth started at shorter incubation times than for the ambient condition.

In addition to long whiskers, localized corrosion was observed on the surface finishes after environmental exposure at high temperature/humidity, as shown in Figure 7.5. Whiskers frequently appeared at or near sites of localized corrosion. Specifically, whisker growth and corrosion are localized and not evenly distributed on a lead or from lead-to-lead. The proximity of the localized corrosion to whiskers and the timing of the appearance of whiskers and corrosion suggest that corrosion may impact whisker growth, but this study alone does not establish a mechanistic relationship.

Based on dew point analysis, it is clear that 93% humidity is difficult or impossible to maintain without creating condensation in a commercially available temperature/humidity chamber, especially when the chamber door is opened. Other research groups have shown that preventing condensation can decrease or eliminate the observation of localized corrosion and associated whiskers. In fact, Osenbach et al. [39] showed that eliminating condensation prevented the formation of whisker clusters near areas of localized corrosion during a 2400-hour exposure. Su [40] observed that decreasing the humidity from 93% RH to 85% RH in a 2000-hour exposure at 60°C reduced the number of leads with whiskers by a factor of 10, decreased the whisker lengths, and prevented localized corrosion. Other researchers observed similar results for 85% RH/55°C test conditions [41].

Condensation was likely present in this experiment because a 1.4°C change in temperature can result in condensation and special precautions were not taken to prevent condensation. As a result of iNEMI testing and other industry work at the 60°C/93% RH condition, the JEDEC test method JESD22A121 adopted conditions

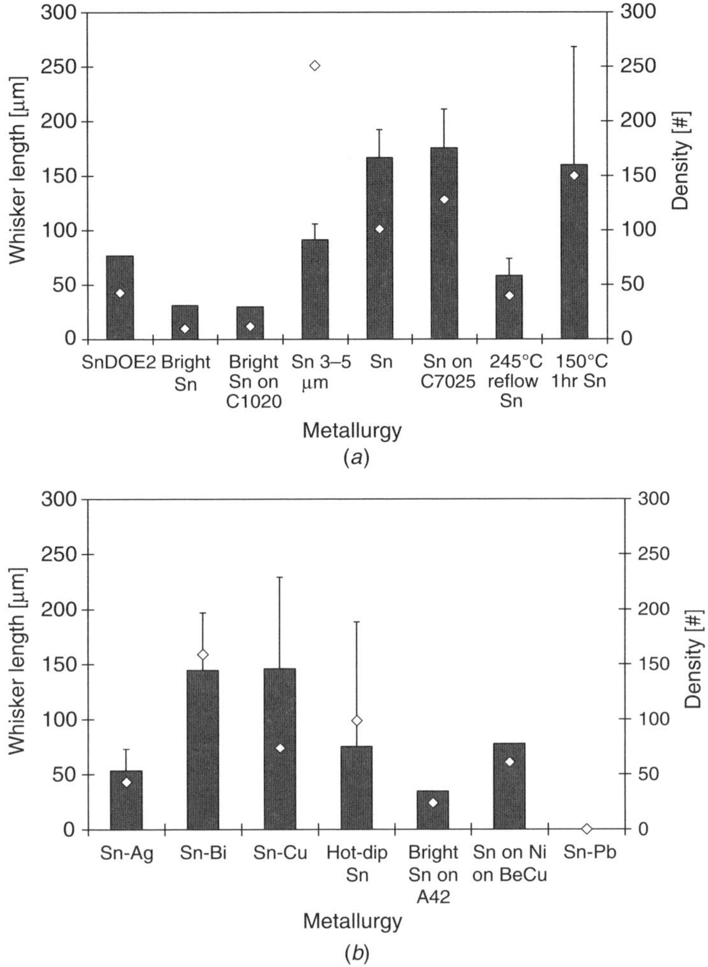

Figure 7.4. Whisker occurrence and length observations for various surface finishes exposed to the high temperature/humidity environmental condition after 8000 hours. *Note*: Whiskers first formed on the Sn–Pb plating after 6000 hours, as shown in Figure 7.6. At the 8000-hour readpoint shown in this figure, the Sn–Pb plating did not exhibit whiskers. Different parts were examined at each inspection interval.

of 60°C and 85% RH and added detailed instructions for reducing the chance of condensation. A relative humidity of 85% provides a high-humidity environment with less risk for condensation during testing. Tin whisker test data at 60°C and 85% RH is now being collected throughout the electronics industry.

In service localized corrosion has not been prevalent for Sn–Pb surface finishes. However, it should be noted that in the current experiment the Pb-free, Sn-based surface finishes exhibited more severe corrosion in testing than did Sn–Pb.

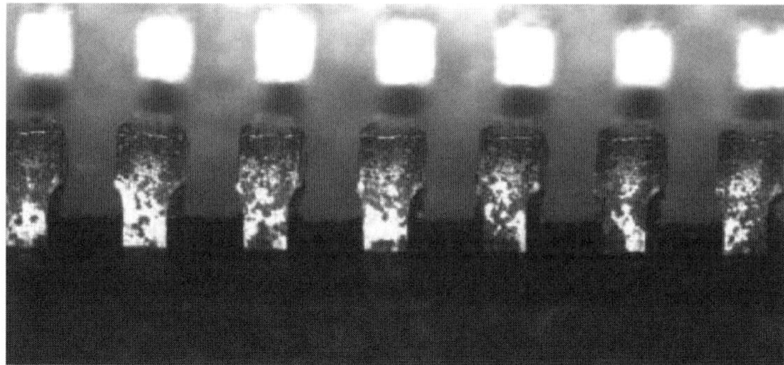

Figure 7.5. A representative optical image of localized corrosion on a tin surface finish.

Corrosion may cause a new whisker mechanism or may alter factors involved in whisker growth, such as stress [41] or diffusion. However, corrosion is not the only explanation for whisker growth, since we observe whiskers in ambient testing and thermal cycle testing, which does not create localized corrosion.

In Figure 7.6, maximum whisker lengths are plotted as a function of the surface finish and the inspection interval. Whisker growth may be stabilizing by 9000 hours with lengths up to 360 μm. However, the corrosion present on the surface was so extensive after 9000 hours that testing was discontinued. This dataset cannot be used to discuss saturation of whisker growth in the absence of localized corrosion.

7.3.2.3.3. Thermal Cycle Condition. During thermal cycling, all Pb-free surface finishes formed whiskers on copper-base lead frames with a maximum whisker length of 60 μm, as shown in Figure 7.7. In fact, whisker growth was initiated on all Pb-free surface finishes, except Sn–Ag, by the first inspection at 500 cycles, as shown in Table 7.7. Whisker growth was initiated on the Sn–Ag sample by the second inspection at 1000 cycles. In contrast, whiskers did not grow on Sn–Pb surface finish up to 3000 cycles. Note that thermal cycle testing was not conducted on samples K–O.

Thermal cycling did not seem to differentiate between the various copper-based substrate cells tested. Due to the greater coefficient of thermal expansion (CTE) mismatch, stress-driven whisker growth theory predicts whisker growth will be greater on Alloy 42 than on a copper lead-frame substrate. Although not demonstrated in the current study, industry data support this prediction [6, 7].

Thermal cycling causes significant damage to the plating, including possibly grain growth and spalling. This damage likely affects the whisker growth behavior and may create saturation of whisker growth during testing. This extensive damage is not expected for the smaller temperature range cycles in many service conditions. However, some electronic products, such as automotive, are exposed to harsher environments and thermal cycles.

Whisker lengths appeared to stabilize by 1500 cycles for the Cu-based substrates tested with average lengths up to 35 μm, as shown in Figure 7.8, and maximum

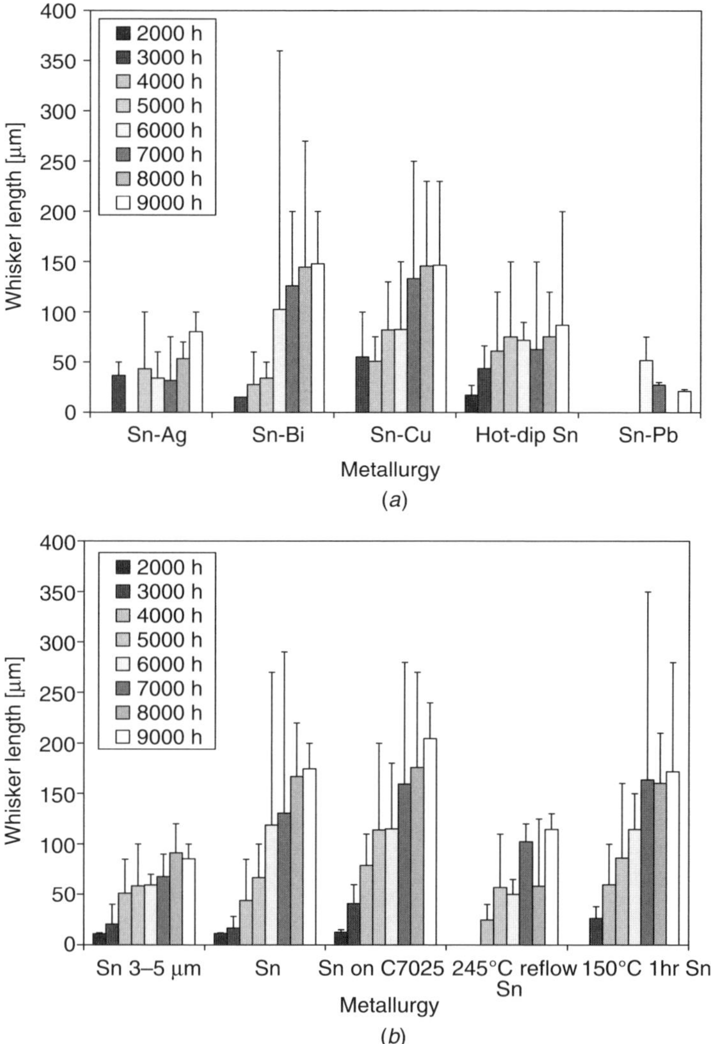

Figure 7.6. Whisker growth over time for various surface finishes exposed to the 60°C/93% RH high-temperature/humidity environmental condition.

lengths up to 60 μm. The observed surface finish damage may contribute to the whisker length saturation. Durations of 1500 cycles for thermal cycle testing are adequate since whisker growth, at least for Sn-based finishes on Cu-based lead frames, stabilizes around 1500 cycles at the tested thermal cycle range.

Overall, the density of whiskers observed in the ambient test condition was significantly less than the densities observed in both the high-temperature/humidity and thermal cycle test conditions. This difference may be due to differences in incubation times at the given test conditions or due to other environmentally induced

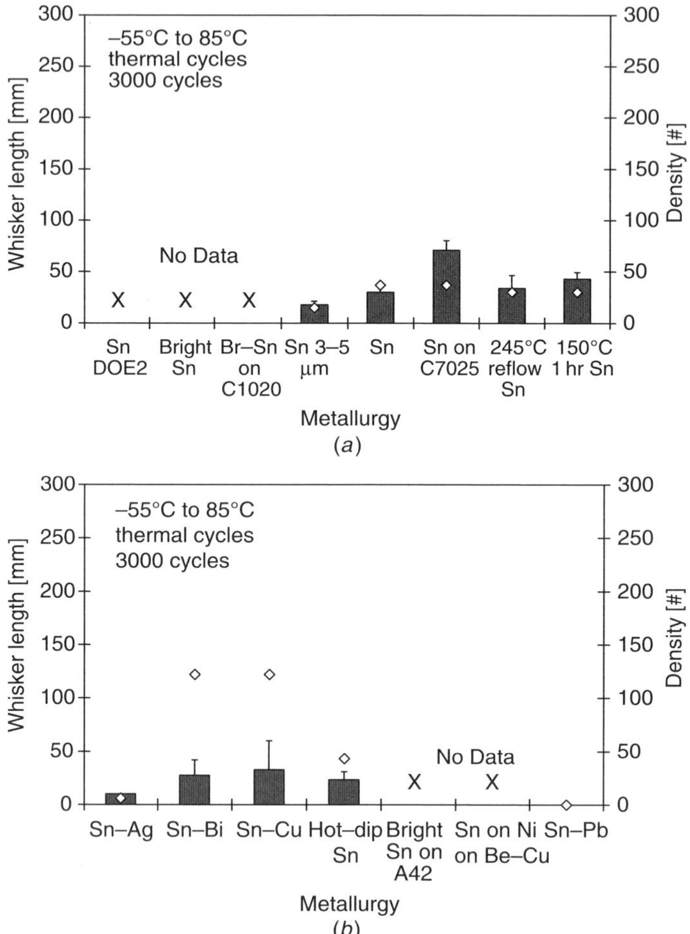

Figure 7.7. Whisker occurrence and length observations for various surface finishes exposed to the thermal cycling test condition from −55°C to +85°C for 3000 cycles.

differences in the finishes. Further study would be required to draw any conclusions with respect to whisker density as a function of test condition.

7.3.2.3.4. Sample Size. Whisker length, density, diameter, growth rate, and incubation time occur across a statistical distribution making the detection of a "true" maximum whisker length difficult. Examples of statistical sampling errors were observed in results for the Sn–Ag and Sn–Bi surface finishes exposed to high temperature/humidity. Specifically, the Sn–Ag alloy finish had whiskers on 1 lead out of almost 200 after 6000 hours, and the maximum whisker length was 100 μm. In the case of the Sn–Bi surface finish, after 5000 hours, the finish had the shortest actual

TABLE 7.7. Incubation Times for Various Surface Finishes for the Thermal Cycling Test Condition

Test Duration (Cycles)	First Appearance of Whiskers by Surface Finish Type (Thermal Cycling)
0	
500	3–5 μm Sn, Sn, Sn on C7025, Sn solder dip, annealed Sn, Sn–Cu, Sn–Bi, reflowed Sn at 245°C
1000	Sn–Ag
1500	
2000	
2500	
3000	
>3000	Sn–Pb

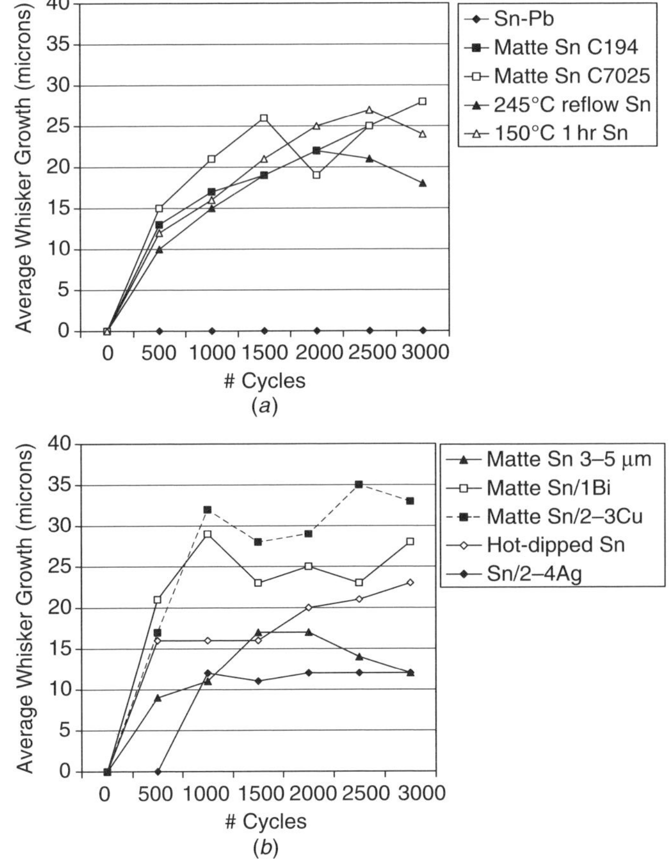

Figure 7.8. Whisker growth over time for various surface finishes exposed to thermal cycling conditions of −55°C to +85°C.

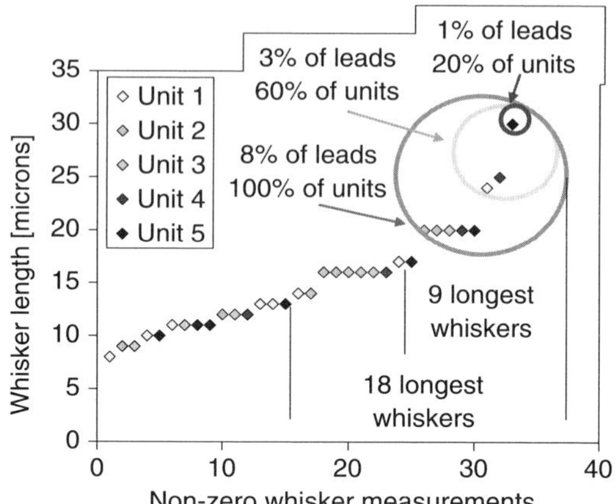

Figure 7.9. Presentation of a complete set of whisker length measurements after 3000 thermal cycles.

maximum whisker of 50 μm compared to the other finishes (A–J), but after 6000 hours it had the longest maximum whisker of 360 μm.

In the case of the ambient condition in this study, whisker growth can often occur sparsely. Even when whiskers are present in higher numbers in the thermal cycling environmental condition, a higher number of leads must be inspected to capture the statistical variation of whisker length, as shown in Figure 7.9. For the data presented in Figure 7.9, all 16 leads on 5 units were inspected (80 leads) after 500 cycles. One hundred percent of units had whiskers. Nonzero whisker measurements were made on 33 of 80 leads. Absolute maximum whisker length measured on 5 units (30 μm) is 50% greater than maximum measured on 1 or 2 units (20 μm).

A sample size of about 200 leads was not large enough to prevent scatter in the absolute maximum and average maximum whisker lengths between inspection intervals. This scatter is apparent in Figure 7.4, for which a different set of samples was inspected at each inspection interval. Moreover, a sample size of 3 random leads that was initially proposed by iNEMI was substantially insufficient to capture statistical variation and sparseness of whisker formation, particularly in isothermal conditions.

It is necessary to look at a large number of leads to make sure a whisker is present (in some cases only 1 out of 50 to 1 out of 200 have been reported). The JEDEC recommendation, to pre-inspect 6 units and 96 leads to identify 18 leads for detailed inspection, is reasonable for high densities of whiskers.

7.3.3. Conclusions: Phases 1, 2, and 3

These experiments resulted in progress toward the objective of developing industry standard test methods for assessing tin whiskers risk. The experimental results from Phase 3 helped to change the JEDEC standard JESD22A121 tin whisker test method by (1) decreasing the humidity in the high-temperature/humidity test, (2) requiring better chamber control and procedures to prevent condensation in the high-temperature/humidity condition, (3) controlling the ambient condition with 60% humidity, and (4) increasing the sample size to a 96-lead inspection.

The Phase 3 work also demonstrated that short-term testing could not be used to predict the relative ranking of finishes based on whisker length. However, incubation time may be a predictor of performance in different environments. Finally, JEDEC inspection intervals of 1000 hours for ambient and high temperature/humidity and 500 cycles for thermal cycle testing and total durations of 4000 hours and 1500 cycles, respectively, are reasonable.

7.3.4. Current and Future Work: Phase 4 and 5 Evaluations

7.3.4.1. Phase 4 Evaluation. The Phase 4 evaluation was originally intended to investigate the effects of electrical bias on the susceptibility of [unwetted] tin finishes to form and grow whiskers. Special printed circuit test boards were designed and assembled to provide bias between adjacent leads on a component. A portion of an actual test board and a schematic illustration are shown in Figure 7.10. For the iNEMI Phase 4 evaluation, bright Sn, matte Sn, and Sn–Pb finishes were tested in two isothermal storage environments: 30°C/60% RH and 60°C/85% RH. Additional variables included package types, assembly solder paste, current, and applied voltage bias. Samples were tested for up to 5000 hours. Inspections were conducted at 1500, 3000, and 5000 hours.

Tin whisker testing conducted on assembled modules can add a number of complicating factors, as illustrated by the inspection of the Phase 4 samples. Complete or nearly complete wetting of the package leads resulted in an insufficient sample size

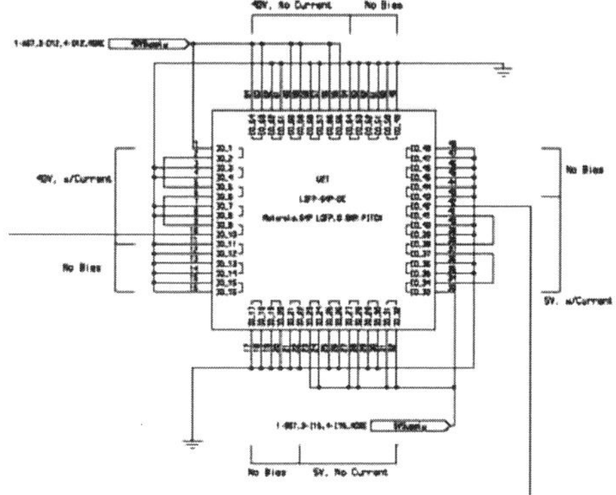

Figure 7.10. Phase 4 evaluation test board and schematic, designed to provide varying electrical bias between adjacent leads after assembly. During inspection, the schematic needed be followed to determine which leads were biased relative to each other.

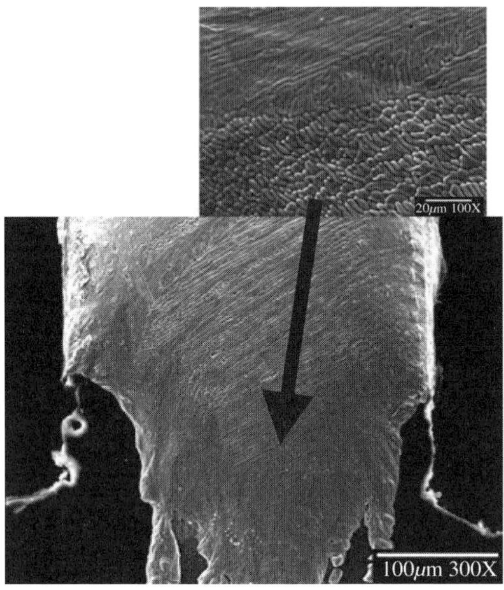

Figure 7.11. Showing complete wetting of a 64LQFP matte Sn lead soldered with Sn–Ag–Cu solder paste. Whisker inspection cannot be performed on an "unwetted" or exposed tin-plated lead.

for inspection of the exposed or unwetted tin finish and, therefore, resulted in inconclusive results regarding the effect of applied bias on the formation and growth of whiskers in the unwetted region of the lead frame. Figure 7.11 shows the 64LFQP with matte Sn finish and Sn–Ag–Cu solder paste. Here the lead finish is completely alloyed with the solder paste.

Although conclusions with regard to whisker growth [on the exposed, unwetted tin finish] as a function of applied bias could not be drawn from the Phase 4 evaluation, a number of interesting observations were made. No whiskers were seen in any of the samples stored at 30°C/60% RH. In the high-temperature/humidity test condition, whiskers were observed in the reflowed/wetted region at the foot of several components assembled with Sn–Ag–Cu paste as illustrated in Figure 7.12. Furthermore, flux residue did not appear to inhibit whisker growth. Additional characterization of the Phase 4 evaluation samples is ongoing, as well as comparison of the observations in Phase 4 with other experimental results found in the literature.

7.3.4.2. Phase 5 Evaluations. Additional testing has been initiated to understand the effect of temperature and humidity over a range of conditions for Pb-free, matte Sn surface finishes on Cu-based substrates. These tests include a matrix of temperature and humidity conditions from 30°C to 100°C and from 30% to 90% RH. Production-plated matte tin samples from three sources with 3-μm-thick and 10-μm-thick finishes are being used in this evaluation. For the

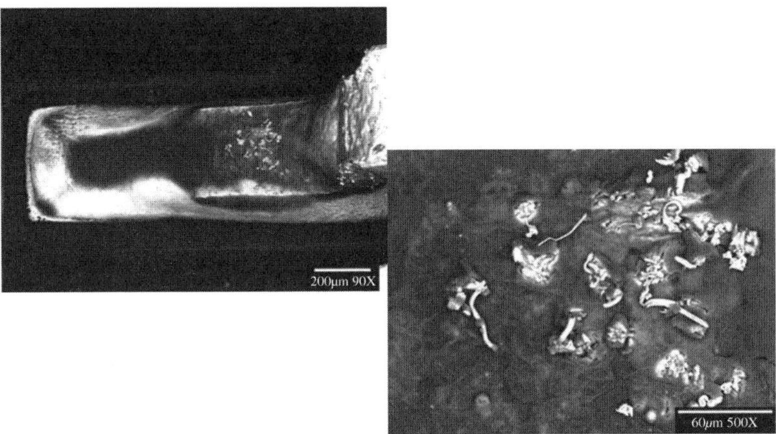

Figure 7.12. Whiskers growing in a region that has been "wetted" by Sn–Ag–Cu solder.

10-μm-thick matte tin, the surface finish is being evaluated in both the as-plated and reflowed (260°C) conditions.

The hypothesis for this work is that whisker presence and/or length, measured in isothermal environments, is a function of temperature and humidity. Whisker length could be discontinuous at a threshold point or could vary as a function of temperature and humidity over the entire tested range. If such a function exists and becomes known, this function could be used to determine optimal whisker test condition(s) and whether whisker behavior measured at accelerated testing conditions can be related to actual storage and/or customer service conditions. At the time of this writing, no significant whisker growth has been observed in the Phase 5 evaluation. The experiment is ongoing.

7.3.5. Closing Remarks on Tin Whisker Test Development

Multiple factors influence whisker growth. Some of these factors are known and under investigation, while other factors are yet unknown. Test methods (as well the theory/model describing it) must take into consideration numerous parameters. Development of an accelerated test may not be possible until the underlying theory of whisker formation and growth is fully understood. In the interim, the recommended whisker test methods will aid the industry in (a) comparing and characterizing whisker propensity of existing Pb-free, Sn-based finishes in preparation for production, (b) evaluating different whisker growth theories, and (c) developing reduced whisker processes and materials.

7.4. SUMMARY

The removal of Pb from Sn–Pb termination finishes in the last few years has significantly increased the risk for failures in the electronic industry due to tin whiskers. The objective of this chapter is to give the reader an introduction to the issues related to tin whiskers including a brief background, recommended mitigation strategies, and an understanding of recent tin whisker test development. With this background, users and suppliers of electronic components, alike, can make informed engineering decisions to reduce the probability of tin whisker failures. A complete resolution of the problems related to tin whiskers will not be attained until the fundamental mechanisms for whisker formation and growth are understood. In the meantime, a strategy that includes use of the recommended mitigation practices, careful control and monitoring of plating processes and ongoing tin whisker testing is the best practice to reduce tin-whisker-related failures in the field.

ACKNOWLEDGMENTS

The authors would like to acknowledge the contributions of former iNEMI Project Chairs V. Schroeder, N. Vo, and S. Prasad and also thank the members of the iNEMI User Group and Test Group for their participation and support.

REFERENCES

1. JEDEC Standard JESD22A121, Measuring whisker growth on tin and tin alloy surface finishes (available at www.jedec.org), May 2005.
2. B. D. Dunn, Whisker formation on electronic materials, *Circuit World* **2**, 32–40, 1976.
3. K. G. Compton, A. Mendizza, and S. M. Arnold, Filamentary growths on metal surfaces—whiskers, *Corrosion* **7**(10), 327–334, 1951.
4. G. T. Galyon, Annotated tin whisker bibliography and anthology, *IEEE Trans. Electron. Packaging Manuf.* **28**(1), 94–122, 2005.
5. R. M. Fisher, L. S. Darken, and K. G. Carroll, Accelerated growth of tin whiskers, *Acta Metall.* 2, 368–372, May 1954.
6. Y. Zhang, C. Fan, C. Xu, O. Khaselev, and J. A. Abys, in *Proceedings of the 2004 APEX Conference*, February 2004.
7. M. Dittes, P. Oberndorff, P. Crema, and V. Schroeder, Tin whisker formation in thermal cycling conditions, in *Proceedings of 5th Electronics Packaging Technology Conference (EPTC)*, Singapore, December 2003, pp. 183–188.
8. K. N. Tu, Interdiffusion and reaction in bimetallic Cu–Sn thin films, *Acta Metallurg.* **21**, 347–354, 1973.
9. B. Z. Lee and D. N. Lee, Spontaneous growth mechanism of tin whiskers, *Acta Metall.* **46**(10), 3701–3714, 1998.
10. M. Dittes, P. Oberndorff and L. Petit, Tin whisker formation—Results, test methods, countermeasures, *Proceedings of the 53rd Electronic Components & Tech Conference*, New Orleans, 2003, pp. 822–826.
11. iNEMI User Group, Tin whisker acceptance test requirements (available at www.inemi.org), July 2004.

12. JEDEC standard JESD 201, Environmental acceptance requirements for tin whisker susceptibility of tin and tin alloy surface finishes (available at www.jedec.org), March 2006.
13. A. Teverovsky, Gold whiskers: Introducing a new member to the family, Internal NASA Goddard Space Flight Center Memorandum, April 2003 (http://nepp.nasa.gov/WHISKER/other_whisker/gold/index.htm).
14. P. Oberndorff et al., JISSO/PROTEC Forum JAPAN 2004 Session B, October 2004.
15. P. Su, M. Ding, and S. Chopin, Effects of reflow on the microstructure and whisker growth propensity of Sn finish, in *Proceedings of the 55th Electronic Components & Technology Conference*, Orlando, May/June 2005, pp. 434–440.
16. Private communication between the authors and industry component suppliers, 2006.
17. J. Haimovich, Hot-air-leveled tin: Solderability and some related properties, in *Proceedings of the 39th Electronic Components Conference*, Houston, 1989, pp. 107–112.
18. P. Oberndorff, M. Dittes, and L. Petit, Intermetallic formation in relation to tin whiskers, in *Proceedings, International Conference on Lead-Free Electronics*, Brussels, 2003.
19. G. Binder, L. Boyer, and J. C. Puyon, A new way of obtaining consistent crimped connections in tinned copper wire, The Wire Association, Stratford-on-Avon, England, 1988.
20. G. Bürstner and E. Fröhlich, The hot-dipped tinning process for lead-free electrical connections, in CARTS USA '97, p. 179.
21. C. H. Pitt and R. G. Henning, Pressure-induced growth of metal whiskers, *J. Appl. Phys. (Commun.)* **35**, 459–460, 1964.
22. J. W. Osenbach, R. L. Shook, B. T. Vaccaro, B. D. Potteiger, A. N. Amin, K. N. Hooghan, P. Suratkar, and P. Ruengsinsub, Sn whiskers: Material, design, processing, and post-plate reflow effects and development of an overall phenomenological theory, *IEEE Trans. Electron. Packaging Manuf.* **28**(1), 36–62, 2005.
23. P. Oberndorff, M. Dittes, and P. Crema, Whisker testing: Reality and fiction, *Proc. IPC/Soldertec Global 2nd International Conf. on Pb-Free Electronics,* Amsterdam, The Netherlands, June 2004.
24. R. Schetty, N. Brown, A. Egli, J. Heber, and A. Vinckler, Lead-free finishes—whisker studies and practical methods for minimizing the risk of whisker growth, in *Proceedings of AESF SUR/FIN Conference*, June 2001, pp. 1–5.
25. J. Kadesch and J. Brusse, "The Continuing Dangers of Tin Whiskers and Attempts to Control Them with Conformal Coating", in *NASA's EEE Links Newsletter* **1**(2), 2001, http://nepp.nasa.gov/WHISKER/reference/tech_papers/kadesch2001-article-dangers-of-tin-whiskers-and-conformal-coat-study.pdf
26. J. Kadesch and H. Leidecker, Effects of uralane conformal coatings on tin whisker growth, in *Proceedings of IMAPS Nordic, The 37th IMAPS Nordic Annual Conference*, September 10–13, 2000, pp. 109–116.
27. C. H. Tung, Process–structure–property relationship and its impact on microelectronics device reliability and failure mechanism, *J. Semicond. Technol. Sci.* **3**, pp. 107–113, 2003.
28. M. E. Williams, C. E. Johnson, K.-W. Moon, G. R. Stafford, C. A. Handwerker, and W. J. Boettinger, Whisker formation on electroplated Sn–Cu, in *Proceedings of the American Electroplaters and Surface Finishers Society (AESF) SUR/FIN 2002 Annual International Technical Conference*, The American Electroplaters and Finishers Society, Inc., Orlando FL, 2002, pp. 31–39.

29. N. Vo, Y. Nadaira, T. Matsura, M. Tsuriya, R. Kangas, J. Conrad, B. Sundram, K. Lee, and S. Arunasalam, Pb-free plating for peripheral/leadframe packages, in *Proceedings of the IEEE Electronic Components Conference*, 2001, pp. 213–218.
30. W. J. Choi, T. Y. Lee, K. N. Tu, N. Tamura, R. S. Celestre, A. A. MacDowell, Y. Y. Bong, L. Nguyen, and G. T. T. Sheng, Structure and kinetics of Sn whisker growth on Pb-free solder finish, in *Proceedings of the 52nd Electronic Components & Technology Conference*, San Diego, CA, 2002, pp. 628–633.
31. K. W. Moon, M. E. Williams, C. E. Johnson, G. R. Stafford, C. A. Handwerker, and W. J. Boettinger, The formation of whiskers on electroplated tin containing copper, in *Proceedings of the 4th Pacific Rim International Conference on Advanced Materials and Processing, Japanese Institute of Metallurgy*, 2001, pp. 1115–1118.
32. R. D. Hilty, Bright tin for whisker mitigated electronics applications, in *IPC/JEDEC 11th Conference on Lead Free*, Boston, 2005.
33. Limitation of hot solder dipping for mitigation of tin whisker formation, http://nepp.nasa.gov/WHISKER/photos/pom/2004april.htm
34. S. Prasad, NEMI Tin Whisker Test Project Update, IPC New Orleans, 2002.
35. I. Boguslavsky, P. Bush, E. Kam-Lum, M. Kwoka, and J. McCullen, and N. Vo, JEITA, May 2003.
36. N. Vo, M. Kwoka, and P. Bush, Tin whisker test standardization, IEEE *Trans. Electron. Packaging Manuf.* **28**(1), pp. 3–9, 2005.
37. J. Osenbach, Sn-whiskers: Preliminary temperature/humidity acceleration factors and risk assessment, iNEMI Test Group Meeting presentation at APEX (Anaheim), February 2005.
38. I. Sakamoto, Whisker test methods of JEITA whisker growth mechanism for test methods, Tin Whisker Workshop/ECTC May/June 2004 and IEEE *Trans. Electron. Packaging Manuf.* **28**(1), 10–16, 2005.
39. J. Osenbach et al., IEEE CPMT, 2004.
40. P. Su, private communication, material included in Accelerated Test Group presentation at iNEMI Tin Whisker Workshop at ECTC, Orlando, 2005.
41. P. Oberndorff, M. Dittes, P. Crema, and S. Chopin, Whisker formation on matte Sn influencing of high humidity, in *Proceedings of the 55th Electronic Components & Technology Conference*, Orlando, May/June 2005, pp. 429–433.

CHAPTER 8

Lead-Free Reflow and Rework

JASBIR BATH

8.1. INTRODUCTION

With tin–lead (Sn–Pb) soldering, there is a long history of soldering experience beginning from hand soldering to wave soldering to surface-mount technology. The development of Pb-free solder manufacturing experience has been a relatively recent occurrence. For reflow soldering and rework the developments that have occurred have not been reviewed in a comprehensive manner and their developments have not been accelerated until recently with increasing legislation against the use of lead. This chapter aims to review what data exist regarding Pb-free surface-mount technology and rework and what future work needs to be done with particular emphasis on Sn–Ag–Cu based Pb-free solder alloys. These alloys are gaining increasing favor and use within the electronics industry. It has probably only been since 1999 that Sn–Ag–Cu has begun to gain wider acceptance as the Pb-free surface-mount alloy of choice. The development work and focus on this alloy since that time has been growing with a wide variety of results and work presented. Manufacturing experience has been less developed; however, with increasing Pb-free production, these data are becoming available.

This chapter will review the lead-free data in comparison with Sn–Pb with emphasis on the materials, processes, and equipment used during surface-mount assembly and rework. The chapter is broken down into sections with a progression of information from the basic Pb-free materials used to process and equipment issues to the inspection of Pb-free solder joints with reliability and yield data presented in relation to the process where available. The important aspect to realize about Pb-free soldering or any type of production process is that certain manufacturing issues may only be noticed when actually manufacturing quantities of product. Experimentation is a useful tool, but experience is only gained when commitment is made by the manufacturing company and end-user to actually build prototype/production boards under actual manufacturing conditions.

Lead-Free Electronics. Edited by Bradley, Handwerker, Bath, Parker, and Gedney

It is hoped that the chapter will provide a good set of building blocks and information for the manufacturer to use when moving to assembling product with Pb-free solder. Movement to Pb-free soldering is underway in Japan. In Europe, this is rapidly occurring; with the increasing globalization of manufacturing, lead-free manufacturing will be implemented in other regions of the world.

8.2. PRINTABILITY OF LEAD-FREE SOLDER PASTES

Printing of the solder paste can be considered one of the most important processes for surface mount assembly operations. Without an adequate, consistent paste volume with correct registration onto the solder pad, solder defect rates will increase and reliability of the reflowed solder joint may be compromised. The factors typically affecting the printability of the solder paste include:

1. Printer machine settings: print pressure, printer speed, stencil release rate
2. Stencil thickness and stencil aperture openings
3. Solder metal alloy, particle size and distribution, solder–flux (binder) ratio (metal content), paste viscosity
4. Stencil and squeegee type
5. For paste-in-hole applications, thickness of the printed circuit board

With a Pb-free Sn–Ag–Cu solder paste, factor (3) will be most affected. Three other factors (1, 2, and 4) may be affected to some extent because they are interlinked. With tin–lead solder paste, because of the history of manufacturing use, there are guidelines in various standards and by solder paste, printer, and stencil manufacturers to allow for effective printing. For Pb-free pastes, manufacturing experience and experimentation will allow the same developments to occur. The following section will review the developments so far, with particular emphasis on Pb-free Sn–Ag–Cu solder paste.

8.2.1. Printability Results with Different Machine Printer Settings

There have been studies to determine the affect of printing parameters on Pb-free solder paste print volumes [1, 2]. Nguty et al. [2] evaluated the printability process window of Sn–3.8Ag–0.7Cu, Sn–57Bi, Sn–3.7Ag–0.7Cu–2Bi, and Sn–0.7Cu no-clean Type 3 size solder pastes by varying print pressure, print speed, aperture width, aperture length, and metal content in comparison with Sn–37Pb paste.

A design of experiment (DOE) was constructed with squeegee speeds of 50, 100, and 150 mm/s, squeegee pressures of 3, 5, and 7 kg, aperture widths of 5, 7, and 9 mils, aperture lengths of 60, 70, and 80 mils, and metal content of 88.5 wt% and 89.5 wt%. The flux formulation was the same for all the pastes tested. A stencil thickness of 4 mils (100 μm) was used, although typically a 5 (125-μm)- or

6-mil (150-μm)-thick stencil would be used to ensure an adequate paste volume for a solder joint. No indication in the work was given of squeegee length, although this will affect the amount of pressure being exerted by the squeegee on the stencil during printing. For a 16-in. squeegee length of blade, 8-kg print pressure is usually used with a rule of thumb of 1-lb print pressure for every in. length of squeegee blade. Low print pressure (giving reduced wear of stencil) and high print speed giving high print volumes are usually considered beneficial print parameters, especially for high-volume and smaller-sized boards such as cell phones.

For 88.5 wt% metal content of the Pb-free solder pastes, the results indicate that for a paste height approaching 80% of the stencil thickness (3.2-mil paste height), the printing process window is wide with squeegee speeds between 50 and 150 mm/s and with print pressure between 4 and 5 kg. For 89.5 wt% metal content of Sn–Ag–Cu, Sn–Ag–Cu–Bi, and Sn–Cu solder pastes, the process printing window was narrowed compared with Sn–57Bi Pb-free paste, which had a much wider process window (no printability data were presented for 88.5 wt% Sn–37Pb solder paste). The only similarity between Sn–37Pb and Sn–57Bi solder paste was that the solder densities were similar (8.4 g/cm^3 versus 8.6 g/cm^3), which were different from those for the other Pb-free pastes (Sn–Ag–Cu, Sn–Ag–Cu–Bi, Sn–Cu) with densities of 7.4–7.5 g/cm^3.

The lower density of the majority of the Pb-free solder pastes (Sn–3.9Ag–0.6Cu) compared with Sn–37Pb would indicate that to obtain the correct solder–binder ratio, you would need a slightly lower metal content of 88 wt% compared with Sn–Pb solder paste, which usually has a metal content of 90 wt%. This wt% would maintain a solder–volume ratio of 50:50. Thus using a slightly lower metal content than is used for tin–lead pastes typically optimizes the printability of Pb-free solder pastes. The property of the flux binder used in the Pb-free solder paste will also vary from solder paste supplier, because of the development of fluxes to meet the increased melting temperature of the solder powder, which will affect the printability of the paste.

In Pb-free paste work by Bath and Crombez [1], a DOE was constructed with squeegee pressure of 6, 12, and 18 kg, squeegee speed of 10, 15, and 20 mm/s, and stencil release rate of 1, 1.75, and 2.5 mm/s. The factors analyzed were print volume and consistency of print volume on 16-mil-pitch pads. The squeegee blade was 16 in. long with a stencil thickness of 6 mils (150 μm). The stencil aperture dimensions of the 16-mil pitch pads were 8 × 80 mils. The aperture-width to stencil-thickness ratio was 1.3. This would be considered a tough test with Type 3 paste (20–45 μm) because typical aspect ratios used are greater than 1.5 to 2, which allows for relatively easy printing.

The printing performance of Sn–4Ag–0.5Cu Type 3 no-clean paste varied with different squeegee print pressures, speed, and stencil release rate with an optimum print volume (2475 cubic mils) and consistency of print (standard deviation of paste volume = 194 cubic mils) found when using a high print pressure (18 kg), medium print speed (20 mm/s), and low stencil release rate (1 mm/s). A very good printability Sn–Pb Type 3 no-clean solder paste would have a print volume approaching 2700 cubic mils on the same 16-mil pitch pads with a standard deviation of print volume of 160 cubic mils with optimum printer settings of 6-kg pressure, 20-mm/s speed, and 1-mm/s release rate. The printability of the Pb-free Sn–Ag–Cu paste is thus not as good as the best Sn–Pb pastes but within the range of the Sn–Pb pastes available on the market. With development work and manufacturing experience, lead-free pastes are improving in printability.

For Sn–58Bi solder paste with a different flux used in the paste than with Sn–Ag–Cu, the optimum printer settings were 15-kg pressure, 10-mm/s speed, and 1-mm/s release rate on 16-mil pitch pads, with the highest print volume of 2265 cubic mils and standard deviation of print volume of 251 cubic mils with these optimized settings. Thus the printability is dependent not only on the alloy but also on the flux used in the paste.

Printability studies by Detert et al. [3] showed that for the Pb-free Sn–Ag–Cu and Sn–3.5Ag solder pastes tested, printability was very good on 14- to 16-mil-pitch pads and good on 8- to 12-mil-pitch pads with 3-mil-thick stencils. The particle sizes of the solder pastes used was not mentioned, and in practice a 3-mil-thick stencil would not be used for the majority of printing applications.

In the NEMI Pb-free work [4], print volumes were measured on five Sn–Ag–Cu and five Sn–Pb Type 3 no-clean solder pastes from various North American solder paste suppliers. Two separate printer settings were used to replicate large and mid-sized board printing: (a) 18-kg pressure, 10-mm/s speed, and 1-mm/s stencil release rate and (b) 6-kg pressure, 20-mm/s speed, and 1-mm/s stencil release rate. The print volumes were measured on 16-, 20-, 50-, and 40-mil-pitch pads and 0402 chip pads on a process test vehicle. The stencil thickness was 6 mils laser cut with apertures of 8×80 mils for 16-mil-pitch pads.

The results indicate that for 16-mil (0.4-mm)-pitch pads there was a difference between print volumes between Sn–Ag–Cu pastes and Sn–Pb pastes. The typical print volume for Sn–Ag–Cu was 2500 cubic mils compared with 2800 cubic mils for Sn–Pb. These actual results translate to a transfer rate for the five Sn–Ag–Cu solder pastes ranging from 55% to 73% compared with transfer rates for the five Sn–Pb pastes ranging from 65% to 74% for the 16-mil pitch pads. Transfer rate is calculated as the actual paste volume divided by the maximum theoretical paste volume (from stencil aperture measurements). In general, the Sn–Pb pastes printed with high volumes and more consistency than Sn–Ag–Cu pastes, but this may have been more due to the development nature of the Pb-free Sn–Ag–Cu pastes evaluated at that time.

This is also evident in the transfer rate print volume results for the Sn–Ag–Cu pastes compared to the Sn–Pb pastes for the 20-mil-pitch QFP, 40-mil-pitch BGA, 50-mil-pitch BGA, and 0402 chip pads. The print transfer rates for 20-mil-pitch QFP pads vary from 71% to 94% for the Sn–Ag–Cu pastes compared with 77–100% for the five Sn–Pb pastes. The transfer rates for 40-mil-pitch BGA pads vary from 56% to 91% for the five Sn–Ag–Cu pastes compared with 60–100% for the five Sn–Pb pastes. The transfer rates for 50-mil-pitch BGA pads vary from 59% to 97% for the five Sn–Ag–Cu pastes compared with 94–100% for the five Sn–Pb pastes. The transfer rates for 0402 chip component pads vary from 47% to 84% for the Sn–Ag–Cu pastes compared with 72–97% for the five Sn–Pb pastes as shown in Table 8.1.

TABLE 8.1. Paste Transfer Rates for Sn–Pb and Pb-Free Sn–Ag–Cu during NEMI Solder Paste Evaluations [4]

Paste Transfer Rates	Sn–Pb	Sn–Ag–Cu
16-mil-pitch QFP	65–74%	55–73%
20-mil-pitch QFP	77–100%	71–94%
40-mil-pitch BGA	60–100%	56–91%
50-mil-pitch BGA	94–100%	59–97%
0402 chip	72–97%	47–84%

It should be noted that there was an area of concern with the print volume results for the 50-mil-pitch pads, with some of the Pb-free Sn–Ag–Cu pastes being lower than for those with Sn–Pb paste, although the stencil aperture dimensions were not considered challenging for printing. The assembled Pb-free printed boards, which had reasonable print volumes at these BGA pads, showed no signs of open or insufficient solder after BGA assembly and after cross sections of these components, which indicates that the printing process window could be relatively wide.

No real details could be obtained as to the benefits of the different printer settings on the print volumes for Sn–Ag–Cu and Sn–Pb solder pastes. Based on the results of these printing experiments and X-ray and solderability evaluations of assembled boards, suitable Sn–Ag–Cu and Sn–Pb pastes were chosen for the NEMI reliability test board builds which had the best combination of soldering, printing, and X-ray inspection scores.

In addition to the use of metal squeegees for printing solder paste, there have been developments in the use of printer head technology whereby the transfer head of the printing machine is enclosed and attached to a pressure mechanism which is connected to the printer's squeegee carrying mechanism. The transfer head remains in contact with the stencil during production, sealing the solder paste from the environment. The advantages are that reduced wastage of solder paste due to extended life of the solder paste from the sealed process and the transfer force of the solder paste through the stencil apertures can be increased, thereby giving advantages for small aperture printing and paste-in-hole applications.

Ashmore and Goldsmith [5] conducted evaluations comparing Pb-free Sn–Ag–Cu with Sn–Pb solder paste using the enclosed print head technology on 0.4-mm-pitch QFP and 0.5-mm-pitch CSP component pads with various stencil aperture dimensions on 4- and 5-mil-thick stencils. The Type 3 size Sn–Ag–Cu solder pastes had metal content varying from 88 to 89.3 wt%, whereas the Type 3 Sn–Pb solder paste had a metal content of 90.25 wt%. The lower metal contents used for Sn–Ag–Cu was to maintain the flux/solder powder volume ratio for good printing results due to the lower density of the Pb-free Sn–Ag–Cu paste (7.4 g/cm^3) compared with Sn–Pb (8.4 g/cm^3).

A DOE was constructed with three factors: print pressure, print speed, and separation speed. Paste volume was measured and paste transfer efficiency was calculated. The results showed that the low print pressure and fast print speed were beneficial for printability for both Pb-free Sn–Ag–Cu and Sn–Pb solder pastes, which are beneficial parameters to use for the printing process. Separation speed was also found to be an important factor to consider during the printing process to provide the best printing results. The experiments showed that the print volumes for Pb-free Sn–Ag–Cu pastes were, in general, slightly lower than those for the Sn–Pb control. This could be remedied by the use of more optimized Pb-free Sn–Ag–Cu solder pastes in terms of printing characteristics and a possible adjustment in Sn–Ag–Cu solder metal content in the paste.

8.2.2. Printability Assessments for Pb-Free Sn–Ag–Cu and Sn–Pb Paste with NEMI Pb-Free Reliability Test Boards

Prior to the NEMI reliability test board builds, the selected Sn–Pb and Pb-free Sn–Ag–Cu paste from previous evaluations [4] were subjected to further printability evaluations [6]. These included print paste studies to determine the color and texture of the Sn–Pb and Pb-free Sn–Ag–Cu paste (Figure 8.1) and the ease of print roll with the machine squeegee blade (Figure 8.2).

The paste volume on CSP169 component board pads was recorded, and measurement of print quality and stencil clogging after 10 printing cycles without stencil cleaning are shown (Figure 8.3).

In all tests there was found to be no difference in printing quality between the Sn–Pb and Pb-free Sn–Ag–Cu, with the following remark being made during the evaluation: "excellent printing characteristics of both solder pastes" for the test boards to be assembled. During the actual NEMI test board builds, all paste printed boards were labeled with a unique board number and measured for paste volume at all pad locations for future analysis using automated laser scanning print inspection equipment. Typical solder paste volumes on the pads measured during the reliability test board builds for each board (package) type are listed in Table 8.2. Six-mil-thick stencils were used for all the builds. The values are calculated for the solder paste on one whole representative board for each case.

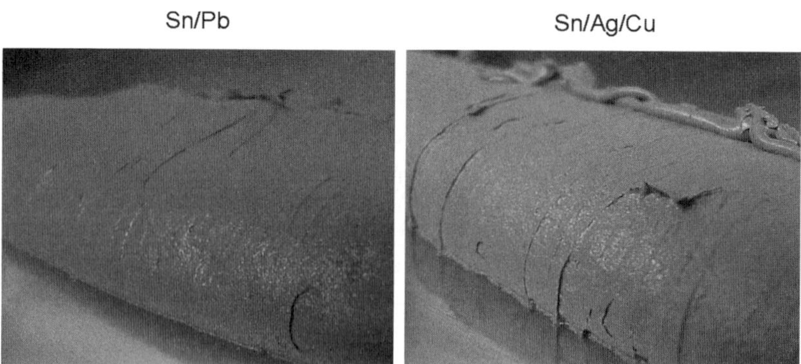

Figure 8.1. Color and texture comparison of the selected NEMI Sn–Pb and Pb-free Sn–Ag–Cu solder paste.

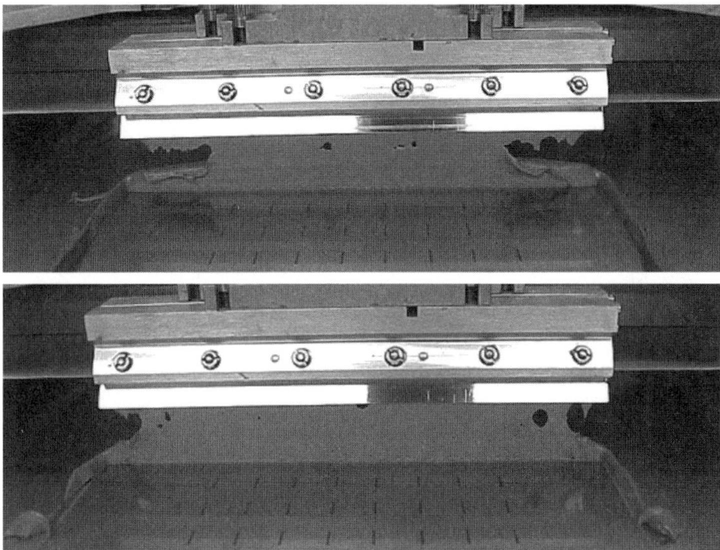

Figure 8.2. Ease of print roll with the machine squeegee blade (upper picture for Sn–Pb paste, lower picture for Sn–Ag–Cu paste).

Paste volumes for the Pb-free Sn–Ag–Cu paste were found to be higher than Sn–Pb paste in all cases using the same printer settings of 5-kg pressure, 15-mm/s print speed and 0.5-mm/s separation speed. The squeegee blade length was 25 cm. This would be due to the good printing characteristics of the selected

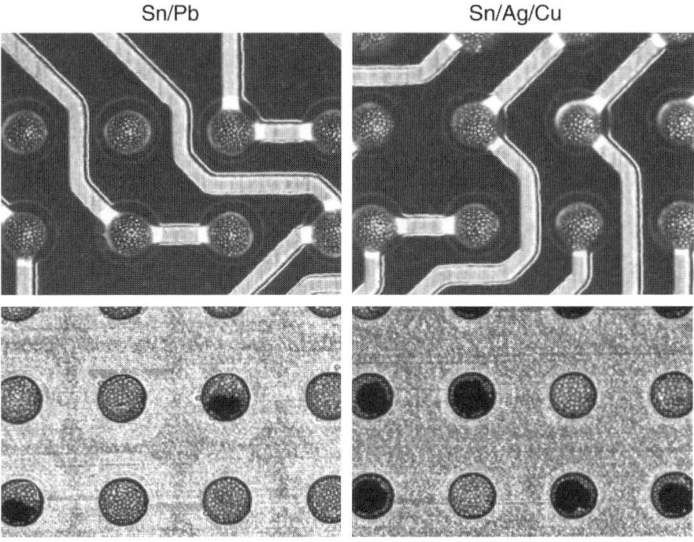

Figure 8.3. Print quality on board pad and evaluation of stencil clogging after 10 print cycles without stencil cleaning for CSP169 test board and stencil.

TABLE 8.2. Typical Pb-Free Sn–Ag–Cu and Sn–Pb Paste Volumes on the NEMI Reliability Test Vehicle Boards

Package	Paste	Average Paste Volume (cubic mils)	Paste Transfer Rate (%)	Average Paste Height (mils)
256PBGA	Sn–Pb	2,410	86%	6.7
	Sn–Ag–Cu	2,677	96%	6.5
169CSP	Sn–Pb	552	56%	4.4
	Sn–Ag–Cu	681	69%	5.5
208CSP	Sn–Pb	522	53%	5.4
	Sn–Ag–Cu	535	54%	5.1
256CBGA	Sn–Pb	3,836	95%	7.0
	Sn–Ag–Cu	4,037	100%	6.9
R2512	Sn–Pb	30,908	100%	7.2
	Sn–Ag–Cu	32,894	100%	7.9
TSOP48	Sn–Pb	2,312	79%	5.4
	Sn–Ag–Cu	2,373	81%	5.6

Sn–Pb and Pb-free SnAg–Cu paste and the relatively coarse pitch of the components being assembled.

There was one printing issue encountered during the start of the TSOP board builds. The stencil aperture-to-pad ratio was 1:1, with too much paste printed on the pads. During placement of the TSOP component with the production placement machine solder bridging was found. This was rectified by having a new TSOP stencil manufactured with a reduced stencil aperture width (reduction of 10%). This reduced the volume of paste printed and removed the solder bridging issue.

All other boards and components were printed with solder paste and placed with components without issues. The machine placement images were similar for the Sn–Pb and Pb-free components for these test boards [6], and no differences in placement accuracy were encountered for Sn–Pb and Pb-free components.

8.2.3. Printability of Type 3 Sn–Ag–Cu Paste Compared with Type 4 Pastes

Evaluations of Type 3 versus Type 4 Sn–Ag–Cu no-clean pastes by Arra et al. [7] have shown that the amount of misprinted pads noticed with Type 4 Sn–Ag–Cu paste seemed to be more than Type 3 Sn–Ag–Cu paste. The stencil thickness was 4 mils with aperture openings of 6, 8, 10, 12, 14, and 16 mils. With a 6-mil circular aperture opening, the aspect ratio is 1.5, which is considered a fairly difficult printing operation. The number of misprints for Type 4 Sn–Ag–Cu paste was, in general, greater than Type 3 Sn–Ag–Cu and Type 3 Sn–Pb pastes on 6-, 8-, 10- and 12-mil aperture openings. This could be a result of the type of flux used but also indicates that Type 3 paste can be used for fine-pitch printing where the maximum solder particle size should be less than a half of the aperture width and the stencil thickness. Aspect ratio considerations need to be taken into account for fine-pitch printing.

8.2.4. Shelf Life/Viscosity of Pb-Free Solder Pastes

The shelf life or storage life of Sn–Pb solder paste is usually 6 months at around 5°C compared with 3 months at 20°C. Solder paste is a mixture of metal and flux with two different densities, and separation of flux and solder metal in a paste jar can occur especially when exposed to excessive heat which can affect the rheological properties of the solder paste.

Because of the relative new development of Pb-free solder pastes, the typical shelf life is recommended to be 3 months initially especially when using newer flux formulations, which are relatively untested in the field. There has been work done by Huang and Lee [8] to evaluate the shelf life of Pb-free solder pastes by monitoring the viscosity of Pb-free solder pastes compared with Sn–Pb paste over a period of one month. A changing viscosity, which would usually be increasing over time, is considered undesirable. A high-viscosity solder paste may lead to an insufficient print, whereas a low-viscosity paste may lead to print slumping.

Viscosity of the solder paste was measured at 1 day, 7 days, and 30 days at a testing temperature of 25°C. The Sn–3.8Ag–0.7Cu solder paste had a small viscosity change over the month testing, which was similar to the results for Sn–Pb solder paste, which is considered desirable.

8.2.5. Tack Force of Pb-Free Sn–Ag–Cu Solder Pastes

An important property of a solder paste prior to reflow once it has been printed on the board is its tack force when a component is placed onto the paste. If the tack force or "tackiness" of a solder paste is insufficient, then the paste will not have sufficient force to hold the components to the board before reflow. The IPC TM 650 specification (2.4.44) measures the tack life of the solder paste over time. Butterfield et al. [9] measured the tack force of the no-clean Sn–Ag–Cu solder paste over a period of 8 hours compared with that for a control Sn–Pb paste. The tack force remained over 0.3 N for both Sn–Ag–Cu and Sn–Pb pastes, which was considered the pass tack force. Work by Huang and Lee [8] reported similar tackiness for the Sn–3.8Ag–0.7Cu solder pastes compared with Sn–Pb over a period of testing of over 3 days. Arra et al. [7] conducted tackiness tests on seven Sn–Ag–Cu solder pastes compared with one Sn–Pb solder paste. Six out of the seven Sn–Ag–Cu pastes had better tackiness than Sn–Pb paste over the 13-hour period of tackiness testing to replicate what could happen during production. It is interesting to note that three different workers used three different types of test to calculate tackiness, which indicates that the current industry standard tackiness test may need to be revised.

8.2.6. Slumping of Pb-Free Solder Pastes

8.2.6.1. Cold Slump. The slump of a solder paste is the collapse of the printed paste creating bridging between adjacent component pads. This may be caused by such factors as excessive paste being deposited (stencil aperture-to-pad ratio too high, e.g., 1:1). The correct solder-paste-metal to flux-weight ratio is required, and a thickening agent may also be added with the flux to maintain solder paste shape after printing. Component contact with the solder paste deposit can also skew the deposit causing the solder paste to bridge. Arra et al. [7] conducted evaluation tests on seven Sn–Ag–Cu pastes and one Sn–Pb paste after printing and storage at ambient conditions for 10–20 minutes. No cold slumping was observed.

8.2.6.2. Hot Slump. Slump can also be attributed to heat. After the paste is printed and components are placed, the board will subjected to the reflow profile, which can cause hot slump. Within the soak preheat zone prior to reflow, the solder paste may slump, causing bridging between pads. Arra et al. [7] conducted tests on hot slump on seven Sn–Ag–Cu pastes. The slump test was based on IPC Test Method 650 (2.4.35). Printed solder paste boards were subjected to 10–15 minutes storage in an oven at 150°C. The Sn–Pb paste and three of the seven Sn–Ag–Cu pastes had no slumping between pads after this storage. The remaining four Sn–Ag–Cu pastes had evidence of slumping with the smallest pad spacing of 0.75 mm (28 mils) and 0.8 mm (32 mils).

Hart et al. [10] conducted work on hot slumping of Sn–Ag–Cu solder pastes after printing. The printed solder pastes were subjected to 5 minutes storage in an oven at 150°C. This was to try and replicate the 2–2.5 minutes that a Sn–Pb solder paste typically encounters in the soak zone at 150–170°C prior to reflow. Five of the 16 Sn–Ag–Cu solder pastes examined after 150°C storage had no slumping down to a 0.33 mm (13 mils) print spacing (highest rating of 2). Ten of the 16 Sn–Ag–Cu solder pastes had slumped with spacing of 0.63 mm (25 mils) and below (middle rating of 1). The remaining Sn–Ag–Cu solder paste had slumping at 0.71-mm (28-mils) spacing and above, which is considered a poor hot slump result (rating of zero). The Sn–Pb water-soluble paste control had no slumping under the 150°C conditions (rating of 2), whereas the Sn–Pb no-clean paste control had a medium rating of 1 (slumping with spacing of 0.63 mm and below).

It is interesting to note that the hot slumping studies were done at 150°C, which is what is used for Sn–Pb solder paste. For Sn–Ag–Cu paste, which has a melting temperature of 217°C compared with 183°C for Sn–Pb, the soak zone may be at a higher temperature than for Sn–Pb with the development of flux formulations rated to the higher temperatures, so the testing that was conducted may need to be done at elevated temperatures.

8.2.6.3. Humidity Slump. Humidity as well as heat can also cause solder paste slumping. Typical humidity conditions on the manufacturing floor are less than 60% RH. On a humid day, humidity can exceed these levels especially in hot humid climates, which can lead to solder paste slump and subsequent bridging after reflow. Humidity typically affects water-soluble pastes more than no-clean Sn–Pb solder pastes. Hart et al. [10] conducted tests on 16 Sn–Ag–Cu solder pastes by subjecting printed solder paste boards to 90% RH and 30°C for 1 hour. Ten of the Sn–Ag–Cu solder pastes had no slumping down to 0.33 mm (13 mils) print spacing (highest rating of 2). One of the Sn–Ag–Cu pastes had slumping with print spacing of 0.63 mm and below (middle rating of 1). The remaining five

Sn–Ag–Cu solder pastes had slumping at 0.71 mm (28 mils) spacing and below considered a poor humidity slump result (rating of zero). The Sn–Pb no-clean solder paste control had no humidity slumping (rating of 2), whereas the Sn–Pb water-soluble paste control had a medium rating of 1 (slumping with spacing of 0.63 mm and below).

8.2.7. Solder Balling of Reflowed Pb-Free Solder Pastes

Solder balling is the formation of very small spherical particles of solder separating from the main body of the solder joint after reflow. In the worst case there could be an artificial solder bridge between two adjacent pads. The problem is typically a concern for no-clean solder pastes and not for water-soluble pastes where the solder balls would be washed away during the cleaning process. Solder balling can occur by an improper solder reflow profile being used which can give rise to solder splattering. Excessive surface oxide on the solder powder can also cause incomplete solder coalescence during reflow.

Arra et al. [7] conducted solder ball testing on seven Sn–Ag–Cu and one Sn–Pb solder paste using IPC Test Method 650 (3.4.43), except a solder mask coated PCB was used. Solder paste was printed and inspected after reflow. Of the Sn–Ag–Cu pastes, three had no evidence of solder balls (preferred IPC category). Three Sn–Ag–Cu solder pastes had some solder balls that were grouped in the IPC acceptable category. One Sn–Ag–Cu solder paste had a large number of solder balls and was termed in the IPC unacceptable category. The Sn–Pb paste had some solder balls that were grouped in the acceptable IPC category.

In work by Huang and Lee [8] the solder balling for Sn–3.8Ag–0.7Cu solder paste was slightly lower than Sn–37Pb paste, with an average of 8 solder balls around a soldered pad for Sn–Ag–Cu compared with 10 solder balls with Sn–Pb with observations at 20× magnification. Hart et al. [10] reflowed printed Sn–Ag–Cu solder paste and subjected the reflowed paste to 90% RH for 1 day. The humidity can cause flux residue to dislodge unattached particles from the coalesced solder sphere. Examination at 30× magnification after humidity testing revealed that 12 of the Sn–Ag–Cu pastes had no solder balls in the flux residue surrounding the solder sphere (highest rating of 2). Three Sn–Ag–Cu pastes had a few small-sized solder balls in the flux residue (medium rating of 1). Only one Sn–Ag–Cu paste had numerous solder balls (rating of zero). The control Sn–Pb water-soluble and no-clean pastes both had no solder balling (rating of 2).

In work by Wada [11] for the assembly of Pb-free mini-disc players, there was a reduction in the number of solder balls when using Sn–3Ag–3Bi solder paste compared with the assembled Sn–Pb-Ag paste boards.

8.3. SOAK VERSUS RAMP TEMPERATURE PROFILES

Once the board has been printed with solder paste and components have been placed, the board will go through a reflow oven to reflow the solder paste, which will cause an electrical and thermal bond between the component and board. For a solder paste, which is a mixture of solder particles and flux, use of the correct reflow profile is important to avoid or minimize solder defects such as opens, bridging, and voiding. The use of the correct reflow profile begins with the preheat stage before the solder particles reflow and coalesce into a solder joint.

The reflow profile can be broken down into four stages for Sn–Pb solder paste similar to that reported by Rajewski [12]:

Initial Preheat Stage (Room Temperature to 120°C): The PCB and components are heated up gradually. Excess solvent is driven off. The ramp up rate is less than 3°C per second to avoid flux splattering which would give solder balling and potentially slumping.

Soak Stage (120–183°C): Flux components are activated and begin to reduce the surface oxides on the solder particles, component leads, and PCB pads. Activated flux continues to keep the metal surfaces from reoxidizing and allows the different thermal mass components on the board to reach the same maximum temperature.

Reflow Stage (183–220°C): Paste is brought to and beyond the Sn–Pb alloy's melting point (183°C). The activated flux reduces the surface tension at the metal interfaces adding the soldering process.

Cool-Down Stage (220–183°C to Room Temperature): Assembly is cooled down as evenly as possible so that excess intermetallics are not formed and the components and boards are not subjected to excessive thermal shock.

For Pb-free Sn–Ag–Cu solder paste, the same four stages will apply but the temperature ranges will change because the melting point for Sn–Ag–Cu is 217°C compared to 183°C for Sn–Pb, a 34°C increase. Thus the four stages of a Pb-free Sn–Ag–Cu profile could be:

Initial Preheating Stage: Room temperature to 170°C

Soak Stage: 170–217°C

Reflow Stage: 217–255°C

Cool-Down Stage: 255–217°C to room temperature

Existing flux systems used with Sn–Pb solder paste may not be rated to these higher soak and reflow profiling temperatures, which is why there has been development of Pb-free solder pastes recently which will continue in the coming years.

The soak stage, which is used for Sn–Pb solder paste, has undergone a change of profiling over recent years. Previously, older style infrared ovens have used a ramp–soak–spike reflow profile which has included a long soak time at the soaking preheat temperature of around 2 minutes in order to achieve optimum results. With the use of more efficient and newer ovens that transfer heat to the assembly through forced convection, the reflow profile has changed to a ramp–spike reflow profile with a very short or minimal soak time at the soaking temperature. The older generation of Sn–Pb pastes, some of which are still used today, have tended to be developed with the use of the ramp–soak–spike profile in mind, whereas the newer generation of pastes may be more suitable with a ramp–spike profile with limited soak time. With large boards with Sn–Pb paste having a large ΔT across the board, a long soak time is still preferred in manufacture in order to ensure the different mass of components on the board reach the same temperature prior to the reflow stage.

However, excessive soak times and temperatures can cause the breakdown of flux activators in the solder paste prior to reflow, which may give solderability issues. The ramp–spike profile typically has a ramp up rate of less than 2°C per second. It is beneficial to use this type of profile when there may be uncertainty of maintaining the flux vehicle throughout the preheat soak stage where the flux is activated and the flux volatiles are driven off prior to the reflow stage.

For Pb-free Sn–Ag–Cu solder paste, because of the higher temperatures being used and the initial use of flux vehicles which may have been developed for Sn–Pb pastes, it is preferred that a sharper ramp–spike profile is used to avoid the potential for flux breakdown at the elevated temperatures. Fluxes are and will need to be developed which will be more rated to the higher Pb-free soldering temperatures and maintain their activity during the soak and reflow process. The development of Pb-free pastes so far seems to have been for smaller-sized consumer boards where the requirement for an extended soak time is considered minimal. There will be an increasing shift to medium- to large-sized boards which will require more development of pastes.

Work has been done on preheat soak temperatures and times for Pb-free soldering by Shina et al. [13] with assembled boards using Sn–Ag–Cu solder paste. The flux used was no-clean with a high flux activity and a large amount of flux residue formed after reflow. There was a soak time of between 150°C and 170°C in the design of experiment for 90–120 seconds corresponding to a soak profile. This was compared with boards assembled using a linear ramp profile with minimum soak time and temperature. The visual inspection results for the Pb-free soldered leaded and leadless devices (QFP, SOIC, chips) indicated no real differences between the use of a soak or a linear ramp profile. Pull test results for SOIC components both before and after thermal cycling indicated no real differences between using a soak or linear profile [14]. The pull test results for Sn–Ag–Cu were equivalent to or better than those for Sn–Pb soldered components both before and after thermal cycling.

Work by Skidmore and Walters [15] also reviewed the affect of soak temperature on the visual appearance of assembled Sn–Ag–Cu boards with three different types of no-clean flux in the solder paste. Flux 1 had a high flux activity and large amount of residue formed after reflow. Flux 2 had a medium flux activity with a low amount of residue formed after reflow. Flux 3 had a medium flux activity with a large amount of residue formed after reflow. From the visual inspection results of the leadframe components, the linear profile with minimal soak tended to give slightly better results compared with the use of a soak profile. The soak profile used a soak time between 150°C and 175°C of 90–120 seconds. The boards and components assembled by Shina [13, 14] and Skidmore [15] were identical with board dimensions of 4×5.5 in., and thus cannot be considered a large-sized assembly board.

8.4. EFFECT OF PEAK TEMPERATURE VERSUS REFLOW PERFORMANCE

8.4.1. Wetting/Solderability

For Sn–Pb soldering, wetting (bonding) of the solder paste or wave alloy to the board and component is essential to form a good and reliable solder joint. Too low a reflow time and temperature indicates insufficient bonding between solder and component or board, whereas too high a reflow time and temperature indicates too much intermetallic compound created from bonding, which can create a less reliable solder joint.

The wetting behavior data available indicate that Sn–Ag–Cu alloys may have a slightly better or equivalent solderability to Sn–3.5Ag with the benefit of a slightly reduced melting temperature (217°C compared to 221°C for Sn–3.5Ag). The Pb-free high-tin-containing solders typically have a higher surface tension than Sn–Pb, and due to this, to have increased wetting angles and a tendency for reduced spread. This does not mean that there is a tendency for reduced reliability as has been shown in reliability testing where the bonding is still occurring between the tin in the solder (either from Pb-free Sn–Ag–Cu or from Sn–37Pb) and the component or board metallization (usually copper or nickel).

Analysis of wetting for evaluation/testing purposes is usually carried out by two methods:

1. Solderability analysis under dynamic conditions with the use of wetting balances to measure wetting times and forces of solders on lead frames/chip terminations or solderable coupons
2. Solderability analysis under more static conditions with measurement of the contact angle and spreading of solder on various solderable surface finish coupons

The use of either of these methods should be used in conjunction with analysis of the solderability of assembled components on the actual production board. The analysis from either of these two wetting methodologies should also be used for relative comparison purposes only between Sn–Pb and Pb-free solder and not as absolute figures. Solderability analysis using the wetting balance test was originally designed for wave soldering where a component which achieved two-thirds of its maximum theoretical wetting force [calculated from the component geometry and the physical properties (surface tension, density) of the solder used] within 2 seconds would be deemed to pass the wetting balance test. The 2-second rule was based on the time that the component/board would typically pass through the molten wave solder during the wave soldering operation.

With the increased movement to surface-mount technology, where typically the time over reflow is around 45–75 seconds, comparisons become even more difficult to make. On one hand, the time over reflow is extended compared with wave soldering (45–75 seconds for reflow compared with 2–4 seconds for wave soldering with

Sn–Pb solder) which is beneficial but on the other hand the actual peak temperature is reduced compared with wave soldering (200–220°C for reflow compared with 250–255°C for wave soldering with Sn–Pb). The increasing use of less active no-clean VOC-free fluxes with little or no halide in the solder paste during surface-mount assembly and the same type of fluxes for wave soldering reduces the wetting ability of the solder alloy further with a tightening of the process window.

8.4.1.1. Wetting Balance Tests. The main data that have been available so far for Sn–Ag–Cu is the wetting balance dynamic test data. Sattiraju et al. [16] conducted tests with Sn–3.9Ag–0.6Cu using copper coupons coated with immersion tin, immersion silver, palladium, nickel–palladium, nickel–palladium–gold, nickel–gold, and OSP. The sheer amount of surface finishes evaluated indicates that standardization of Pb-free surface finish will be essential in order to more forward in the same way as Sn–Ag–Cu solder is becoming the standard Pb-free surface mount alloy.

Reviewing the various board surface finishes, the solder has to simply dissolve the tin, silver, or OSP into the solder to bond to copper substrates or dissolve the gold and/or palladium to bond to the nickel substrate. The measurements of maximum wetting force and time before producing a positive wetting force (time for the solder to be at an angle of 90° to the test coupon) for the surface finishes were compared for Sn–Ag–Cu and Sn–Pb solder using a no-clean and water-soluble flux. In general the Sn–Ag–Cu had similar maximum wetting forces and time taken for a positive wetting force to Sn–Pb with the majority of the surface finishes. The Sn–Ag–Cu has a slightly better wetting result with immersion tin surface finish, whereas the Sn–Pb had a slightly better wetting result with immersion silver surface finish. The pot temperature used for Sn–Ag–Cu was 235°C, whereas for Sn–Pb it was 201°C (both were 18°C over the respective liquidus temperature). The wetting results with water-soluble flux were, in general, better than no-clean flux for both Sn–Ag–Cu and Sn–Pb, which was probably related to the improved flux activity of the water-soluble flux.

Previous work by Heiser et al. [17] on Sn–Pb solder using a combination of wetting force evaluations and assembly production trials has shown that a positive wetting balance force (i.e., maximum wetting force greater than 0 mN) will produce a good solder joint. The work above would indicate that the Sn–Ag–Cu should give adequate wetting when used in production. A good approach to testing would be to conduct a similar experiment with Pb-free Sn–Ag–Cu solder as was conducted by Heiser et al. [17] to confirm that wetting balance findings for Sn–Ag–Cu would give good results in production.

In continuation of the work by Sattiraju et al. [18], the same tests were conducted with Sn–Ag–Cu and Sn–Pb solder using the wetting balance with the same fluxes and similar surface finishes (Sn, Ni–Au, Pd, OSP, Ag). The only difference was that a pot temperature of 210°C (liquidus + 27°C) was used with Sn–Pb instead of 201°C used previously, with the Sn–Ag–Cu pot temperature remaining the same at 235°C (liquidus + 18°C). In this case, the wetting times were, in general, quicker for the Sn–Pb at 210°C compared with Sn–Pb at 201°C and better than the Sn–Ag–Cu solder for all surface finishes, with the maximum wetting force for the Sn–Pb solder remaining the same. For the Sn–Ag–Cu, when the solder pot temperature was increased from 235°C to 255°C, there was no significant improvement in the maximum wetting force with Ni–Au surface finish with water-soluble flux, but there was a small improvement using no-clean flux. For Sn–Ag–Cu with immersion tin surface finish, there was a more marked increase in wetting force from a pot temperature of 230°C to 235°C with water-soluble flux and a marked increase in wetting force from 235°C to 240°C with no-clean flux. Above 240°C the improvement in wetting force with the immersion tin surface finish was less noticeably for either flux with Sn–Ag–Cu.

The work done by Johnson et al. [19] when evaluating no-clean and water-soluble fluxes using the Sn–Ag–Cu solder indicated that acceptable wetting can be achieved at peak temperatures of 230°C and above for most surface finishes with the benefit of improved wetting times (marginal wetting with OSP at 230°C). The use of 235°C pot temperature improved wetting times significantly compared with 230°C with a further more modest improvement at 240°C with no significantly improvements above 240°C with no-clean flux. 235–240°C peak temperature will be a good minimum peak temperature to aim for based on the solderability experiments.

Work by Kwoka and O'Brien [20) using Sn–Ag–Cu solder with a wetting balance to evaluate the solderability of Ni–Au and OSP surface finish coupons indicated that there were significant improvements in solderability of Sn–Ag–Cu when increasing solder pot temperatures from 235°C in increments of 10°C up to 265°C. An experimental no-clean flux developed for use with the wetting balance was used in the experiments which was based on pure rosin with 0.2 wt% halide. Positive wetting forces for Sn–Ag–Cu were being achieved at 3–4 seconds with a pot temperature of 235°C compared with less than 1 second for a pot temperature of 265°C for both Ni–Au and OSP surface finishes. Tin–lead solder wetting balance tests were conducted at the same pot temperatures as Sn–Ag–Cu with the same surface finishes. The results indicated positive wetting forces being achieved within 1 second at all pot temperatures (235°C, 245°C, 255°C, 265°C). It would have been preferred to use lower pot temperatures with Sn–Pb solder (i.e., 201°C, 211°C, 221°C, 231°C) to allow a better comparison with the results with Sn–Ag–Cu, especially for assessing surface-mount soldering.

Work was conducted by Fan et al. [21] on the solderability of lead-frame finishes with Sn–Ag–Cu using solder pot temperatures of 235°C, 245°C, 260°C, and 280°C. The surface finishes included pure Sn, Sn–0.7Cu, Ni–Pd, and Ni–Pd–Au. One non-activated and one mildly activated flux was used in the wetting balance tests. In general, positive wetting forces were being achieved within 5–8 seconds at 235°C solder temperature for Sn–Ag–Cu with the pure Sn and Sn–0.7Cu lead-frame finish, within 2–5 seconds at 245°C solder temperatures for these two finishes and less than 1 second for all the four lead-frame finishes at 260°C and 280°C. Steam aging for the four lead-frame finishes for 8 hours followed by wetting balance tests at 260°C Sn–Ag–Cu solder temperature also indicated positive wetting forces were being achieved close to or less than 1 second.

Work by Lal et al. [22] evaluating the solderability of Sn–Ag–Cu on as-received pure Sn- and Sn–Cu-coated coupons showed that positive wetting forces were being achieved within 0.8 seconds using a solder pot temperature of 245°C using no-clean flux. After 4.5 seconds of continuous immersion in the solder, there was still a positive wetting force indicating no evidence of dewetting of the as-received coupons. With 8-hour steam aging of the coupons, positive wetting forces were achieved within 0.5–5 seconds with Sn–Ag–Cu solder for the pure-Sn-coated and Sn–Cu-coated coupons with a pot temperature of 245°C. After 4.5 seconds of immersion in the solder, the majority of the pure-Sn-coated and Sn–Cu-coated steam-aged coupons tested still had a positive wetting force with some evidence of dewetting related to certain specific Sn/Sn–Cu-plating chemistries used (2 out of 7 chemistries had negative wetting forces at 4.5-second immersion times after steam aging). The use of active compared with nonactive no-clean fluxes improved the wetting ability of steam-aged samples by being more aggressive in removing surface oxide aiding solderability.

Work by Romm et al. [23] conducted wetting balance tests on lead-frames coated with Ni–Pd, Ni–Pd–Au, and Sn–Pb surface finishes with a Sn–Ag–Cu solder globule at a temperature of 250°C. Positive wetting forces were being achieved within a time of less than 0.5 seconds, and the time to reach two-thirds of the maximum wetting force was less than one second. Hunt [24] conducted wetting balance tests with Pb-free and Sn–Pb lead-frame finishes using a Sn–Ag–Cu solder globule. The temperature of the Sn–Ag–Cu globule was 267°C (liquidus plus 50°C) with a water-based flux used in the tests. The components were aged for 4 days at 155°C prior to testing to replicate a storage condition. In general the lead-frames for all surface finishes (Ni–Au, Ni–Pd–Au, Sn–Pb) has a positive wetting force after 2 seconds with Sn–Ag–Cu. The wetting force with Sn–Pb surface finish was usually higher than with Ni–Pd or Ni–Pd–Au surface finish when using Sn–Ag–Cu solder. The wetting forces when using aged Sn–Pb, Ni–Pd, and Ni–Pd–Au surface-finish lead frames with a Sn–Pb globule at a solder temperature of 233°C (liquidus plus 50°C) were always greater than when using the aged lead-frame finishes with the Sn–Ag–Cu globule at 267°C.

Work by Harrison et al. [25] investigated the solderability of Sn–Ag–Cu using a wetting balance with cleaned copper coupons using a wetting balance with a no-clean flux based on pure rosin with 0.5 wt% halide. Tests were conducted at pot temperatures of 242°C (liquidus plus 25°C), 250°C, and 267°C (liquidus plus 50°C) for Sn–Ag–Cu. For Sn–40Pb, pot temperatures of 208°C (liquidus plus 25°C), 218°C (liquidus plus 35°C), 233°C (liquidus plus 50°C), and 250°C were used. Maximum wetting forces for Sn–Ag–Cu were similar to Sn–Pb at all pot temperatures. The time to reach two-thirds of the maximum wetting force for Sn–Ag–Cu was similar to Sn–Pb when the same temperature over liquidus was used for Sn–Pb and Sn–Ag–Cu. With an increase in solder pot temperature over liquidus, the time to reach two-thirds of the maximum wetting force was reduced for both Sn–Ag–Cu and Sn–Pb from 1 second at the lowest pot temperatures (242°C for Sn–Ag–Cu and 208°C for Sn–Pb) to less than 0.6 seconds at the highest pot temperatures (267°C for Sn–Ag–Cu and 233–250°C for Sn–Pb).

The increased activity of the no-clean flux used by Harrison (0.5 wt% halide) compared with Kwoka (0.2 wt% halide) [20] had a beneficial affect in terms of improved wetting for Sn–Ag–Cu. Harrison et al. [25] also conducted the same experiments with pure rosin flux containing no halide with much more scatter in the wetting balance data and difficulty in determining conclusions.

Evaluations of wetting balance results from Albrecht and Wilke [26] indicate that the wetting time to reach two-thirds of the maximum wetting force for cleaned copper wires using Sn–Ag–Cu solder with no-clean rosin flux containing 0.5 wt% halide (same flux as used by Harrison [25]) will decrease with increasing solder pot temperature. Solder pot temperatures of 225°C were shown to have longer wetting times (2–8 seconds), whereas for a pot temperature of 235°C these wetting times were reduced down to 1.5–2.5 seconds. With Sn–Ag–Cu solder pot temperatures of 245°C and 260°C the wetting times were around or below 1 second. The general conclusion from Albrecht and Wilke [26] was that reflow soldering temperatures between 230°C and 240°C temperatures could be used for Sn–Ag–Cu.

8.4.1.2. Area of Spread/Wetting Angle Test. An additional test to the wetting balance test method is to determine the extent of spread of the solder on a board surface finish. This involves measuring the amount of spread of a fluxed solder pellet heated on a copper coupon usually with a determination of wetting angle between the solder and the coupon. The lower the wetting angle the more beneficial in terms of good wetting. A wetting angle of 0–10° for Sn–Pb is usually considered perfect wetting, with 10–20° excellent wetting, 20–30° very good wetting, and greater than 55° poor wetting. Too low a wetting angle (close to 0°) is not considered beneficial because the solder will tend to spread too thin on the board and component, thereby reducing the amount of solder forming the actual joint. The surface tension of Sn–Ag–Cu solder is higher than for Sn–Pb so the area of spread will be less with a corresponding increase in wetting angle. This does not mean it will have a reduced solder joint bond integrity with the board and component termination.

Vincent et al. [27] conducted work to measure the contact angle of fluxed Sn–Ag–Cu pellets on different coupon surface finishes. Wetting angles were lowest for a Sn–0.7Cu-coated coupon with an 18° wetting angle with Sn–Ag–Cu pellets compared with 17° with Sn–Pb pellets on the same surface finish. With immersion silver finish, the wetting angle with Sn–Ag–Cu solder was 24° compared with 13° with Sn–Pb solder. With a cleaned copper coupon the wetting angle was 43° with Sn–Ag–Cu compared with 12° with Sn–Pb solder. The wetting angles for Sn–Pb solder for all surface finishes were less than 20°, which can be considered excellent wetting, whereas for Sn–Ag–Cu solder the wetting angles were 18–43°, which can be considered from very good to adequate wetting.

Huang and Lee [8] evaluated spreading on OSP surface finish comparing Sn–Ag–Cu paste with Sn–Pb paste. The spreadability was measured in terms of the extent of spreading on a pad where there was only 30% of the pad printed with solder paste. Reflow profiles of liquidus plus 15°C and liquidus plus 30°C were used. In general, for both a 198°C and a 213°C peak Sn–Pb reflow profile, there was close to 100% solder coverage of the exposed OSP pad after Sn–Pb solder paste reflow. For the Sn–Ag–Cu solder paste for both a 232°C and 247°C peak reflow profile, there was approximately 50% solder coverage of the exposed OSP pad after Sn–Ag–Cu paste reflow.

Ludwig et al. [28] evaluated spreading of Sn–Ag–Cu paste compared to Sn–Pb paste on different surface finishes of OSP, Ni–Au and Ni–Sn. The spreadability of Sn–Pb solder paste was in general better than Sn–Ag–Cu for OSP and Ni–Sn surface finishes and similar to Sn–Ag–Cu solder paste reflowed on Ni–Au surface finish. The reflow peak temperature for Sn–Pb solder paste was 213°C and for Sn–Ag–Cu solder paste was 247°C. The spreadability of Sn–Ag–Cu solder paste was found to be slightly better with Ni–Au than OSP or Ni–Sn surface finishes after reflow.

Nowottnick et al. [29] evaluated spreading of Sn–3.5Ag–0.7Cu solder paste compared with Sn–36Pb–2Ag solder paste after screen printing and reflow soldering on a variety of surface finishes in a vapor-phase reflow oven. The vapor-phase oven temperature was 240°C for Sn–Ag–Cu and 203°C for Sn–Pb–Ag. With immersion tin, OSP and Ni–Pd–Au surface finishes, the Sn–Ag–Cu solder paste spread much less than with Sn–Pb–Ag. The immersion silver surface finish had slightly increased spreadability with Sn–Ag–Cu solder paste than with Sn–Pb–Ag. The spreadability of Sn–Ag–Cu solder paste increased on the surface finishes in the following order: OSP < Sn < Ni–Pd–Au < Ag.

8.4.2. Voiding

Voiding in a solder joint occurs during the solder reflow process. Large voids in a solder joint can lead to reliability issues. There are many theories as to why voids are created. One theory indicates that it is due to the difficulty in releasing out gassing flux material entrapped during the reflow process (Hance and Lee [30]). The primary concern for voids for the assembled components are for BGA/CSP components where these "covered" component joints are less likely to expel the void compared with a more uncovered component joint such as lead-frame devices.

Some of the factors influencing the incidence and amount of voiding include:

Flux material used in the solder paste

Reflow profile used

Solderability of the pad, component, and solder paste

Surface oxide for the solder paste particles

Reflow atmosphere used

For Pb-free solder pastes, there seems to be more incidence of voiding than for Sn–Pb pastes. Some of the reasons are that the flux materials used with Pb-free solder pastes are less developed and thus have not been optimized to reduce the amount of voiding. In addition the Sn–Ag–Cu pastes will be reflowed at a higher temperature than Sn–Pb pastes with an increased tendency to have more oxidized surfaces on the component, pad and solder particles. The solderability of Pb-free Sn–Ag–Cu is also less than Sn–Pb with a reduced tendency to wet to the surfaces of the component termination and board pad. The reflow atmosphere will help to offset the oxidation tendency at higher reflow temperatures and times and will help to reduce voiding. The surface tension of the Sn–Ag–Cu paste may also have a part to play in that its higher surface tension compared with Sn–Pb may increase the tendency for entrapped flux gases to remain within the solder joint.

The affect of reflow profile needs to be considered. Li et al. [31] found that there was an increased trend for voiding if the peak solder joint temperature was higher and the time over reflow was increased for Sn–Ag–Cu printed paste for flip-chip wafer bumping on a Ni–Au metallization. Voiding was decreased to zero from 95% when no Ni–Au underbump metallization was used on a silicon wafer. When an electroplated Sn–0.7Cu metallization was used there was again no voiding which indicates that the primary cause for voiding was the result of a less solderable surface potentially as a result of surface impurities.

For Sn–Pb pastes, the increased peak temperature also tends to increase the amount of voiding [32], especially when using lower boiling point flux solvents [33]. This is thought to be due to the fact that increased reflow temperatures and times tend to dry out the volatile flux solvent more readily than a cooler reflow. This will leave a more viscous flux remnant that will be more difficult to expel from the molten solder. The voiding situation with higher-temperature Pb-free Sn–Ag–Cu solders will likely be increased, especially with the use of existing flux solvents primarily developed for Sn–Pb solder pastes and lower Sn–Pb reflow temperatures.

Considering soak temperatures and times, a slower heating profile with a longer preheat stage may help to reduce the tendency for entrapped flux prior to paste coalesce [30]. Primavera [32] also conducted some work on Sn–Pb pastes to show that a longer soak time can help to reduce voiding.

Bath and Crombez [34] found that Sn–Pb BGA components tended to void more with Pb-free Sn–Ag–Cu paste than when using Sn–Pb solder paste. This is due to the Sn–Pb ball reflowing sooner than the Pb-free paste with an increased tendency for outgassed flux to remain in the molten solder ball. Work done with various Sn–Ag–Cu solder pastes has indicated that with the correct choice of flux material in the solder paste, voiding can be reduced but in general not to the levels associated with Sn–Pb solder pastes because more flux optimization development work is required.

As mentioned previously, the incidence of voiding can give rise to a reduction in mechanical integrity of the BGA/CSP joint. In general a small percentage (5–10%) of voids are not considered an issue [32]. Large voids or a collection of smaller voids at the board or component interface can be more problematic. In general, ATC failures in a BGA/CSP solder joint occur at the component side, where there can be a stress concentration introduced from the use of solder mask defined pads which induces a crack. One benefit of voiding is that it may actually enhance the reliability of a solder joint because it acts to reduce the crack propagation energy, in effect acting as a crack blunter.

In the work so far with Pb-free Sn–Ag–Cu pastes, although there has been voiding encountered to a greater extent than with Sn–Pb paste, the data so far indicate that the reliability has not been compromised compared with Sn–Pb and in some aspects may be enhanced (please see Section 8.10). The amount of voiding and the size of the voids and their position near the interfaces are all considerations on their effect on reliability, and further work is needed to define how these factors affect reliability.

8.4.3. Solder Joint Integrity

The affect of peak reflow temperature to give a visually acceptable and reliable solder joint is one of the relatively least explored areas for Pb-free Sn–Ag–Cu solder. As mentioned previously, there needs to be an adequate amount of time over the melting temperature of the solder in order to ensure that an intermetallic is formed, which is an indication of a reliable solder joint.

For Sn–Pb solder, which has a melting temperature of 183°C, the reflow profile would typically have a peak solder joint temperature of 200–220°C and time over the melting temperature of Sn–Pb of 30–90 seconds with 45–75 seconds over 183°C and 205–215°C peak, a typical time and temperature over reflow to aim for. This has been developed from manufacturing experience and guidelines from solder paste and oven manufacturers. If this was translated for Sn–3.9Ag–0.6Cu with a melting point of 217°C, the peak solder joint temperature would be 234–255°C and time over 217°C of 30–90 seconds. This would have two main significant changes:

(i) The need for solder paste to be developed to reflow at these higher temperatures
(ii) The need for components and boards rated to the higher soldering temperatures

Because of the cost implications of rating components to the higher reflow temperatures, various studies have and will be undertaken to determine if the minimum peak reflow temperature for Sn–Ag–Cu can lowered to give a visually acceptable and reliable joint. One of the main published works has been undertaken during the European Community "IDEALS" project [35]. Investigations were undertaken to determine the minimum solder joint temperature used for Sn–Ag–Cu solder paste that can give a visually acceptable and reliable solder joint.

The board used was epoxy FR4 62-mil thickness with QFP and chip resistors placed on the board. Assembly was done in air using an eight-zone convection oven using a high-solids containing Sn–Ag–Cu no-clean solder paste. Visual inspection was conducted on Pb-free pure-tin-coated 1206 chip and Ni–Pd or pure-tin coated QFP100 components using a company-based visual inspection standard, not the IPC610 visual inspection standard that would be considered to have a wider following in the industry. The minimum Sn–Ag–Cu solder joint peak temperature found to create a visually acceptable joint was 225–227°C with a minimum time over reflow of 25 seconds. With Sn–15Pb-coated QFPs, this could lower the peak temperature further as the lead in the lead frame would lower the melting point of the Sn–Ag–Cu. As we are migrating to Pb-free coatings for the terminations, it is wiser to consider the case for Pb-free terminations only when dealing with Sn–Ag–Cu solder paste. Because the 225°C is only 8°C above the melting temperature of Sn–Ag–Cu, an assumption was made that it would be more practical to consider a minimum peak temperature of 230°C for Sn–Ag–Cu (13°C ΔT), which would be equivalent to having a reflow peak temperature for Sn–Pb solder of 196°C.

Reliability data were obtained using different peak temperatures for the pure-tin-coated chip components reflowed with Sn–Ag–Cu paste. Reflow peak solder joint chip temperatures used were as low as 226°C with 29 seconds over 217°C and as high as 251°C for 57 seconds with a preheat temperature of either 160°C or 175°C for 100 seconds. Thermal shock from −25°C to 125°C for 3000 cycles (8 minutes per cycle) followed by shear testing showed no differences between Sn–Ag–Cu and Sn–Pb solder paste reflowed chip samples on immersion Sn or Ni–Au board surface finishes. No corresponding reliability data were obtained for the QFP soldered components. With data developed, there always needs to be caution attached to it if extrapolating to specific manufacturing circumstances with different fluxes used in the solder paste.

Work in the NEMI group with Pb-free CSP, PBGA, CBGA, and TSOP and 2512 chip resistor components soldered with Sn–Ag–Cu no-clean solder paste used peak reflow temperatures of 239–248°C and time over 217°C of 78–90 seconds reflowed in a nitrogen atmosphere [36]. For the NEMI reliability test builds [6] a 10-zone convection oven with a nitrogen atmosphere of less than 30 ppm O_2 was used for all builds. Table 8.3 shows the reflow measurements taken when building the reliability test hardware.

There was little difference in the manufacture of boards assembled with eutectic Sn–Pb and Pb-free Sn–Ag–Cu solder paste during printing, placement, and reflow apart from the increase in reflow temperature in the reflow oven for Pb-free.

The reliability results for Pb-free paste soldered with Pb-free components, which are developed in more detail in Chapter 6, indicated a reliability equivalent to or better than that of Sn–Pb control soldered components after ATC temperature cycling to 50% failure from 0°C to 100°C or −40°C to 125°C. Visual inspection of some of the Pb-free soldered joints [6] may not have strictly passed the visual inspection standards in terms of adequate wetting on the chip components requiring a rework hand soldering operation in manufacturing, but a reliable solder joint was made.

Work by Shina [14] on assembled boards with Sn–Ag–Cu no-clean solder paste showed that with a peak solder joint temperature of 235°C and time over reflow between 60 and 120 seconds, the pull test results for the SO14 Ni–Pd-coated soldered components were similar to or better than Sn–Pb paste soldered components. Pull testing was conducted before, during, and after thermal cycling up to 2000 cycles from 0°C to 100°C (1-hour cycles). The results were similar with air or

TABLE 8.3. NEMI Reliability Test Board Reflow Measurements for the Three Test Cells

Component Board Type	Specifications	Pb-Free (Sn–Ag–Cu) Paste/Pb-Free Component	Pb-Free (Sn–Ag–Cu) Paste/Sn–Pb Component	Sn–Pb Paste/ Sn–Pb Component
256CBGA	TAL (reflow time)	64–66 s	64–66 s	84–86 s
	Peak temperature	240–241°C	240–241°C	209–210°C
256PBGA	TAL	61–69 s	61–69 s	84–88 s
	Peak temperature	239–243°C	239–243°C	209–214°C
208 CSP	TAL	78–81 s	78–81 s	83 s
	Peak temperature	244–246°C	244–246°C	217°C
169 CSP	TAL	80 s	80 s	88 s
	Peak temperature	244°C	244°C	215°C
48TSOP	TAL	87 s	87 s	85–86 s
	Peak temperature	247°C	247°C	217–218°C
2512 resistor	TAL	90 s	90 s	87–88 s
	Peak temperature	247–248°C	247–248°C	217–220°C

Pb-free paste = Sn–3.9Ag–0.6Cu (mp 217°C).
Sn–Pb paste = Sn–37Pb (mp 183°C).
TAL, time above liquidus.

nitrogen reflow (20 ppm O_2) and with a linear or soak preheat on both Ni–Au and OSP surface finish boards.

Work by Romm et al. [37] using Sn–Ag–Cu no-clean solder paste soldered to Ni–Pd-coated SOIC20 components showed no significant difference in pull test force between joints soldered at a peak temperature of 235°C and 260°C for Sn–Ag–Cu both before and after thermal shock testing from −40°C to 125°C for 1000 cycles (10 minute cycles) in comparison with Sn–Pb–Ag paste soldered components. The Sn–Pb–Ag paste was reflowed at a solder joint peak temperature of 225°C with a reflow time of 65 seconds over 179°C. The 235°C Sn–Ag–Cu reflow profile had a preheat between 140°C and 180°C of 100 seconds with a reflow time of 45 seconds over 217°C, and the 260°C Sn–Ag–Cu profile had a preheat between 140°C and 200°C of 100 seconds with a reflow time of 75 seconds over 217°C. The board surface finish used was OSP and the nitrogen reflow atmosphere in the convection oven had an oxygen impurity level of 500–1000 ppm.

Additional pull test work by Romm et al. [23] using Sn–Ag–Cu soldered to Ni–Pd–Au- and Ni–Pd-coated SOP20, TQFP176, and TQFP208 using a peak temperature profile of 235–238°C compared with these same components soldered with Sn–Pb–Ag solder paste using a peak temperature profile of 225–228°C indicates no significant difference both before and after thermal shock testing for 1000 cycles from −40°C to 125°C (10 minute cycles). The board surface finishes were both OSP and Ni–Au and had no significance on the results. The same convection oven was used with nitrogen reflow atmosphere with an oxygen impurity level of 500–1000 ppm.

Work by Albrecht and Wilke [26] on assemblies soldered with Sn–Ag–Cu paste showed that pure-tin-coated 1206 chip components assembled at 235°C peak temperature had adequate shear force results up to 2000 cycles of temperature cycling from −40°C to 125°C (no real indication given of thermal cycling conditions used). Sn–Ag–Cu CSP components were also assembled with Sn–Ag–Cu paste with a 240°C peak temperature and reflowed in air. There were no reliability issues after thermal cycling from −40°C to 125°C for 1000 cycles for these CSP components.

Tanaka et al. [38] conducted work on the reliability of Pb-free Sn–Ag–Cu soldered pager boards using a minimum solder joint peak processing temperature of 230°C. They found that there were no issues in terms of reliability passing all product testing and long-term reliability test requirements.

From the results available so far, a minimum peak solder joint temperature for Sn–Ag–Cu appears to be around 230–240°C with 235°C being a good target solder joint peak temperature. More work needs to be done to define the minimum peak temperature on a fuller range of components using accelerated thermal cycling, other reliability test methods, and assembly trials. The minimum peak temperature and time over reflow must take into account the criteria for a visually acceptable, reliable, and inspectable solder joint. If the process window is made too narrow, then assembly operations will become more difficult to maintain at the production levels used for Sn–Pb soldering.

8.5. EFFECT OF REFLOW ATMOSPHERE ON SOLDERABILITY OF LEAD-FREE SOLDER

There have been relatively few papers comparing the solderability of Sn–Ag–Cu pastes using air and nitrogen reflow atmosphere in the reflow oven. The benefits of using nitrogen are that it helps to reduce oxidation of the solder paste, aiding the wetting of the solder to the board and component. The negative in its usage is its consumption and subsequent cost. With Sn–Pb soldering, there are variations across companies in electronics manufacturing regarding how much and what purity of nitrogen is used in reflow ovens. The use of nitrogen is not only dependent on the solder alloy used but on the type of flux used with the alloy.

With the increased movement to no-clean and less active solder pastes in electronics assembly, nitrogen is considered to aid solderability and in general improve the process window for Sn–Pb soldering. For Pb-free Sn–Ag–Cu, the wettability and process window is already reduced compared to Sn–Pb, so nitrogen would be considered an option to aid solderability and improve the process window.

Melton [39] found that Sn–3.5Ag solder had improved wetting times to pure-tin-coated chip resistors in nitrogen atmosphere compared with air. Hunt et al. [40] conducted experiments with Sn–Cu and Sn–Ag–Bi-based alloys and found that nitrogen benefited wetting during globule tests with solder pellets and flux. Nitrogen atmosphere levels of 5000 ppm O_2 were found to be beneficial with nitrogen atmospheres of 50 ppm O_2 further benefiting wetting in terms of allowing the use of lower reflow temperatures for the solder.

Tests conducted by Dong et al. [41] showed that the temperature at which a fluxless solder preform started to spread on a Ni–Au-coated substrate was related to temperature and nitrogen purity. For Sn–0.7Cu the solder preform spread at a temperature of 230°C with a nitrogen purity level of 10 ppm O_2 compared with a temperature of 245°C with a nitrogen purity level of 1000 ppm O_2. For Sn–3.5Ag, the solder spread occurred at 230°C with 10 ppm O_2 and 240°C with 1000 ppm O_2. No spread occurred for both alloys when a nitrogen purity level of 10,000 ppm O_2 was used which was the same result as with Sn–37Pb solder. If the results were estimated for Sn–Ag–Cu solder (melting point of 217°C), then solder spread would occur at

227°C with a nitrogen purity level of 10 ppm O_2
232°C with a nitrogen purity level of 100 ppm O_2
237°C with a nitrogen purity level of 1000 ppm O_2

Typically, nitrogen purity levels used now for a reflow oven would be from 50–500 ppm O_2. The amount of tin oxide film (Sn–O) formed in a separate experiment by Dong et al. [41] with Sn–0.7Cu, Sn–3.5Ag, and Sn–37Pb solder preforms in air at 140°C above the solder melting temperature was found to be of the same magnitude in terms of an initial thickness [20–30 Å (10^{-18} m)], but over time the thickness of the oxide film for Sn–Pb grew more than Sn–0.7Cu or Sn–3.5Ag.

Miric and Grusd [42] conducted reflow tests on QFP components assembled on boards with Sn–Ag–Cu paste and found that wetting was improved when using nitrogen atmosphere compared with air. Albrecht and Wilke [26] indicated that for Sn–Ag–Cu on assembled boards, nitrogen atmosphere would allow a minimum solder joint temperature of 230°C to be used on a board for good solder wetting whereas with air atmosphere the minimum solder joint temperature could be 240°C.

Shina et al. [13, 14] found that test boards assembled with Sn–3.8Ag–0.7Cu no-clean paste had less soldering defects when using nitrogen atmosphere (20 ppm O_2 impurity level) in terms of solder joint fillets, wetting, bridging, and solder balls than when reflowing in air atmosphere using a 10-zone convection oven. These results followed the same trend as with Sn–Pb solder paste, although the number of visual defects with Pb-free Sn–Ag–Cu paste was increased. The board surface finishes used were OSP and Ni–Au. The Pb-free components used were pure-tin-coated 0402, 0805, 1206 chips and Ni–Pd-coated LQFP120, LQFP100, SO14, and SO16. The minimum pull strength result recorded for Pb-free Sn–Ag–Cu-soldered Ni–Pd-coated components before and after 2000 cycles of ATC testing (0–100°C) showed no difference in value for components assembled in air or nitrogen reflow atmosphere [14].

For Pb-free Sn–Ag–Cu-based solders, which have higher surface tension and are slower to wet than Sn–Pb, the development in the fluxes used will help to increase the process window for Pb-free soldering. Developments of no-clean fluxes in Pb-free solder pastes which have increased flux activity have to be balanced by the fact that the flux activity cannot be excessive to cause reliability issues from the no-clean flux residues remaining on the board. Nitrogen atmosphere will be one option to improve the process window somewhat for Pb-free solders, especially when newer flux formulations are still being developed.

The decision as to the use of nitrogen may be determined during production evaluations for the boards to be assembled with Pb-free. The purity level of the nitrogen may play a role in determining whether it is used for Pb-free soldering due to cost. If lower purities of nitrogen are used (100–1000 ppm O_2), the flow rates and control of the purity of nitrogen through the reflow oven could also be more difficult to control.

8.6. CONVECTION VERSUS IR REFLOW OVENS

KAREN WALTERS

8.6.1. Heat Transfer Via Forced Convection

In forced convection reflow ovens, gases are typically electrically heated and pressurized in plenums via a blower motor or fan and forced through orifice holes onto the product. Factors that affect heat transfer in a convection oven are film coefficients, gas flow rates, surface area, and changes in temperature between the heated gas and product in the oven. As temperature set points increase within the oven the convection rate becomes limited and hence, the heat transfer efficiency goes down. A means to counter this effect is closed loop convection control in multiple zones to maintain constant convection rates independent of temperature. This feature promotes more efficient heat transfer between the furnace gas and the product.

8.6.2. Heat Transfer Via an Infrared Radiation

In an infrared (IR) radiation oven, energy transfer to the product consists of thermal energy being transferred across an open space between two bodies of different temperature via electromagnetic waves. All radiant heat transfer is a function of infrared wavelengths. The factors that affect heat transfer in IR furnaces include emissivity of the heater and product, surface area, Stephan–Boltzman's constant as applied to the difference in temperature between the radiating element or heat source, and product in the oven [43]. The emissivity of a product can be an issue with an IR oven operation. Very dark surfaces have emissivity closer to 1 and, hence, absorb or emit more energy than a shiny or light color object whose emissivity is closer to 0. When a product contains both high- and low-emissivity areas, they will heat at different rates and can lead to variances in product temperature.

8.6.3. Infrared Radiation Versus Forced Convection Oven Comparison

In forced convection reflow, the temperature profile of the product depends primarily upon its mass, the mass flow rate, and the temperature of the process gas.

However, where IR radiation is the primary means of heat transfer, there are more factors that can play more of a role in a product's thermal profile. These factors include sensitivity toward uneven thermal mass distribution, difference in color and material type of board or parts, and component shadow effects. As a result, a larger temperature gradient can exist across a product in a radiant furnace. Therefore, it is more probable that hot spots will develop in some areas, resulting in potential heat damage to components or the substrate itself; and in other locations, cold spots may develop, potentially resulting in cold solder joints or poorer wetting. In general, for an IR furnace a larger reflow process window is required to obtain acceptable solder interconnects. Forced convection ovens are generally more efficient in the temperature range where reflow is accomplished and hence give a tighter process control.

With Pb-free solders, the process window will likely be reduced and more tighter temperature uniformity will become more important, especially with the higher-temperature excursions that the board will likely see and its detrimental affect on the components. For larger products with more uneven mass distribution, Pb-free solder reflow will become an increasing challenge. Previously, a soak was introduced into the profile on larger boards to achieve component temperature equilibrium prior to elevating component leads to peak temperature. For Pb-free soldering, the optimized profiles so far have been a straight ramp-to-peak temperature. This profile would help to prevent burn-off of the flux activators because the Pb-free solder pastes tested typically used fluxes in the paste which were rated to Sn–Pb soldering temperatures and not the elevated temperatures used for Pb-free soldering.

In effect, for Pb-free soldering, an efficient oven will be required that is not sensitive to color, board materials, thermal mass distribution, reflectivity, and component shadowing. The furnace must produce tight thermal uniformity in order to meet Pb-free profile requirements. A means to ensure constant convection rates independent of process temperature is a key benefit, if not a requirement, to reflowing Pb-free solders. Forced convection ovens tend to produce tighter temperature uniformity.

All convection ovens accomplish some level of heat transfer via radiation, be it direct or indirect. Generally, forced convection ovens that possess the lowest levels of radiant heat transfer also tend to yield products with the lowest temperature gradient. During developments of Pb-free products which consist of larger boards with large temperature variations across the board, it has been suggested that in order to reduce the ΔT across the board, IR radiation could be used in conjunction with convection heating [44]. Aluminum electrolytic capacitors, which are temperature-sensitive, tend to be more heat conductive, and increases in temperature of this component during reflow can be rapid. By using the fact that the aluminum could actually reflect IR heat, a combination of convection and infrared heating could be used to keep the temperature-sensitive components such as electrolytic capacitors lower in temperature while increasing the temperature of adjacent components and the board.

Work carried out by Baggio et al. [45] on a medium-size eight-layer FR4 board with thickness of 1.6 mm (63 mils) and board dimension of 23 cm (9 in.) × 30.5 cm

(12 in.) indicated that a combination of hot air convection and IR technology coupled with thermal masking material to distribute the heat more evenly could reduce the temperature variation across a board from 40°C (min temperature 213°C, max temperature 253°C) to 21°C (min 215°C, max 236°C) with an Sn–Ag–Bi (melting point of 210°C) soldered assembly. Ovens with this type of capability at this stage are developmental, and significant demand would be needed to make these ovens used broadly across the global industry.

8.7. REFLOW TEMPERATURE DELTA ON BOARDS AND COMPONENTS

One of the biggest concerns for Pb-free soldering is the higher reflow temperatures used during processing, which increases the temperatures at which the board and components are subjected.

To determine the extent of the Pb-free reflow temperatures on the boards and components will be subjected, there have been increasing studies on production boards.

The NEMI Pb-free component group conducted a study on reflow temperatures during Pb-free soldering to cover the following issues:

1. What is the ΔT across a typical circuit board and the components?
2. What are the typical processing tolerances for reflow ovens?

The "practical" peak temperature that could be used for a good cross section of products was defined as the minimum required solder joint temperature + ΔT (board and components temperature range) + process tolerance + measurement error. {The ΔT for board and components would be different for every reflow oven and board combination.}

Several solder paste manufacturers indicated for Sn–Ag–Cu the minimum solder joint temperature was 232°C to 235°C for an actual process, for the current generation of fluxes. There are those who indicate it could be as low as 225°C to 226°C, but this clearly would not be a uniformly accepted limit in a large-scale process with large board assemblies. To assess the impact that reflow equipment had on ΔT, two oven manufacturers ran experiments.

One oven manufacturer determined the maximum ΔT across a circuit board assembly for two product boards and one test board. Board S was 92 mils thick with dimensions 18 × 12 in., and the thermal mass of the board was 1.1 kg. Board I was 60 mils thick with dimensions 12 × 8 in. and the thermal mass was 0.5 kg. Board H (test vehicle) was 93 mils thick with dimensions 12 × 18 in. and 1-kg thermal mass. The initial efforts were based on the premise that 240°C was to be the target maximum peak temperature. The best ΔT achieved, with an oven that had closed-loop convection control, was in the order of 9 to 10°C as shown in the Table 8.4. This feature maintained constant convection rates independent of temperature, voltage, and atmospheric pressure variations.

TABLE 8.4. ΔT Temperature for the Three Boards

Board Tested	ΔT Achieved
Board S	10°C (estimated value with fixture)
Board H (test vehicle board)	9°C
Board I (motherboard)	9°C

The best ΔT achieved for Board S with a Pb-free profile would be 10°C for a fixtured and fully populated assembly with a reduction in conveyor speed from an Sn–Pb profile of around 20%. The best ΔT for the Board I motherboard would be 9°C for a fully populated assembly. The best ΔT for Board H test vehicle would be 9°C for a partially populated assembly. A maximum ΔT of 3°C was noticed between the component body and Pb–solder joint during profiling for Board I with a 10-zone oven.

Several DOEs were run to test the product orientation and profile types. The best results were obtained by reducing the conveyor speed, which had the effect of allowing more time for the larger thermal mass parts to reach the desired temperature and in some cases (Board S and I) to change the orientation of the board so that the shorter side would be parallel to the direction of the conveyor belt [46–48]. Linear versus soak profiles were found to have minimal affect on reducing ΔT.

Work to determine process and thermocouple tolerances was conducted and results shown in Table 8.5 [49].

The reflow peak temperatures must be increased to allow for the total tolerance listed in Table 8.5. The furnace repeatability values will be unique to each board assembler's equipment. A value of 7°C would need to be added to the reflow profile stack-up to determine actual set points desired.

Inputting the three temperatures that were established into the reflow temperature equation yielded the following results: 235°C {minimum joint temperature} + 10°C ΔT {board and components temperature range} + 5°C {process tolerance} + 2°C {measurement error} = 252°C. Basically, there is a 17°C ΔT between the minimum solder joint temperature and the maximum solder joint temperature across a board.

From this work, we can see the importance of understanding the reflow profile and profiling the board to find the coldest and hottest joints to determine ΔT. The narrower the ΔT of the solder joints on the assembly, the lower the peak temperature

TABLE 8.5. Reflow Temperature Tolerances

	Typical	Worst Case
Furnace repeatability	±0.4	±0.6
Furnace to furnace	±1.5	±2.3
Load vs. no load	±1.4	±1.8
Thermocouple	±1.1	±2.2
Total	±4.4	±6.9

needs to be. Every assembly/reflow oven combination will be unique. Detailed temperature profiling for each product would allow the lowest temperature profile to be used, reducing the thermal stress on the components.

Assemblers would still want and need as much process window as possible and as such would prefer the components be qualified to Pb-free temperatures (260°C). But the component manufacturers are concerned with the component's integrity if subjected to the hotter profiles and the cost of implementation of components rated to higher temperatures. As development moves forward, it is possible that more robust technologies may present themselves and higher temperature capability will be possible.

In a separate study, another oven manufacturer conducted Pb-free profiling work from a different perspective with a 10-zone convection oven [50]. The impact of various Pb-free reflow profiles was studied for the following five scenarios: line speed and throughput, line speed and ΔT, profile type (shape) and ΔT, time above liquidus and ΔT, and power consumption of Pb-free versus Sn–Pb baseline. This work was done using Board IK motherboard. The board was 60 mils thick with dimensions 13 × 16 in. and thermal mass 1.25 kg.

These scenarios attempted to minimize the ΔT for each situation. The ΔT between large and small components on the board was 20°C for the Pb-free profile. Using four zones instead of three for reflow, the ΔT was reduced to 16°C from 20°C. By reducing conveyor speed by 22%, the ΔT was reduced to 11°C. By using a ramp–soak–spike profile instead of the straight ramp profile, the ΔT was reduced to 16°C. The orientation of the board through the convection oven was found to have minimal affect on reducing ΔT.

Considering again the minimum solder joint temperature of 235°C, the range of the temperatures across a board is 11–20°C ΔT {board and components temperature range} + 5°C {process tolerance} + 2°C {measurement error} = 18–27°C. Thus the peak temperature for the smallest component could be between 253°C and 262°C. Achieving an optimized Pb-free profile is totally dependent on the equipment used and the circuit board and components being processed and the ability of the process engineer to define and develop the optimal reflow profile. The ability to optimize all the control parameters would have a direct impact on processing Pb-free assemblies. The trends presented should aid in making the compromises necessary to obtain a profile with the minimum ΔT.

The oven power consumption comparison suggested that a maximum increase in the range of 11.5% would be required for Pb-free reflow using a ramp–soak–spike reflow profile from a straight ramp profile. There could be a reduction in power consumption of 1% compared to the Sn–Pb baseline if the conveyor speed was reduced by 22% for Pb-free reflow. However, cycle time or throughput reduction can be expected for the Pb-free profiles with reduced line speed. An increase in power consumption of 4% was found from the Sn–Pb baseline to the Pb-free baseline profile and when moving from a Sn–Pb four-zone reflow profile to the Pb-free four-zone reflow profile.

Factors such as the total number of oven zones can affect the ability to reduce ΔT across a board. In work by Johnson et al. [19], the minimum solder joint Pb-free temperature on the board was 233°C with ΔT of 10°C using a six-zone reflow oven. By having additional zones, the ΔT was reduced to 7°C for this double-sided automotive controller board with dimensions 5.5 × 7.5 in. During second-pass (topside) reflow of this board, the less densely populated bottom-side components reached 241°C.

In a separate component temperature study for Sn–Pb and Pb-free assemblies, four production boards were selected to cover a large range of components and board sizes [51]. The four FR4 epoxy boards selected were 36 mils, 48 mils, 78 mils, and 92 mils thick with OSP surface finish.

The 36-mil-thick-disk drive board was 5×4-in. dimensions with components ranging from 292I/O PBGA, 32-lead connector, TQFP100, SOIC8, SOT-23, and 0603 components. The 48-mil-thick double-sided laptop board had dimensions of 8 × 8 in. with components ranging from 433 I/O PBGA, TQFP144, PLCC68, TSOP56, 136-lead connector, SC-70, and 0603 chips.

The 78-mil-thick server board had dimensions of 20 × 15 in. with components ranging from 1657 CCGA, 331 I/O PBGA, PQFP160, PLCC44, TSOP54, SOT223, and 0805 components. The 92-mil-thick server board had dimensions of 10 × 6 in. with components ranging from 1247 CCGA, 153 I/O PBGA, LQFP100, PLCC44, TSSOP8, SOT223, and 0805 chips.

The minimum solder joint temperature for Pb-free Sn–Ag–Cu soldering was kept between 232°C and 235°C based on internal company development work which indicated this was the minimum temperatures to use for Sn–Ag–Cu while still producing a reliable solder joint.

For the 32-mil thick disk drive board, the ΔT during a Sn–Pb profile was 9°C between a PBGA292 (213°C) and TQFP48 (222°C) solder joint. For the Pb-free Sn–Ag–Cu reflow profile the temperature ΔT was 6°C between the same PBGA (233°C) and TQFP (239°C) solder joint. There was a difference of 3°C between the PBGA joint and component body during Sn–Pb processing, which was only 1°C during Pb-free Sn–Ag–Cu processing.

For the 48-mil thick laptop board, the ΔT during an Sn–Pb profile on the bottom side of the board (first pass) was 2°C between a PBGA241 (205°C) and TSSOP56 (207°C) solder joint. For the Pb-free Sn–Ag–Cu reflow profile the ΔT was 4°C between the same PBGA (234°C) and TSSOP (238°C) solder joint. There was a difference of 6°C between the PBGA joint and component body during Sn–Pb processing, which was 10°C during Pb-free Sn–Ag–Cu processing (component body hotter than solder joint). This large difference between the solder joint and component body temperature was related to the type of "closed" surface-mount fixture used (for paste in hole processing), which mainly shielded the board from being heated from one side, allowing heating by air in direct contact with the reflow side and little heat transfer through the board itself. The SC-70 component package (package volume of only 2.3 mm^3) reached 253°C and the board reached 246°C on the bottom side of this board during Pb-free processing, which highlighted the need for temperature mapping of Pb-free production boards to ensure that temperature limits of the parts are adhered to during processing.

For the 78-mil-thick server board, the ΔT during an Sn–Pb profile was 6°C between a CCGA626 (206°C) solder joint and the board (212°C). For the Pb-free

Sn–Ag–Cu reflow profile the temperature ΔT was 4°C between the same CCGA (233°C) solder joint and the board (237°C).

For the 92-mil-thick server board, the ΔT during an Sn–Pb profile was 11°C between a CCGA1247 (207°C) solder joint and the board (218°C). For the Pb-free Sn–Ag–Cu reflow profile the ΔT was 6°C between the same CCGA (232°C) solder joint and the board (238°C).

The results indicate that it is not necessarily the largest and thicker boards which have the largest spread of temperatures during Pb-free processing. The fixtures required for the 48-mil-thick laptop board for the paste-in-hole process caused larger ΔT across the board [19°C for Pb-free bottom-side (first pass), 14°C for Pb-free topside (second pass)] than for 36-mil-thick (6°C ΔT), 78-mil-thick (8°C ΔT), or 92-mil-thick boards (6°C ΔT), which used more open surface-mount fixtures. Assuming that the minimum solder joint temperature for the Sn–Ag–Cu is 235°C, then the maximum Pb-free solder joint temperature can range from 241°C to 254°C for the boards tested during surface-mount reflow processing.

The results also indicate that component body temperatures during Pb-free processing can be grouped into two package volume categories as with Sn–Pb soldering, those being $\geq$350 mm^3 termed as "large" components and $<$350 mm^3 termed as "small" components for the purposes of component temperature qualifications. From the results, component packages remained below 245°C (227–244°C) for large-sized components $\geq$350 mm^3 package volume and around or below 250°C (233–253°C) for small-sized components $<$350 mm^3 package volume. It should be borne in mind that each product board is unique, and it is necessary to ensure that thermal profiling is done on Pb-free boards to ensure that these specific temperatures are not exceeded during reflow soldering.

8.8. VISUAL INSPECTION OF LEAD-FREE SOLDERED JOINTS

The purpose of this inspection is to detect manufacturing nonconformities by visual assessment of the solder joint appearance—that is, bridging, insufficiency, misalignment, opens, non-wetting. Recorded details may include visible soldering defects, flux residues, and general appearance of solder joints for components in the assembled and/or reworked condition.

The criteria for inspecting visual defects is usually by industry standards such as IPC610. Shiny/dull joints are acceptable but reduced wetting is not. The relevant IPC standards have been under review to determine how to incorporate Pb-free inspection criteria with the release of a new standard revision.

The Pb-free solder joints can be slightly duller when compared with Sn–Pb soldered joints and the solder wetting/spreading for Pb-free soldered joints can be reduced compared to Sn–Pb soldered joints. For the NEMI reliability test board builds [6], the six different reliability test boards used were:

1. CSP169 test vehicle
2. CSP208 test vehicle

3. TSOP48 test vehicle
4. 2512 chip test vehicle
5. PBGA256 test vehicle
6. CBGA256 test vehicle

For each component test vehicle, there were three distinct test cells built:

1. Sn–Pb paste boards with Sn–Pb component
2. Pb-free Sn–3.9Ag–0.6Cu paste (NEMI alloy) boards with Sn–Pb component
3. Pb-free Sn–3.9Ag–0.6Cu paste boards with Pb-free component

For the CSP169 and CSP208 soldered joints there was not an obvious visual difference between Sn–Pb paste soldered components and Pb-free paste soldered components across the three distinct test cells (Figure 8.4). For the PBGA256

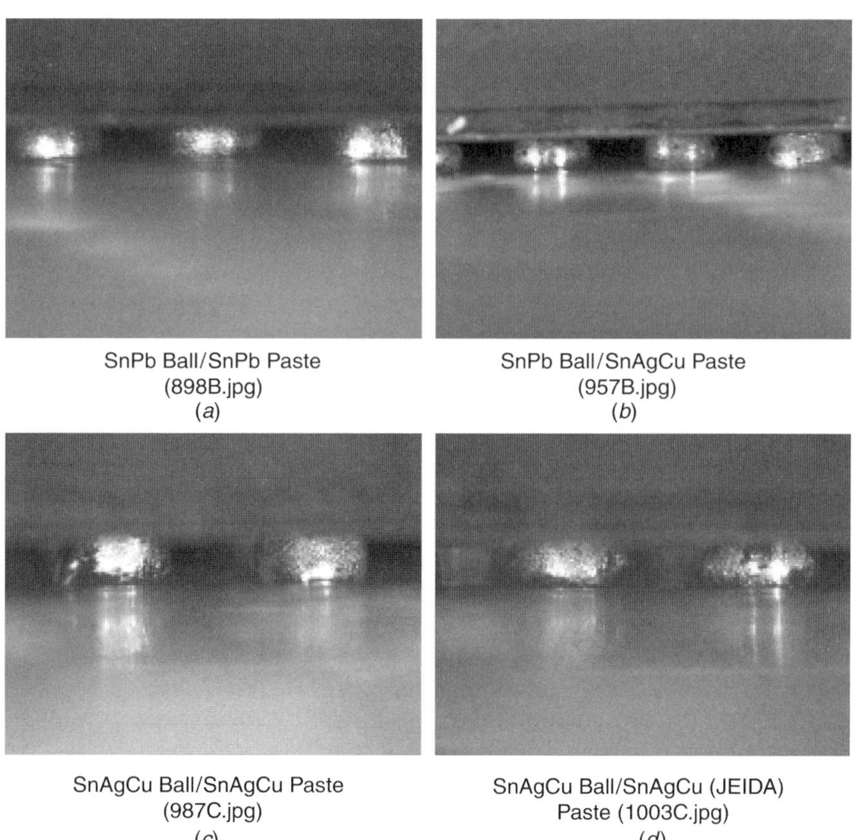

Figure 8.4. CSP208 visual appearance.

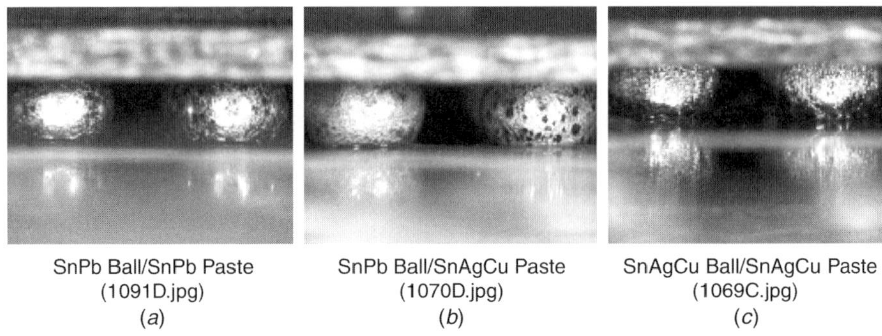

SnPb Ball/SnPb Paste (1091D.jpg) (*a*)

SnPb Ball/SnAgCu Paste (1070D.jpg) (*b*)

SnAgCu Ball/SnAgCu Paste (1069C.jpg) (*c*)

Figure 8.5. PBGA256 visual appearance.

boards (Figure 8.5), the Pb-free Sn–Ag–Cu paste soldered Sn–Ag–Cu BGA component showed a rougher solder ball surface appearance compared to the other two PBGA test cells.

For the CBGA boards (Figure 8.6), the Sn–Pb paste with Sn–Pb–Ag CBGA component showed shiny solder joints, which was expected. The Pb-free Sn–Ag–Cu paste with Sn–Pb–Ag CBGA component showed dull solder joints. The Pb-free Sn–Ag–Cu paste with Pb-free Sn–Ag–Cu CBGA component showed solder joints with a cratered appearance.

For the TSOP48 soldered joints there was not an obvious difference between Sn–Pb paste soldered components and Pb-free paste soldered components (Figure 8.7). For the Sn–Pb paste soldered Sn–Pb-coated 2512 resistor and the Pb-free Sn–Ag–Cu paste soldered Sn–Pb-coated 2512 resistor good wetting was observed on the chip resistors (Figure 8.8), whereas for the Pb-free Sn–Ag–Cu paste soldered to a pure Sn component, reduced wetting was observed.

In summary, the major difference that was noticed from the builds between Sn–Pb and Pb-free Sn–Ag–Cu paste assembled boards were indications of different visual solder appearance for Pb-free soldered components compared with Sn–Pb soldered components, especially CBGA, PBGA, TSOP, and resistor components.

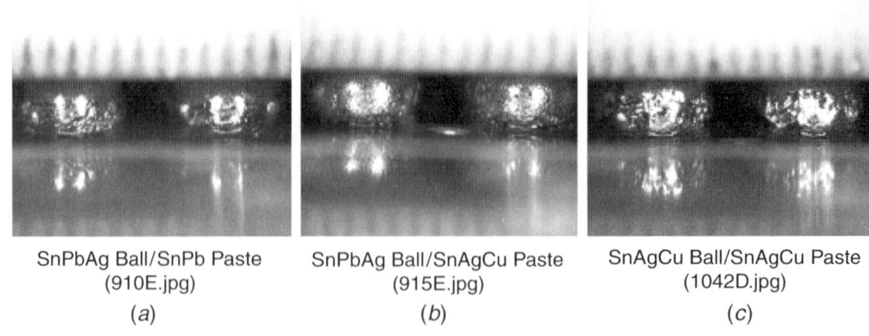

SnPbAg Ball/SnPb Paste (910E.jpg) (*a*)

SnPbAg Ball/SnAgCu Paste (915E.jpg) (*b*)

SnAgCu Ball/SnAgCu Paste (1042D.jpg) (*c*)

Figure 8.6. CBGA256 visual appearance.

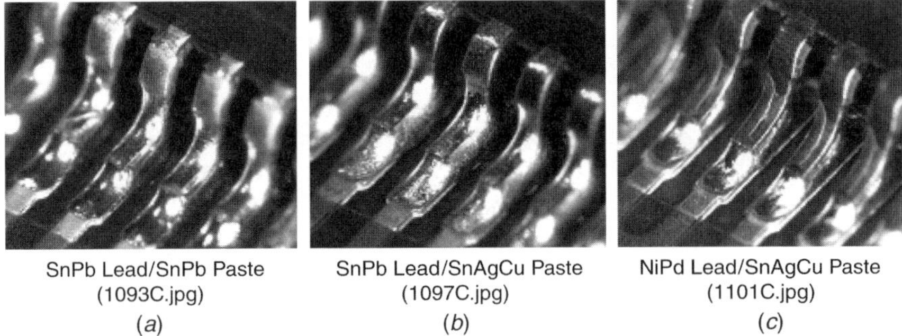

SnPb Lead/SnPb Paste (1093C.jpg) (*a*) | SnPb Lead/SnAgCu Paste (1097C.jpg) (*b*) | NiPd Lead/SnAgCu Paste (1101C.jpg) (*c*)

Figure 8.7. TSOP48 visual appearance.

SnPb Termination/SnPb Paste (872H.jpg) (*a*) | SnPb Termination/SnAgCu Paste (971C.jpg) (*b*) | Sn Termination/SnAgCu Paste (974G.jpg) (*c*)

Figure 8.8. 2512 resistor visual appearance.

This difference in visual appearance did not affect the reliability of the test boards; but during manufacturing, this may cause operators to rework more components, especially in the transition period from Sn–Pb to Pb-free soldering.

8.9. AUTOMATED OPTICAL INSPECTION (AOI)

Since the Pb-free soldered joints may be dull, less wetted, less smooth, and potentially more cratered when compared with Sn–Pb soldered joints, the same AOI criteria used for Sn–Pb soldered joints may not be used for solder joint quality. But when possible the system can be taught separately for Pb-free and Sn–Pb for good/marginal/bad joint quality. The NEMI AOI work [52] was conducted on soldered TSOP and 2512 chip resistors to determine what affect the potential difference in appearance of Pb-free soldered joints may have compared to Sn–Pb soldered joints in relation to inspection criteria, methodology, and shape of Pb-free versus Sn–Pb soldered joints.

Figure 8.9. Sn–Ag–Cu paste soldered pure-tin-coated 2512 chip resistor.

For the Sn–Ag–Cu paste-soldered Sn–Pb resistor components, the solder joint appearance was similar to the Sn–Pb paste soldered Sn–Pb resistors. For the Sn–Ag–Cu paste soldered pure-tin resistors, there was more evidence of differences in wetting at the solder joints. There was evidence of a lack of wetting on the board pad (Figure 8.9), and some evidence of surface cracking at the solder joint which may be due to the way the solder joint cooled. The ATC reliability results for these boards which was conducted later did not give any indication that there was any mechanical integrity issues with the Pb-free soldered joints even though the wetting/spreading of the soldered joints may have been reduced.

For the TSOP48 boards, comparing the three test cell solder joint appearances (Sn–Pb paste with Sn–Pb component, Sn–Ag–Cu paste with Sn–Pb component, Sn–Ag–Cu paste with Pb-free component), there was little difference noticed in the solder joint shapes. There was some evidence of bad wetting on the Sn–Pb coated components mainly in the Sn–Pb paste/Sn–Pb component test cell due to exposed copper at the lead-frame toe. There was no real evidence that the reduced wetting at the toe of the some of the Sn–Pb coated lead frames was causing a reliability issue from subsequent ATC thermal cycling studies. There was some evidence of a slightly rougher solder joint shape with the Sn–Ag–Cu paste soldered components (Figure 8.10) compared with Sn–Pb paste soldered components (Figure 8.11).

This study was used to determine what the visual appearance of Pb-free soldered joints was in relation to Sn–Pb and if there was a requirement to reprogram AOI equipment for Pb-free soldered joints. The study highlighted that there were some differences in visual appearance across the test cells for the resistor and TSOP boards. A greater variety of soldered lead-frame and chip components would need to be tested to truly evaluate the differences between Sn–Pb soldered and Pb-free soldered components. With the increased variety of solder joint shapes seen with Pb-free solder paste especially when having a mix of Pb-free and Sn–Pb-coated components on the board, this could potentially cause more reprogramming requirements for AOI systems.

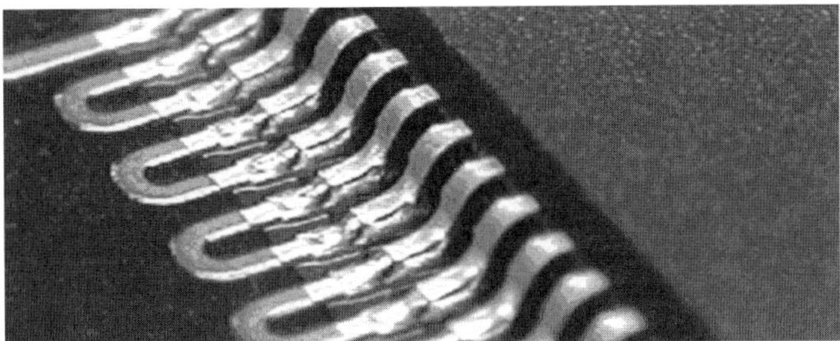

Figure 8.10. Sn–Ag–Cu paste with Pb-free Ni–Pd-coated TSOP.

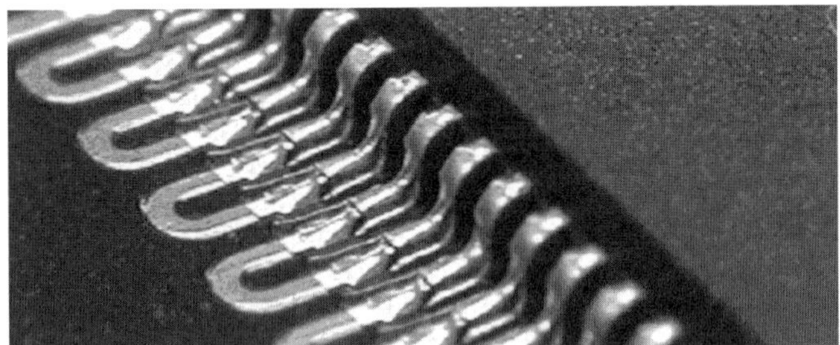

Figure 8.11. Sn–Pb paste with Sn–Pb-coated TSOP.

8.10. X-RAY INSPECTION OF LEAD-FREE SOLDERED JOINTS

8.10.1. Manual X-Ray Inspection

Manual X-ray inspection is typically performed using radiography or transmission X-ray technology. The wetting quality of solder joints will impact the X-ray images, but dull-looking solder joints have minor impact on the X-ray image quality. Manual X-ray inspection may have very small impact on the images taken on Pb-free and Sn–Pb soldered joints.

8.10.2. Automated X-Ray (Laminography/Cross-Sectional) Inspection

The purpose of this inspection is to quantitatively assess solder joint quality using the X-ray laminography/cross-sectional slice technology. Numerous measurements such as solder volume and fillet length are reported for each solder joint. From these measurements, the defects are detected and reported. This technology is especially well-suited for detecting hidden joint defects such as area array packages, where visual inspection has limited access.

The samples are exposed to the X-ray system and inspection performed on the programmed solder joints. The solder joints are pronounced good/marginal/bad based on the thresholds taught to the system in the program development stage.

The wetting quality of solder joints will impact the X-ray images, but dull-looking solder joints have minor impacts on the X-ray image quality. Although it is not obvious, the coefficient of absorption of the Pb-free alloys are different from that for Sn–Pb to potentially create difference in X-ray images. The corresponding calibration coupons can be used to teach the X-ray system. The NEMI reliability test assembled boards were inspected by X-ray laminography [53].

The Pb-free assembled boards had only slightly adjustments needed in the equipment settings. The information regarding adjustments needed could be requested directly from the X-ray machine manufacturer. Fazzio [54] showed that the stopping power of the high-Sn-containing Pb-free solders such as Sn–3.2Ag–0.5Cu was about 88% of that for Sn–37Pb. The stopping power indicates the relative amount of X-rays that are attenuated in the material. Thus the images of solder joints with the higher Sn-containing solders will have slightly less contrast than Sn–Pb, but in practice there will be little or no difficulty imaging the Pb-free soldered joints.

The X-ray inspection equipment takes X-ray "slices" at two or three levels for a CSP/BGA component (center solder joint solder, board side solder joint slice, component side solder joint slice) to inspect for X-ray defects such as voiding, insufficient solder, and X-ray diameter measurements. For TSOP/2512 chip-soldered components, one X-ray "slice" was taken to inspect for (a) X-ray defects such as insufficient solder and (b) X-ray measurements such as solder heel and solder joint thickness.

Typically, X-ray inspection analysis at the time of inspection measured voiding with a 20% area criteria based on internal X-ray development work and void diameter measurements for a minimum of 10-mil diameter based on repeatability and accuracy of measurement limits for the equipment. During the X-ray inspection analysis, general observations were that most of the solder joints inspected and passed by electrical continuity testing after assembly were passed as good solder joints during X-ray inspection.

Overall, there were more X-ray defects observed in moving from Sn–Pb paste soldered components to Pb-free Sn–Ag–Cu paste soldered components. These X-ray defects were mainly in the form of increased voiding for Pb-free Sn–Ag–Cu paste with Sn–Pb or Pb-free BGA/CSP components and some indications of reduced wetting with the Pb-free Sn–Ag–Cu paste soldered Pb-free TSSOP component combination.

The voiding was attributed to the use of the specific flux in the Pb-free solder paste, the reduced solderability of the Pb-free Sn–Ag–Cu alloy compared with Sn–Pb, and the elevated temperatures used during Pb-free solder paste processing. More development work would be needed to optimize the fluxes in the Pb-free Sn–Ag–Cu solder paste to operate at the higher reflow temperatures. Improvements are being made over time. The reduced wetting with Pb-free Sn–Ag–Cu paste with Pb-free TSSOP components was more related to the reduced solderability of the Sn–Ag–Cu solder alloy compared with the Sn–Pb solder alloy which was already observed in the Pb-free solder paste evaluation development work for Pb-free Sn–Ag–Cu solder paste on OSP board surface finish compared with

TABLE 8.6. Thickness Measurements of Sn–Pb and Pb-Free Soldered PBGA Component Joints by X-Ray Analysis

NEMI PBGA256 X-Ray Measurement	Sn–Pb Paste		Sn–Ag–Cu Paste		Sn–Ag–Cu Paste
Slice 1 midball	**Sn–Pb PBGA**	**Slice 1 midball**	**Sn–Pb PBGA**	**Slice 1 midball**	**Sn–Ag–Cu PBGA**
Average diameter (mils)	32.82	Average diameter (mils)	33.17	Average diameter (mils)	32.83
St dev	0.8	St dev	0.6	St dev	0.58
Min	29.54	Min	25.17	Min	29.54
Max	35.16	Max	35.51	Max	37.49
Slice 2 pcb pad		**Slice 2 pcb pad**	**PBGASACPB**	**Slice 2 pcb pad**	**PBGASAC**
Average diameter (mils)	24.78	Average diameter (mils)	26.57	Average diameter (mils)	26.21
St dev	0.73	St dev	0.65	St dev	0.58
Min	22.06	Min	21.62	Min	22.89
Max	27.88	Max	29.52	Max	30.75
Slice 3 component pad		**Slice 3 component pad**	**PBGASACPB**	**Slice 3 component pad**	**PBGASAC**
Average diameter (mils)	28.24	Average diameter (mils)	27.5	Average diameter (mils)	27.28
St dev	0.52	St dev	0.56	St dev	0.54
Min	25.89	Min	22.98	Min	24.45
Max	30.28	Max	30.21	Max	31.67

TABLE 8.7. X-Ray Defect Summary (Percent of Total Solder Joints with a Specific Defect)

	Test Cell (Paste/Component)		
Component Type	Sn–Pb/Sn–Pb	Sn–Ag–Cu/Sn–Pb	Sn–Ag–Cu/ Sn–Ag–Cu
256CerBGA	No defects (0)	22% V (−)	3% V (+)
256PBGA	0.1% In (0)	4% V (0)	3% V (0)
208 CSP	0.2% In (0)	0.5% In, 0.8% V (0)	5% V, 0.2% In (+)
169 CSP	5% V (0)	18% V (+)	8% V, 1% In (+)
48TSSOP	No defects (0)	0.5% In (−)	3% In, 0.3% Ex (0)
2512 chip	No defects (0)	No defects (0)	No defects (0)

In, insufficient solder; Ex, excess solder (bridging).
V, voids evident >20% of the area of the solder joint and minimum void diameter of 10 mils at board level solder joint slice or center of ball.
V, voids with minimum diameter of 10 mils at board level solder joint slice or center of ball.
(O), ATC testing result equivalent to Sn–Pb paste soldered Sn–Pb component.
(−), ATC testing result worse than Sn–Pb paste soldered Sn–Pb component.
(+), ATC testing result better than Sn–Pb paste soldered Sn–Pb component.

Sn–Pb [4]. The solder ball diameters measured and solder joint thicknesses were similar across a specific component type for the 3–4 test cells within the component group. Table 8.6 shows the results for the PBGA soldered component with measurements taken at the PCB side, the middle of the solder sphere and close to the component pad side.

There were indications of reduced wetting with Pb-free Sn–Ag–Cu paste and Pb-free TSSOP components shown in Table 8.7, which was not evident with the Sn–Pb paste and Pb-free paste soldered Sn–Pb coated TSOP test cells.

8.11. ACOUSTIC MICROSCOPY INSPECTION OF COMPONENTS BEFORE AND AFTER LEAD-FREE REFLOW

During the NEMI reliability test builds TSOP48, CSP169, CSP208, PBGA256 and CBGA256 components were evaluated before and after Sn–Pb and Pb-free reflow processing using acoustic microscopy equipment to determine and compare the amount of internal damage the components may incur from Sn–Pb and elevated Pb-free reflow temperatures. Typically, the higher the reflow temperature, the more damage that could be sustained in a component plastic package as a result of delamination such as "popcorning" caused by rapid release of moisture contained in the package.

The CSP169, CSP208, PBGA256, and CBGA256 components were manufactured in a relative short space of time before the Pb-free experiments (<3 to 6 months), whereas the TSOP48 samples were over 2 years old. The components were typically received in JEDEC trays. The CSP169, CSP208, PBGA256, CBGA256, and TSOP48 were all rated to Sn–Pb reflow temperatures, but none had been tested or rated to elevated Pb-free soldering temperatures with the exception of the CSP208 and PBGA256 components. Although the MSL level of these daisy-chained components was not noted, the MSL of the packages would typically be MSL 1, 2, or 3 at Sn–Pb reflow temperatures. For the CSP208 (1.2 mm), CSP169 (1.2 mm) and TSOP48 (1 mm) the package thickness was less than 1.4 mm. For the PBGA the package thickness was between 2 and 4.5 mm (2.4 mm). A total of 2037 components were evaluated prior to reflow assembly using reflection and through transmission modes of acoustic microscopy (Phase 1). Each component was individually numbered. Data for each component was recorded and analyzed.

Components were then shipped for assembly on the various test boards and their corresponding test cells. The components were pre-baked at 125°C for 12 hours prior to assembly and stored in a nitrogen storage cabinet. After pre-baking, the components were removed from the nitrogen storage cabinet only for assembly. The Sn–Pb paste and Pb-free paste reflow profiles used for each component test cell are shown (Table 8.3).

CBGA256 and PBGA256 component boards were subjected to peak temperatures of 239–243°C during Pb-free solder paste reflow compared with peak temperatures of 209–214°C during Sn–Pb paste reflow. CSP169 and CSP208 component boards were subjected to peak temperatures of 244–246°C during Pb-free solder paste reflow compared with peak temperatures of 215–217°C during Sn–Pb paste reflow. TSOP48 component boards were subjected to peak temperatures of 247°C during Pb-free solder paste reflow compared with peak temperatures of 217–218°C during Sn–Pb paste reflow. After reflow assembly on test boards, the daisy chained boards were electrically continuity tested and shipped back for acoustic microscopy analysis using reflection mode (Phase 2) since transmission mode acoustic microscopy could not be used for mounted components on boards. The level of internal damage for each of the assembled components was recorded and compared with the results for the same individually numbered components from Phase 1 (pre-reflow condition).

The analysis of the components by acoustic microscopy was done using the guidelines in IPC/JEDEC J-STD-020 (Moisture/Reflow Sensitivity Classification for Non-Hermetic Solid State Surface Mount Devices). Any component showing internal damage that did not meet the acceptance criteria for JSTD020 was noted [55].

For the TSOP48 components analyzed in Phase 1 (pre-reflow), 8% of Sn–Pb parts had some internal component damage whereas 58% of Pb-free parts had some internal damage. This damage was relatively minor and not sufficient to cause nonacceptance according to J-STD-020. After reflow (Phase 2), 34% of the Sn–Pb TSOP parts which had been subjected to a lower Sn–Pb reflow profile had some internal damage that was relatively minor in nature and not causing failure according to J-STD-020 criteria. Eighty-eight percent of the TSOP parts assembled with a Pb-free reflow profile had internal damage, some of which would have been unacceptable according to J-STD-020 criteria.

There are two reasons why this level of unacceptable damage for the TSOP was noticed. The component itself would probably have only been rated to Sn–Pb reflow process temperatures, and the amount of bake-out of the component (12 hours at 125°C) prior to reflow to remove the moisture in the component was insufficient in time duration according to the JSTD-033 document (Handling, Packaging, and Shipping of Moisture/Reflow-Sensitive Surface-Mount Devices).

If the TSOP48 was a MSL 5a part (withstanding 48 hours under normal floor life conditions), the previous JSTD033 specification lists 14 hours whereas a later revision lists 16 hours at the baking time of 125°C to remove the moisture in the package which has just exceeded its floor life. The bake-out time for this part during the experiment was only 12 hours at 125°C.

For the CBGA256 components, there was no die, so the condition of the ceramic substrate was monitored. There was a minimal amount of damage observed pre-reflow and no significant increase in damage post-reflow for Sn–Pb or Pb-free processing temperatures. The minimal damage reported was found to be acceptable according to J-STD-020 criteria.

For the CSP169 components analyzed in Phase 1 (pre-reflow), none of the Sn–Pb or Pb-free parts had internal component damage. After reflow (Phase 2), all the Sn–Pb and Pb-free parts had some internal damage, but it was relatively minor in nature and well within J-STD-020 criteria. For the CSP208 components analyzed in Phase 1 (pre-reflow), none of the Sn–Pb parts had internal component damage whereas 94% of Pb-free parts had some internal damage. This damage for the Pb-free parts was relatively minor and not sufficient to cause nonacceptance according to J-STD-020. After reflow (Phase 2), all the Sn–Pb and Pb-free parts that had been subjected to the Sn–Pb and Pb-free reflow profiles had some internal damage that was relatively minor in nature and not causing failure according to J-STD-020 criteria.

For the PBGA components analyzed in Phase 1 (pre-reflow), all of the components had die attachment anomalies. These die attachment anomalies made comparison of the data between pre-reflow and post-reflow difficult because die attach does have an influence on damage mechanisms in PBGA components. After reflow (Phase 2), 58% of the parts subjected to the Sn–Pb reflow profiles had some internal damage which was relatively minor in nature and not causing failure according to J-STD-020 criteria. After reflow (Phase 2), 61% of the parts subjected to the elevated Pb-free reflow profiles had some internal damage that was relatively minor in nature and not causing failure according to J-STD-020 criteria.

In conclusion, the NEMI acoustic microscopy study for pre-reflow (Phase 1) and post-reflowed (Phase 2) components showed in most cases an increase in internal component damage with elevated Pb-free peak temperature reflow profiles. In most cases this would not cause rejection under the classifications of Moisture/Reflow Sensitivity of Non-Hermetically Sealed Surface Mount devices in J-STD-020 for the components tested. The need for increased bake-out times for components was highlighted along with the requirement for components with plastic molding compounds which were less moisture-sensitive and rated to the higher Pb-free processing temperatures. After thermal cycle reliability tests, the soldered parts were again analyzed by C-SAM (Phase 3) and no indications of further increases in delamination were found.

8.12. LEAD-FREE REWORK OF BGA/CSP SOLDERED JOINTS

During the NEMI reliability test build evaluation, five Pb-free CSP208 and five Pb-free PBGA256 assembled test boards together with some Sn–Pb paste assembled boards with Sn–Pb CSP208 and PBGA256 components were set aside to conduct rework experiments. The work on these boards is discussed below.

8.12.1. NEMI PBGA Rework

For Pb-free Sn–Ag–Cu paste soldered 1.27-mm-pitch Pb-free Sn–Ag–Cu PBGA component test boards, a peak solder joint temperature of 244°C was targeted to achieve component removal [56]. For Sn–Pb paste soldered Sn–Pb PBGA component boards, a peak solder temperature of 205°C was targeted to achieve component removal. These profiles were both implemented successfully.

After component removal, site preparation was conducted to remove residual solder and prepare the site for a new component to rework. This was done with a device that removed solder from the pads by applying heat on the paste fluxed surface and using a vacuum to remove molten solder. The "scavenged" Sn–Ag–Cu and Sn–Pb PBGA test pads had a uniform flat solder-coated appearance.

Paste flux was applied liberally to both component and board surfaces and new Pb-free PBGA components were placed on the redressed component sites. The peak solder joint temperature was targeted at 244°C (time over 217°C of 66 seconds) for the Pb-free Sn–Ag–Cu rework. The top of the Pb-free component body would normally be expected to be around 15°C higher than the solder joint temperature. A method known as "positive air flow" was used to apply cooler temperature air over the top of the component to reduce its temperature while the component solder joints were being heated from the four sides by the heater nozzle. This had the affect of keeping the component body temperature to a maximum peak of 201°C.

Three PBGA boards were sent for thermal cycling. No early reliability defects were observed on the reworked PBGA boards (−40 to 125°C thermal cycling range).

Further work was conducted with the PBGA256 test boards in terms of Sn–Pb and Pb-free Sn–Ag–Cu rework temperature profiling with and without the use of positive air flow. For a Sn–Pb rework profile, the PBGA solder joint was 216°C and time over 183°C was 75 seconds. The temperature at 100 mils away from the component on the board was measured at approximately 172°C, which was within the limits acceptable to prevent secondary reflow of adjacent components. No positive air flow was used for the Sn–Pb profiling.

For the Pb-free Sn–Ag–Cu PBGA rework profiling, the PBGA solder joint had a peak temperature of approximately 246°C and the time over 217°C was 74 seconds.

When positive air flow was used, the top of the component reached only 231°C and the time over 217°C was 72 seconds. When positive air flow was not used, the top of the PBGA component reached 265°C and time over 217°C was 75 seconds. The temperature at 100 mils away from the component on the board when not using positive air flow was measured at approximately 193°C, which would be within the limits acceptable to prevent secondary reflow of adjacent Pb-free soldered Pb-free components.

Typical component spacings on a board are 150 mils or greater. For high-density component boards these spacings may be smaller down to 120 mils. This can cause concerns during BGA/CSP rework operations with secondary reflow of adjacent components causing reliability issues. The typical minimum keep out spacings for BGA/CSP components is usually 150 mils distance or greater from lead-frame components to avoid localized secondary reflow of adjacent components during rework operations.

8.12.2. NEMI CSP208 Rework

For the 0.8-mm-pitch CSP208 component test boards, the thermal processes were developed in a manner similar to that used for the PBGA256 component test boards. For Pb-free Sn–Ag–Cu paste soldered Pb-free CSP208 component test boards, a peak solder temperature of around 246°C was targeted to achieve component removal. The Pb-free component removal profile was implemented successfully.

After component removal, site preparation was conducted to remove residual solder and prepare the site for a new component to rework. For the CSP component boards, this was done with a typical solder wick and hot soldering iron method compared to the scavenging type device used for the PBGA test boards. It was observed that the solder wicking method resulted in partial filling of vias on the board with solder and more nonuniform solder-coated component pads.

New CSP components were placed on the redressed component sites which had paste flux applied liberally to both component and board surfaces for the Pb-free CSP component attachment to aid soldering. The top of the component body would normally be expected to be around 15°C higher than the solder joint temperature. The same method known as "positive air flow" was used to apply cooler temperature air over the top of the Pb-free component to reduce its temperature while the component solder joints were being heated from the four sides by the heater nozzle. The peak solder joint temperature was targeted at 246°C (time over 217°C of 68 seconds) for the Pb-free Sn–Ag–Cu rework profile while the top of the CSP component reached no higher than 231°C for 19 seconds over 217°C. This had the affect of keeping the component body temperature much lower than would be expected.

The X-ray images showed evidence of a columnar solder joint structure which may indicate that the solder joints had not reached their melting point or had not gone over the melting point for a sufficient temperature and time. The CSP208 reworked boards shipped for thermal cycling prior to ATC reliability testing did not show any opens using electrical continuity measurement. Visual inspection of the outer CSP rows showed no apparent defects. The reworked CSP208 boards were thermally cycled alongside the reworked PBGA256 boards. Early ATC reliability defects were observed for the reworked CSP boards.

The development nature of the CSP208 and PBGA256 Pb-free rework part of this process and the small sample sizes of boards reworked and tested gave an indication of some of the issues that may be faced during Pb-free rework. Subsequent Pb-free rework projects would help to increasing identify and determine solutions to issues encountered.

8.12.3. Other Pb-Free BGA/CSP Component Rework Projects

Other development rework projects include work on a network interface card assembled with Pb-free Sn–Ag–Cu paste and Sn–Pb paste in production for rework and reliability testing trials [57]. This board is shown in Figure 8.12. The only PBGA component on the board was intentionally reworked to assess the Pb-free and Sn–Pb rework temperature profiles for this component. The rework equipment used was standard production BGA/CSP rework equipment.

The board material was Epoxy FR4, 4 layer with T_g of 130–140°C and thickness of 63 mils. The board surface finish for the Sn–Pb paste assembled boards was Sn–Pb HASL. The BGA components for the Sn–Pb paste assembled boards had Sn–Pb spheres. Paste flux was used for the BGA rework process. The Sn–Pb rework profile for the Sn–Pb BGA had a solder peak temperature of 208°C, time over 183°C of 62 seconds, time between 130°C and 160°C of 49 seconds, and component body temperature peak of 219°C. This is shown for the Sn–Pb rework profile curve in Figure 8.13.

For the Pb-free Sn–Ag–Cu board assembly, the board surface finish was immersion silver. The BGA components for the Pb-free paste (Sn–3.9Ag–0.6Cu) assembled boards had Sn–Ag–Cu spheres. Paste flux (flux material contained in the paste) was used for the Pb-free BGA rework process. The Pb-free rework profile for the Sn–Ag–Cu BGA had a solder joint peak temperature of 240°C, with time over 217°C for the solder joint of 48 seconds as shown in Figure 8.14, and component body temperature peak of 255°C with time over 217°C for the component of 54 seconds.

There were no issues noticed for the reworked Sn–Pb or Pb-free BGAs using X-ray inspection. The reworked boards passed ICT and functional testing and underwent reliability testing with no defects encountered. The reliability testing on the Sn–Pb and Pb-free assembled and reworked boards included ATC testing (−40°C to 85°C for 1000 cycles), unpackaged shock and vibration tests, and temperature and voltage operating and nonoperating tests.

In work conducted by Gowda et al. [58–60], Pb-free Sn–3.2Ag–0.5Cu CSP46 0.75-mm-pitch packages were assembled with Sn–3.5Ag–0.7Cu solder paste on 62-mil-thick FR4 Ni–Au surface finish boards and reworked with Pb-free Sn–Ag–Cu solder paste and Pb-free CSP components. The peak temperature of the center solder joint during rework was 243°C with time over reflow of around 90 seconds. The top of the component reached a temperature of 262°C. An adjacent

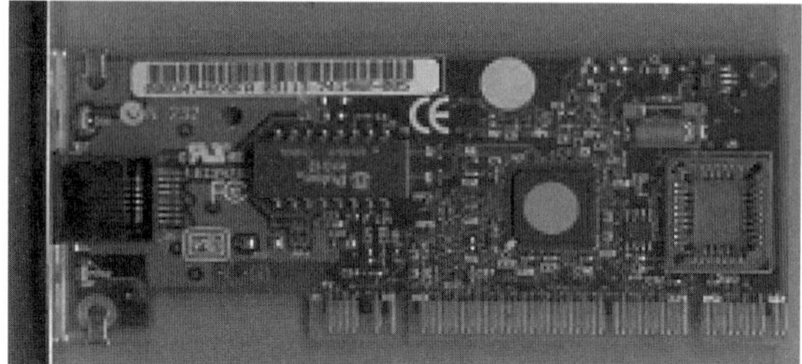

Figure 8.12. Image of fully assembled Pb-free network interface card.

component 180-mil distance from the component being reworked did not exceed 160°C. There was a 21°C ΔT between the top of the component and the solder joint during rework. If the solder joint was kept at 235°C, the top of the component would have reached 256°C. The reliability testing of as-assembled and reworked Pb-free CSP components was found to have no issues after 2500 air-to-air thermal cycles from 0°C to 100°C in 20-minute durations for each cycle.

In addition, Pb-free rework was conducted on Pb-free Ni–Pd–Au leadless micro lead-frame (MLF) 0.5-mm-pitch CSP48 components assembled with Pb-free Sn–3.5Ag–0.5Cu solder paste on OSP-coated 93-mil-thick boards. This showed rework profiles with 245°C peak at the solder joint with a top component peak surface temperature of 267°C. A location on the board 100 mils from the exterior of the rework nozzle reached 185°C peak temperature while an adjacent component 250 mils from the component being reworked reached a maximum of 151°C. The ΔT of 22°C between the solder joint and top of the rework component would be equivalent to having a minimum solder joint peak of 235°C and the top of the component reaching 257°C.

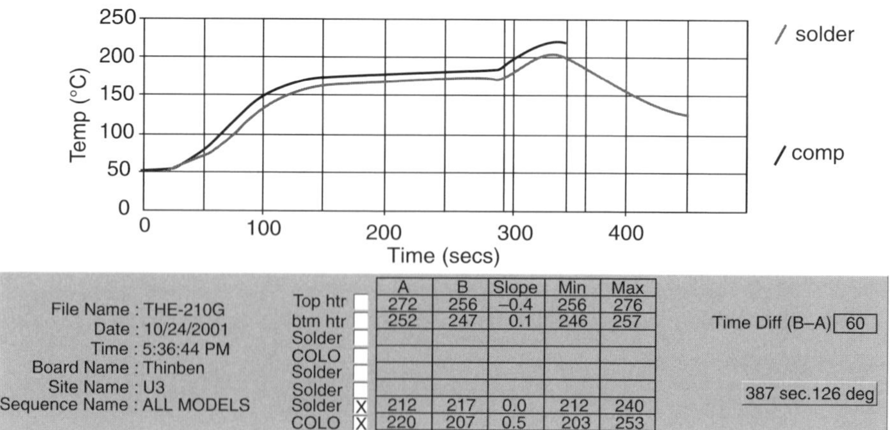

Figure 8.13. Sn–Pb BGA rework profile.

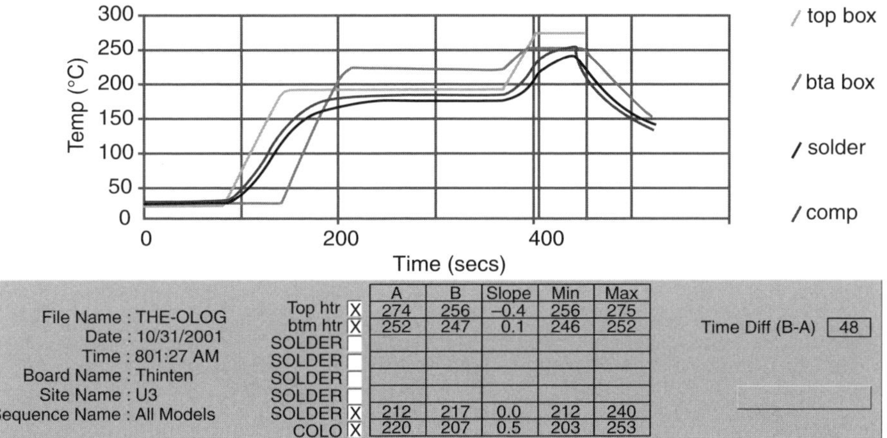

	A	B	Slope	Min	Max
Top htr X	274	256	−0.4	256	275
btm htr X	252	247	0.1	246	252
SOLDER					
SOLDER					
SOLDER					
SOLDER					
SOLDER X	212	217	0.0	212	240
COLO X	220	207	0.5	203	253

Figure 8.14. Sn–Ag–Cu (SAC) Pb-free rework profile for the BGA solder joint and component body.

In work by Tsang and Szymanowski [61], Pb-free BGA256 (Sn–Ag–Cu) and QFP208 (Ni–Pd) components were soldered onto Ni–Au and Immersion Silver FR4 boards with Sn–Ag–Cu solder paste and then hot air gas reworked at 2 different solder joint peak temperatures of 230°C and 250°C. Time over 217°C (60 seconds or 90 seconds), the use of different atmospheres (nitrogen or air), number of reworks (1× or 2×) were varied. The use of nitrogen during rework tended to give a more shinier solder joint and reduced the tendency for voiding. There was no difference between the 60 seconds or 90 seconds over reflow temperature in terms of joint quality. Results conducted on components undergoing 2× rework were not as good as those for 1× rework. The rework joint peak temperature of 230°C gave similar results to that for 250°C peak. The typical die (component body) temperature is usually 15–20°C higher than the joint temperature during rework. It would be beneficial to keep the joint as close to 230–235°C in order to reduce the temperature that the component body reached. With a minimum joint temperature of 235°C, the top of the component would still likely see a temperature of 250–255°C.

8.12.4. BGA/CSP Results and Discussion

Current BGA/CSP rework equipment can be used for Pb-free rework operations. The major areas of development that need to be conducted for BGA/CSP rework equipment for Pb-free processing is equipment rated to higher rework temperatures which can keep the ΔT between the solder joint and top of the component as low as possible. The usual ΔT is around 15–25°C between the solder joint and the top of the BGA/CSP component.

The solder paste to use for BGA/CSP rework would be Sn–3.9Ag–0.6Cu or the paste flux used in this solder paste. For Pb-free rework soldering, the minimum solder joint peak temperature will be around 230–235°C. This would mean that the top of a BGA/CSP using a Pb-free rework profile will reach 245–260°C. BGA/CSP components may have reliability concerns at these temperatures due to the delamination defects, referred to as "popcorning," for moisture-sensitive plastic components.

Typically components for Sn–Pb soldering are rated to either 220–225°C for large body package volumes or 235–240°C for smaller body package volumes (refer to testing temperatures for Sn–Pb plastic component packages in J-STD020). The testing temperatures currently set in JSTD020 for Pb-free soldering are 245°C for large package volumes and 260°C for smaller package volumes for lead-free rework, a component not rated to 260°C would need to pass a 1× rework profile at 260°C to simulate lead-free BGA/CSP rework.

A "wish" list recommendation to component manufacturers is to have large package volume components rated to 255°C peak temperature and small package volumes rated to 265°C peak temperature to cover the assembly and rework temperatures that could be faced for Pb-free assembly. Thus 260°C is an acceptable Pb-free component peak temperature that components should be tested and rated to.

The tendency for internal component package delamination will be increased when reflowing or reworking at higher peak temperatures. Component manufacturers are addressing the concerns and are developing materials rated to the higher Pb-free temperatures. This will take time, and component costs may increase initially with the use of newer, more moisture-resistant molding compounds that are less sensitive to internal delamination at the higher temperature due to popcorning or outgassing of moisture in the part. There may be more need for baking of components in order to remove moisture in the component prior to Pb-free assembly and rework operations.

Typically, a plastic component that has a moisture sensitivity level of 1 (unlimited storage life) at Sn–Pb processing temperatures (220°C) may have a higher moisture sensitivity level when processed at higher Pb-free temperatures of 240°C (e.g., moisture sensitivity level of 3, which means that the storage life in the factory is 7 days before baking is required). A typical rule of thumb is that for every 5–10°C increase in peak reflow temperature, the component loses one level of moisture sensitivity. With the use of higher temperature rated moisture resistant component materials, this "rule of thumb" can be offset to some extent.

The component supplier should always be consulted prior to Pb-free processing to find out what temperature the component is rated to and what effect that higher Pb-free reflow/rework temperatures will have on the reliability of the component.

Education will need to be given to customers, component suppliers, and the industry of the temperature concerns that are faced during BGA/CSP rework. Developments in rework equipment will help to reduce the concerns, but this will be a gradual development process.

The movement to Pb-free has occurred on smaller consumer-type products where reliability concerns will not likely be as great as for large network boards. Collaboration between the manufacturer, the component supplier, and the end-user will address the concerns and work towards developing solutions. Most component manufacturers in internal testing seem to be targeting 260°C peak reflow temperature as a test rating for their components, which is prudent. The development of these higher temperature-rated Pb-free components has accelerated due to customer demand for these types of components.

With increasing Pb-free implementation, concerns are being addressed in the Pb-free manufacturing production environment or with development projects such as the NEMI Lead-Free Hybrid Assembly and Rework Project [62], where rework evaluations and reliability testing have been conducted on 62-mil, 93-mil, and 135-mil-thick boards on various Pb-free components such as CBGA, PBGA, CSP, TSOP, and connectors using two board surface finishes (immersion silver, Ni–Au). These are discussed in more detail in Chapter 9.

8.13. LEAD-FREE HAND-SOLDERING REWORK

As the assembly materials for surface mount are Sn–3.9Ag–0.6Cu and for wave soldering are Sn–0.7Cu or SnAgCu, the hand-soldering rework material can be Sn–Ag–Cu or Sn–0.7Cu or Sn–3.5Ag.

Experiments were conducted on the three Pb-free wire materials (Sn–Ag–Cu, Sn–Ag, Sn–Ag–Cu) and one Sn–Pb wire material with no-clean liquid rework flux to determine the best Pb-free solder wire composition to rework Sn–Ag–Cu surface-mount soldered joints [63]. The same hand-soldering rework equipment was used for Pb-free and Sn–Pb rework. The settings on the hand-soldering equipment were raised by 1–2 settings above those usually used for Sn–Pb rework because the melting temperature of the Pb-free solder wire is around 30°C above that for Sn–Pb (217°C for Sn–Ag–Cu, 221°C for Sn–3.5Ag, 227°C for Sn–0.7Cu, 183°C for Sn–Pb).

The visual solder joint and flux appearance are shown for some sample reworked boards in the photos below for Sn–3.5Ag (Figure 8.15) and Sn–37Pb (Figure 8.16) solder wire reworked boards and rated in terms of visual solder joint and flux rating appearance in Table 8.8. The locations reworked in the photos shown are QFP100 (Sn–Pb-coated), U34 SOL20 (Ni–Pd-coated), and 1206 chip (Pure-tin-coated).

From Table 8.8, all three Pb-free rework wire alloys (Sn–3.5Ag, Sn–3.9Ag–0.6Cu, Sn–0.7Cu) have a higher (and better) visual and flux rating than the Sn–Pb reworked control boards. The Sn–3.9Ag–0.6Cu Pb-free rework wire had a slightly higher rating than the Sn–0.7Cu or Sn–3.5Ag. As either of the three Pb-free rework wires can be used, the Sn–3.5Ag was suggested as the Pb-free rework wire of choice to rework both Sn–Ag–Cu surface-mount soldered boards and Sn–0.7Cu or SnAgCu wave soldered boards. Sn–3.5Ag solder wire has a history of good use in the industry. Work from the IDEALS Pb-free European project has also shown Sn–3.5Ag wire to be effective for reworking Pb-free Sn–Ag–Cu surface-mount boards, with the joints looking less "grainy" than when rework was conducted with Sn–Ag–Cu solder wire [64].

Figure 8.15. Sn–3.5Ag solid wire Pb-free rework with no-clean liquid flux pen (Sn–Ag–Cu soldered OSP-coated board).

Work by Tsang and Szymanowski [61] evaluated Pb-free hand-soldering rework with Sn–4Ag and Sn–3.8Ag–0.7Cu cored wire on Pb-free fine-pitch QFP (Ni–Pd-coated), 1206 chip (pure-tin-coated), SOIC20 (Ni–Pd-coated), and PLCC44 (Ni–Pd-coated) on immersion silver and Ni–Au surface finish FR4 boards. The variables investigated were the Pb-free rework alloy (Sn–4Ag or

Figure 8.16. Sn–37Pb solid wire rework board with no-clean liquid flux pen (Sn–Pb soldered OSP-coated board).

TABLE 8.8. Visual and Inspection Rating for Pb-Free and Sn–Pb Reworked Boards

Alloy	Rating
Sn–37Pb	2.3
Sn–3.5Ag	2.5
Sn–3.9Ag–0.6Cu	3.2
Sn–0.7Cu	3.1

Sn–3.8Ag–0.7Cu), the solder iron tip temperature (500°F, 600°F, 700°F, 800°F), the flux volume (small or large flux core size), and the rework atmosphere (air or nitrogen). When Sn–4Ag and Sn–3.8Ag–0.7Cu were evaluated against one another, both could be used for rework. The Sn–Ag–Cu showed a grayer appearance of solder joints in certain rework situations. Because the Sn–4Ag rework wire had a history of successful use in the industry, it was recommended as the rework material of choice.

The solder iron tip temperatures of 500°F (260°C) and 600°F (316°C) were found to be unacceptable for rework with either Sn–4Ag or Sn–3.8Ag–0.7Cu cored solder wire with a minimum solder iron tip temperature of 700°F (370°C) suggested without a measurable change in cycle time compared with Sn–Pb solder wire rework. Increasing the volume of flux used tended to give better hand-soldering results for Pb-free rework for both Pb-free alloys. Nitrogen was found to benefit the rework process in terms of cosmetic appearance of the solder joint, especially when reworking the fine-pitch QFP component. Lead-free reworked Ni–Pd 20-mil-pitch QFP components were found to have pull-strength results equivalent to or better than those for Sn–Pb reworked components on both immersion silver and Ni–Au board surface finishes.

The Pb-free solder iron tip temperature suggested by Tsang and Szymanowski [61] was similar to results found by Tsunematsu et al. [65] on Pb-free cored wire rework evaluations with Sn–Ag–Cu. The Pb-free solder spreading times for Sn–Ag–Cu were found to be similar to Sn–Pb when a solder iron tip temperature of 716°F (380°C) was used. The obvious problem would be that these tip temperatures are slightly higher than those normally used for Sn–Pb hand-soldering rework. As the rework hand-soldering temperatures are usually localized around the lead-frame termination tip, the temperature of the component will not rise in the same way as for convection hot air gas rework such as for BGAs/CSPs, so only very temperature-sensitive components will most likely be affected by this increase in tip temperature.

8.13.1. Hand-Soldering Results and Discussion

Hand-soldering rework will likely be less of a concern compared with Pb-free BGA/CSP rework because of the lower peak temperatures and times over reflow exerted on the components in relation to BGA/CSP rework. The choice of Pb-free rework wire would be Sn–3.5Ag.

8.14. IN-CIRCUIT TESTING AND FUNCTIONAL TESTING (ICT/FT) OF SOLDERED JOINTS

8.14.1. ICT (In-Circuit Testing)

The objective of conducting in-circuit testing for Pb-free no-clean soldered joints would be to measure the probeability of different paste (flux) types with the test probes. The more difficult the flux residues are to probe, the more false rejects call will occur during ICT on assembled boards. The use of the higher Pb-free soldering temperatures will increase the likelihood of the flux residue hardening, which reduces probeability. Two paste types (with the same flux formulation), one representing the Sn–Pb paste and the other representing the Sn–Ag–Cu paste, were tested by ICT with high/low-temperature peak reflow profiles, and after aging (i.e., post reflow time) and for reflow tested with air or nitrogen atmosphere [66].

The results showed that Sn–Pb no-clean soldered paste had a lower number of probing problems with each of probe types used (chisel and crown type) compared with the Sn–Ag–Cu soldered no-clean paste. This was likely due to the hardening of the no-clean flux residue on the board, especially at higher soldering temperatures for Pb-free. Nitrogen atmosphere use showed fewer probing problems when compared with air reflow environment. Regardless of the reflow atmosphere used, the aging factor (post reflow age) showed that the longer the aging, the more the probing defects for the specific pastes used which contained the same flux material.

In work by Arzadon et al. [67] on no-clean Pb-free Sn–4Ag–0.5Cu solder paste, ICT pin probe testing was conducted on soldered Sn–Ag–Cu test pads on OSP board surface finish boards. The three reflow conditions used for the Sn–Ag–Cu solder paste were:

(i) Peak temperature 230°C, time over liquidus of 50 seconds, 1× reflow
(ii) Peak temperature 240°C, time over liquidus of 80 seconds, 1× reflow
(iii) Peak temperature 240°C, time over liquidus of 80 seconds, 2× reflow

The worse probe test results were obtained with the 230°C peak and 1× reflow (i). The second worse probe test results were obtained with 240°C peak and 2× reflow (iii). The best probe test results were with 240°C peak and 1× reflow (ii). At the coolest reflow profile, a moderate amount of flux residue was picked up by the probe tips that lead to an increase in contact resistance of over 10 ohms for 5% of the probes tested. With the hotter single reflow profile, half the amount of flux residue was picked up by the probe tips that lead to a reduced incidence of probe testing issues (3% of probes tested had contact resistances of over 10 ohms). With the hotter profile undergoing 2× reflow, there was almost no residue at the probe tips but the flux that was remaining was less penetrable after 2× reflow compared with 1× reflow (4.7% of probes tested had contact resistance of over 10 ohms). These probe test results were found to be comparable with those of the most pin-testable Sn–Pb no-clean solder pastes, which also had less than 5% of probes tested with contact resistances over 10 ohms.

Board surface finish can also play a part in probe-testability results. Wong and Nestor [68], during production trials with Pb-free Sn–Ag solder paste, noticed that the changeover from Sn–Pb HASL-coated boards to Pb-free immersion silver finish led to less penetration on the surface of immersion silver test pads compared with Sn–Pb HASL, which was remedied by the use of a higher spring force and three-sided chisel test probes. The actual testing probes could also wear out quicker if they were in contact with the harder Pb-free solder alloys [69]. These were all considerations to bear in mind when conducting Pb-free assembly trials. In a majority of cases, there may be no issues with ICT testing because the flux residues may not flow onto the test probe contacts on the board.

8.14.2. Functional Testing

There are no indications of differences between functional testing for Pb-free soldered boards compared with Sn–Pb soldered boards at this time.

8.15. YIELD DATA

Yield data information for lead-free has been relatively scarce to date.

Wada [11], who conducted development and production for Pb-free mini-disk players, indicated a reduced number of solder balling defects with Pb-free Sn–Ag–Bi solder paste compared with the previous Sn–Pb–Ag soldered assemblies. Godavarti [70] conducted work on Pb-free cell phones and found that the yield for production with Sn–Ag–Cu solder paste was better than for Sn–Pb. In Pb-free production and reliability trials conducted on a network interface card (Figure 8.12; see Ref. 57), the incidence of defects for Pb-free Sn–Ag–Cu soldering was slightly decreased compared to Sn–Pb production due to reduced wetting evident on a soldered LED diode center lead termination. However, even though the Pb-free defects levels were slightly increased relative to Sn–Pb, they were still within the expected range for Sn–Pb soldering and were deemed acceptable for production. The reliability of these Pb-free soldered network boards were equivalent to Sn–Pb.

Companies who have been most active in Pb-free implementation have learned from the example of increasing production development experience to overcome Pb-free production issues in order to make the Pb-free yields equivalent or better than Sn–Pb assembly.

8.16. SURFACE-MOUNT FILLET LIFTING AND RELIABILITY OF REFLOWED SOLDERED JOINTS

There have been various reports of the potentially harmful affect on reliability of using Bi and Pb in solder joints when assembling with Sn-based Pb-free solder [71, 72]. The effect has typically been found mainly for wave soldered joints where the combination of low-melting-point phases formed when combining Sn, Pb, and Bi, and the geometry of the joint has caused separation in soldered joints, termed fillet-lifting or "lift-off."

The low-melting-point phase of Sn–Bi–Pb, which melts at 96°C, concentrates at the last area of the solder joint to solidify which is typically at the interface of the board and solder joint. Because of the specific geometry of the wave soldered through-hole joint, the stresses during solidification act to cause pull away the solder from the board while it is still molten at the board interface, causing fillet lifting. The geometries of a surface-mount joint are less susceptible than the wave soldered joint, but there is still some susceptibility particularly when Bi, Sn, and Pb are present in the solder joint.

There is a benefit of using lower-melting-temperature Pb-free solder alloys because as the reflow profiles can be reduced in temperature this enables the use of components rated to lower reflow and wave soldering temperatures. The benefit of using Bi additions to Sn–Ag and Sn–Ag–Cu alloys is that it lowers the melting temperature while enhancing solderability. The negative is the reliability implications, particularly when Pb is still present in the system from Sn–Pb-coated component terminations and printed circuit boards or rework operations.

Akinade et al. [73] showed potential problems with the use of Pb-free Sn–7.5Bi–2Ag–0.5Cu Pb-free solder paste containing Bi when assembling on 63Sn–37Pb HASL-coated boards and 70–95%Sn–5–30%Pb-coated components. Reliability testing showed failures after thermal cycling from −10°C to 100°C for 510 cycles due to "lifted" surface-mount connectors, whereas no failures were noted when using Pb-free Sn–3.5Ag or eutectic Sn–Pb solder paste with the same Sn–Pb-coated components.

Nakatsuka et al. [74] reported work on Pb-free Sn–3Ag3–15Bi and Sn–37Pb solder paste assembled with Sn–10Pb-coated 1608 chip resistors and 0.5-mm QFP components on OSP coated Cu boards. Shear testing of the soldered chip components after thermal shock testing (500 cycles from −40°C to 85°C) with subsequent analysis of the fracture surfaces showed a migration of fracture from the bulk solder joint with Sn–37Pb and Sn–3Ag–3Bi to fracture at the solder to printed circuit board land interface with Sn–3Ag–6Bi and Sn–3Ag–15Bi. Bismuth concentrations at the solder–land interface were noted mainly with the 6%Bi and 15%Bi containing solders. For the soldered Sn–Pb-coated QFP components, which were subjected to thermal shock cycling testing (500 cycles from −40°C to 85°C) followed by pull testing, there was a decrease in pull strength for Sn–3Ag–6Bi and Sn–3Ag–15Bi soldered components compared to Sn–3Ag–3Bi and Sn–37Pb soldered components, probably due to the concentration of the bismuth at the interface, which would be a easy path for crack propagation due to the brittle structure of bismuth. As a result, Sn–3Ag–3Bi was used for production of the Mini-Disc Pb-free product rather than a higher Bi-containing solder.

The use of Pb-free Sn–Ag–Cu–Bi soldered surface-mount components has also been studied in terms of defects caused by Pb-free wave soldering from the bottom side of the board [74]. The surface-mount solder fillet has been found to separate from the heat through the board generated by wave soldering. From the evaluations, there appears to be separation at the solder joint–land interface on large Sn–Pb-coated surface-mount QFP devices when subjected to a simulated bottom-side wave soldering process. The higher the Bi content of the surface-mount paste usually from 7%Bi and higher (Sn–3Ag–Cu > 7Bi), the more likely the separation in the solder joint. This was found to be due to the redistribution of the Bi to the last area of the solder joint to solidify, which was at the solder–land interface.

When the minimum peak temperature of the printed circuit board surface reached 215°C from the wave soldering process, separation was estimated to occur with a 1 wt% Bi Pb-free solder alloy for the surface-mount Sn–Pb-coated QFP joint, whereas the temperature at the printed circuit board for separation to occur with a 3 wt% Bi-containing alloy would be only 185°C. For a 7 wt% Bi-containing alloy, separation would be estimated to occur when the temperature at the printed circuit board surface reached 135°C. Separation was found to occur due to the decreased strength of the solder joints but also due to the warping of the board during processing.

Comparing the results for the large-sized QFP mentioned above and a small-sized TSOP Sn–Pb coated component, the TSOP had no instances of any issues because of the lower thermal capacity of the TSOP component, indicating that there was less thermal gradient across the device during cooling and thus a lower potential for Bi redistribution in the solder joint giving fewer areas of stress concentration or crack propagation paths. By reducing the thermal gradient from the top of the QFP package to the surface of the printed circuit board by heating the top of the component during a simulated wave soldering process, the incidence of separation in the solder joint was reduced. In practice this technique may be difficult to control during wave soldering. In other work by Baggio et al. [45] using Sn–Ag–Bi wave solder baths with lead impurities down to 0.2 wt% in the solder bath, the authors found reductions in pull strength for wave soldered wires through 63-mil-thick paper phenolic PCBs, and this was the reason to avoid the use of Sn–Ag–Bi solder for wave soldering.

Not only can joint quality deteriorate with a surface-mount top-side and wave soldered bottom-side operation, but also with a double-sided reflow operation where Bi is used in combination with Pb and Sn in the solder joint. Yamaguichi et al. [75] investigated the effect that aging could have on a Bi-containing Pb-free solder joint in terms of solder joint integrity at the interface. After assembling an Sn–Pb-coated QFP and chip component with Sn–3Ag–2.5Bi–2.5In solder on an OSP-coated board, reheating the solder joint at 180°C was found to cause a segregation layer of Sn–Bi–Pb (melting point of 96°C) at the solder–land interface, which could be a source for solder joint deterioration.

No segregation layer of Sn–Bi–Pb was observed when assembling the Sn–Pb-coated QFP with Sn–3Ag–0.5Bi–3In or Sn–3Ag–0.5Cu solder due to the reduced or zero presence of bismuth in the solder joint. The segregation layer was also found to be due to the consumption of Sn as Cu–Sn intermetallic during aging which would leave a concentration of Bi at the solder joint interface contributing to reduced reliability.

Surface-mount assembly evaluations by Bath and Crombez [34] with Sn–Ag–3Bi, Sn–3Ag–5Bi and Sn–58Bi solder pastes using Sn–Pb-coated components also revealed some issues. Sn–3Ag–5Bi soldered Sn–Pb-coated QFP components on OSP-coated boards showed separation at the solder–land board interface after assembly. The pull forces for the Sn–3Ag–3Bi and Sn–3Ag–5Bi soldered QFP component both before and after thermal cycling (−40°C to 125°C, 1000 cycles) were lower than the Sn–Pb and Sn–3.9Ag–0.6Cu and Sn–3Ag–0.8Cu soldered Sn–Pb-coated QFP component pull-force results. This suggested that there was evidence of weak interfaces for both the Sn–3Ag–3Bi and Sn–3Ag–5Bi soldered QFP components. After thermal cycling, lead from the component coating was more evident at the solder joint–land PCB interface for all the Pb-free solders evaluated, which indicated that the lowest-melting-temperature phase in the solder joints, which could be Sn–Pb (183°C) or Sn–Pb–Ag (179°C) or Sn–Bi–Pb (96°C), was distributed close to the board interface, which was the last area of the solder joint to solidify. For the Sn–58Bi soldered Sn–Pb-coated QFP components, the as-assembled pull test results were so low that they could not be measured. During thermal cycling from −40°C to 125°C, the Sn–58Bi soldered Sn–Pb QFP components were reflowing and falling from the test board because of the Sn–Bi–Pb phase, which melts at 96°C. For Sn–3Ag–3Bi and Sn–3Ag–5Bi soldered Sn–Pb-coated 1206, 0603, and 0402 chip component on OSP-coated boards, there was no evidence of a reduction in shear strength both before and after thermal cycling compared with the Sn–Pb soldered chip components.

Tanaka et al. [38] conducted evaluations of bulk Pb-free solder bar (Sn–3.5Ag–0.7Cu, Sn–3.2Ag–3Bi–1.1Cu–Ge, Sn–3.5Ag–0.5Bi–3In) in tensile and impact testing. The Sn–3.5Ag–0.7Cu and Sn–3.5Ag–0.5Bi–3In had equivalent or better tensile and impact test results compared to Sn–37Pb solder, whereas the 3%Bi-containing Pb-free alloy had a reduced impact strength. Shear test results on Sn–3.2Ag–3Bi–1.1Cu–Ge soldered Sn–Pb-coated 1005 chip components on Ni–Au-coated printed circuit boards were worse than those for Sn–3.5Ag–0.7Cu and Sn–3.5Ag–0.5Bi–3In soldered chip components.

Not only would we need to consider the reliability of solder joints for bismuth added to a Pb-free solder when lead is introduced from the component lead or board surface finish, we would also need to evaluate the case where Pb interacted with Pb-free solder in the absence of Bi.

Zhu et al. [76] conducted work on Sn–3.5Ag and Sn–5Sb Pb-free solder pastes with various additions of lead ranging from 5 wt%, 10 wt%, 15 wt%, and 20 wt% to simulate what would happen if Pb was introduced from the board surface finish or component surface finish. The solder pastes with various Pb additions were soldered to pure-Sn-coated 1206 chip components on an OSP-coated board surface finish. Shear testing of the chip components was conducted at room temperature and at 125°C. The results showed similar shear forces with no Pb additions and Pb additions to the Sn–3.5Ag and Sn–5Sb soldered components during room temperature testing. Cross sections showed that the Pb was found to be at the grain boundaries of the pure Sn grains.

There was a reduction in shear force, however, when shear testing was conducted at 125°C for the Pb-containing Sn–3.5Ag and Sn–5Sb solders (greater than or equal to 5 wt% Pb). The fracture surfaces of the Pb-containing solders that had been subjected to 125°C shear testing showed increased concentrations of Pb phase at the fracture surface, which suggested that failures occurred through the lead phase. The aging may have acted to increase the concentration of Pb phase by depleting areas of Sn as Cu–Sn intermetallic.

To predict the amount of Pb in a solder joint contributed from a component lead-frame, board surface finish or the solder paste would be difficult. For a Sn–Pb solder joint, estimated contributions of Pb in a lead-frame solder joint have indicated that 5% of the Pb contribution is from the lead-frame finish, 75% of the Pb contribution is from the solder paste, and 20% of the Pb contribution is from the Sn–Pb HASL board surface finish.

For a Pb-free solder paste joint, for a lead-frame only contribution (Sn–10 wt% Pb) to the Pb in the solder joint, the estimated amount of Pb dissolved in a 3-mil-thick (0.075-mm) Pb-free solder joint would be around 0.4–1.6 wt%, dependent on the thickness of the lead-frame coating (3 μm [connector]–12 μm [lead-frame]) (Table 8.9). For a Sn–Pb HASL board (Sn-37 wt% Pb), the estimated amount of Pb dissolved in a Pb-free solder joint would be around 0.7–2.3 wt% (1.5 μm–5 μm). For a Sn–Pb HASL board and Sn–Pb component-coated lead-frame, the estimated amount of Pb dissolved in a Pb-free solder joint would be around 1.1–3.9 wt%. For an Sn–37Pb BGA component, the estimated amount of Pb dissolved in a Pb-free solder joint would be around 34 wt%.

Choi et al. [77] conducted work with Sn–3.5Ag where situations where Pb contamination from the board surface finish and/or the lead-frame finish could be studied. Three Pb concentrations (2 wt%, 5 wt%, 7.5 wt%) were evaluated with Sn–3.5Ag Pb-free solder. For all three Pb concentrations, there was found to be the formation of Sn–36Pb–1.5Ag ternary phase with a melting temperature of 179°C. This lower-melting-temperature phase may be swept to the last area of the solder joint to solidify, which in the case of a surface mount solder joint would be at the joint/board surface finish interface increasing the concentration of Pb in this region. As the melting temperature of the ternary eutectic would be close to the operating temperature desired in under-the-hood automotive applications of 150°C, this may have an effect on reliability of the solder joint during elevated temperature operation.

TABLE 8.9. Estimated wt% of Pb in Pb-Free Sn–Ag–Cu Solder Paste Joints from Sn–Pb Component or Board Surface Finish Mixing

Component/ Board Type	Sn–Pb Solder Alloy	Thickness Range (μm)	Pb wt% in Sn–Ag–Cu Paste Soldered Joint
Component Pb frame	90Sn–10Pb	3–12	0.4–1.6
HASL board finish	63Sn–37Pb	1.5–5	0.7–2.3
Component Pb frame plus board finish			1.1–3.9
BGA sphere	63Sn–37Pb		34

Seelig and Suraski [78] found that a Pb-free Sn–Ag–Cu soldered joint with an Sn–Pb-coated lead-frame component caused a concentration of Pb at the joint/board surface interface, which was the last area of the solder joint to solidify. The Sn–Pb–Ag with a melting temperature of 179°C could be present in this area. The resulting solder joint was found to have a weak interface.

Oliver et al. [79] conducted experiments on temperature-cycled Pb-free Sn–3.5Ag–0.7Cu solder joints which had Pb contamination introduced from the component termination and/or the board surface finish. The estimated Pb content in the solder joint was 2–6 wt%. The Pb phase was found to exist around the Sn grains at the grain boundaries. After 500 thermal shock cycles from −15°C to +125°C, micro-cracks were observed through the Pb phases at the Sn grain boundaries which would have a detrimental effect on reliability. Tests on model ring and plug solder joints using mechanical cycling at room temperature to replicate the stresses occurring during thermal cycling on Pb-free Sn–3.5Ag–0.7Cu solder with various additions of Pb (1, 2, 5, 10, 15, 20 wt%) were also conducted. Data showed that additions of 2–5 wt% of lead in Sn–3.5Ag–0.7Cu caused a drop in fatigue life compared with Sn–3.5Ag–0.7Cu, with a fatigue life similar to that of Sn–37Pb solder. Low or zero Pb additions (0, 1 wt%) and high lead additions (10 wt%, 15 wt%, 20 wt%) to Sn–3.5Ag–0.7Cu had an improved fatigue life compared to Sn–37Pb for the model solder joints.

From the observations above, we would need to ensure that we are aware of the surface finishes of the component and board materials used with Pb-free solder. To reduce the Pb content as much as possible would be the route that should be taken as much as possible. The most obvious and easiest steps would have Pb-free solder paste, bar, rework materials, Pb-free board surface finish, and Pb-free BGA/CSP. Of more difficulty to transition may be to lead-frames because of the testing requirement criteria such as tin whisker testing for Pb-free pure Sn-coated components. These requirements were being developed in order to facilitate a complete Pb-free solution.

8.17. CONCLUSIONS

The printability of Pb-free Sn–Ag–Cu solder pastes has been shown to be equivalent to that for Sn–Pb in most cases. Improvements were being made to the Pb-free pastes with more production experience and feedback to the solder paste suppliers.

There have been no placement issues associated with assembling Pb-free components compared with Sn–Pb. The main difference occurring during SMT processing was the increase in reflow temperature for Sn–Ag–Cu. This would require more development of Pb-free solder pastes to be rated to the higher temperatures.

The wettability of Sn–Ag–Cu was found to be reduced compared with Sn–Pb but this was not found to reduce the reliability of the solder joint. Voiding was found to occur to a greater extent in Pb-free soldered joints. This was mainly due to the increased temperatures used and the need to develop the fluxes for these temperatures. The reliability of these joints was not found to be reduced.

The minimum solder joint peak temperature needed for Sn–Ag–Cu surface-mount reflow was found to be in the range 230–240°C, with 235°C being a good minimum peak temperature to aim for.

Reflow profiling of boards was conducted to determine the ΔT across the board during Pb-free processing. The ΔT across the board was found to be 10–15°C. Taking into account process tolerances across difference reflow ovens used in production, the ΔT across boards could be as small as 5–10°C for small mobile phone-type boards and as high as 20°C for larger boards requiring components rated to 255–260°C. Thermocouple profiling of Pb-free boards would become an increasing requirement to determine what component temperatures were actually being reached.

The visual appearance of Pb-free Sn–Ag–Cu solder joints was found to be different from that for Sn–Pb joints. This did not mean a reduction in reliability but a need to develop and refine the visual inspection standards to take account of Pb-free soldered joints which has been done in the IPC 610 standard.

In most cases, existing manufacturing tools and equipment could be used for Pb-free processes. X-ray, acoustic, and AOI (automated optical inspection) equipment could be used to evaluate Pb-free assembly processes.

AOI would need to be reprogrammed more to take account of the visual differences for Pb-free solder joints. X-ray inspection showed increased voiding with Pb-free Sn–Ag–Cu paste soldered Pb-free CSP/BGA components and some reduced wetting with Pb-free paste soldered Pb-free TSOP components, but this was not found to affect reliability compared with Sn–Pb soldered components. Components examined by acoustic microscopy showed internal defects induced by the higher-temperature Pb-free reflow, but most of the defects were well within component specifications.

Lead-free BGA/CSP rework had been assessed and the temperatures being experienced during this rework were found to be higher for the component than when using reflow ovens. The peak temperature of the component during rework was found to be 15–25°C above that of the solder joint. More development would be needed in this area. A good testing temperature for components for Pb-free soldering would be 260°C.

For Pb-free hand-soldering rework, typical soldering iron tip temperatures were found to be around 700°F (370°C), which was higher than those used for Sn–Pb. The hand-soldering tip would usually apply localized heating so that the component would less likely increase in temperature sufficiently to cause concern compared with BGA/CSP rework.

The probeability of flux residues for reflowed Pb-free no-clean solder pastes was reduced compared with Sn–Pb paste flux residues during in-circuit testing. This was found to be due to the higher temperature used during Pb-free processing, which would harden the flux residue more. Developments would be made over time by solder paste suppliers. No difference in functionally testing was found for Pb-free and Sn–Pb soldering.

Lead-free yield data for products built so far indicated initial reductions in yield by a moderate amount; but with increasing production experience, yields would reach and in some cases exceed Sn–Pb levels.

Surface-mount fillet lifting with a mixture of Sn, Pb, and Bi in the solder joint was assessed. This was found to be due to the low melting point phase of Sn–Bi–Pb (melting point of 96°C). With a reduced amount of Bi the effect was reduced. During the transition period where Sn–Pb components could still be used, it was advisable to not use any Bi in the board surface finish, component finish and solder paste when transitioning to Pb-free soldering.

Lead introduction from lead-frame coatings, BGA/CSP spheres, or Sn–Pb HASL board finishes with Pb-free Sn–Ag–Cu solder paste could also create some problems with lower melting point phases (Sn–36Pb–2Ag: melting point of 179°C) but not to the extent that bismuth would. It was advisable to reduce the Pb content introduced into a solder joint as much as possible. Having a Pb-free BGA/CSP and Pb-free solder paste and board surface finish could reduce the Pb content in a solder joint by over 95%. Work had been conducted to develop Pb-free Sn-based component coating standardization tin whisker testing.

With more production experience and evaluations, lead-free solutions were and should be found to overcome any issues which occur. There was no substitute for Pb-free production experience and assembly and reliability trials on actual product in the implementation phase.

8.18. FUTURE WORK

There are many areas where development can take place for Pb-free soldering. New developments would occur with Pb-free Sn–Ag–Cu solder pastes rated to the higher soldering temperatures to perform as good as or better than existing Sn–Pb pastes.

The increase in temperature from Sn–Pb to Pb-free Sn–Ag–Cu imply a 30°C increase in processing temperature. Reflow ovens are being developed which will aim to reduce the ΔT across a board, but this will be a gradual process. Components will need to be rated to the higher Pb-free soldering temperatures.

Much more work is needed in the areas of Pb-free BGA/CSP rework where the temperature that the components had to withstand is typically higher that for reflow soldering. Larger-sized boards with thicknesses of greater than 100 mils an area where there was a limited experience of Pb-free soldering where the demands on the components in terms of temperature ratings will increase.

The increase in production for Pb-free will highlight where there needs to be development work done.

ACKNOWLEDGMENTS

The author would like to acknowledge the work from the various persons and companies involved within the iNEMI Pb-free process group and task force who contributed much of the data presented in this chapter.

The persons and companies involved in the iNEMI assembly builds and inspection testing and writing reports included Len Poch, Maurice Davies, Jeff Schake, Denis Barbini, Oliver Bast, Jeremy Jessen, Steve Martell, Mark Walz, and Terry Leahy. The author acknowledges the persons who contributed to the various Pb-free papers listed in the references section who made the task of writing a chapter on Pb-free reflow and rework relatively easier.

The author would also like to thank Karen Walters for her contribution to Section 8.6, as well as Karen, Denis Barbini, Rich Parker, Edwin Bradley, and Kay Nimmo for reviewing this chapter.

Finally, the author would like to thank my mum and dad, my wife Piyanoot, and my three little girls Palm, Mint, and Surinder for the patience they have shown when writing this chapter.

REFERENCES

Section 8.2

1. J. Bath and E. Crombez, Evaluation of the printability of no-clean and water-soluble solder pastes, NEPCON West, February 2000, Anaheim, CA.
2. T. A. Nguty et al., Understanding the Process Window for Printing Lead-Free Pastes, ECTC 2000.
3. M. Detert et al., Screen Printing, Placement and Joining Technologies with Lead-Free Joining Materials, ECTC 2000.
4. J. Bath, NEMI Lead-Free and Tin-Lead Paste Evaluation Report, March 2001.
5. C. Ashmore and R. Goldsmith, Investigating Mass Imaging Lead-Free Materials Using Enclosed Print Head Technology, Nepcon West, December 2002, San Jose, CA.
6. L. Poch et al., NEMI Lead-Free Reliability Build Report, March 2002.
7. M. Arra et al., Performance Evaluation of Lead-Free Solder Pastes, SMI 2001.
8. B. Huang and N.-C. Lee, Prospect of Lead-Free Alternatives for Reflow Soldering, IPC Works 1999, Minneapolis, MN.
9. A. Butterfield et al., Lead Free Solder Flux Vehicle Selection Process, ECTC 2000.
10. P. Hart et al., Characteristics of Some Lead-Free Solder Pastes for Microelectronics Package Assembly, SEMICON West 2000, San Jose, CA.
11. Y. Wada, Lead-Free Soldering Technology Overview, February 8, 1999.

Section 8.3

12. K. Rajewski, Kester Solder, SMT Process Recommendations Defect Minimization Methods for a No-clean SMT process, www.metcal.com/kester/smtrecs.html
13. S. Shina et al., Design of Experiments for Lead-Free Materials, Surface Finishes and Manufacturing Processes of Printed Wiring Boards, APEX 2001, San Diego, CA.
14. S. Shina et al., Process and Material Selection for Zero Defects and Superior Adhesion Lead-Free SMT Soldering, SMI 2001.
15. T. Skidmore and K. Walters, Lead-Free Research: Optimizing Solder Joint Quality, Circuits Assembly, April 2000.

Section 8.4

16. S. V. Sattiraju et al., Wetting Performance of Several Lead-Free Board Surface Finishes and Solder Alloys, IPC Works 2000.
17. E. A. Heiser et al., Development of Internal Solderability Standard: Building on External Specifications and Assembly Experience, SMI 1995, San Jose, CA.
18. S. V. Sattiraju et al., Wetting Performance Versus Board Finish and Flux for Several Pb-Free Solder Alloys, IEEE/CPMT International Electronics Manufacturing Technology Sympoisum, 2000.
19. W. Johnson et al., Improved Thermal Process Control for Lead-Free Assembly, APEX 2001, San Diego, CA.
20. M. Kwoka and G. O'Brien, Lead-Free Components FOCUS Group Evaluation—Status Report 1, IPC Works 2000, Miami, FL.
21. C. Fan et al., Solderability of Lead-Free Finishes Using Lead-Free Solders, SMI 2001.
22. S. Lal et al., Lead-Free Development of Electronics Connectors, APEX 2002, San Diego, CA.
23. D. Romm et al., Evaluation of Nickel/Palladium/Gold-Finished Surface-Mount Integrated Circuits, Application Report SZZA026, July 2001.
24. C. Hunt, Benchmarking Palladium and Gold Component Termination Finishes, IPC Works 2000, Miami, FL.
25. M. Harrison, J. Vincent and H. Steen, Lead-Free Reflow Soldering for Electronics Assembly, Soldering and Surface Mount Technology, 2001.
26. H.-J. Albrecht and K. Wilke, Board Level Reliability of Lead-Free Soldered Interconnects on Conventional and Area Array Components, SMI 2001.
27. J. Vincent et al., Lead-Free Reflow Soldering in Electronics Assembly, IPC Works 2000, Miami, FL.
28. R. Ludwig et al., Evaluation of Two Novel Lead-Free Surface Finishes, APEX 2002, San Diego, CA.
29. M. Nowottnick et al., Investigation of Lead-Free Solder Processing, SMI 2001.
30. W. B. Hance and N.-C. Lee, Voiding Mechanisms in SMT, China Lake 17th Annual Electronics Manufacturing Seminar, 1993.
31. L. Li et al., Pb-Free Solder Paste Reflow Window Study for Flip Chip Wafer Bumping, 2001 Symposium on Advanced Packaging Materials.
32. A. Primavera, Factors that Affect Void Formation in BGA Assembly, IPC/SMTA Electronics Assembly Expo, 1998, Rhode Island.
33. W. O'Hara and N.-C. Lee, Voiding mechanism in BGA assembly, in Proceedings, 1995 ISHM, Los Angeles, CA.
34. J. Bath and E. Crombez, Surface Mount Evaluations of Lead-Free Solder Pastes, NEPCON East, June 2000, Boston, MA.
35. M. Harrison and J. Vincent, Improved Design Life and Environmentally Aware Manufacturing of Electronics Assemblies by Lead-Free Soldering, SMART Lead-Free conference, UK, 1999.
36. J. Bath, A Manufacturable Lead-Free Surface Mount Process, Circuits Assembly, January 2003.
37. D. Romm et al., Evaluation of Nickel/Palladium-Finished ICs with Lead-Free Solder Alloys, Application Report SZZA024, January 2001.

38. Y. Tanaka et al., Lead-Free Soldering Technology for Mobile Equipment, 2000 International Symposium on Microelecronics.

Section 8.5

39. C. Melton, Nitrogen Atmosphere Processing in Lead-Free Soldering, Nepcon West 1995, Anaheim, CA.
40. C. Hunt et al., Evaluation of the Comparative Solderability of Lead-Free Solders in Nitrogen, APEX 2002, San Diego, CA.
41. C. C. Dong et al., Effects of Atmosphere Composition on Soldering Performance of Lead-Free Alternatives, Nepcon West 1997, Anaheim, CA.
42. A. Z. Miric and A. Grusd, Lead-Free Alloys, Soldering and Surface Mount Technology, 1998.

Section 8.6

43. J. P. Holman, *Heat Transfer*, McGraw-Hill, New York, 1986.
44. K. Suetsugu, Development and Application of Lead-Free Solder Bonding Technology, InterNepcon Japan, January 1999.
45. T. Baggio et al., Challenges and Solutions for Lead-Free Soldering of Large PCB Assembly, APEX 2000, Long Beach, CA.

Section 8.7

46. K. Walters, Board S Lead-Free Applications NEMI Report, May 2001.
47. K. Walters, Board I Lead-Free Applications NEMI Report, July 2001.
48. K. Walters, Board H Lead-Free Applications NEMI Report, July 2001.
49. R. Parker, NEMI Lead-Free Component Group Executive Summary Report, September 2002.
50. P. Bourgelais and D. Barbini, Profiling Trade-Offs for a Lead-Free Process, August 2001.
51. M. Kelly et al., Component Temperature Study on Tin-Lead and Lead-Free Assemblies, SMTA International, Chicago 2002.

Section 8.9

52. O. Bast, NEMI Optical Appearance of Solder Joints Using Different Solder Alloys, June 2001.

Section 8.10

53. J. Bath, NEMI Lead-Free Reliability X-Ray Inspection Report, April 2002.
54. R. Shane Fazzio, Effects of Lead-Free Solders on Imaging Characteristics of Laminographic X-Ray Inspection System, X-Ray User's Group Conference, 1998.

Section 8.11

55. S. Martell, Interim Acoustic Microscopy (AM) Analysis Rework for NEMI Lead-Free Solder Study: Phase 1 (Pre-Reflow) and Phase 2 (Post-Reflow) Evaluations, February 2002.

Section 8.12

56. M. Walz and T. Leahy, NEMI Lead-Free BGA/CSP Rework Study, December 2001.
57. J. Furnanz et al., Manufacturing and Reliability Evaluation of a Lead-Free Electronics Network Card, IPC Annual Meeting, November 2002, New Orleans.
58. A. Gowda, H. Srihari and A. Primavera, Lead-Free Rework Process for Chip Scale Packages, NEPCON East, June 2001, Boston, MA.
59. A. Gowda et al., Rework of Lead-Free Surface Mount Components, APEX 2002, San Diego, CA.
60. A. Gowda et al., Challenges in Lead-Free Rework, Pan Pacific Conference, February 2002, Hawaii.
61. M. Tsang and R. A. Szymanowski, Rework Processes for Lead-Free Assembly, APEX 2002, San Diego, CA.
62. C. Reynolds and J. Gleason, NEMI Lead-Free Hybrid Assembly and Rework Project Statement of Work, October 2002.

Section 8.13

63. J. Bath and S. Sethuraman, Lead-Free Solder Wire Rework Evaluation, Internal Report, November 2002.
64. M. Warwick, Implementing Lead-Free Consortium Research, SMTA-I Conference, San Jose, CA, September 1999.
65. Y. Tsunematsu et al., Evaluation of Pb-Free Solders for Adaptability to Various Soldering Processes, in *Proceedings of EcoDesign '99*, Tokyo, Japan, Feb. 1999, pp. 610–614 .

Section 8.14

66. J. Bath and J. Nguyen, Lead-Free Solder Paste ICT Evaluation, Soletron Internal Report, November 2002.
67. B. Arzadon et al., Lead-Free Solder Paste with ICT Probe Penetrable Residues, APEX 2002, San Diego, CA.
68. S. Wong and M. Nestor, High Volume Lead-Free Production, SMI 2001.
69. T. Baggio, The Lead-Free Mini Disk Player: Turning a New Leaf in a Lead-Free Market, IPC Works 1999, Minneapolis, MN.

Section 8.15

70. V. Godavarti, Lead-Free Cell Phone Manufacturability Evaluations, IPC/JEDEC Lead-Free Conference, April 2002, Santa Clara, CA.

Section 8.16

71. P. Harris and M. Whitmore, DTI-UK Research Program on Lead-Free Solders, *Circuits World* **19**, 25–27, 1993.
72. NCMS Lead-Free Solder Project Final Report, August 1997.

73. K. Akinade et al., Lead-Free Solder Paste Evaluations at Transmission Products Division, Soldering and Surface Mount Technology, 1995.
74. T. Nakatsuka et al., Reliability of Pb-Free Solder Joints of Surface-Mounted LSI Packages after Flow (Wave) Soldering, 2000 International Symposium on Microelectronics.
75. A. Yamaguichi et al., Development of Lead-Free Reflow Process Technology—Influence of Heat History on Solder Joint Quality, Pan Pacific Conference, 2002, Hawaii.
76. Q. Zhu et al., The Effect of Pb Contamination on the Microstructure and Mechanical Properties of SnAg/Cu and SnSb/Cu Solder Joints in SMT, Soldering and Surface Mount Technology, 2000.
77. S. Choi et al., Effects of Pb Contamination on the Eutectic Sn–Ag Solder Joint, Soldering and Surface Mount Technology, 2001.
78. K. Seelig and D. Suraski, Advances Issues in Assembly, Part 1: Lead contamination in lead-free assembly, *SMT Magazine*, October 2001.
79. J. Oliver et al., Fatigue Properties of Sn3.5Ag0.7Cu Solder Joints and Effects of Pb Contamination, SMTAI Conference, October 2002, Chicago, IL.

CHAPTER 9

Case Study: Pb-Free Assembly, Rework, and Reliability Analysis of IPC Class 2 Assemblies

JERRY GLEASON, CHARLIE REYNOLDS, MATT KELLY, JASBIR BATH, QUYEN CHU, KEN LYJAK, and PATRICK ROUBAUD

9.1. INTRODUCTION

The 2002 International Electronics Manufacturing Initiative (iNEMI) Roadmap acknowledged that the first iNEMI Pb-free project had laid the foundations for Pb-free manufacturing processes, including the selection and recommendation of the Sn–3.9Ag–0.6Cu solder alloy. However, it was recognized that more development work was needed for rework and wave soldering and to extend manufacturing process development to larger, higher-thermal-mass printed circuit assemblies. To accomplish this, a new project was initiated that included manufacturing-level studies on the assembly and rework of large, complex, high-thermal-mass component board assemblies, representing IPC Class 2 second-level assembly manufacturing.

For Pb-free assembly processing to be successful in the electronics industry, a highly capable manufacturing process must be demonstrated that includes multiple high-temperature component exposures during assembly and rework. Also, the repeatability and reliability of completed assemblies must be assessed. The electronics industry today is challenged to provide the highest reliability in products at competitive costs while meeting regulatory requirements. The tin (Sn)–silver (Ag)–copper (Cu) (SAC) solders have demonstrated the most promise for Pb-free assembly, including the iNEMI established formulation of Sn–3.9Ag–0.6Cu, which was utilized in this work. Similar work has been reported for baseline SMT assembly of large Pb-free assemblies [1]. As part of iNEMI's charter to establish manufacturing principles in the electronics' industry, a team of 19 companies and one university formed a project workgroup to evaluate the Sn–Ag–Cu alloy as a viable manufacturing

Lead-Free Electronics. Edited by Bradley, Handwerker, Bath, Parker, and Gedney

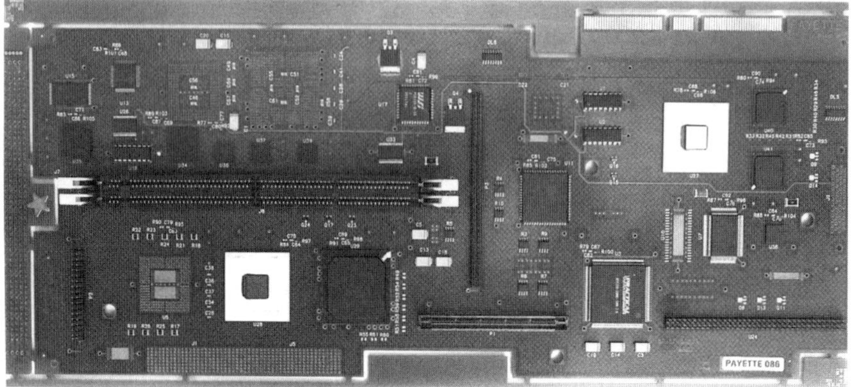

Figure 9.1. iNEMI project Phase 3 test vehicle (Payette).

alternative solder material to Sn–Pb solder. For reference, the top side of the 7-in. × 17-in. IPC Class 2 test vehicle named Payette is shown in Figure 9.1.

9.2. APPROACH AND STRATEGY

9.2.1. Assembly Strategy—SMT (Surface-Mount Technology) Focus

The main approach used for the assembly development team was to (1) focus on IPC Class 2-type assemblies, (2) utilize current process flows and equipment, (3) study, develop, improve, and implement new Pb-free SMT processes including printing, placement, and reflow operations, and (4) ensure PCB laminate survivability throughout the two passes of SMT followed by PTH and rework processes. Deliverables were to include the following:

A. Assembly Process

1. Define assembly process window for IPC Class 2 assemblies.
2. Determine the absolute minimum solder joint temperature during Sn–Ag–Cu reflow processes.
3. Quantify any differences between trace surface finish and board thickness effects.
4. Multiple iterations to minimize ΔT ($T_{max} - T_{min}$) across IPC Class 2 board assemblies.

B. Laminate Performance Assessment

1. Develop a pre-conditioning method of 260°C maximum peak temperature with six reflow passes (6×),
2. Perform IST (interconnect stress testing) and CITC (current induced temperature cycling) testing after 6× Pb-free laminate pre-conditioning.

3. Compare IST and CITC test methods.
4. Assess one potential laminate material for survivability.

9.2.2. Rework Strategy—Area Array Package Focus

Rework development was conducted on the iNEMI Payette reliability test vehicle on both top-side and bottom-side components. The top-side rework included micro ball grid array (μBGA) packages, plastic ball grid array (PBGA) packages, ceramic ball grid array (CBGA) packages, and dual inline packages (DIP). The bottom-side rework included a PBGA544 component and both thin, small outline packages (TSOP) and 2512 chip components. The rework objectives were as follows:

1. Develop a rework process for area array packages using conventional hot gas rework system.
2. Investigate a pin-through-hole (PTH) component attachment rework process.

The following parameters were considered in area array package and PTH rework development:

- Array Packages
  - Conventional hot gas rework equipment
  - Board thickness: 0.093 in. and 0.135 in.
  - Components: μBGA, PBGA, CBGA
  - Surface finish: immersion Ag and electrolytic Ni–Au
  - No-clean solder paste: Sn–3.9Ag–0.6Cu and Sn–37Pb
- DIP16:
  - Mini-pot solder fountain equipment
  - Board thickness: 0.135 in.
  - Rework nozzle: 0.48 in. × 0.96 in.
  - Board finish: Ni–Au
  - Solder alloy: Sn–Ag–Cu and Sn–37Pb
  - Wave flux: no-clean water-based VOC-free

9.2.3 Process Robustness Test Strategy—ATC Focus

Essential to this iNEMI project was the evaluation of the reliability of first pass and reworked Pb-free component solder joints on thick, high-thermal-mass Sn–Pb and Pb-free PCBs. To evaluate this reliability, two tests were chosen:

A. Accelerated thermal cycling (ATC) tests were used to measure solder joint thermal fatigue resistance using continuous *in situ* daisy chain resistance measurements.

TABLE 9.1. Parameters Evaluated During the ATC and the Bend Test Experiments

Parameter	ATC	Bend Test
Metallurgy of the solder joint (Sn–Pb or Pb-free)	Yes	Yes
Rework operation	Yes	Yes
Thickness of the PCB (0.093 in. or 0.135 in.)	Yes	Yes
Nature of the PCB surface finish (Ni–Au or immersion Ag)	Yes	No

B. A four-point bend test was used to measure solder joint strength and PCB structure. Four parameters were evaluated, as shown in Table 9.1.

In addition, the joints were analyzed using X-ray images, cross sectioning, dye and pry, optical microscopy, and SEM.

The iNEMI Payette board used as a test vehicle has many types of components. Seven different components were selected to represent classes covering a wide variety of solder joint types including area array, leadframe, through-hole components, and passive components. More details on the reliability test strategy can be found in Ref. 2.

9.3. OBSERVATIONS AND RESULTS

9.3.1. SMT Assembly

Figure 9.2 shows the effects of elevated lead-free processing temperatures when compared with current Sn–Pb reflow and hot gas rework processes [3]. The Pb-free Sn–Ag–Cu-based alloy system evaluated has a near-eutectic melting point (217°C)

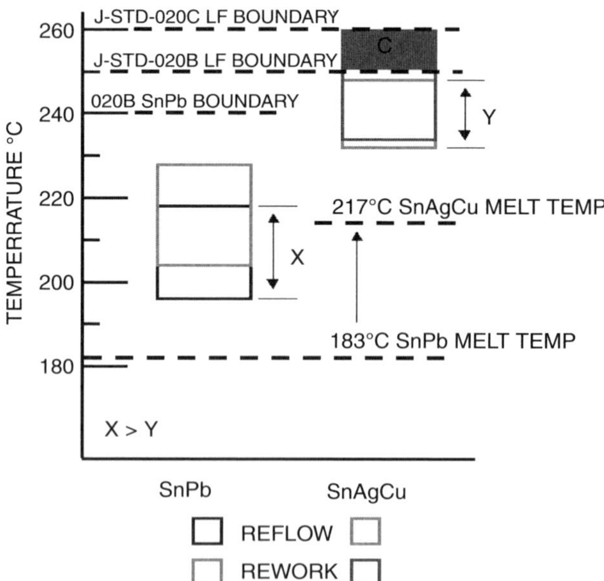

Figure 9.2. Lead-free versus Sn–Pb process windows.

approximately 34°C higher than the Sn–37Pb eutectic (183°C). Maximum package body temperature limits, which are now set by the newly revised IPC/JEDEC J-STD-020C Moisture/Reflow Sensitivity Classification specification for components, are 245°C, 250°C, or 260°C, depending on the package volume and thickness.

During Pb-free reflow soldering, maximum peak component body temperatures of 245–250°C were consistently measured in the iNEMI Payette board trials. The hottest measurements were found on small form factor components, usually passives, while the coolest temperatures were recorded in center solder joints of BGA-area array devices which were consistent with other studies [4]. These temperatures were attained on the 7-in. × 17-in. Payette board using both the 0.093-in. and 0.135-in. thicknesses. A Pb-free SMT reflow process was developed to assemble more than 100 Payette test boards conforming to J-STD-020C specifications for component temperature exposure limits.

Higher lead-free processing temperatures placed greater stress on components and boards and amplified the cumulative heat exposure effect. Figure 9.3 illustrates the cumulative effect of reflow and rework heat cycles on a single assembly. A typical assembly, such as the iNEMI Payette test board shown in Figure 9.1, would be subjected to

1. Bottom-side reflow (join bottom side mounted components)
2. Top-side reflow (join top side mounted components)
3. Wave solder (join pin in hole components)
4. First (1X) rework (repair faulty components)
    a. Remove component (unsolder)
    b. Dress board by reflowing and removing excess solder
    c. Place and resolder component

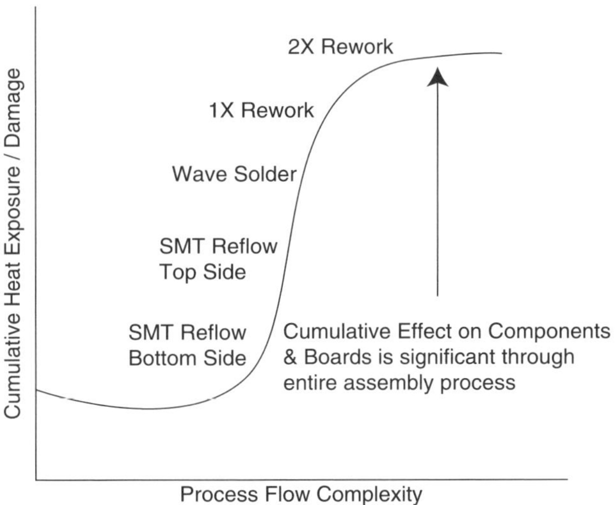

Figure 9.3. Cumulative heat exposure/damage.

5. A second (2×) rework. (repair another faulty component)
   a. Remove component (unsolder)
   b. Dress board
   c. Place and resolder component

Therefore, an assembly could be exposed to between five and nine thermal heat excursions: two reflow passes, a wave solder pass, and two hot gas rework passes (SMT, 2; wave, 1; BGA rework (1×), 2–3 (remove, replace, and potentially non-contact redress where the whole board is heated from the bottom side), BGA rework (2×): 2–3). By surviving at least five full thermal excursions, the board can be considered fairly robust in today's manufacturing environment where customers demand high-value hardware with rework capability.

Internal package structures within components and PCBs must survive all processes and still provide long-term reliability. Component effects included increased moisture sensitivity levels (MSLs) and resulting shorter exposed floor life, while PCB laminates must withstand internal layer delamination, via cracking and board warpage.

Linear ramp-to-peak Pb-free profiles were successfully developed for 0.093-in.-thick and 0.130-in.-thick constructions of Payette, and are shown in Figure 9.4. The immersion Ag surface finish was used only on 0.093-in.-thick Payette boards, while the Ni–Au surface finish was used on both thicknesses. Extensive metallurgical analysis clearly showed that properly formed intermetallics were achieved at the

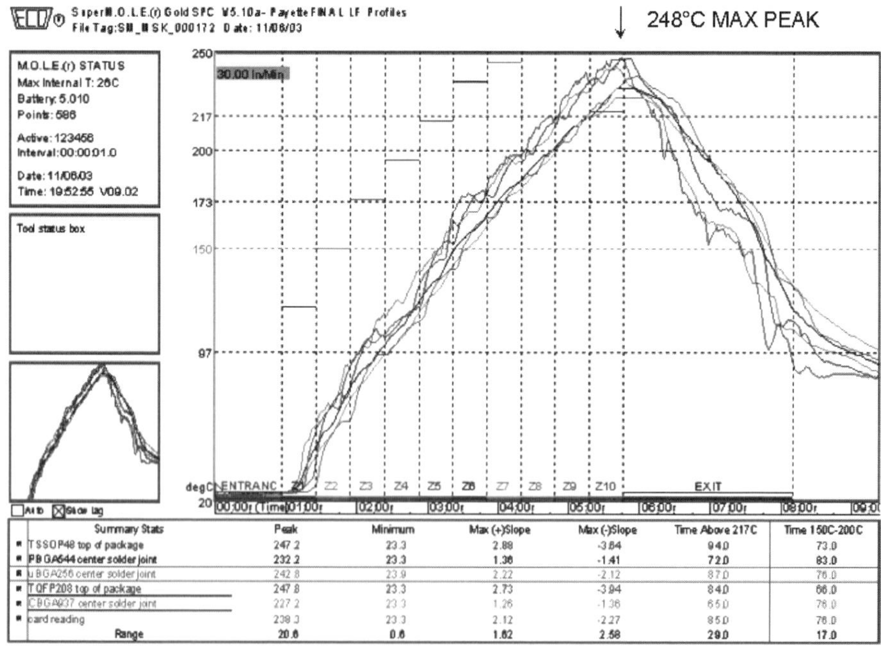

Summary Stats	Peak	Minimum	Max (+)Slope	Max (-)Slope	Time Above 217C	Time 150C-200C
TSSOP48 top of package	247.2	23.3	2.88	-3.84	94.0	73.0
PBGA544 center solder joint	232.2	23.3	1.36	-1.41	72.0	83.0
uBGA256 center solder joint	242.8	23.9	2.22	-2.12	87.0	76.0
TQFP208 top of package	247.8	23.3	2.73	-3.94	84.0	66.0
CBGA937 center solder joint	227.2	23.3	1.26	-1.36	65.0	76.0
board reading	238.3	23.3	2.12	-2.27	85.0	76.0
Range	20.6	0.6	1.62	2.58	29.0	17.0

Figure 9.4. Example of Pb-free SMT temperature profile used on iNEMI Payette board.

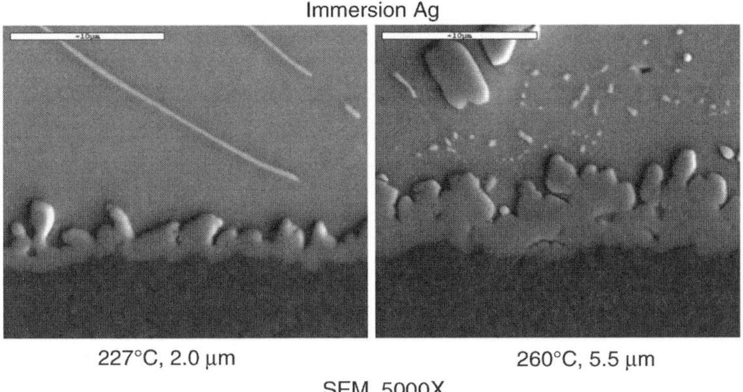

Figure 9.5. Immersion Ag joint formation.

minimum peak temperature of 227°C for Sn–Ag–Cu. Figures 9.5 and 9.6 show proper joint formation at this temperature on both electrolytic Ni–Au and immersion Ag surface finished PCB pads. The thicknesses of the intermetallic layers are dependent on the peak temperature: The higher the peak temperature, the thicker the intermetallic layer. Intermetallic layer thickness also depends on the surface finish: The Cu–Sn intermetallic layer formed on an immersion Ag surface finish is thicker than the Ni–Sn intermetallic layer formed on an Ni–Au surface finish.

All final Payette TQ (technology qualification) assemblies were SMT reflowed in an air atmosphere. Early development work indicated that the use of nitrogen during reflow increased solder joint aesthetics significantly, but did not help to increase solder joint performance based on pull and shear tests. Below in Figures 9.7 and 9.8, aesthetic appearance differences are shown for Sn–Ag–Cu soldered resistor and lead-frame components on two different board surface finishes.

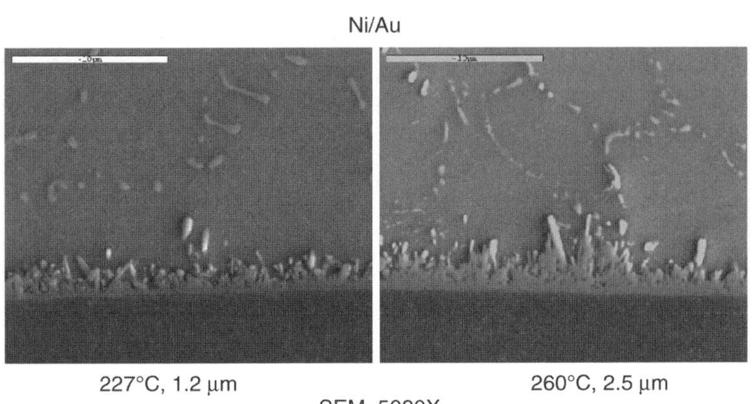

Figure 9.6. Ni–Au joint formation.

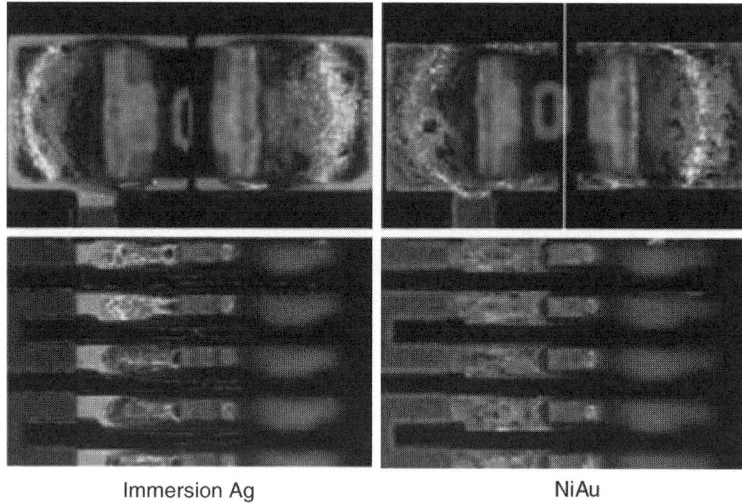

Figure 9.7. Air reflow joint aesthetic appearance.

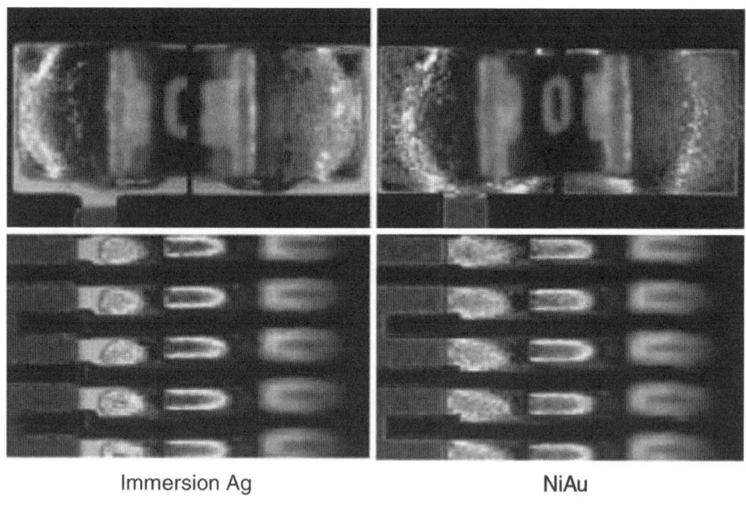

Figure 9.8. Nitrogen reflow aesthetic joint appearance.

9.3.2. Laminate Performance Assessment

Many of the laminate materials found in today's mid-to-high end reliability applications were originally formulated, designed, and qualified for eutectic Sn–Pb compatible materials and processes. These same laminate systems, when subjected to higher temperature Pb-free processes showed signs of blistering, delamination,

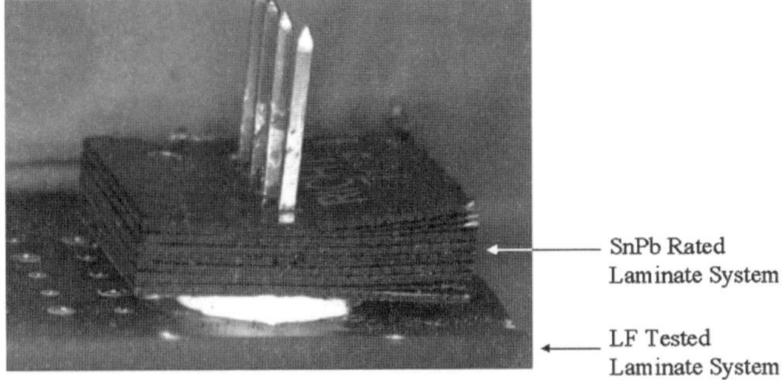

Figure 9.9. FR-4 Laminate header used for IST connections.

inner layer separation, micro-hardening, and laminate/barrel cracking. It is recommended that (1) more effort be focused on testing and qualifying new laminate systems suitable for elevated temperatures and (2) more effort be focused on qualifying fabricators who build these types of boards. A reliable PCB would require a good laminate material and a qualified fabricator who can demonstrate proper fabrication using these new Pb-free laminate materials. Figure 9.9 shows laminate performance differences: On top is a regular SnPb compatible laminate system, and below shows characteristics of a Pb-free laminate.

Based on the observations above, the assembly team worked on creating a laminate survivability test method, to help ensure that the selected laminate would survive the entire Pb-free assembly process. Interconnect stress testing (IST) and IBM's current-induced temperature cycling (CITC) test methods in combination with a "pre-conditioning" reflow profile were developed to help tackle this problem. Results showed that a 260°C 6× preconditioning profile followed by subsequent IST or CITC testing allowed the team to measure Pb-free laminate survivability through the entire process. Examples of different IST coupons from different

Figure 9.10. Examples of IST coupon versions from different test boards.

board thicknesses and surface finishes of the IST coupon design are shown in Figure 9.10.

9.3.3. Rework

9.3.3.1. Through-Hole Rework Development. Preliminary rework evaluations were conducted on a through-hole soldered PDIP component. An Sn–Pb through-hole rework process was used as a baseline. Acceptable hole fill results were obtained for Sn–Pb reworked PDIPs. For the Pb-free Sn–Ag–Cu rework process, different solder pot temperature settings were used: 500°F (260°C), 525°F (274°C), 550°F (300°C) at 5- and 10-second contact times with and without board preheat. Top board preheat of 120°C was provided with an external BGA rework machine because the mini pot used did not have the board preheat capability.

Removal of Pb-free assembled DIP16 component was achieved with 525°F (274°C) pot temperature. Re-soldering of the component could be accomplished at 525°F (274°C) pot temperature only by preheating the board to 120°C top side. A representative photograph is shown in Figure 9.11.

It was found that re-soldering (rework) was much more difficult than first-pass wave assembly or removal. Longer dwell times and preheat are needed for Pb-free mini-pot rework to obtain adequate solder wetting and hole fill.

Sn–Pb and Sn–Ag–Cu solder rework were compared for the DIP16 components on the 135-mil-thick boards, Sn–Pb solder appeared to have much more fluidity than the Pb-free Sn–Ag–Cu solder during the mini-pot rework operations. The time required for soldering and removing of Sn–Pb parts with Sn–Pb solder was much shorter than Pb-free parts with Sn–Ag–Cu solder. The Sn–Pb solder also gave better top-side soldering than did the Pb-free Sn–Ag–Cu solder. A preheat setup is required for Sn–Ag–Cu rework to achieve results similar to those of Sn–Pb

Figure 9.11. Representative photograph of reworked lead-free PDIP component.

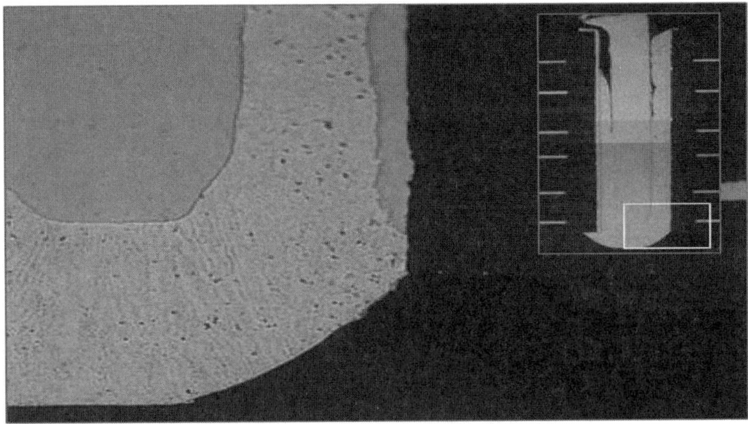

Figure 9.12. Cross section of reworked PDIP on a 0.135-in.-thick board using Sn–Ag–Cu solder. *Note*: Copper pad/barrel has been dissolved in the minipot during rework operation (274°C, 30-second dwell time).

rework. However, the use of a preheat stage is not typically common in a production environment, which would necessitate equipment upgrades.

Once a rework process was developed, PDIPs were reworked with Sn–Ag–Cu. Visual, X-ray, and cross-sectional analysis were performed. Using cross-sectional analysis, it was found that part of the copper pad barrel had dissolved into the solder pot. A representative picture is shown in Figure 9.12. Additional work is needed to define a workable process to characterize the integrity of the rework for through-hole solder joints while minimizing copper dissolution.

9.3.3.2. BGA Rework. Rework development evaluated the site redressing, paste printing, and more critical rework reflow profiles for three area array components: μBGA, CBGA, and PBGA. The rework was performed using current production rework equipment and tools. The previous IPC/JEDEC J-STD-020B specification was followed during the rework development, which was current during most of the development stage of the project.

9.3.3.2.1. Site Redressing. Two methods for site redressing, contact and non-contact, were conducted. The contact operation used was a traditional method that required a soldering iron making contact with the board through a copper braid. The second method used a vacuum scavenging system that sucks up the residual solder leaving behind a semi-flat surface for solder print and part placement. The techniques for both systems were found to be adequate for Pb-free site redressing. However, for the scavenging method, the filtration life was found to be about 30% shorter than Sn–Pb site redressing. Though not directly compared, the scavenging method appeared to be less likely to cause damage to the solder mask during site redressing but the cycle time may be longer.

9.3.3.2.2. Paste Printing. Three paste printing approaches were used for depositing the solder paste: (1) paste printing on component, (2) paste printing on board, and (3) paste dispensing on board. All three approaches were found to yield adequate paste deposition. Paste printing is more similar to the primary assembly screen printing process, using similar paste and stencil apertures.

9.3.3.2.3. Profile Development. Multiple rework trials were performed before developing satisfactory profiles for the PBGA, CBGA, and μBGA. Typical Sn–Pb and Sn–Ag–Cu PBGA544 rework profiles are shown in Figure 9.13.

For the Pb-free PBGA544 profile, the minimum solder joint temperature was approximately 230°C while the maximum package temperature was approximately 245°C. The Pb-free rework temperatures for three of the five reworks conformed to J-STD-020B (245–250°C peak), but this work was done with optimized rework equipment and nozzles with the active support and co-development from rework equipment suppliers. In addition, Pb-free rework profile runs using thermocoupled boards were conducted over a period of several months to develop and verify the best rework profiles. In production, the time spent developing rework profiles is typically hours or days, and this can be done only if thermocoupled profile boards are available. J-STD-020C, which has higher temperature limits (245–260°C peak), allows a much needed wider Pb-free process temperature window. Overall rework

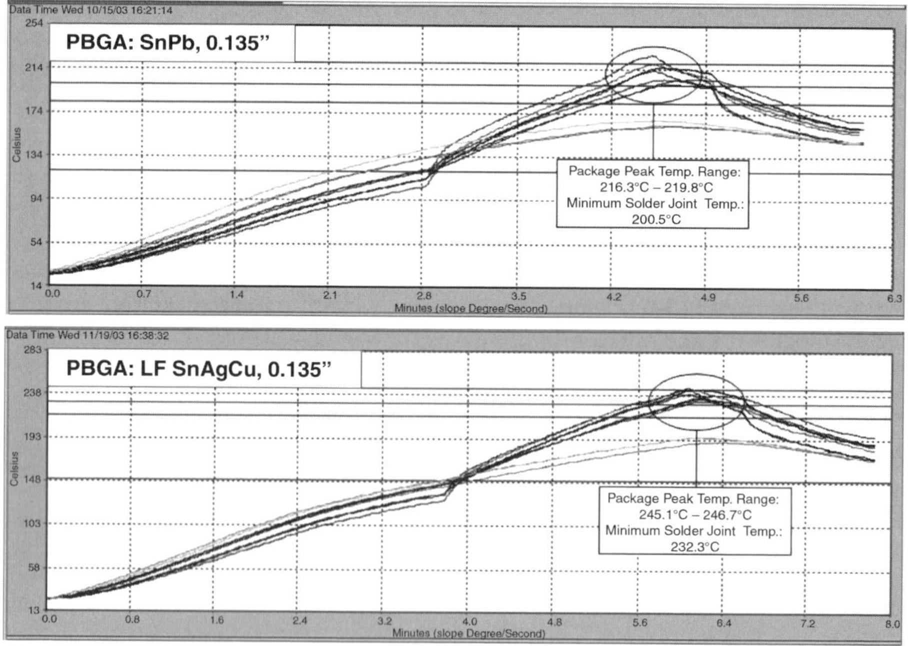

Figure 9.13. Representative Sn–Pb and LF Sn–Ag–Cu PBGA544 rework reflow profiles for iNEMI Payette Board.

time was approximately 8 minutes for Pb-free rework and 6 minutes for the Sn–Pb rework profile. In most cases, the board temperature 150 mils away from the rework component was above the liquidus reflow temperatures (for Sn–Pb and Pb-free rework). More details can be found in Ref. 5. During the profile developments, three key challenges were encountered:

- Minimizing top package temperature while allowing sufficient heat to form good solder joints
- Adjacent and bottom-side components exposed to unintended reflow temperatures
- Pb-free reflow parameters near the limits of solder paste and package specifications

With Pb-free rework, it was found that the bottom-side heater set point needed to be elevated compared to Sn–Pb reflow profiles. This was required to keep the top heaters from over heating the top of the package beyond the JEDEC 020C package temperature limitation. An even higher heat was applied to the more thermally challenging 0.135-in.-thick boards.

Increasing the bottom-side heaters to compensate for reduced heat from the top nozzle was found to have an adverse effect on bottom-side and adjacent components in terms of exceeding liquidus temperatures. During the μBGA rework, it was observed that the nearby CBGA was affected by this heat, which resulted in open solder joint connections on the CBGA. However, this was not observed for an adjacent μBGA that was also spaced at a similar distance to the CBGA. It was believed that component construction and size contributes to the differences observed.

Shielding of the CBGA during rework of the μBGA helped somewhat to reduce subsequent opens observed post rework but not completely. However, reliability was found to have decreased, and this was believed to be due to the adjacent rework process. Once all rework was performed on test boards for reliability testing, a side experiment was performed to better understand the thermal characteristics of adjacent heating.

Preliminary results showed that the adjacent CBGA had joint temperatures ranging from 211°C to 223°C, with the 223°C being closest to the reworked μBGA. Thermocouples were placed at the bottom side of the PCB corresponding to the CBGA joint locations above registered temperatures ranging from 237°C to 245°C. The adjacent μBGA had a solder joint temperature of 245°C. Table 9.2 and Figure 9.14 show the temperature results with locations of the thermocouples. The Pb-free CBGA had certain solder joints which melted and some which had not. The same issue was observed with the Sn–Pb CBGA where the adjacent Sn–Pb μBGA was reworked. Additional work is needed to help reduce bottom-side and adjacent component temperatures.

The higher temperature of the PCB and the adjacent μBGA observed was believed to have been caused by the direct bottom-side heating. The bottom-side heating element is a chamber with the top plate having a series of holes for hot air to flow and warm up the board (denoted in the Figure 9.14 as brown hashed

TABLE 9.2. TAL and Peak Temperature of μBGA-CBGA from Adjacent Rework Study

TC Location	Time Above Liquidus (seconds)	Peak Temperature (°C)
Reworked μBGA Joint	103.4	232.9
Adjacent μBGA Joint	149.4	245.5
CBGA 1 Joint	56.6	223.0
CBGA 1 Bottom PCB	124.1	237.2
CBGA 2 Joint	21.7	218.7
CBGA 2 Bottom PCB	121.4	237.3
CBGA 3 Joint	0.0	211.3
CBGA 3 Bottom PCB	153.4	245.1

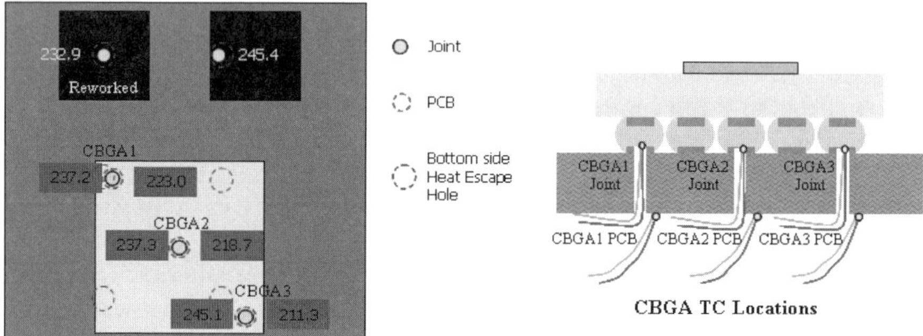

Figure 9.14. Thermocouple placement location of μBGA–CBGA Adjacent Rework Study.

circles). Depending where the board is positioned and how far away to the bottom heater, the PCB could experience different temperatures. Figure 9.15 illustrates how hot air flows from the bottom-side heater and concentrating on a given area.

After reviewing the Pb-free solder rework times for the PBGA, CBGA, and uBGA, the time above liquidus was found to be close to 90 seconds on many occasions. This leads to increased solder voiding for the solder paste used. More

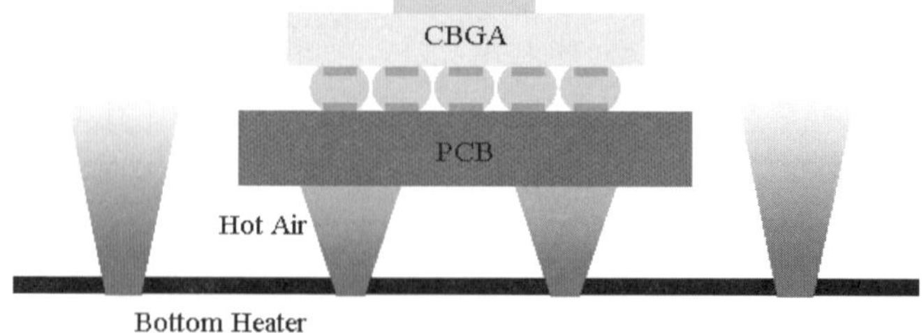

Figure 9.15. Diagram of bottom-side heating.

solder paste development work is needed to support the elevated Pb-free solder temperature profiles.

9.3.4. Process Robustness Assessment—Thermal

9.3.4.1. ATC Fatigue Resistance. Close to 6000 thermal cycles were applied to both as-assembled and reworked test boards. The thermal profile settings followed JEDEC JESD22-A104B standard recommendations with a minimum temperature of 0°C, a maximum temperature of 100°C, 11-minute ramp times, and 10-minute dwell times.

The parameters studied were the solder joint metallurgy (Sn–Pb or Pb-free Sn–Ag–Cu), the board thickness (0.093 in. or 0.135 in.), and the PCB surface finish (Ni–Au or immersion Ag). The test matrix of this experiment is shown in Table 9.3. On each PCB the electrical resistance of the following component daisy chains was continuously monitored using two automatic data acquisition systems:

- CBGA 937 (2 per board)
- Micro BGA 256 (2 per board)
- CSP 81 (3 per board): Not reworked component (first-pass assembly only)
- DIP 16 (2 per board): Not reworked component (first-pass assembly only)
- TSOP 48 (4 per board)
- PBGA 544 (2 per board)

In total, the electrical resistances of 952 components were individually and continuously monitored during this ATC experiment.

At the end of the experiment, three components types had sufficient data points (failed parts) to develop Weibull plots: the CBGA 937, the Micro BGA 256, and the CSP81. The other components (DIP 16, TSOP 48, and PBGA 544) did not

TABLE 9.3. Test Matrix for the ATC Experiment

Cell	Paste	Thickness (inch)	Rework	Surface Finish	Number of Boards
1	Sn–Pb	0.135	No	Ni–Au	4
2	Sn–Pb	0.135	Yes	Ni–Au	4
3	Sn–Pb	0.093	No	Ni–Au	4
4	Sn–Pb	0.093	Yes	Ni–Au	4
5	SAC	0.135	No	Ni–Au	8
6	SAC	0.135	Yes	Ni–Au	8
7	SAC	0.093	No	Ni–Au	8
8	SAC	0.093	Yes	Ni–Au	8
9	SAC	0.093	No	Immersion Ag	4
10	SAC	0.093	Yes	Immersion Ag	4
Total					56

fail in sufficient numbers for the generation of meaningful Weibull plots, indicating the excellent accelerated thermal fatigue lives obtained for these components.

9.3.4.1.1. Impact of the Solder Joint Metallurgy. The first-pass Pb-free Sn–Ag–Cu soldered parts did have longer fatigue life than their Sn–Pb soldered counterparts in our experiment. This trend is illustrated on the comparative Weibull graph in Figure 9.16.

9.3.4.1.2. Impact of PCB Thickness. For the as-assembled condition (no rework) the impact of the PCB thickness appeared to be small in our experiment. The Sn–Pb soldered components assembled on the 135-mil-thick boards tended to have the shorter accelerated fatigue lives. The trend was opposite for the Pb-free soldered components. The Pb-free soldered components assembled on the thicker board tended to have the longer accelerated fatigue lives as shown in Figure 9.16.

9.3.4.1.3. Impact of Rework. After rework, the same trend was observed. The Pb-free reworked parts had longer accelerated fatigue life than the reworked Sn–Pb parts in our ATC experiment.

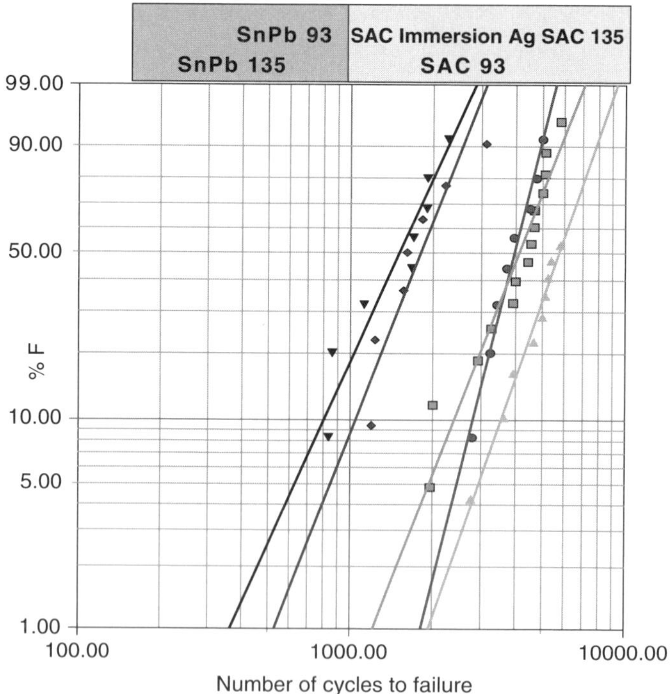

Figure 9.16. Weibull plots for the MicroBGA 256. The Pb-free components performed better than the Sn–Pb ones. The same trend was observed for the other component types.

In most cases the reworked components (both Sn–Pb and Pb-free) assembled on 0.093-in.-thick boards performed comparably to the as-assembled ones. However, the rework on thicker boards (0.135 in. thick) negatively impacted the accelerated fatigue life of many components. The high thermal exposures used to rework thick boards may have been responsible for this degradation. Excessive thermal exposures can damage the PCB material and create thicker intermetallic compounds. This degradation were was noticeable only on the μBGA solder joints, probably because of their smaller solder joints.

These excessive thermal exposures could have also negatively impact components adjacent to the reworked area, probably by inducing a reflow of some of their solder joints. This effect was observed on our test boards for several components. The CSP81 on Pb-free 0.135-in. boards, for example, had a degraded resistance to accelerated thermal fatigue performance potentially due to the rework of adjacent components (see Figure 9.17).

We observed that some CBGA solder joints reflowed during the rework of an adjacent μBGA. This reflow led to electrical opens and to very marginal solder joints for these CBGAs. The chart in Figure 9.18 illustrates the percentage of CBGAs that failed for different test cells. (The "R" prefix refers to Reworked

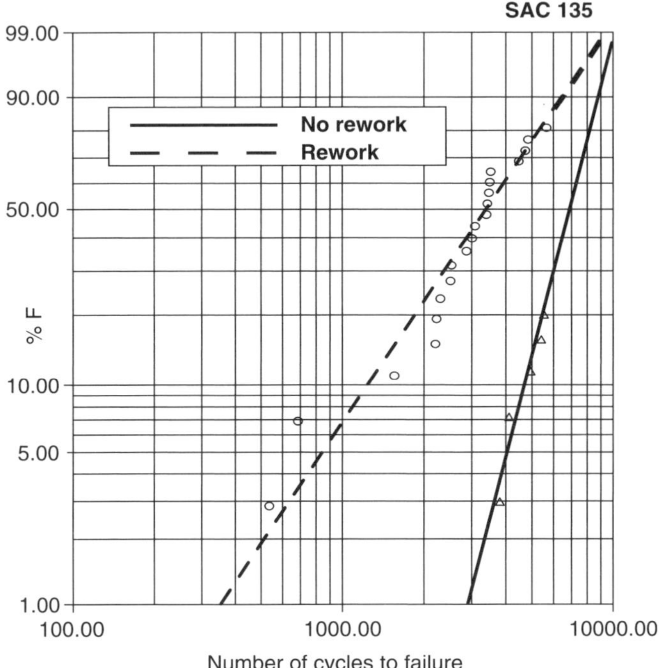

Figure 9.17. The resistance to thermal fatigue for the CSP81 on the 0.135-in.-thick boards with Ni–Au surface finish appeared to be negatively impacted by the rework of adjacent components. The CSP 81 themselves were *not* reworked.

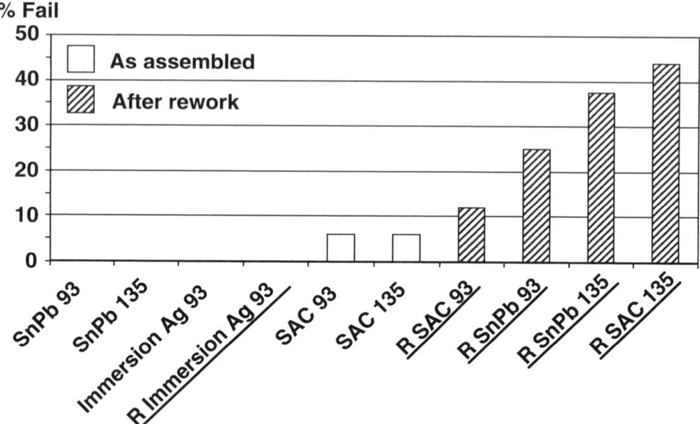

Figure 9.18. Many CBGA failed after the rework of adjacent μBGAs. There were no failures for the immersion Ag boards (fourth cell from the left) because in this case the CBGA was reworked after the μBGA.

components. The others are as-assembled.) One can see that test cells for the thicker boards (0.135 in.) following rework are the ones with the highest percentage of failed CBGAs. The reason is because higher thermal exposures had to be applied to these boards during the rework of adjacent μBGAs (this is true for both Sn–Pb and Pb-free reworked boards). We suggest performing a controlled experiment to understand the exact mechanism of this failure mode as a follow-on study. The rework profiling conducted is described in the previous BGA rework section.

9.3.4.1.4. Impact of the PCB Surface Finish. The surface finishes (Ni–Au or immersion Ag) had no significant impact on the resistance to thermal fatigue. (Immersion Ag was used only on the 0.093-in.-thick boards.)

9.3.4.1.5. Accelerated Thermal Fatigue Test Conclusions. The Pb-free components had acceptable results in terms of resistance to accelerated thermal fatigue. On average, they had longer accelerated thermal fatigue lives than their Sn–Pb soldered counterparts. It can be noted that the high thermal exposures used to rework the Pb-free components assembled on the thicker boards (0.135 in. thick) degraded the resistance to accelerated thermal fatigue of the components. The rework operation also led to premature failures leading to lower yields. The nature of the surface finish (Ni–Au or immersion Ag) had no significant impact on the failure rate.

It appeared that the rework process of Pb-free parts on thick boards (0.093 and 0.135 in.) was achievable, but is not a mature process at this time. More development work is needed to be done to transform it into a robust industrial process by improving yield and controlling the thermal exposures transmitted to the adjacent components in the reworked areas. Improving the temperature resistance of PCB laminate should also be explored.

9.3.4.2. Failure Analysis. This work presents the investigation of the as-assembled and reworked test boards.

In general the solder joints appeared to be acceptable except for the smallest BGA joints in the μBGA256. In this case the rework process appeared to be marginal because it created large levels of voiding, thicker layers of intermetallic compounds, and degradation in the solder mask material.

9.3.4.2.1. As-Assembled Solder Joints. Visual observations of the Pb-free and Sn–Pb solder joint shape revealed mostly good solder joints and some isolated defects. Cross-sectional analysis of the solder joints revealed typical microstructures for both the Sn–Pb and the Sn–Ag–Cu samples.

The level of voiding was higher in the Pb-free Sn–Ag–Cu soldered joints. For the smallest joints (such as the μBGA), this level was measured above 20% (by area).

For both the Sn–Pb and the Sn–Ag–Cu soldered joints the intermetallic compound (IMC) thicknesses was typically under 3 μm, which would not create a reliability concern. As expected, the IMC was thicker when the joints were formed on the immersion silver surface finish due to Cu–Sn intermetallic formation compared with Ni–Sn intermetallic for Ni–Au boards.

9.3.4.2.2. Impact of Rework. The microstructures observed after rework are typical for these kinds of solder joints. The lead-free reworked solder joints had more voiding than the as-assembled ones. For the smallest joints such as the μBGA, the level of voiding exceeded 20% (see Figure 9.19).

The intermetallic layers were slightly thicker after rework compared with first pass. In this case of the small solder joints (μBGAs) formed on the immersion silver surface finish, the thickness was measured above 6.5 μm.

Figure 9.19. Illustration of high level of voiding in reworked Sn–Ag–Cu μBGA solder joints.

Solder penetration under the solder mask in the vicinity of the μBGAs was observed. The higher thermal exposure imposed during rework of these components may have induced this defect.

9.3.4. Process Robustness Assessment—Mechanical

PCBs with Pb-free Sn–Ag–Cu and eutectic Sn–Pb CBGAs and PBGAs were subjected to four-point bend tests. The test matrix was set up such that the effect of type of package, board thickness (0.093 in. versus 0.135 in.), and effect of rework (reworked versus non-reworked) on robustness during the 4-point test mode could be ascertained. The test matrix is described in Table 9.4. Only boards with Ni–Au surface finish were tested.

Both load and deflection were significantly lower (worse) for Sn–Ag–Cu solder than for eutectic Sn–Pb solder. This result is illustrated on Figure 9.20. The average load to failure was 0.41 kN for Sn–Pb packages and only 0.23 kN for Sn–Ag–Cu packages.

Almost all failure modes observed in the samples were either within the PCB laminate or between the intermetallic compound and the nickel underlayer on the PCB land.

Measurements showed that the micro hardness of the laminate under the PCB pads significantly increased after the Pb-free Sn–Ag–Cu solder assembly processes. This implies some embrittlement of the laminate when subjected to the higher Sn–Ag–Cu reflow and rework temperatures.

To explain the lower robustness of Pb-free CBGAs and PBGAs in the four-point bend test experiment, two possible root causes were proposed:

- The higher Sn–Ag–Cu solder stiffness subjected more mechanical stress into the laminate material.
- The laminate material became more brittle after being subjected to the higher Sn–Ag–Cu reflow and rework temperatures.

The higher sensitivity of the Pb-free solder joint/pad-board structure (Sn–Ag–Cu solder joint with higher stiffness combined with more brittle laminate material) to

TABLE 9.4. Test Matrix for the Bend Test Experiment

Cell	Paste	Thickness (inch)	Rework	Surface Finish	Number of Boards
1	Sn–Pb	0.135	No	Ni–Au	3
2	Sn–Pb	0.135	Yes	Ni–Au	3
3	Sn–Pb	0.093	No	Ni–Au	3
4	Sn–Pb	0.093	Yes	Ni–Au	3
5	SAC	0.135	No	Ni–Au	3
6	SAC	0.135	Yes	Ni–Au	3
7	SAC	0.093	No	Ni–Au	3
8	SAC	0.093	Yes	Ni–Au	3
Total					24

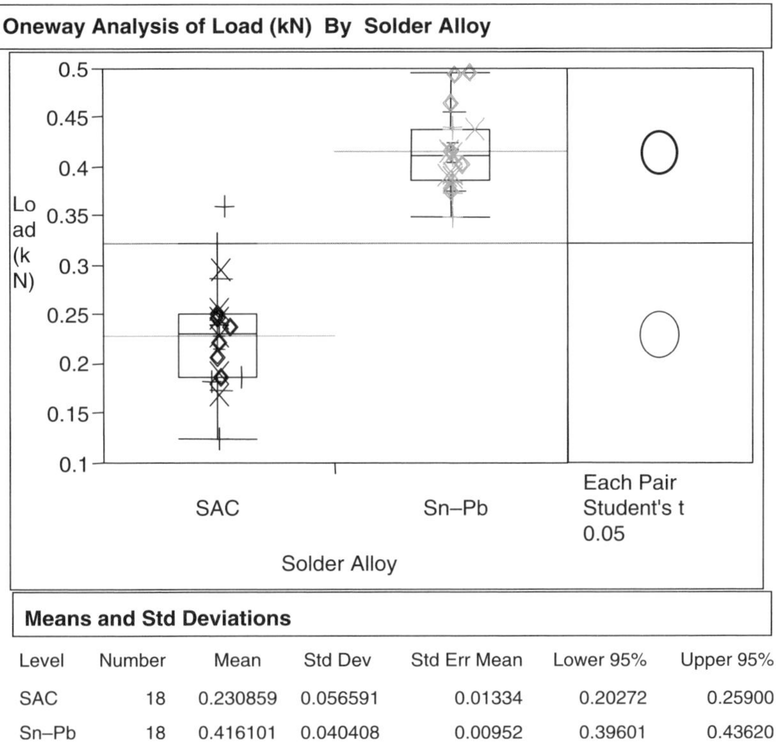

Level	Number	Mean	Std Dev	Std Err Mean	Lower 95%	Upper 95%
SAC	18	0.230859	0.056591	0.01334	0.20272	0.25900
Sn–Pb	18	0.416101	0.040408	0.00952	0.39601	0.43620

Figure 9.20. Four-point bend test results for the 0.093-in.-thick test boards. The Sn–Pb packages had significantly higher loads to failures than the Sn–Ag–Cu ones.

bending suggests that failures may happen during manufacturing operations like electrical testing where mechanical stress is applied to the board. Handling and testing procedures used with Sn–Pb assemblies might need to be modified when switching to lead-free assemblies in order to minimize risk of damage.

9.3.5. Shadow–Moiré Measurements

Shadow–moiré measurements were performed to measure the amount of deformation induced by the high temperatures reached during the Pb-free assembly process. Both the PCB itself and a set of components were measured with this technique. Cross sectioning of a board was conducted from the same board lot to check for delamination or other defects. A set of selected components was also cross-sectioned to get an understanding of the solder joint microstructure. Finally, some components were checked for internal delamination using a CSAM analysis.

The boards exhibited a permanent deformation (which was still within IPC specifications) after being submitted to a Pb-free reflow cycle similar to that used during Pb-free assembly or rework of the boards (see Figure 9.21). The PCA flatness was measured at 3.6 mils/inch. It still does meet the standard IPC-A-610D (Section 10. 2.7, Note), which recommends a flatness specification of 7.5 mils/inch or less.

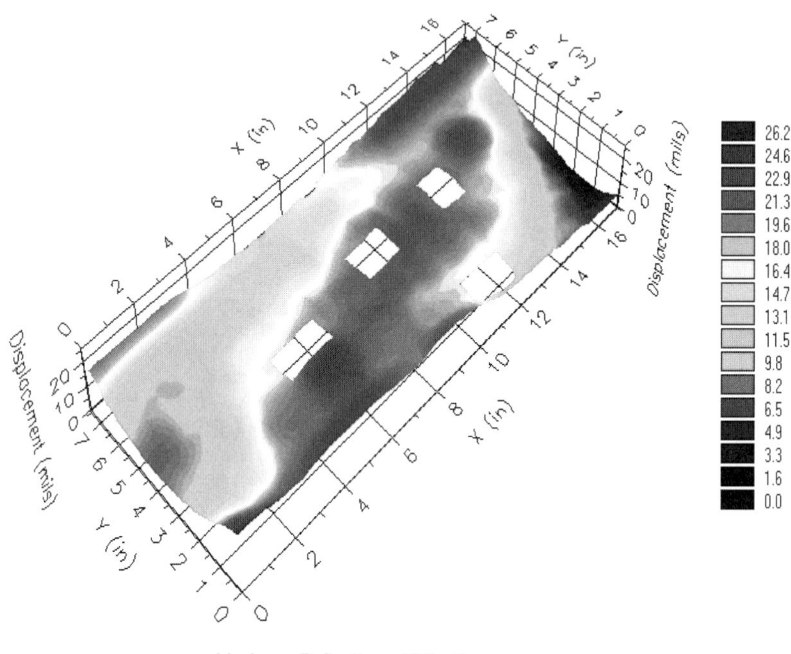

Figure 9.21. Thermo moiré of a board after a thermal excursion following the thermal profile used for Pb-free assembly. The thermal excursion induced a permanent deformation of the board.

Neither delamination nor other structural defect was observed during the cross-section study and CSAM analysis.

9.4. CONCLUSIONS

9.4.1. SMT Assembly

1. IPC Class 2 Pb-free SMT reflow process windows for Sn–Ag–Cu will shrink when compared to current Sn–Pb processes. The Pb-free Sn–Ag–Cu-based alloy systems have a near-eutectic melting point approximately 34°C higher than the Sn–Pb eutectic point. Utilizing current 10-zone convection reflow ovens within the Pb-free process means that assemblies are required to get much hotter in the same length of oven, without violating many of the key variables including time above liquidus, maximum peak temperature, soak temperature/time, and

heating/cooling ramp rates. Therefore, greater care will be required during profiling efforts to ensure that all specified targets are met while ensuring that throughput rates are not significantly reduced.

2. Multiple heat cycles cause laminate and via damage. The higher lead-free processing temperatures placed greater stress on components and boards and amplified the cumulative heat exposure effect. Internal package structures within components and PCBs need to survive all processes and still provide long-term reliability. Component effects included increased moisture sensitivity levels (MSL) resulting in shorter exposed floor life. PCB laminates must suppress internal layer delamination, via cracking and board warpage.

3. Based on iNEMI test results, Pb-free SMT first-pass reflow processing typically had the following characteristics:

- Four- to six-minute cycle times.
- Temperatures ranging from 230°C to 250°C for joints and body temperatures.
- ΔT ranges from 5°C to 20°C on a single assembly.
- Time above liquidus ranges from 60 to 90 seconds.
- Linear ramp to reflow profile shapes (paste supplier recommendation).

4. SMT reflow using an air atmosphere was shown to produce properly formed solder joints. All final iNEMI Payette assemblies were run in air only. The study shows that reflow processing in air was acceptable to produce reliable solder joints.

5. The use of nitrogen was shown to promote wetting, and it created shinier-looking solder joints (improved aesthetics). If there are aesthetic concerns, the use of nitrogen would ensure that Pb-free soldered joints "look better." The study between air and nitrogen showed no significant difference in microstructure formation and final reliability performance (pull/shear tests).

6. Trials and testing recommended that the coldest solder joint on Any Pb-free assembly should be no less than 230°C. This would be the lowest recommended temperature at any location on an assembly. This coldest temperature would be expected at the center-joint of the largest BGA package on the assembly (or equivalent). Metallurgy studies conducted within this program show that solder joint metallurgy was still acceptable at processing temperatures of 227°C. It is not recommended to process solder joints at this 227°C temperature because of process tolerances. The work completed helped to indicate the lower limit of temperature requirements needed to make reliable lead-free solder joints.

9.4.2. SMT Assembly Future Work

1. It was found that many PCB laminate systems are not surviving Pb-free primary attachment or rework processes and should be the subject of future studies. Elevated Pb-free SMT and rework processing temperatures can cause

significant quality and reliability concerns when using existing PCB laminate systems (originally designed for conventional Sn–Pb processes).

2. IST and CITC tests were shown to help in assessment of survivability of laminates involving elevated Pb-free temperatures. Lead-free pre-conditioning reflow excursions were used to help expose the laminate to temperatures up to 260°C. Laminates were first pre-conditioned and were then run through subsequent IST or CITC test protocols. More work is recommended to further develop pre-conditioning methods and to help continue assessment of the reliability of laminates for use with Pb-free second-level assembly processes.

3. The majority of challenges during processing were recorded when using 0.135-in. stack-up designs. The study shows that continued focus is needed on process control to ensure high quality and high reliability for "thick" assemblies. Reliable Pb-free primary attachment is possible using 0.135-in. board constructions; however, processing windows are small, and tight process control is necessary during assembly.

9.4.3. Rework-Area Array Package Focus

Based on the rework done, a set of Pb-free area-array component best practices is listed below:

1. Check solder paste and component specifications to verify maximum allowable rework temperature limits for Pb-free. The new J-STD-020C standard (with body package temperature maximum limits of 260°C) will provide a larger window for the rework profile developer to create rework profiles.
2. When developing Pb-free profiles, bottom heaters may need to be set ~50°C higher than Sn–Pb operations.
3. One needs to consider the following when developing the rework reflow profile:
   a. Peak temperature ranging from 230°C to 255°C for joints and body temperature.
   b. Time above liquidus may range from 45 to 90 seconds, and potentially over 90 seconds. For large packages on thicker boards, this needs to be controlled.
   c. Linear ramp to peak temperature to help minimize ΔT across component body.
   d. Attach thermocouples to monitor at least two solder joint temperatures and two package body temperatures (center and corner for both) during rework to ensure good thermal profile characterization.
4. When starting a new batch of rework, re-verify rework profile developed.
5. Confirm that the laminate system being used can withstand up to five Pb-free thermal excursions. This is the minimal number to guarantee for a robust process that demands reworkability.

9.4.4. Rework Future Work

This should be concentrated in two areas indicated in the following sections:

BGA Rework

1. Minimize adjacent component and bottom-side temperatures. The board temperature (150 mils away from the reworked component) was found to exceed the melting temperature of the soldering alloy for Sn–Pb and Pb-free soldering. A particular issue was also noted when reworking the μBGA256. It was found that the adjacent CBGA had undergone a partial double reflow which could have weakened its mechanical solder joint integrity. This needs to be understood in terms of process optimization and/or implications for design guidelines for Sn–Pb and Sn–Ag–Cu.
2. Thermal controls of the rework machines were developed for Sn–Pb soldering, and there is a need for better and higher temperature capability and control with Sn–Ag–Cu solder.
3. It was found that bottom-side heat and thermal uniformity was critical to bring the board up to proper Pb-free rework temperatures, but increased bottom heating may impact the reliability of the board material. This needs to be assessed.
4. Use of a retrofit heat shroud over the reworked board could add the benefit of reducing bottom preheat setting used, especially for thicker boards. The use of retrofit heat shroud needs to be assessed further.
5. Rework equipment suppliers need to develop their equipment more for higher-temperature Pb-free soldering with an emphasis on optimized rework profiles and optimized machine tool sets. At this time, limited data exist on rework equipment temperature tolerances and repeatability.

Through-Hole Component Rework (PDIP). Improved hole fill is needed with Sn–Ag–Cu solder with emphasis on development of the rework process on thicker boards such as 135 mils thick. Copper dissolution was observed on copper traces/pads while performing the through hole rework with Sn–Ag–Cu solder due to increased pot temperatures and soldering times used. Some success was achieved in reducing pot temperatures and times by preheating the board. Moving forward, development work should continue on reworking through-hole components on thick boards without using external preheat.

9.5. SUMMARY

Overall the as-assembled Pb-free assembled components showed good reliability results using accelerated thermal fatigue testing.

The high-temperature exposures necessary for Pb-free or Sn–Pb rework on high thermal-mass boards can induce partial reflow of adjacent component solder joints. This lowered the yield and reduced the accelerated thermal fatigue resistance.

Degradation was measured on non-reworked adjacent components (CSP 81) after accelerated thermal cycling. This likely reflects the influence of unintended joint reflow as was noted for the CBGA component during μBGA rework.

High-temperature exposures during Pb-free assembly and rework on Ni–Au boards had a detrimental impact on the mechanical deflection sensitivity and the board resin material used in this study.

High-temperature exposures during Pb-free rework had a detrimental impact on board solder mask adhesion and integrity. Also, via structures can be impacted.

ACKNOWLEDGMENTS

The authors gratefully thank all the participants of the iNEMI Pb-free assembly and rework project. Appreciation is also expressed to the following companies for the management support provided: Agilent, Celestica, Cisco, CMAP, Cookson, Dell, Delphi, Endicott Interconnect, HP, IBM, Intel, Jabil, Lace, Nortel, Sanmina-SCI, Solectron, Teradyne, Texas Instruments, and Vitronics-Soltec. Finally, the entire project team would like to thank the iNEMI Council, Secretariat, and support staff, especially Mr. Ron Gedney, for their untiring help and assistance in the quest to complete this project in a timely manner.

REFERENCES

1. E. Hernández, et al., Development of a lead-free surface mount manufacturing process for high complexity electronic assemblies, in *IPC/SOLDERTEC, First International Conference on Lead-Free Electronics*, Brussels, June 2003.
2. P. Roubaud, et al., Development of baseline lead-free rework and assembly processes for large printed circuit assemblies, in *Proceedings, IPC/SOLDERTEC 2nd International Conference on Lead-Free Electronics*, Amsterdam, Netherlands, June 2004.
3. M. Kelly, Q. Chu, and J. Bath, Pb-free reflow and rework (Cover Story), *Circuits Assembly Magazine*, November 2004.
4. M. Kelly, D. Colnago, K. Lyjak, J. Bath, et al., Component temperature mass study on tin–lead and lead–free assemblies, *J. Surf. Mount Technol.* **15**(4), 11–22, 2002.
5. J. Bath, et al., Lead-free and tin–lead rework development activities within the NEMI lead-free assembly and rework project, *Proceedings, SMTAI Conference*, September 2004.

CHAPTER 10

Implementing RoHS and WEEE-Compliant Products

JIM McELROY and CYNTHIA WILLIAMS

10.1. INTRODUCTION

The European Union's RoHS (Restriction on the use of certain Hazardous Substances) Directive has had a dramatic impact on the electronics industry, governing the material content of many types of "electrical and electronic equipment" (EEE) sold in Europe.

The RoHS Directive, which went into effect July 1, 2006, severely restricts the use of six substances, including lead (Pb), in EEE. The precedents set by this directive have triggered legislative activities in all major geographic regions to address management, reporting, or elimination of hazardous substances.

The vast majority of this book is associated with the technical challenges of RoHS compliance with the bulk of the work focused on the removal of Pb and the related issues of implementing alternative materials in a manufacturing environment. As this chapter will indicate, however, there is much more to this conversion than the resolution of technology gaps. In today's distributed manufacturing environment, cooperation across the value chain is a necessity—from product design through end-of-life disposition—in order to achieve RoHS compliance. While the focus of this chapter is on the deployment issues associated with a particular set of regulations, the concepts described here would generally apply to any major regulation-driven technology change that is broadly adopted by industry, and thus these observations will remain relevant for future applications.

We will begin by untangling the complexity of the RoHS Directive by reviewing a decision tree for compliance. This will help you understand where your company's products fit within the various categories identified in the regulations and their related exemptions. Next we will provide a proposed management structure for working the transition issues across the functions of your company and your supply base (driven by the complexity of the work). RoHS compliance puts

Lead-Free Electronics. Edited by Bradley, Handwerker, Bath, Parker, and Gedney

additional demands on inventory identification (differentiating between those products that contain lead and those that do not) and will also create new information requirements across the supply chain. To address these needs, we will review the justification for distinct part numbers for RoHS-compliant components and will explore the alternatives for efficient mechanisms for exchanging compliance data between trading partners (as one aspect of due diligence programs).

Standards are playing a critical role to help ensure that conversion to RoHS compliance is done in a consistent, cost-effective, and reliable manner. We will discuss a number of standards that have been updated, along with new standards that have been created to help the industry comply.

High-reliability products provide some special requirements to be considered. While many of these products enjoy a Pb exemption, they must meet other materials restrictions and they must contend with a component supply base that is driven by high-volume segments. These special needs—along with solutions—will be discussed.

Conversion to RoHS-compliant products is a massive undertaking with significant business and supply chain implications. We will discuss some of these issues and how industry is coping with the associated financial implications.

The RoHS landscape continues to change: Legislation is revised, updated, or amended; standards are updated to reflect process changes; and the European Commission is continuing to clarify its position regarding the directive's intent. Readers should check for the most current information in all of the areas discussed in this chapter.

10.2. ARE YOUR PRODUCTS WITHIN THE SCOPE OF THE EU RoHS? [1]

For the vast majority of products, the determination of whether they must meet the RoHS Directive's substance restrictions is reasonably straightforward. However, there is a significant minority of products (particularly in specialized or industrial sectors), where there may be considerable uncertainty. Figure 10.1 provides an example of a "decision tree" that can be used to help determine whether products come within the scope of the RoHS Directive. Note that this is only a guideline and it may be advisable to seek an independent legal opinion for a final decision.

The criteria discussed below can be used to help assess "gray area" products. The products referenced here are outside the scope of the RoHS Directive. These guidelines come from the UK's Department of Trade and Industry (DTI) and are reflected in that Member State's regulations.

EEE Intended to Protect National Security or to Be Used for Military Purposes. The WEEE (Waste Electrical and Electronic Equipment) Directive states that equipment connected with national security or military purposes is excluded from the scope of the directive. The RoHS Directive does not specifically

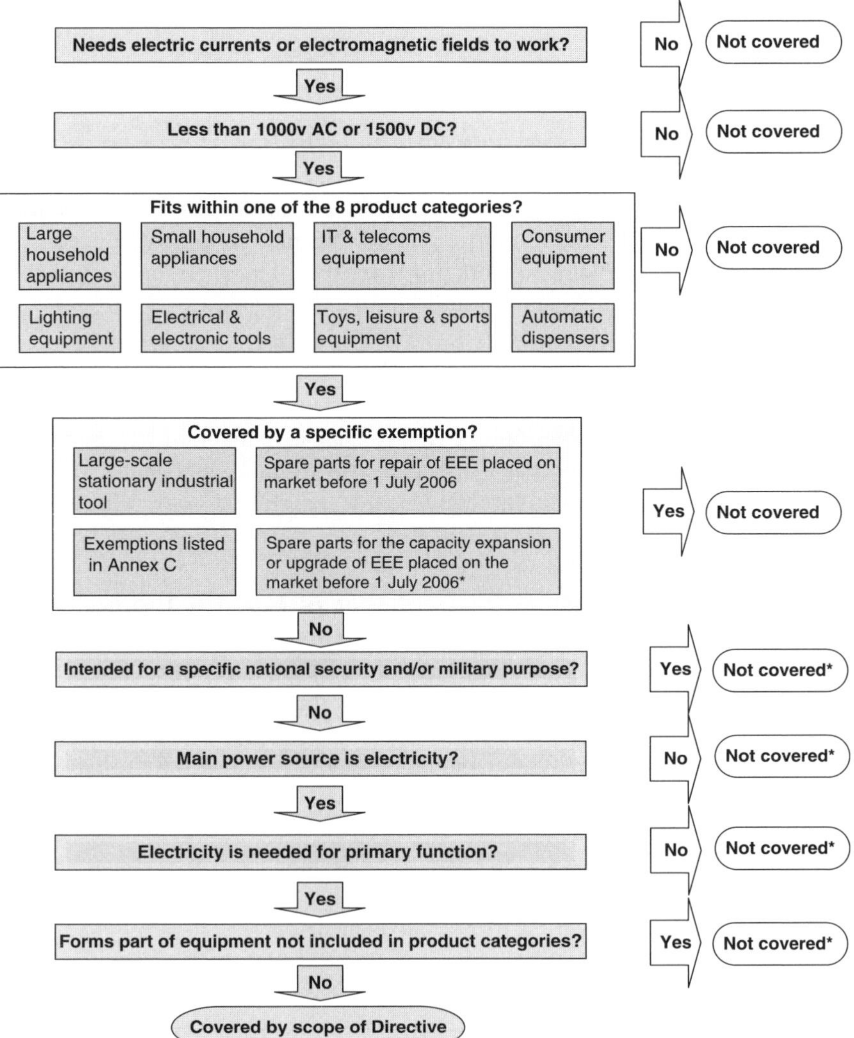

Figure 10.1. This decision tree helps clarify what products are within the scope of the RoHS Directive. *While these exclusions are not expressly provided for in the Directive, it is the DTI view that they apply. It should be noted, however, that a definitive legal interpretation is only available from the court. Producers should rely on independent legal advice on compliance. (*Source*: DTI RoHS Regulations, Government Guidance Notes, Annex B November 2005.)

provide for a similar exemption but, since the RoHS Directive draws its coverage from the WEEE Directive, the exemption applies equally to products covered by RoHS. (This coverage is confirmed by the Frequently Asked Questions document published by the European Commission to clarify the intentions of the directives, which can be downloaded from http://europa.eu.int/comm/environment/waste/pdf/faq_weee.pdf) The exemption for military products does not apply to dual-use products (i.e., used for both military and civilian purposes).

Products for Which Electricity Is Not the Main Power Source. Many products contain electrical and electronic components, either for additional functionality or as peripheral parts (e.g., a combustion engine with an electronic ignition). The definition of EEE extends only to those products that are dependent on electric currents or electromagnetic fields to work properly, such that when the electric current is switched off, the product cannot fulfill its primary function.

Products Where the Electrical or Electronic Components Are Not Needed to Fulfill the Primary Function. This category includes products that contain electrical or electronic elements that add functionality but that can still fulfill their primary function without the electronic components. This would include, for example, musical greetings cards or stuffed toys with electronic components.

Electrical and Electronic Equipment that is Part of Another Type of Equipment not Within the Scope of the Directive. The WEEE Directive excludes EEE that is part of another type of equipment that does not fall within the scope of the Directive. Again, since EEE is defined in identical terms under RoHS, this exclusion typically extends to EEE under the RoHS Directive. Examples of such equipment would be lighting or entertainment equipment for use in vehicles, trains, or aircraft.

Batteries. The RoHS Directive's restrictions on the use of the named hazardous substances do not apply to batteries. This includes batteries that are permanently fixed into the product, as well as disposable batteries.

10.2.1. Exemptions

There are several product and application exemptions listed in the original Annex of the RoHS Directive. Additional exemptions were approved in October 2005 and February 2006. By early 2006, approximately 60 more proposed exemptions were under review. The exemptions included in the original Annex are as follows:

- Large-scale stationary industrial tools consisting of a combination of equipment, systems, products, and/or components, each of which is designed, manufactured, and intended to be used only in fixed industrial applications. In early 2006 there was discussion among the Member States regarding the interpretation of "large-scale stationary industrial tools," and several Member States were asking the European Commission to clarify this issue further in the next update of the FAQ.

- Spare parts for the repair of EEE placed on the market before July 1, 2006. The European Commission and Member States agreed to extend this exemption to include parts that expand the capacity of and/or upgrade EEE placed on the market before the RoHS deadline, provided that the EEE concerned is not put on the market as a new product.
- The reuse of EEE placed on the EU market before July 1, 2006.
- Specific applications of lead, mercury, cadmium, hexavalent chromium, PBB and PBDE as set out in Schedule 2 of the Regulations. These include, for example, mercury in straight fluorescent lamps for special purposes; lead in the glass of cathode ray tubes, electronic components, and fluorescent tubes; and hexavalent chromium as an anti-corrosion of the carbon steel cooling system in absorption refrigerators.
- The original exemptions outlined in the Annex to the RoHS Directive were amended by two Commission decisions published in October 2005. Additional exemptions include, among other things, Deca BDE in polymeric applications and Pb in lead–bronze-bearing shells and bushes.

The RoHS Directive, including the exemptions specified under the directive, will likely change over time. Just a few months before the directive was set to go into effect, there were several proposed exemptions being considered and many outstanding issues that still needed to be defined. Therefore, the conclusions listed here need to be understood and/or modified based on the latest regulations.

10.3. TEN STEPS TO RoHS COMPLIANCE [2]

Perhaps one of the greatest challenges for companies selling EEE products in the EU is how the RoHS Directive will be enforced from state to state. Each Member State has transposed the directive through national legislation, and there are differences between laws.

In early 2006, there was discussion among Member States of developing a guidance document that would help establish a common enforcement approach across the EU. This document, which would be purely advisory rather than legally binding, would outline the principles to support RoHS enforcement, such as the documentation that "producers" should keep, how enforcement bodies might use such documentation to check for RoHS compliance, and how/when sample preparation and analytical testing might be employed.

The approach most likely to be followed is one of risk-based market surveillance and seeking documentation before proceeding to testing if necessary—in other words, a "due diligence" approach.

Once the guidance document is agreed to by the Member States, it will be published and made publicly available to all stakeholders. Rather than expecting manufacturers to prove that every part they buy is free of the RoHS-defined substances, Member States will likely expect manufacturers to establish documented and auditable systems to prevent noncompliant products from entering the EU market. Assuming that this approach will be taken, development, implementation, and proof of a corporate RoHS compliance strategy is essential. Following are 10 steps that companies can take to develop and implement a strategy for achieving RoHS Directive compliance.

1. Determine Company's Legal Exposure

The first step associated with RoHS compliance is to determine the company's legal exposure. It is important that senior management support this initial exposure assessment and that someone within the company take responsibility for directing the effort. This person should initiate a company dialogue to determine whether the company is subject to the directive or whether its products are covered by an exemption. An in-house attorney or outside legal counsel should make a quick assessment of whether—and to what extent—the company must comply with requirements.

2. Form a Corporate-Wide Compliance Team

The next step is to assign ownership of the task to a corporate-wide compliance team. The RoHS Directive's requirements are complex and their impacts are far-reaching, affecting product design, manufacture, distribution, procurement, and sale. Virtually all corporate departments must be aware of the directive and its potential impacts. It is important that people from various departments be involved in the compliance team and that they meet regularly to assess the impact on current and future operations.

3. Develop a Corporate RoHS Directive Compliance Statement

The team's first task should be to develop a corporate compliance statement articulating the company's commitment to, and goals for achieving, RoHS compliance. It can be a simple one-page document stating that the company intends to "comply with all regulations worldwide," or a more detailed, substance-by-substance discussion of its compliance status. It should include a date (or dates) for compliance and might also outline supplier requirements, such as methods for verification of compliance (i.e., testing, documentation). The objective is to have a formal document to respond to customer inquiries about RoHS compliance. Without this document, customers may perceive the company as an ineligible supplier, which could lead to lost sales.

4. Develop Internal RoHS Compliance "Roadmap"

The company should develop an internal document to help guide its plans for compliance. This roadmap should define the concrete actions the company will take, including an implementation schedule. It should identify which products will be offered as lead-free with timelines and deadlines for sample availability and production availability. It might also set deadlines for the discontinuation of "leaded" products, identify plans for part renumbering (e.g., separate part numbers for compliant versus non compliant products), and discuss substitute materials (e.g., preferred alloys to replace Sn–Pb solder). The roadmap is a living document that will adapt and change as the company's RoHS preparedness evolves.

5. Assess Company's Supply Chain Exposure

The company needs to look very closely at the components and parts that it purchases from suppliers to determine where RoHS substances may exist within its covered product lines. There are the obvious places, such as lead in solders and surface finishes; but there are also less apparent high-risk components, such as lead and cadmium in cables and banned flame retardants in plastics, housings, cables, connectors, fans, and components. A company's restricted substance specification, updated to include the RoHS substance MCVs (maximum concentration values), is an appropriate compliance metric for suppliers.

Product-specific assessments are necessary to identify (1) components at greatest risk for compliance and (2) key suppliers of those components. A risk-based supply chain compliance program should combine documentation of components' RoHS compliance status with limited testing requirements.

6. Qualify Suppliers

Companies should qualify suppliers to determine their level of RoHS preparedness. A questionnaire may help begin this process. Questions may include: Are you aware of/familiar with the RoHS Directive? Does your company have a person responsible for environmental compliance? Is your company ISO 14001 certified? This information, combined with the product risk assessment described above, will help companies identify suppliers that will require the greatest attention because they provide high-risk components and/or they require more education and supervision to ensure that they can deliver RoHS-compliant components. It may be useful to develop a compliance "checklist" for high-risk components to both inform procurement personnel and provide documentation that RoHS requirements were addressed in procurement decisions.

7. Establish a Supply Chain Material Declaration Process

Although RoHS requires manufacturers to know what substances are *prohibited* in their products, it's also important to know what substances and materials are *used* in their products. A material declaration questionnaire is typically used to obtain this information. A good starting point is the *Joint Industry Guide for Material Composition Declaration*, developed by EIA, JEDEC, and JGPSSI (http://www.eia.org/new_policy/jig_download.phtml). This guide sets an industry standard for the materials and substances that, when present in products and subparts above certain thresholds, must be disclosed (i.e., "declared").

Companies should establish a database for the information that they will receive from their suppliers so they can use it to determine compliance. This database may be as simple as a corporate spreadsheet; however, software vendors, seeing the market potential generated by the RoHS Directive, are busy developing proprietary and/or standards-based solutions to help companies manage and use supply chain disclosure data.

8. Perform Limited Testing and Validate Results

Educating suppliers and procurement managers and managing RoHS compliance through SDoCs (supplier declarations of conformity) is the "first line" of defense in a compliance strategy. Additionally, testing may be performed to supplement supply chain material declarations. Companies may also choose to test products directly to determine compliance. It may not be necessary to test every component, but companies should have a legitimate risk-based testing process using standardized methods. The evolution of RoHS substance limits for products has created the need for a reference standard to objectively define analytical testing procedures for use in compliance determinations. An IEC (International Electrotechnical Commission) Working Group (WG 3) has been established under the Technical Committee "Environment" (TC 111) to address this need. The WG 3 has published several widely acclaimed draft standards already, with the Committee Draft for Voting (111/54/CDV) released in May 2006.

The standard on *Procedures for the Determination of Levels of Six Regulated Substances (Lead, Mercury, Hexavalent Chromium, Polybrominated Biphenyls, Polybrominated Diphenyl Ethers) in Electrotechnical Products* (IEC 62321) provides test procedures as one option that will allow the electrotechnical industry to determine the levels of the regulated substances Pb, Hg, Cd, Cr VI, and their compounds and PBB and PBDE (except decabrominated diphenyl ether, DecaBDE) in electrotechnical products on a consistent, global basis. Additionally, a general guidance to obtain representative samples from finished electronic products to be tested for the determination of levels of regulated substances is given in the Annex A of the standard.

Table 10.1 lists the RoHS substance limits (maximum concentration values, MCVs) and test methods suggested by the IEC standard.

Instead of dealing with the inherent difficulties in digestion and analysis of product materials using wet chemical methods, screening using X-ray fluorescence (XRF) is proposed as an alternative compliance approach. Screening may be carried out either by directly measuring the sample (nondestructive sample preparation) or by crushing the sample to make it uniform. A screening of representative samples of many uniform materials (such as plastics, alloys, glass) may be done nondestructively, while for other, more complex samples (like a component), mechanical sample preparation is necessary. Because XRF can only detect elemental substances, it must be used in conjunction with other analytical methods in most cases to achieve a full compliance determination. The XRF analysis technique has limitations to its use and the applicability of the results obtained, although fast and resource-efficient analysis has its merits, particularly for the demands of the

TABLE 10.1. MCVs and Suggested Test Methods

<table>
<tr><th>Substance</th><th>RoHS MCV Limits</th><th>Chemical Sample Preparation</th><th>Test Methods</th></tr>
<tr><td>Lead</td><td>1000 ppm[a]</td><td rowspan="3">• Microwave digestion
• Acid digestion
• Dry ashing
• Solvent extraction</td><td rowspan="2">• ICP-OES
• ICP-MS
• AAS
• XRF (screening)</td></tr>
<tr><td>Cadmium</td><td>100 ppm</td></tr>
<tr><td>Mercury</td><td>1000 ppm</td><td>• CV-AAS
• AFS
• ICP-OES
• ICP-MS
• XRF (screening)</td></tr>
<tr><td>PBB/PBDE</td><td>1000 ppm</td><td>n/a</td><td>• GC-MS
• XRF (screening, for elemental Br only)</td></tr>
<tr><td>Hexavalent Chromium</td><td>1000 ppm</td><td>n/a</td><td>• Spot-test procedure/ boiling-water-extraction procedure or
• Alkaline digestion/ colorimetric method
• XRF (screening, for elemental Cr only)</td></tr>
</table>

[a] *Note*: A California Prop 65 ruling limits lead to 300 ppm in external cables to avoid product warning labels.

electrotechnical industry. Caution is especially advised when using portable XRF equipment, mainly due to uncertainties in numerical results and low sensitivity for cadmium. The final steps of any testing plan are to document, incorporate, and periodically evaluate the quality of information received. Testing and compliance documentation should be auditable to ensure credibility and effectiveness.

9. Exchange RoHS Compliance Data with Customers

Once companies compile material content data, they need to be able to manage the data and exchange it with their customers. The most efficient approach is to standardize the data exchange process and formats, which is what the new IPC 1750 series of standards does. IPC-1751, *Generic Requirements for Declaration Process Management*, describes all generic requirements, including company information, and IPC-1752, *Materials Declaration Management*, establishes uniform electronic data formats and standardized forms to simplify the way industry collects, tracks, and discloses material content information. These standards are intended to reduce the cost and complexity of RoHS compliance while increasing data quality and decreasing response times. See Section 10.5 for a more detailed discussion of the IPC 1750 standards.

10. Incorporate Compliance Strategy into Company-Wide Operations

An RoHS compliance strategy is complex and requires commitments from several departments. Once in place, it must be implemented company-wide. Engineers must update design specifications, procurement officers must understand which suppliers are "high-risk," and the corporate team must be aware of changing requirements and legal determinations (such as new restrictions or changes in concentration levels or exemptions). Specifications must reflect the appropriate RoHS substance limits and be cited in manufacturing and procurement contracts. Procurement needs to monitor the supply chain closely and communicate to suppliers the importance of compliance. Inventory managers must be aware of deadlines (noncompliant products must satisfy the RoHS "placed on the market" condition in the EU prior to July 1, 2006). The sales team must be aware of the company's program so that they can respond to customer requests.

10.4. PART NUMBERING IMPORTANT FOR DIFFERENTIATING LEAD-FREE FROM TIN–LEAD COMPONENTS AND BOARDS [3]

Despite the RoHS ban on lead, some electrical and electronic products will continue to be manufactured and assembled using the traditional Sn–Pb process. Several high-reliability applications are exempted under RoHS or have delayed deadlines. These applications include (a) lead in solders for servers, storage, and storage array systems, (b) lead in solders for network infrastructure equipment for switching, signaling and transmission as well as network management for telecommunication, and (c) lead in electronic ceramic parts (e.g., piezoelectronic devices). Aerospace and defense applications are also exempt and will continue to use Sn–Pb assemblies.

So, at least for some period of time—depending on how long exemptions remain in force—manufacturers will be using both Pb-free and Sn–Pb processes and will require boards and components for both. Rework facilities will also be running both processes. The higher temperatures and tighter process windows required for Pb-free processing necessitate segregation of Pb and Pb-free parts; also, in order to prevent manufacturing errors and defects, the assembly shop floor and rework facilities must be able to identify those components, solders, and so on, that contain lead and those that do not. This same information will also be important for end-of-life product disposition to meet the WEEE Directive's recycling and reuse requirements. Therefore, it is important to have separate part numbers to differentiate between Sn–Pb and Pb-free parts.

10.4.1. A Question of Numbers

In the year leading up to the RoHS deadline of July 1, 2006, the issue of part numbers was a hot one, and the camps were divided as might be imagined: The OEMs, EMS providers, and distributors were asking for separate part numbers for

Pb-free parts, and the suppliers wanted to differentiate with lot numbers, date codes, and so on. On the manufacturers' side, there were concerns about being able to differentiate parts when ordering, on the shop floor, in inventory systems, and in manufacturing resource planning (MRP) systems. There was also concern about returned products and end-of-life recycling. Suppliers, however, cited the expense of implementing new part numbers and the strain on their IT systems (anticipating that the number of parts in their databases would nearly double).

Well before the RoHS deadline, many of the leading OEMs and EMS providers had established policies requiring separate part numbers, and several went on record, calling for separate part numbers. In addition, consortia and trade associations, such as iNEMI, NEDA (National Electronic Distributors Association), and the EMS Forum (Celestica, Flextronics, Jabil, Plexus and Sanmina-SCI), stated support for separate part numbers. Although suppliers were at first resistant, it appears that customer demands have pushed many suppliers to provide separate part numbers.

A 2004 survey of suppliers and EMS providers conducted by Avnet, working with Technology Forecasters Inc., indicated that only 52% of suppliers planned to assign new Pb-free part numbers. A 2005 follow-up study showed that number rising to 71%. Also in the 2005 survey, almost 58% of suppliers said they planned to designate compliance by providing information on their packaging, which was up from 37% in the previous year. In addition to providing separate part numbers, the majority of suppliers have said they will communicate compliance through online databases and a lesser number will provide information in paper catalogs.

10.4.2. Marking Schemes and Protocols

JEDEC and IPC Symbols. JEDEC standard JESD97 (published in May 2004) and IPC-1066 (published in January 2005) use the same distinctive symbol and labeling format for components and boards, respectively. These standards are combining to create a joint standard, J-STD-609, *Marking, Symbols and Labels of Lead-Free and Leaded Terminal Finish Materials Used in Electronic Assembly.* The combined standard is gaining wide acceptance in the United States and Europe, and is supported by the U.S. military and the aerospace industry.

IPC-1066 and JESD97 establish a Pb-free symbol (Figure 10.2) that can be used to replace the phrase "Pb-free" or "lead-free" on labels or wherever practical on components/devices, boards or assemblies. They also define a Pb-free identification

Figure 10.2. This symbol (from J-STD-609) can be used to replace the words "lead-free" on labels.

Figure 10.3. This symbol (from J-STD-609) is used only when components or devices, boards and/or assemblies are totally lead-free.

label (Figure 10.3) that is used only when components/devices and/or board assemblies are totally Pb-free, according to the RoHS definition. This label should be affixed to intermediate boxes or other containers not otherwise identified as lead-free.

Figure 10.4 shows the label used to indicate that the second-level interconnect terminal finish, component materials, and/or the solder paste/solder used in board assembly are Pb-free. Note that this label provides information about the "category" and maximum safe component temperature. The category is determined by the type of alloy used and is defined by one of seven designations: e1 through e7 (see Table 10.2 for definitions).

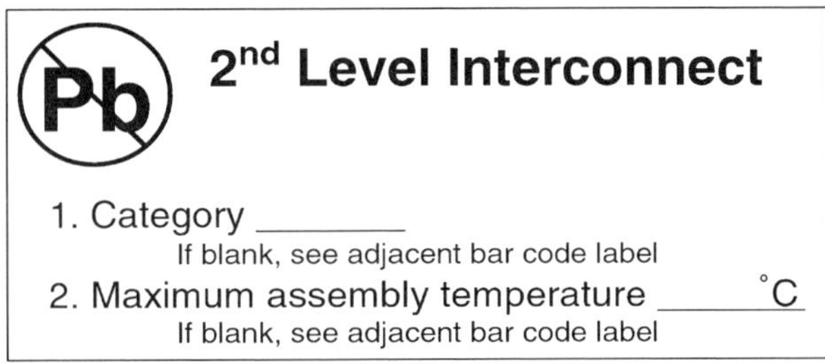

Figure 10.4. This label (from J-STD-609) is recommended for second-level interconnect.

TABLE 10.2. Categories Used to Describe Pb-Free Second-Level Interconnect Terminal Finish/Material or Components and/or the Solder Paste/Solder Used in Assembly

Category/Symbol	Alloy
e1	SnAgCu (not included in category e2)
e2	Sn alloys with no Bi or Zn, excluding SnAgCu
e3	Sn
e4	Precious metals (Ag, Au, NiPd, NiPdAu), no Sn
e5	SnZn, SnZnx (no Bi)
e6	Contains Bi
e7	Low-temperature solder ($\leq 150^\circ C$), containing In (no bi)

Note: Categories e0, e8, and e9 are unassigned.

Figure 10.5. Category symbols such as these are to be used on components and describe the specific finish, material, or solder used in assembly.

The iNEMI Component and Board Marking Project provided inputs to, and supported development of, the original JEDEC and IPC standards, JESD97 and IPC-1066. The identification team considered the models in these standards comprehensive enough to meet the needs of manufacturing. The iNEMI team also developed three standard vocabulary terms—Pb-free, Pb-free second-level interconnect, and RoHS-compliant—which are now included in the RosettaNet dictionary of terms RNTD, version 4.0. Creating consensus on definitions helps develop common language for communication of Pb-free transition status.

These category symbols (e1–e7) are also to be used on components, if space permits, with the category designation enclosed within a circle, ellipse, or parentheses, as shown in Figure 10.5.

JEITA Symbols. JEITA has two documents that outline standardized symbols for marking Pb-free and RoHS-compliant products. The first is a technical report, published in June 2004 (JEITA ETR-7021: *Guidance for the Lead-Free Marking of Materials, Components and Mounted Boards Used in Electronic and Electric Equipment*). These guidelines cover marking for Pb-free products, solders, and assembled boards.

The second is a standard: JEITA ET-7001, *The marking for presence of the specific chemical substances in materials, and mounted boards use in electrical and electronic equipment* (published July 2005). This standard considers materials used in solders, pre-flux, adhesive, ink, and similar materials composing the assembled board, and it provides for markings related to all substances listed in Annex 1 of the RoHS Directive.

Harmonization. Although there have been several meetings between representatives of JEDEC, IPC, and JEITA, the groups have not been able to agree on a single marking standard. JEITA's approach may be more complex and difficult to implement. Their marking scheme includes symbols for the exact composition of solder used, phase of conversion, and indication of presence or absence of RoHS Annex 1 substances.

JEDEC/iNEMI/IPC believe that it is only necessary to identify second-level interconnect (board to component joint) and the solder family used and that lead and halogens are the only components that need to be identified since this is the information needed for manufacturing rework and recycling. The information required for designating RoHS compliance will be readily available in materials declarations. Any environmental markings used for marketing purposes should be OEM-specific and not subject to a standard.

10.4.3. Download Standards

1. *JESD97*: Marking, Symbols, and Labels for Identification of Lead (Pb) Free Assemblies, Components, and Devices http://www.jedec.org/download/search/JESD97.pdf
2. *IPC-1066*: Marking, Symbols and Labels for Identification of Lead-Free and Other Reportable Materials in Lead-Free Assemblies, Components and Devices http://webvision.ipc.org/scripts/mgrqispi.dll?APPNAME=IPC-WEB&PRGNAME?=TOCFRAME&ARGUMENTS=-N,-N,-A,-A,-N50
3. *J-STD-609*: Marking, Symbols and Labels for Identification of Lead Free and Other Reportable Materials in Lead Free Assemblies, Components and Devices; this was a proposed standard for ballot in December 2006. Check the IPC website (www.ipc.org) for most recent draft or published standard
4. *JEITA ETR-7021*: Guidance for the lead-free marking of materials, components and mounted boards used in electronic and electric equipment http://www.jeita.or.jp/english/standard/html/6_42.htm
5. *JEITA ET-7001*: The marking for presence of the specific chemical substances in materials, components and mounted boards use in electrical and electronic equipment (July 2005) http://www.jeita.or.jp/english/standard/html/6_42.htm

10.5. A STANDARDS-BASED APPROACH TO MATERIALS DECLARATION [4]

The RoHS and WEEE directives are driving electronics manufacturers to manage and exchange product content information across the supply chain and throughout a product's life cycle. For RoHS, manufacturers must be able to demonstrate "due diligence" in ensuring that their suppliers provide compliant components and bulk materials (i.e., do not contain more than the allowable amounts of the six restricted substances, unless under specific exemption). For WEEE, manufacturers are expected to provide information about "substances of concern" and the location of any hazardous substances in their products for end-of-life handling and disposition requirements.

Content information is necessary throughout the supply chain, and companies need a means of exchanging data. As shown in Figure 10.6, material declarations are not considered sufficient on their own to ensure RoHS compliance, but provide a key building block of an overall compliance model. Past approaches have generally centered on sending requests using proprietary documents, most often via email. However, the volume of requests that suppliers receive is unmanageable without standardized data exchange processes and formats.

Without a standardized approach to collecting and exchanging item and product content information, there are not only problems with duplication of efforts—leading to higher costs—but also problems with data quality and little or no hope

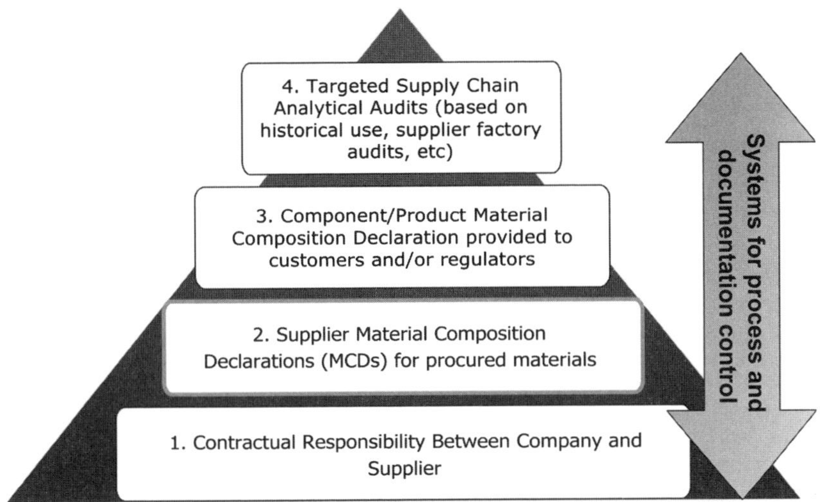

Figure 10.6. General RoHS compliance model.

of automating any part of the process. For OEMs, lack of standardization means that they will likely have gaps in their compliance data, whether they know it or not. Data quality and completeness will vary, depending on the compliance risk management strategies used—that is, whether OEMs collect material content data for all product BOM items themselves, rely on support from their EMS providers to manage compliance, or simply collect signed Certificates of Compliance from direct suppliers. All of these strategies are currently being followed by top tier electronics OEMs, and each has its own associated costs and risks.

Suppliers are being asked for information, but typically in a different format from each customer, and often they are even being asked for different information. Most companies are using some sort of spreadsheet and email to exchange data, but each company uses its own format, and spreadsheets are filled out manually. Most companies don't yet have a searchable database of material content that can be integrated for B2B exchange. In fact, most do not even have a way of tracking what they sent to whom, are not able to ensure that information or requests were received, and sometimes cannot confirm if they have the most recent version of a material declaration.

The IPC-1750 series provides a standardized approach for materials information exchange. IPC-1751, *Generic Requirements for Declaration Process Management*, describes all generic requirements, including company information. IPC-1752, *Materials Declaration Management*, establishes uniform electronic data formats and standardized forms to simplify the way industry collects, tracks, and discloses material content information. It supports small and mid-sized companies, where the interaction is more likely to be manual, as well as large multi-nationals, where IT budgets can support automation and direct business-to-business transactions. It addresses the multiple business requirements needed to support RoHS compliance, all in a single format. For example, the standard provides a format to declare substances used in an item and also supports supplier/manufacturer certification, provides information about alternate "green" parts, and supports the exchange of information related to lead-free manufacturing parameters.

Development of the IPC-1750 series began with two iNEMI projects that brought together users and solution providers interested in shaping standards-based processes for communicating materials content data. The specifications developed by these projects were handed off to IPC's 2-18 Subcommittee (Declaration Process Management), for development and standardization. Altogether, more than 50 of the electronics industry's largest OEMs, EMS providers, and component manufacturers helped define the business requirements and develop specifications for these standards.

10.5.1. Pieces in the Materials Declaration Puzzle

There continues to be considerable confusion in the industry related to regulations, guidance documents, standards, formats, and what may be loosely termed "best practices." There is currently no requirement within the RoHS Directive for producers or suppliers to provide material composition declarations (MCDs). However, it is generally recognized that MCDs are a reasonable way to manage compliance risk, support supplier liability claims, and perform product level compliance analysis. Furthermore, there are references made to "technical documentation" and material declarations within the United Kingdom's RoHS Guidance documents published by their Department of Trade and Industry (DTI) and also reflected in the online guidance provided by their National Weights and Measures Laboratory, which has been tasked with RoHS enforcement in the United Kingdom.

According to the UK DTI Guidance, producers are expected to collect and maintain "technical documentation" that will support that they have conducted "due diligence" in ensuring that their suppliers are providing compliant parts. Upon request or audit, these data are to be made available within 28 days of request and must be retained for a period of four years.

A key element in managing compliance risk is establishing liability across the multiple tiers of the supply chain. This requires an audit trail for information collected or referenced that can be legally attributed to a supplier or manufacturer and referenced to the part numbers (and potentially lot numbers or date codes) used in production.

Figure 10.7 provides a framework of related directives and standards. IPC-1752 directly references the work already done by EIA, JGPSSI, and JEDEC in the Joint Industry Guide (JIG), which defines the reportable substances and threshold levels. IPC-1752 also integrates work done on electronic forms and exchange standards by RosettaNet. The IPC standard's XML scheme is aligned with RosettaNet's 2A13 and 2A15 PIPs.

The IPC-1752 standard is a key part of the materials declaration proposal that has been submitted to the International Electrotechnical Commission (IEC). Along with IEC's Publicly Available Specification (PAS) document 61906, which provides high-level requirements for material declarations, IPC-1752 and JIG 101 form the

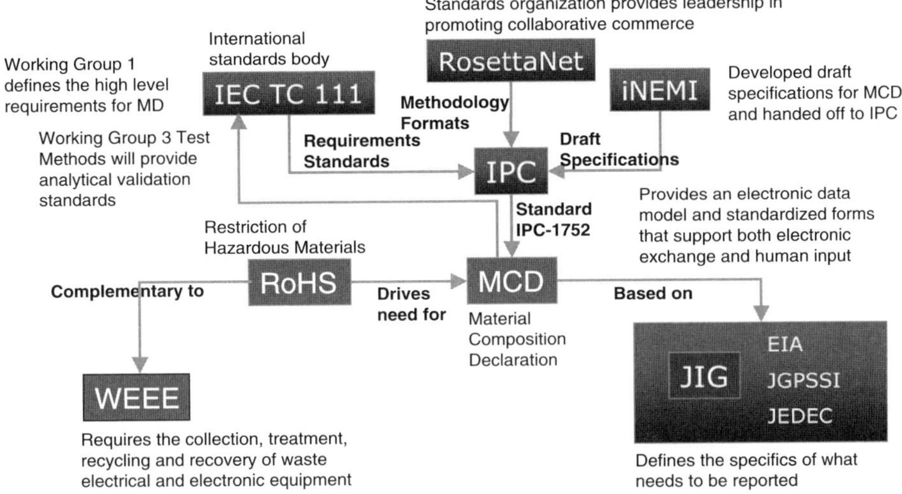

Figure 10.7. Regulatory and standards framework.

basis for the material declaration standard that has been proposed to the IEC's Technical Committee 111. (This committee is responsible for standards supporting environmental stewardship of electronic products.) While there are other formats for material declarations being used or developed in the industry besides company proprietary ones (examples include Japan's JGPSSI spreadsheet and the automotive industry's Compliance Connect spreadsheet), none has the international support and momentum that IPC-1752 is now receiving.

10.5.2. The Flavors of IPC-1752

IPC-1752 meets key business requirements while providing the flexibility needed to support different elements of the supply chain. It provides several user configuration options, is platform independent (Windows, Linux, Solaris, Mac) and, because the forms are PDF-based, does not require suppliers to purchase specific software to complete declarations. Some of the key features and capabilities of IPC-1752 include:

- Enables easy data import and export.
- Provides for bulk material, component, subassembly, and product-level reporting.
- Provides explicit inclusion of all JIG A & B substances, with CAS numbers, to improve data quality.
- Allows auto-generation of requests and responses while also providing a simple human interface.
- Supports "low-tech" data exchange (i.e., email) as well as "high-tech" systems (i.e., XML-based).

- Supports request and response (customer → supplier → customer), as well as publish (supplier → customer) data exchange processes.
- Allows part family declarations.
- Provides for optional inclusion of "legal" statements and allows attachment of supporting documents (i.e., third-party lab analysis).
- Enables electronic signatures, which provide supplier identity validation for certification and support liability requirements.
- Allows inclusion of information required to determine compatibility with lead-free assembly processes.
- Allows requestor and supplier data to be "locked."

IPC-1752 is a single standard, which includes a single data model and supporting XML scheme, but there are two available forms, IPC-1752-1 and IPC-1752-2. The forms, based on Adobe™ PDF forms technology, use a UML data model developed with the assistance of the National Institute of Standards and Technology (NIST). The forms can be printed or can be exchanged via electronic means, ranging from e-mail and Web interfaces to advanced and secure business-to-business methods. Table 10.3 shows how the two forms support the various reporting levels.

The standard supports three levels of declaration:

1. Item-level RoHS declaration in yes/no format, with identification of any applicable exemptions (supported by both forms).
2. RoHS and JIG level A & B and other substances requested, declared at the item level (except for RoHS restrictions) with identification of any applicable

TABLE 10.3. Six Classes of Reporting

Class	Description	Form Type
Class 1	RoHS declaration in yes/no format, with identification of any RoHS exemptions	IPC-1752-1 IPC-1752-2
Class 2	Same as Class 1, with the addition of manufacturing process information	IPC-1752-1 IPC-1752-2
Class 3	RoHS declaration in yes/no format, with identification of any RoHS exemptions + JIG A & B substance reporting with ability for requestor to include additional substances of interest	IPC-1752-1
Class 4	Same as Class 3, with the addition of manufacturing process information	IPC-1752-1
Class 5	RoHS declaration in yes/no format, with identification of any RoHS exemptions + Substance reporting at the homogeneous level. JIG A & B substance list provided, other substances can be added	IPC-1752-2
Class 6	Same as Class 5, with the addition of manufacturing process information	IPC-1752-2

exemptions, as well as providing substance ppm or weight if above thresholds. (Requestors would choose IPC-1752-1 form if they want this level.)

3. Declaration of RoHS, JIG A & B, and other substances at the homogeneous material level, up to full declaration and also including identification of any applicable exemptions. (Requestors would choose IPC-1752-2 form if they want this level.)

While there are technically six classes of reporting, there are really only the original three required levels of declaration, but with the ability to include manufacturing process information, such as lead finish, maximum reflow temperature, and moisture sensitivity level. Obviously, these do not apply to all components, and the supporting fields are optionally enabled by the user only as needed.

The reporting class that a company chooses depends on the overall compliance strategy and level of information required to support that strategy. If a company is comfortable with collecting part-level RoHS compliance declarations only and is not currently concerned with the additional substances in the JIG, then Class 1 or 2 declarations will meet their needs. However, one of the limitations here is that there is no detail on substance amounts, only "yes" or "no" to RoHS compliance, although there is support for identification of any RoHS exemptions that may be applicable. More OEMs will likely opt for Class 3 or 4, as they provide explicit data on the JIG substances, which not only helps manage RoHS compliance, but also supports identification and roll-up of hazardous substances for end-of-life reporting needs. Class 3 and 4 support material and component level reporting as well as subassemblies and products. Some OEMs and EMS providers prefer to use Class 5 and 6, because they provide the most detail, up to and including full disclosure. The primary argument for 100% disclosure is that if you have identified all substances and their levels, there should be no need to go back and request new declarations if the list of restricted substances or allowable thresholds change. But nothing comes without a price. First, many suppliers are either unwilling to disclose everything (for intellectual property reasons) or do not have that level of information available. Second, you need to be able to store and manage the data: The amount of raw data to describe 100% of every substance in every homogeneous material contained in a typical laptop computer is enormous and, many would argue, unnecessary to meet RoHS requirements. Due to this added complexity, Class 5 and 6 declarations are primarily used at the material or component level.

10.5.3. Conclusion

IPC-1752 integrates and leverages several industry efforts, establishing a common solution that is shaped not only by regulatory guidelines but also by industry needs and requirements. It can help eliminate the costly, and burdensome, use of multiple material declaration formats. At the same time, however, it provides flexibility, allowing users to select from several options in terms of the level and scope of the data that is requested. Few would argue with the benefits of having a standard approach to exchange material declaration data. While no standard is perfect, rapid adoption of IPC-1752 represents the best available option for the industry to efficiently meet the business and information needs of the RoHS/WEEE directives.

The IPC-1750 standards and users guide are available as a free download from IPC at http://www.ipc.org/175x.

10.6. STANDARDS [5]

Pb, in the form of Sn–Pb solder, is one of the most basic and well-studied of the six materials impacted by the RoHS Directive. Change to a solder alloy composition requires additional material and process changes to ensure the manufacture of reliable electronic products. The interconnection industry has realized that standardization is essential to understanding and implementing RoHS and other Pb-free restrictions.

In addition to the material declaration standards discussed in the previous section, several key standards have been, or are being, revised and developed to help industry clarify the requirements for RoHS-compliant materials and processes so that manufacturers can effectively compete in the global marketplace. The discussion below reports the status of several standards as of April 2006. For the most current information, go to the IPC (**www.ipc.org**) and JEDEC (**www.jedec.org**) websites.

10.6.1. Component Standards

Component and process compatibility is the first concern of the assembler as companies convert to Pb-free processes. This is because of the higher melting temperatures of commonly used Pb-free alloys. Sn–Pb eutectic solder melts at 183°C with typical processing temperatures of 205–220°C. Compare that to one commonly selected Pb-free alloy of Sn–Ag–Cu, SAC 305, which has a melting point of 217°C and typical processing temperatures of 235–250°C.

MSL. These higher processing temperatures require that component suppliers reclassify moisture sensitive devices. IPC/JEDEC J-STD-020, *Moisture/Reflow Classification for Nonhermetic Solid State Surface Mount Devices* was revised to address the changing requirements for Pb-free processing. The most current revision C of J-STD-020 specifies that the temperature for moisture sensitivity level (MSL) testing is selected according to the thickness, volume and mass of the component, and also raises the maximum MSL testing temperature to 260°C (for certain sizes of component and for Pb-free array area component rework).

Termination Finish Testing. IPC/EIA J-STD-002, *Solderability Tests for Component Leads, Terminations, Lugs, Terminals and Wires* is being updated to include testing of Pb-free component termination finishes.

Tin Whisker Standards. Many component manufacturers have changed component termination finishes from various Sn–Pb alloys to pure-tin and high-tin

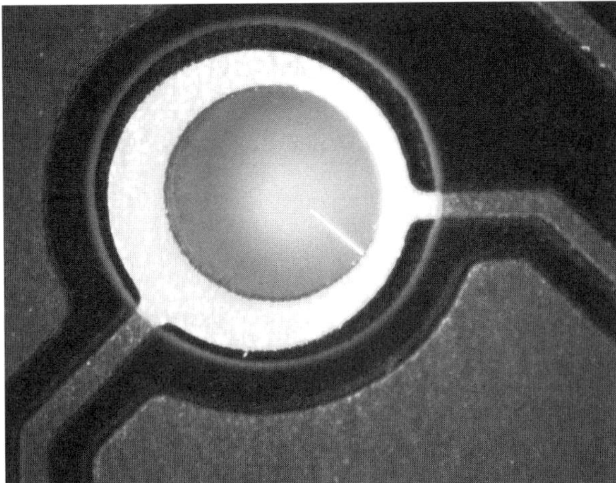

Figure 10.8. Major tin whisker in plated via.

(>95%Sn) Pb-free coatings. These coatings raise potential concerns about tin whiskers (see Figure 10.8). The occurrence of tin whiskers is being actively studied by a number of researchers. However, in order to make comparisons between studies, there is a need for a standardized accelerated tin whisker test. Manufacturers also need test methods for evaluating devices with tin finishes and acceptance criteria. IPC and iNEMI cooperated with JEDEC to develop JESD22-A121, *Test Methods for Measuring Tin Whisker Growth on Tin and Tin Alloy Surface Finishes*. This standard specifies the test conditions to be used for evaluating tin whisker growth and defines how whiskers should be inspected.

JEDEC, again with significant contributions from iNEMI and IPC, developed a second standard, JESD201, *Environmental Acceptance Requirements for Tin Whisker Susceptibility of Tin and Tin Alloy Surface Finishes*. This acceptance specification is intended to be used with JESD22-A121.

Finally, JEDEC and IPC released a joint publication, JP002, *Current Tin Whiskers Theory and Mitigation Practices Guideline*. Also based on iNEMI work, this guide helps tie together all that is known about tin whiskers. It must be noted that this guideline will be frequently revised as further data are generated and incorporated.

10.6.2. PWB Design Standards

Many companies ask how IPC design standards will change to address Pb-free soldering, particularly in the area of SMT land patterns. The answer is that the design standards are not currently changing because of Pb-free requirements. The land pattern designs, identified in IPC-7351, *Generic Requirements for Surface Mount Land Pattern and Design* (superseding IPC-SM-782), have been shown to have similar requirements for surface mount assembly with either Sn–Pb or Pb-free solders. When IPC-7351 design rules are applied, Pb-free SMT assembly does not differ from that experienced with eutectic Sn–Pb soldering, and dimensional modifications are not required.

For through-hole wave soldering, holefill will be more of a concern for Pb-free solders, especially on thick boards that may require more consideration of the use of pressfit components.

10.6.3. Bare-Board Standards

Surface Finishes. For printed wiring board surface finishes, fabricators have many options from which to choose, and nearly all will work for Pb-free assemblies. Choice of finish is dependent on the application and preference. Hot air, Pb-free solder leveling (HASL), immersion tin, immersion silver, organic solderability preservative (OSP), and electroless nickel/immersion gold (ENIG) have all been used successfully in Pb-free assemblies. IPC has published two finish standards, IPC-4552, *Specification for Electroless Nickel/Immersion Gold (ENIG) Plating for Printed Circuit Boards*, for specifying ENIG, and IPC-4553, *Specification for Immersion Silver Plating for Printed Circuit Boards*, for defining immersion silver. A third specification, IPC-4554, *Specification for Immersion Tin Plating for Printed Circuit Boards*, addressing immersion tin was nearing completion in 2006, and the fourth standard in the series, IPC-4555 (OSP), was in development. IPC/EIA J-STD-003A, *Solderability Tests for Printed Boards* was updated to include testing for Pb-free board finishes.

Laminates. IPC's PWB laminate standards are also addressing the application of Pb-free soldering. Again, the higher processing temperatures and the potential impact on the plated through-holes and vias are of major concern. Historically, the glass transition temperature (T_g) has been the parameter utilized to study thermal robustness. T_g can be evaluated per test methods within IPC-TM-650. The requirements for a specific laminate are called out in the individual specification sheets within IPC-4101, *Specification for Base Materials for Rigid and Multilayer Printed Boards*.

However, companies are finding that laminate T_g is not necessarily a complete and accurate indicator of Pb-free process compatibility. It is now recognized that a better predictor of a laminate's temperature stability is needed to show its ability to withstand these higher temperatures and/or longer exposure times.

In addition to T_g, three other thermal parameters are being proposed and investigated as better predictive indicators of a laminate's ability to perform in a Pb-free application. These are: (1) coefficient of thermal expansion (CTE), (2) time to delamination (by TMA), and (3) decomposition temperature (T_d). A lower CTE, particularly in the Z-axis or thickness dimension, is known to increase PTH reliability. The ability of a laminate to withstand TMA dwells for specified periods of time at 260°C, 288°C, and 300°C is a definitive metric showing the material's suitability for Pb-free assembly. T_d is a newer test that uses thermogravimetric analysis (TGA) to determine the temperature at which a laminate material decomposes to a predetermined 2% or 5% weight loss. Additionally, data was gathered via a round-robin test program to properly establish T_d requirements for Pb-free applications.

To gather data and comments from users for the next revision of IPC-4101, these three metrics are being examined and have been included in proposed specification sheets.

10.6.4. Electrical Properties

Changes in PWB materials necessary to improve thermal resistance can alter electrical properties. However, subtle polymeric changes for Pb-free assembly may not align with designs that will require the maximum electronic performance of traditional epoxy materials, particularly in PWBs for high-speed designs. Permittivity or dielectric constant (D_k or ε_r) as well as dissipation factor or loss tangent (DF or tan δ) testing should be run on any new laminate materials designed for Pb-free applications to quantify the need for electrical design adjustments. IPC-4101 provides requirements for the electrical properties of laminates in its specification sheets.

10.6.5. Pb-Free Assembly Materials Qualification

IPC is revising a number of joint standards related to the materials used in electronic assembly. In addition to the solderability standards, revision is in progress for J-STD-004, *Requirements for Soldering Fluxes*. IPC-HDBK-005, *Guide to Solder Paste Assessment* (published) supports J-STD-005, *Requirements for Soldering Pastes* and incorporates details of Pb-free materials. Revision of J-STD-005 is underway and will include additional Pb-free data. An updated version of J-STD-006, *Requirements for Electronic Grade Solder Alloys and Fluxed and Non-Fluxed Solid Solders* addresses Pb-free alloys.

10.6.6. Pb-Free Assembly Standards

IPC-A-610D, *Acceptability of Electronic Assemblies*, and IPC J-STD-001D, *Requirements for Soldered Electrical and Electronic Assemblies*, contain requirements for Pb-free solder connections. At the basic level, there are *no differences* in the solder connection wetting and fillet requirements. However, there are some differences in visual appearance. Often, acceptable Pb-free and SnPb connections exhibit similar appearances (Figures 10.9 and 10.10) but Pb-free alloys are also more likely to have surface roughness (grainy or dull) and potentially different wetting coverage. It is also common for Pb-free to exhibit a greater contact angle, but, again, the 90° angle requirement for Pb-free is no different than that for Sn–Pb.

There are some very unique requirements acceptable only to Pb-free, such as fillet lifting (separation of the bottom of the solder and the top of the land) and hot tearing or shrink holes. There is also concern that BGA voiding may increase, although there is not much data to correlate connection failure to voiding.

Figure 10.9. Pb-free Sn–Ag–Cu solder, water-soluble flux.

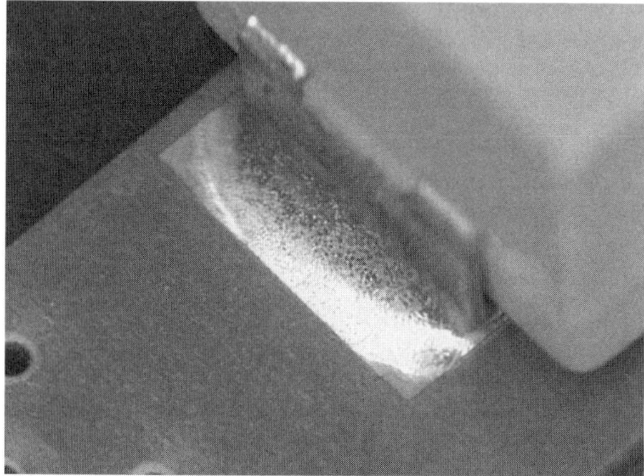

Figure 10.10. Sn–Pb solder, water-soluble flux.

10.6.7. Conclusion

By developing standardized processing requirements, test methods, labeling, and more, the electronics industry can ease the transition to Pb-free. Several industry standards have already been updated to reflect Pb-free processing changes, and several more are currently in development or undergoing revisions. There are still others that will need to be updated, as discussed in the next section.

For IPC standards and tools, go to http://leadfree.ipc.org/RoHS_3-4.asp, http://leadfree.ipc.org/RoHS_3-3.asp, or http://www.ipc.org/onlinestore.

For JEDEC standards, go to http://www.jedec.org/download/.

10.7. HIGH-RELIABILITY REQUIREMENTS [6]

Conversion to Pb-free alloy systems—typically Sn–Ag–Cu (SAC) alloys—raises many questions about the long-term reliability of new Pb-free materials as compared to the standard Sn–Pb formula. Of course, many of the companies producing high-reliability products will take the Pb exemption in the short term, but will still be driven to comply with the Pb restriction at some point in time (either by supply chain considerations or elimination of exemptions).

The broad component mix, board thermal properties, and resulting thermal gradients across complex assemblies pose many thermal challenges and push current capabilities to the very edge of allowable limits. New soldering materials, maximum qualified component temperatures, and primary attach and/or rework equipment are all sources of concern when trying to maintain reliability in the face of new Pb-free requirements.

The iNEMI High-Reliability RoHS Task Force consists of electronic product manufacturers (OEMs and EMS providers) whose products are characterized by long service life and high-reliability requirements. For these companies, maintaining product reliability is absolutely critical to survival. The task force developed the following recommendations to address the primary areas of concern for Pb-free processing of complex, thermally challenging electronic assemblies.

10.7.1. Areas of Concern and Recommended Solutions

Components. For nonhermetic solid-state surface-mount devices, it is imperative that component suppliers demonstrate compliance with J-STD-020C, *Moisture/Reflow Sensitivity Classification for Nonhermetic Solid State Surface Mount Device*, to the maximum exposure temperatures for primary attach and rework specified in Table 4.2 of the standard. (These temperatures range from $245 + 0°C$ to $260 + 0°C$, depending on package thickness and volume.)

For other non-IC, MCM and SiP devices, iNEMI recommends that the procedures outlined in J-STD-020C be used for establishing MSL ratings and peak temperature survivability. In addition, certain industry specifications, namely IPC-9503, *Moisture Sensitivity Classification for Non-IC Components*, and IPC-9504, *Assembly Process Simulation for Evaluation of Non-IC Components (Preconditioning Non-IC Components)*, must be updated to reflect Pb-free processing requirements. Representatives from the iNEMI task force will work with the appropriate standards committees to provide input and rationale for the proposed changes.

Small surface-mount devices (SMDs) that will be attached to the bottom side of a printed circuit board and then passed through a wave solder or solder dip machine (full-body immersion) should be qualified to meet the requirements of JESD22A111, *Evaluation Procedure for Determining Capability to Bottom Side Board Attach by Full Body Solder Immersion of Small Surface Mount Solid State Devices*. Components frequently wave-soldered in this manner include (but are not limited to) small discrete components, SOT-23, and lower I/O leaded devices with pitch ≥1.27 mm/0.050 in. JESD22A111 specifies a maximum exposure temperature of 260°C (+5/−0°C) for a maximum of 10 (±1) seconds for Sn–Pb wave soldering. However, it has not been updated to reflect requirements of Pb-free wave soldering.

Not only do SAC alloys melt at higher temperatures, they also do not spread as well as Sn–Pb. Minimum hole-fill requirements for PTH devices (per IPC-A-610, *Acceptability of Electronic Assemblies*) require increased solder pot temperature. The iNEMI task force is recommending that maximum exposure temperature be increased to a minimum of 265°C to accommodate the new alloys used for Pb-free wave soldering. To help offset this hotter temperature, the time required at 265°C can be reduced to 5 seconds minimum (laminar and chip wave combined). [This requirement of 265°C for 5 seconds minimum is based on a wave solder pot temperature of 260° ±5°C, a minimum wave solder conveyor speed of 3.25 feet/minute (0.65 inches/second), and a maximum wave contact distance (chip wave plus laminar wave) of 3.25 inches.] The task force recognizes, however, that some thermally sensitive components will require special handling or fixtures if used on the bottom side.

For BGAs or other area array packages, suppliers must meet the co-planarity requirements (at room temperature) for the appropriate package outline specifications. Suppliers must also assure assembly capability when packages are subjected to the maximum reflow temperatures specified in Table 4.2 of J-STD-020C.

Laminate and PWB Materials. While laminate materials used for Sn–Pb assembly have been used successfully in less complex assemblies at Pb-free processing temperatures, there are issues when using these materials for complex, thermally challenging products. Moving to higher temperatures increases materials sensitivities, and several parameters—such as T_d, T_g, and CTE(x–y), CTE(z)—must be reexamined for reliability issues such as the potential for delamination, peeling, and warpage. More testing and evaluation is required before new laminate materials can be considered "Pb-free compatible" for all applications in high-complexity systems.

After initial material selection, the manufacturer will need to validate acceptable performance of the materials in the use condition. It is imperative that manufacturers make sure that laminates can withstand both process temperatures and stress and will not be "invisibly weakened" in any way that would show up later in the field. Table 10.4 summarizes the test methods recommended to validate performance, including CAF and thermal/temperature cycling.

Equipment. The capabilities of primary attach and rework equipment are strained by higher thermal mass PWBs and assemblies, and it is difficult to stay within the minimum and maximum temperature limits imposed by J-STD-020C and JESD22A111. Process speed, peak temperatures, flux chemistry, solder pot contamination, and soldering gas atmosphere are key areas of concern, and product and process designers need to work with equipment providers to understand and resolve these issues.

TABLE 10.4. Parameters for Pb-Free Materials

Characteristic	Test Method	Comments/Suggested Value[1]
Decomposition temp, T_d, (5% weight loss by TGA)[2,3]	IPC-TM-650.2.3.40	≥335°C
Glass transition temperature (T_g), °C, by TMA[4]	IPC-TM-650.2.4.24c	• Tg >140°C for all products • >165°C for products with >10 layers, >6:1 aspect ratio, or containing BGAs
In-plane coefficient of thermal expansion CTE(x-y), ppm/°C,	IPC-TM-650.2.4.24c	Solder joint stress depends on peak processing temperature, component CTE
Out-of-plane coefficient of thermal expansion — CTE(z), ppm/°C, α_1 (above T_g) and α_2 (below T_g)[5]	IPC-TM-650.2.4.41	Via/PTH barrel and land stress depends on peak processing temperature, PTH copper ductility
	Secondary Parameter	
Time to delamination (T-260)	IPC-TM-650.2.4.24.1	≥30 minutes
Time to delamination (T-288)	IPC-TM-650.2.4.24.1 modified per paragraph 6.1 to 288°C	≥5 minutes
Test Method, Special		
Copper ductility—PTH barrel	IPC TM 650, 2.4.2	Depends on peak processing temperature, and PTH aspect ratio
	Product Level Validation	
Solder float at 288°C (6X)	Similar to IPC Test Method TM650, 2.4.13 except loaded with SAC solder	First article cross-sections must pass
Conductive anodic filament (CAF) testing	Pre-conditioning + IPC 9691, IPC-TM-650, Method 2.6.25	Pass

Note 1: Suggested values are highly dependent upon the product being assembled. The values suggested are based on thermally complex high layer count boards that have high material resin content, require multiple Pb-free soldering processes at or near the limits of the J-STD-020C profile requirements, and have long life requirements. Specific values will vary with the requirements of the individual products.
Note 2: TGA is Thermo-gravimetric analysis.
Note 3: T_d, a characteristic determined by a standard test method and evaluated at 5 minutes, is substantially higher than delamination temperatures evaluated at 30 and 5 minutes. T_d should be used to compare similar materials, and not used as an absolute value in isolation.
Note 4: Thermo-mechanical analysis (TMA) is preferred over DSC and DMA in determining T_g because total expansion from room temperature to the maximum processing temperature is a critical product parameter and because TMA reports the expansion of the material as a function of temperature.
Note 5: The z-axis CTE's Z-Axis Expansion (%) per IPC TM 650, 2.4.41 (50–260°C), both below and above T_g, are important to long-term reliability. Users should ensure that the materials specified and the associated plated-through-hole copper wall thickness and copper ductility will meet long-term reliability requirements of the products.

Reflow profiling studies using SAC solder for various thermally challenging products have demonstrated that the reflow oven speed should be slowed in order to guarantee a good solder joint (i.e., stay below the upper component body temperature limits defined by J-STD-020C while simultaneously staying above the minimum reflow temperature). However, this modification extends the overall processing time by 20–30% and puts significant additional strain on the soldering materials. Key issues from a materials perspective include:

- Ability of flux to handle higher temperatures at longer pre-heat times (135–200°C for 3–4 minutes).
- Total profile times of 8–9 minutes.

Rework of large assemblies also challenges the equipment's ability to effectively heat the module being reworked while, at the same time, not *over*heating adjacent components. Techniques to improve reworkability include:

- Whole board preheat to reduce heatsinking effects of PWB power and ground planes.
- Improved adjacent component "shielding" from hot gas rework temperatures to prevent secondary reflow.

Preheating PWBs is typically needed when reworking PTH components for Pb-free solders using a mini-pot solder fountain. As discussed previously, the higher tin content combined with the slower wetting/spreading of Sn–Ag–Cu result in the need for higher solder pot temperatures and longer contact times to achieve the same degree of hole fill. This, in turn, increases Cu dissolution of the barrels and traces on the PWB and, therefore, requires tighter control of rework processes and/or equipment.

Manufacturers should seriously consider adding process modeling, thermal and process profiling, and/or some type of overall process equipment verification instrumentation for all process equipment steps (e.g., placement, wave or reflow, and rework). These measures can help ensure that process steps are within the control parameters needed to remain within the extremely tight thermal process window posed by Pb-free process materials.

10.7.2. A Challenge that is not Going Away

The successful conversion to Pb-free alloy systems for products that require high-reliability performance over the long term depends upon the entire supply chain doing its part to create, test, and guarantee materials, components, and process systems. Initial focus of Pb-free manufacturing has understandably been concentrated on high-volume consumer products due to the EU ROHS requirements. As high-reliability products begin to make the conversion to Pb-free, the industry needs to meet the remaining challenges that are encountered when manufacturing these complex assemblies. The iNEMI High-Reliability RoHS Task Force developed these recommendations to provide direction to these efforts and is committed to working with the appropriate standards bodies to ensure that formal standards are updated to support evolving manufacturing needs.

10.8. BUSINESS IMPACT OF SUPPLY CHAIN CONVERSION

Conversion of the electronics manufacturing supply chain to be RoHS-compliant has been a massive undertaking. Some have estimated that the electronics industry has invested on the order of $5B to make this happen. Unfortunately, this investment has not generally translated into increased revenue at the product level. Some companies have seen increased business due to the more stringent requirements for the RoHS-compliant materials, processes, and manufacturing equipment, but the majority have not. Engineering investments have also increased to make all of the necessary product and process changes.

What is missing, however, is an ongoing process for dealing with new environmental regulations as they are introduced. Most firms do not yet have a routine way to deal with these kinds of regulations, although other regulations (such as safety requirements from UL/CSA/IEC or EMI requirements) are built into today's development, manufacturing, and logistics programs. Industry needs to think about environmental regulations in the same way, and firms need to develop an ongoing process/function to deal with an ever increasing set of requirements that are being introduced.

10.9. SUMMARY

RoHS/WEEE is the beginning—not the end—of the environmental regulations story for the electronics industry. While removal of some materials of concern has been relatively straightforward, reduction/elimination of Pb has caused significant disruption to the electronics manufacturing supply chain. While this conversion will be made successfully, it has increased the risk for premature product failures and degraded performance. Use of Sn–Pb solder has over 50 years of manufacturing and field experience and knowledge. Today's Pb-free solutions are nowhere near as mature or as well understood. The next few years will see a dramatic increase in knowledge as we learn from our collective experiences. In the future, our industry must do more to develop the scientific and engineering knowledge of alternative materials well before regulations are created.

REFERENCES

The majority of these articles were part of a series called "Lead-Free Watch: Countdown to July 1, 2006," which ran monthly in *Circuits Assembly* and *Printed Circuits Design & Manufacture*. The following lists the citations for the *Circuits Assembly* articles.

1. S. Andrews, Implementing RoHS in the U.K., *Circuits Assembly* **17**(2), 88–89, 2006.
2. H. Evans and J. Johnson, 10 Steps towards RoHS compliance, *Circuits Assembly* **16**(2), 68–70, 2005.
3. V. Gupta, Marking & part numbering for lead-free components & boards, *Lead-Free Electronics* (a supplement to *Advanced Packaging*, *SMT* and *Connector Specifier*) **July**, 18–20, 2005.
4. R. Kubin, A common solution for materials declaration, *Circuits Assembly* **17**(4), 41–43, 2006.
5. T. Newton, D. Bergman, and J. Crawford, Lead-free standards update, *Circuits Assembly* **16**(9), 56, 2005.
6. J. Smetana and T. Sack, High-complexity, thermally challenging Pb-free product recommendations, *Circuits Assembly* **17**(3), 56–57, 2006.

INDEX

Page numbers followed by an *f* or *t* indicate figures and tables.

Lead-Free Electronics. Edited by Bradley, Handwerker, Bath, Parker, and Gedney